New Trends in Pharmacokinetics

NATO ASI Series

Advanced Science Institutes Series

A series presenting the results of activities sponsored by the NATO Science Committee, which aims at the dissemination of advanced scientific and technological knowledge, with a view to strengthening links between scientific communities.

The series is published by an international board of publishers in conjunction with the NATO Scientific Affairs Division

A	**Life Sciences**	Plenum Publishing Corporation
B	**Physics**	New York and London
C	**Mathematical and Physical Sciences**	Kluwer Academic Publishers
D	**Behavioral and Social Sciences**	Dordrecht, Boston, and London
E	**Applied Sciences**	
F	**Computer and Systems Sciences**	Springer-Verlag
G	**Ecological Sciences**	Berlin, Heidelberg, New York, London,
H	**Cell Biology**	Paris, Tokyo, Hong Kong, and Barcelona
I	**Global Environmental Change**	

Recent Volumes in this Series

Volume 217—Developmental Neuropathology of Schizophrenia
edited by Sarnoff A. Mednick, Tyrone D. Cannon,
Christopher E. Barr, and José M. LaFosse.

Volume 218 — Pharmaceutical Applications of Cell and Tissue Culture to Drug Transport
edited by Glynn Wilson, S. S. Davis, L. Illum, and Alain Zweibaum

Volume 219— Atherosclerotic Plaques: Advances in Imaging for Sequential
Quantitative Evaluation
edited by Robert W. Wissler

Volume 220—The Superfamily of *ras*-Related Genes
edited by Demetrios A. Spandidos

Volume 221—New Trends in Pharmacokinetics
edited by Aldo Rescigno and Ajit K. Thakur

Volume 222—The Changing Visual System: Maturation and Aging in the
Central Nervous System
edited by P. Bagnoli and W. Hodos

Volume 223—Mechanisms in Fibre Carcinogenesis
edited by Robert C. Brown, John A. Hoskins, and Neil F. Johnson

Series A: Life Sciences

New Trends in Pharmacokinetics

Edited by

Aldo Rescigno

University of Parma
Parma, Italy

and

Ajit K. Thakur

Hazleton Washington, Inc.
Vienna, Virginia

Plenum Press
New York and London
Published in cooperation with NATO Scientific Affairs Division

Proceedings of a NATO Advanced Study Institute
on New Trends in Pharmacokinetics,
held September 4–15, 1990,
in Erice, Sicily, Italy

Library of Congress Cataloging-in-Publication Data

New trends in pharmacokinetics / edited by Aldo Rescigno and Ajit K.
Thakur.
      p.   cm. -- (NATO advanced science institutes series. Series
A, Life sciences ; v. 221)
   Includes bibliographical references and index.
   ISBN 0-306-44089-X
   1. Pharmacokinetics.   I. Rescigno, Aldo. II. Thakur, Ajit K.
III. Series.
RM301.5.N49 1992
615'.7--dc20   91-37923
                                                        CIP

ISBN 0-306-44089-X

DIRECTOR

Aldo Rescigno, School of Pharmacy, University of Parma, Italy

ORGANIZING COMMITTEE

Giuliano Mariani, Associate Professor of Internal Medicine, University of Pisa, Pisa, Italy

Aldo Rescigno, Professor of Pharmacokinetics, School of Pharmacy, University of Parma, Parma, Italy

James T. Stevens, Director of Toxicology, Agricultural Division, Ciba-Geigy Corp., Greensboro, North Carolina, U.S.A.

Ajit K. Thakur, Principal Scientist and Biostatistician, Hazelton Washington, Vienna, Virginia, U.S.A.

PREFACE

The last decade or so has witnessed tremendous progress in methodology in the field of drug development in general and pharmacokinetics in particular. Clinical pharmacokinetics is using new tools for probing into the "black box" once being accessible only partly through experimental techniques and mostly through mathematical and computer means. Development of computerized scanning, positron emission tomography (PET), stereoselectivity and other techniques are now enabling investigators to have better pictures of the systems they are studying. Mathematical models through computer simulation and statistical estimation, mostly due to easy access because of inexpensive yet powerful personal computers, are enabling us to investigate ultrastructures and their functional connectivity in more detail. As a consequence, new hypotheses are being formed and tested in various related fields.

In clinical pharmacokinetics, mostly due to mathematical modeling, more accurate interspecies scaling of pharmacokinetic parameters and dosimetry can be done now-a-days. The concept of "a human is a bigger rat" does not necessarily fly as a consequence. Pharmacokinetic concepts are becoming powerful tools in meaningful carcinogenic and toxic risk extrapolation of different chemicals in humans. New dose delivery designs are being formulated using pharmacokinetic techniques for different pharmaceutical compounds. Investigations continue in the academia, research institutions, pharmaceutical, biotechnological, and agricultural industries in developmental and physiological aspects of different chemicals for the benefit of mankind.

The idea of a school on "New Trends in Pharmacokinetics", from which the present publication was made possible, took shape over almost a year. The organizing committee, consisting of Drs. Aldo Rescigno, Ajit K. Thakur, James T.Stevens, and Giuliano Mariani, spent many hours and days worth of efforts to gather experts in various fields of clinical, experimental, and computational pharmacokinetics. The idea was to have these experts from various research environments to teach in this intensive workshop in September, 1990 in Erice,Sicily. The historical background and natural serenity of this island paradise provided the exact atmosphere needed for such an international exchange of ideas under the auspices of the Ettore Majorana Centre.

Of course, none of this could have happened if no money were available for the workshop. Students, investigators, and the speakers had to be supported with funds. The organizing committee relentlessly pursued many different organizations for funds. The ultimate success was due to the generous contributions with funds from the North Atlantic Treaty Organization, the Farmindustria (Rome, Italy), C.N.R. (Rome, Italy), Sigma Tau (Pomezia, Italy), the Italian Section of the Bragg Creek Institute (Parma, Italy), the National Science Foundation (Washington, U.S.A.), Ciba-Geigy Corporation (Basel, Switzerland), and Dr. Ronald Sawchuk of the School of Pharmacy, University of Minnesota (U.S.A.). Thanks to the above, the organizing committee did not have to sell their houses to pay for all expenses, after all!

The lectures and the materials were excellent. Many of the participants took active roles in discussing their research topics with their peers from different countries. Such an international gathering is always enriching from cultural standpoint as well. That was clearly evidenced in the course of the workshop. Several students, post-doctoral fellows, and senior investigators from various countries also presented some of their works in leisurely fashion. The present book is the result of culmination of extensive works by many individuals who, at times, must have wished that they had never seen the faces of the Editors or heard their voices on the telephone! Let the glory be all theirs.

Aldo Rescigno and Ajit K. Thakur

CONTENTS

PHARMACOKINETICS: UNFOLDING OF A CONCEPT

Aldo Rescigno and Bianca Maria Bocchialini

School of Pharmacy
University of Parma
Parma, Italy

INTRODUCTION

The word "Pharmacokinetics", coined from the Greek words $\phi\alpha\rho\mu\alpha\kappa o\nu$, drug, and $\kappa\iota\nu\eta\tau\iota\kappa o\varsigma$, moving, was used for the first time in 1953 by F. H. Dost, a German pediatrician, in his book "Der Blutspiegel" [Dost, 1953], but the concept had been around for a long time before that. The object of Pharmacokinetics is the study of absorption, distribution, and elimination of drugs; but, since the existence of Pharmacology, it has been known that drugs are absorbed, distributed, and eliminated from the organism, and that the rates of absorption, distribution, and elimination are fundamental in determining the effects on the organism they are administered to. Pharmacokinetics as such can be therefore considered a new discipline only since more sophisticated methods have been introduced to study the kinetic properties of drugs. These quantitative methods have been offered by Analytical Chemistry, by Physical Chemistry, and by Applied Mathematics.

In the following few pages we shall try to show how the different concepts used in Pharmacokinetics have unfolded in recent years.

THE INVARIANT QUANTITIES

The volume of distribution

It is not easy to decide which is the earliest paper dealing with the quantitative solution of a pharmacokinetic problem. An important pioneering study is due to Widmark [1919], who in 1919 published in Sweden a paper about the elimination of ethanol and acetone from blood. Widmark observed that, in its final phase, ethanol is eliminated according to an exponential law. He introduced the concept of what we now call the *volume of distribution*.

If a drug is introduced intravenous, let's call D the dose administered and c(t) the concentration in the plasma measured at time t. If we ignore the short interval of time necessary for the drug to distribute uniformly in the plasma, then

$$\lim_{t\to 0} c(t) = D/V,$$

New Trends in Pharmacokinetics, Edited by A. Rescigno and A.K Thakur
Plenum Press, New York, 1991

i.e. the ratio D/V represents the concentration of the drug in the plasma before a fraction of it has been eliminated or has been distributed to other organs. The quantity V must be the volume of the plasma, and can be calculated from the formula above. This observation may seem obvious, but when we measure in an experiment the values of c(t) and D, and compute V using the formula above, sometimes we get values much different from the expected plasma volume. There may be several reasons for this discrepancy; for instance, the drug may be bound to some tissues before being distributed in the plasma. Nowadays we call the ratio

$$V = D / \lim_{t \to 0} c(t)$$

the *apparent volume of distribution*. This is one of the fundamental concepts in Pharmacokinetics, but, as any fundamental concept, it took a while before becoming part of our basic concepts.

Naturally if the biological condition of the subject does not change, with a different dose D the concentration c(t) will change in the same proportion; therefore, the ratio computed with the formula above will not change. The quantity V therefore is called *invariant*. Within the limits of validity of the hypotheses incorporated in the equations used for its computation, an invariant quantity does not depend on the particular experimental conditions.

The time of maximum concentration

Another invariant quantity is t_{max}, i.e. the interval of time necessary for a drug to reach its maximum concentration in the blood, when it has been injected as a bolus. This concept was introduced in 1933 by Gehlen [1933]. He showed that the t_{max} of a drug, other conditions being equal, does not change with the dose. This observation too, apparently obvious today, is a fundamental concept of Pharmacokinetics. It is important to remember, though, that all invariance properties depend on some very specific hypotheses. In the case of t_{max} in particular, the required hypothesis is the linearity of the biological system. Indeed the invariance of t_{max} is commonly used to check the linearity of the system under observation.

The eigenvalues

The second contribution, in chronological order, to the formation of fundamental pharmacokinetic concepts, is due to Biehler [1925], who in 1925 described the elimination of ethanol from blood with a bi-exponential function. But the first systematic treatment of pharmacokinetic problems with exponential functions is due to Teorell [1937a, b], generally considered the originator of Pharmacokinetics. In 1937 Teorell published the paper "Kinetics of Distribution of Substances Administered to the Body;" in that paper Teorell, starting from some general hypotheses, built the equations of what we would call today a "compartmental model". The solutions of Teorell's equations were sums of exponential functions, exactly as used today in most pharmacokinetic models. The exponents of those exponential functions are the *eigenvalues* of the matrix formed with the coefficients of the differential equations [Rescigno, Lambrecht and Duncan, 1983] describing the biological system observed, and do not depend upon the experimental initial conditions. The eigenvalues therefore are invariant quantities, independent not only of the dose, but also of the mode of administration. In other words the eigenvalues do not change if the drug is administered in a single dose or in multiple doses or by continuous infusion.

The following year, 1938, another fundamental paper appeared [Artom, Sarzana and Segré, 1938]. It was written by a physiologist, a histologist, and a physicist who combined their expertises to do a very innovative piece of work. Using the isotope ^{32}P prepared by Ernest Lawrence at the University of California in Berkeley, they studied the synthesis and distribution of phospholipids in rats after administration of inorganic Phosphorus. Theirs is probably the first paper dealing with the use of radioactive isotopes for the solution of pharmacokinetic problems. It is worth noting that, even though they did not use the term "compartment" explicitly, they were the first authors to use this concept in a precise and consistent way.

The half-life of a drug

Finding the eigenvalues from the experimental data is not always an easy problem. Most of the times the observation errors propagate in such a way as to invalidate most of the numerical procedures towards this goal. In general, the easiest eigenvalue that can be computed is the smallest one in absolute value.

Suppose that a particular drug in a particular organism is characterized by three eigenvalues; in other words, the function representing the concentration of that drug in the plasma is a sum of three exponential functions; then

$$c(t) = A_1 e^{-\alpha_1 t} + A_2 e^{-\alpha_2 t} + A_3 e^{-\alpha_3 t}, \tag{1}$$

where $-\alpha_1, -\alpha_2, -\alpha_3$ are the three eigenvalues. Suppose also that

$$\alpha_1 > \alpha_2 > \alpha_3.$$

When t increases, the first two exponential functions decrease faster than the third, so that after a sufficiently long time

$$c(t) \cong A_3 e^{-\alpha_3 t},$$

and as a consequence,

$$\ln(c(t)) \cong -\alpha_3 \cdot t + \ln(A_3); \tag{2}$$

The plot of ln c(t) versus t is a straight line of slope $-\alpha_3$, therefore α_3 can easily be determined by plotting c(t) as a function of t on a semilogarithmic scale and extrapolating for $t \to \infty$.

The interval of time necessary for c(t) to decrease 50%, in the range of t where the approximation of equation (2) is valid, is called $t_{1/2}$ or *half-life* of that drug. Clearly

$$t_{1/2} = \ln 2/\alpha_3 = 0.693 \cdot 1/\alpha_3.$$

We mentioned earlier that this particular eigenvalue is easy to determine, but this is not always the case. There are at least two cases when this determination is difficult and inaccurate. If A_3 is very small, the approximation in equation (2) is still valid, but only for values of c(t) correspondingly small, that is for measurements of c(t) taken for large values of t, when experimental errors are more likely. This difficulty sometimes can be overcome. In fact, α_3 is invariant, but A_3 is not; if the initial conditions are modified appropriately, for instance, by using a continuous infusion, A_3 may sufficiently increase while α_3 stays constant.

Table 1

	$\kappa=2$	$\kappa=3$	$\kappa=4$
$\tau=2$	14~41	2~9	0~2
$\tau=3$	5~20	0~2	0~0
$\tau=4$	2~9	0~0	0~0
$\tau=5$	1~4	0~0	0~0
$\tau=6$	0~2	0~0	0~0

Table 1 shows the relative errors in % committed when assuming

$$A_3 e^{-\alpha_3 t} \cong A_1 e^{-\alpha_1 t} + A_2 e^{-\alpha_2 t} + A_3 e^{-\alpha_3 t} \, ;$$

the entries of the table are $\kappa = \alpha_2/\alpha_3$ and $\tau = \alpha_3 t$; the range of values indicated for the error corresponds to the different possible values of α_1; the larger the difference $\alpha_1 - \alpha_2$, the smaller the error.

Table 2 shows the corresponding errors when the drug is given with a continuous infusion until steady state is reached, then $c(t)$ is measured and extrapolated as above. The reduction of the errors in considerable.

Another case when the determination of α_3 is difficult is when $\alpha_2 = \alpha_3$; in this case equation (1) must be substituted by

$$c(t) = A \cdot e^{-\alpha_1 t} + (B \cdot t + C) e^{-\alpha_2 t};$$

if $\alpha_1 > \alpha_2$, for t sufficiently large, the approximations

$$c(t) \cong (B \cdot t + C) e^{-\alpha_2 t};$$

$$\ln(c(t)) \cong -\alpha_2 \cdot t + \ln\left(B \cdot t + C\right)$$

are valid, but this last one is not the equation of a straight line.

Of course the probability of two eigenvalues being exactly equal is extremely small. Suppose that

$$\alpha_1 > \alpha_2, \quad \alpha_2 - \alpha_3 = \varepsilon > 0,$$

where ε is small; the coefficients of equation (1) are given by [Rescigno and Beck, 1972]

Table 2

	$\kappa=2$	$\kappa=3$	$\kappa=4$
$\tau=2$	7~24	1~3	0~2
$\tau=3$	2~11	0~1	0~0
$\tau=4$	1~5	0~0	0~0
$\tau=5$	0~2	0~0	0~0
$\tau=6$	0~1	0~0	0~0

$$A_1 = \frac{c(0)}{(\alpha_1 - \alpha_2)(\alpha_1 - \alpha_3)}, \quad A_2 = \frac{c(0)}{(\alpha_2 - \alpha_1)(\alpha_2 - \alpha_3)}, \quad A_3 = \frac{c(0)}{(\alpha_3 - \alpha_1)(\alpha_3 - \alpha_2)};$$

therefore

$$c(t) = \frac{c(0)}{(\alpha_1 - \alpha_2)(\alpha_1 - \alpha_3)(\alpha_2 - \alpha_3)}\left((\alpha_2 - \alpha_3)\,e^{-\alpha_1 t} - (\alpha_1 - \alpha_3)\,e^{-\alpha_2 t} + (\alpha_1 - \alpha_2)\,e^{-\alpha_3 t}\right)$$

$$c(t) = \frac{c(0)}{\varepsilon\cdot(\alpha_1 - \alpha_2)(\alpha_1 - \alpha_3)}\left(\varepsilon\cdot e^{-\alpha_1 t} - (\alpha_1 - \alpha_3)\,e^{-\alpha_2 t} + (\alpha_1 - \alpha_2)\,e^{-\alpha_3 t}\right).$$

For t very large we may use the approximation

$$c(t) \cong \frac{c(0)}{\varepsilon\cdot(\alpha_1 - \alpha_3)}\,e^{-\alpha_3 t},$$

but the error of this approximation does not decrease very rapidly with t; in the best case, i.e. when $\alpha_1 \gg \alpha_2$, the relative error is given by $e^{-\varepsilon t}$. Table 3 shows some typical values for this error.

Table 3

εt	2	3	4	5
$e^{-\varepsilon t}$	13.5%	5.0%	1.8%	0.7%

The area under the curve

The area under the curve measuring the concentration of a drug in the plasma as a function of time, often abbreviated AUC, depends in a simple way upon the fraction of the drug reaching the systemic circulation and upon its clearance therefrom.

If $c(t)$ is the concentration of a drug in the plasma and Cl the volume eliminated per unit time, then $Cl\cdot c(t)\cdot dt$ is the amount of drug eliminated during the interval of time from t to t+dt. Suppose that the drug present in the plasma will be eliminated completely in due time, then

$$F\cdot D = \int_0^\infty Cl\cdot c(t)\cdot dt, \tag{3}$$

where D is the *dose* or amount of drug administered, and F is the fraction reaching the systemic circulation.

If Cl is constant, then equation (3) can be written in the form

$$F\cdot D = Cl\cdot\int_0^\infty c(t)\cdot dt. \tag{4}$$

This equation is known as the "Stewart-Hamilton principle," often written in the form

$$AUC/D = F/Cl$$

where

$$AUC = \int_0^\infty c(t) \cdot dt.$$

Evidently the ratio AUC/D is another invariant quantity, provided our hypothesis is valid, i.e. Cl is constant.

The transfer time

Another important invariant quantity is the *transfer time* of a drug from a compartment to another compartment, as introduced in 1961 by Rescigno and Segre [1961a]. If $c_a(t)$ and $c_b(t)$ are the concentrations of a drug in compartments a and b respectively, and a is the precursor of b, the transfer time from a to b is the difference

$$T_{ab} = \frac{\int_0^\infty t \cdot c_b(t)dt}{\int_0^\infty c_b(t)dt} - \frac{\int_0^\infty t \cdot c_a(t)dt}{\int_0^\infty c_a(t)dt}. \tag{5}$$

This quantity does not depend upon the dose or the mode of administration of the drug.

We shall say more about the transfer time in the sections about compartments and moments.

The transfer function

Consider a system where $X(t)$ is the amount of drug present at time t in the compartment where it was first supplied, while $Y(t)$ is the amount present at time t in another compartment. We call the first compartment the *precursor* of the second, and the second the *successor* of the first [Rescigno and Segre, 1961b]. If all the processes involved in the transfer of the drug from precursor to successor are linear and do not change in time, the relationship between $X(t)$ and $Y(t)$ can be described by the integral equation

$$Y(t) = \int_0^t X(\tau)\, g(t-\tau)\, d\tau. \tag{6}$$

The integral above is called a *convolution*, and the function g(t) is called the *transfer function* [Rescigno, 1960] between $X(t)$ and $Y(t)$. The transfer function is not, strictly speaking, an invariant quantity, but it is a characteristic of the system and many invariant quantities can be derived from it.

The actual process of determining function g(t) is not a simple one because small experimental errors in $X(t)$ and $Y(t)$ propagate non-linearly in the numerical computation of g(t) and may become very large. Nevertheless, there are a number of properties of the transfer function that are important and can be easily observed.

Consider the ratio

$$\frac{Y(t)}{t\cdot X(t)} = \frac{\displaystyle\int_0^t X(\tau)\, g(t-\tau)\, d\tau}{t\cdot X(t)}, \tag{7}$$

which can easily be computed for a number of values of t; for $t = 0$ this ratio is indeterminate, but using L'Hospital's rule one gets

$$\lim_{t\to 0}\frac{Y(t)}{t\cdot X(t)} = \lim_{t\to 0}\frac{\displaystyle\int_0^t X(\tau)g'(t-\tau)d\tau + X(t)g(0)}{X(t) + t\cdot X'(t)}.$$

Now, if $X(0) \neq 0$, then

$$\lim_{t\to 0}\frac{Y(t)}{t\cdot X(t)} = g(0).$$

If $X(0) = 0$, we can use L'Hospital's rule once more to get

$$\lim_{t\to 0}\frac{Y(t)}{t\cdot X(t)} = \lim_{t\to 0}\frac{\displaystyle\int_0^t X(\tau)g''(t-\tau)d\tau + X(t)g'(0) + X'(t)g(0)}{2X'(t) + t\cdot X''(t)}.$$

Now, if $X'(0) \neq 0$, then

$$\lim_{t\to 0}\frac{Y(t)}{t\cdot X(t)} = \frac{g(0)}{2}.$$

Proceeding in the same way we find that, if

$$X(0) = X'(0) = X''(0) = \cdots = X^{(n-1)}(0) = 0,$$

but

$$X^{(n)}(0) \neq 0,$$

then

$$\lim_{t\to 0}\frac{Y(t)}{t\cdot X(t)} = \frac{g(0)}{(n-1)!}.$$

Both n and g(0) are invariant quantities [Beck and Rescigno, 1964] and frequently can be determined with ease.

The precursor order

The number n shown above is called the *precursor order* between $X(t)$ and $Y(t)$ [Rescigno and Segre, 1961b]. If the precursor order is one, there are no intermediate products between the precursor and the successor; in general n is the minimum number of transformations involved in the process.

The value g(0) itself is another important invariant quantity; its properties were described by Rescigno and Segre [1965]. Let us observe more closely its physical meaning.

Equation (6) can be interpreted in the following way. Call $X(\tau)$ the probability that a particle of a given drug is present in the precursor at time τ, $Y(t)$ the probability that the same particle is present in the successor at time t, with $t \geq \tau$, and $g(t-\tau)d\tau$ the probability that a particle that left the precursor in the interval of time $\tau, \tau + d\tau$ will be present in the successor at time t. It follows that

$$\int_0^\infty X(\tau) \cdot g(t-\tau) \cdot d\tau$$

is the probability that a particle will be in the successor at time t if it was in the precursor any time before. The initial value g(0) of the transfer function is the rate at which the particles leaving the precursor enter the successor. In the case of an order-one precursor, this is simply the fractional transfer rate from precursor to successor.

If $g(0) = 0$, there is no direct transfer from precursor to successor; the precursor order is two or more.

Additional information can be gathered by plotting the ratio (7) versus t; the slope of that function is

$$\lim_{t \to 0} \frac{d}{dt} \frac{Y(t)}{t \cdot X(t)} = \lim_{t \to 0} \frac{d}{dt} \frac{\int_0^t X(\tau) \, g(t-\tau) \, d\tau}{t \cdot X(t)} .$$

Using L'Hospital's rule twice we get

$$\lim_{t \to 0} \frac{d}{dt} \frac{Y(t)}{t \cdot X(t)} = \frac{g(0)}{2} \left(\frac{g'(0)}{g(0)} - \frac{X'(0)}{X(0)} \right).$$

From this identity, knowing $g(0)$, $X(0)$, and $X'(0)$, one can compute $g'(0)$; this value is the rate of exit of the drug from the successor, or its turnover rate [Rescigno, Thakur, Brill, and Mariani, 1990].

COMPARTMENTAL ANALYSIS

The Rutherford equations

The first compartmental models were used in Physics for the description of radioactive decay. After Becquerel [1896] discovered radioactivity, Rutherford and Soddy [1902] found experimentally that Thorium X decays in time according to an exponential law, i.e. that the number of radioactive atoms decaying per unit time is proportional to the number of radioactive atoms present. If $X(t)$ is the quantity of radioactive substance present at time t, the law of radioactive decay is

$$\frac{dX}{dt} = -KX, \tag{8}$$

whose integral is

$$X(t) = X(t_0) \cdot e^{-K(t-t_0)}.$$

Later Rutherford [1904] developed the theory of successive radioactive transformations. If A is transformed into B, B is transformed into C, and so forth, X_a, X_b, X_c, ... be the amounts of A, B, C, ... present at any given time; and K_a, K_b, K_c be the rates of such transformations. In analogy with equation (8), he wrote

$$\frac{dX_a}{dt} = -K_a X_a,$$

$$\frac{dX_b}{dt} = +K_a X_a - K_b X_b, \tag{9}$$

$$\frac{dX_c}{dt} = +K_b X_b - K_c X_c,$$

$$\cdot\ \cdot\ \cdot\ \cdot\ \cdot\ \cdot\ \cdot\ \cdot\ \cdot\ \cdot\ \cdot\ \cdot\ \cdot\ \cdot\ \cdot\ ,$$

and by integration, provided all K_i's are different,

$$X_a(t) = X_a(t_0) \cdot e^{-K_a(t-t_0)},$$

$$X_b(t) = \frac{K_a}{K_b - K_a} \cdot X_a(t_0) \cdot e^{-K_a(t-t_0)} + \left(X_b(t_0) + \frac{K_a}{K_a - K_b} \cdot X_a(t_0)\right) \cdot e^{-K_b(t-t_0)},$$

$$X_c(t) = \frac{K_a K_b}{(K_b - K_a)(K_c - K_a)} \cdot X_a(t_0) \cdot e^{-K_a(t-t_0)} +$$

$$+ \left(\frac{K_b}{K_c - K_b} \cdot X_b(t_0) + \frac{K_a K_b}{(K_a - K_b)(K_c - K_b)} \cdot X_a(t_0)\right) \cdot e^{-K_b(t-t_0)} +$$

$$+ \left(X_c(t_0) + \frac{K_b}{K_b - K_c} \cdot X_b(t_0) + \frac{K_a K_b}{(K_b - K_c)(K_a - K_c)} \cdot X_a(t_0)\right) \cdot e^{-K_c(t-t_0)}$$

and so forth.

Many experimental observations have shown that this compartmental model is consistent with the behavior of all radioactive substances, thus confirming the hypothesis incorporated into equations (8) and (9), i.e. that radioactive decay is a first order process.

The compartmental equations

Let us go back to Widmark [1919], who studied both theoretically and experimentally the kinetics of distribution of several narcotics, in particular, acetone. He studied the concentration curve of acetone in the blood after a single dose administration, and assumed that the fall of the curve was due principally to elimination from the lungs and to chemical metabolism. The mathematical model used by Widmark was

$$\frac{dx}{dt} = -ax - bx \qquad x(0) = x_0$$
$$\frac{dy}{dt} = +ax \qquad y(0) = 0 \tag{10}$$
$$\frac{dz}{dt} = \qquad +bx \qquad z(0) = 0$$

where x, y, z are the amounts of acetone in the body, exhaled, and metabolized, respectively; x_0 is the amount administered initially.

The solution of the above equations is

$$x(t) = x_0 \cdot e^{-(a+b)\,t}$$
$$y(t) = x_0 \cdot \frac{a}{a+b}\left(1 - e^{-(a+b)\,t}\right) \tag{11}$$
$$z(t) = x_0 \cdot \frac{b}{a+b}\left(1 - e^{-(a+b)\,t}\right)$$

By observing that it is possible to determine a and b in such a way that the values of, say, x(t) computed from equations (11) correspond to the values measured from an actual experiment, Widmark concluded that the hypotheses implied by equations (10) were acceptable.

From a knowledge of the time behavior of the concentration c(t) of the acetone in blood and of the so-called "reduced body volume", m, where m = x/c, Widmark computed the time behavior of x, y, z in several experimental conditions.

Later Widmark and Tandberg [1924] derived the equation of a model where there is a constant rate of administration, and also when the drug is administered with rapid intravenous injections repeated at uniform intervals of time.

Another important contribution has been given by Gehlen [1933] who derived some theoretical expressions for what we would now call a two-compartment system.

Widmark [1932] also studied the elimination of ethanol and developed in this context, what we would now call a zero-order compartmental model.

The later works of Teorell [1937 a, b] and of Artom, Sarzana and Segré [1938], and of many other authors start almost always from differential equations of the type

$$\frac{dX_i}{dt} = \Sigma_j \, k_{ji}X_j - K_iX_i, \quad i=1,2,...,n, \tag{12}$$

where X_i is the amount of drug present in compartment i, the constant k_{ji} is the fraction of drug in compartment j that is transferred to compartment i per unit time, and the constant K_i is the fraction of drug leaving compartment i per unit time.

The solution of equations (12) in general is a sum of exponential functions, i.e. it has the form

$$X_i(t) = \Sigma_j a_j \cdot e^{-\alpha_j t}. \tag{13}$$

Usually we cannot measure the amount of drug in an organ, but only its concentration, therefore we transform equations (12) by dividing both sides by V_i, the volume of compartment i, to get

$$\frac{dc_i}{dt} = \Sigma_j \frac{k_{ji} V_j}{V_i} \cdot c_j - K_i c_i, \quad i=1,2,...,n, \tag{14}$$

where c_i is the concentration of the drug in compartment i. Equations (14) differ from equations (12) only for the values of their coefficients, therefore their solutions still have the form of (13).

Equations (12) are based on some fundamental hypotheses that is worth recalling; they are:

a) The rate of transfer of the drug from one compartment to another is proportional to the amount of drug present in that compartment; this implies that the process causing this transfer is a process of order one.
b) The coefficients k_{ji} and K_i are constant.

In the case of equations (14) there is another fundamental hypothesis:

c) The concentrations measured in one point of a compartment are representative of the amount of drug present in the whole compartment, i.e. that compartment is homogeneous.

Definition of compartment

From the hypotheses described above ensues, even if this is not always explicitly declared, that a compartment must be a homogeneous set of particles defined by a physical boundary and by a chemical structure, such that they all have the same probability, constant in time, of being transferred or transformed (*transition probability*). Failing these conditions, the differential equations (12) and (14) cease to be valid.

Nevertheless many experimental data generated by systems certainly non-compartmental, can be fit with exponential functions like (13). This fact has been known for a long time and has been the subject of many discussions; one valuable reference on this matter is a paper by Bergner [1962] almost thirty years old, but still timely.

Coding and modulation of experimental data

To discuss the merits and the limits of compartmental analysis it is necessary to introduce here a few definitions [Rescigno and Beck, 1987].

All our knowledge originates from observations, but the data generated therefrom must be organized to become part of our consciousness. To this end we modify the experimental data we collect to represent them in some efficient way.

Table 4

t	1	2	3
c(t)	1	4	7

For instance the data of Table 4 can be represented in the form

$$c(t) = 3{\cdot}t - 2, \quad t = 1, 2, 3 \tag{15}$$

or in the form

$$c(t) = 3{\cdot}t - 2, \quad t > 0 \tag{16}$$

or graphically as in Figure 1 or Figure 2.

Equation (15) and Figure 1 convey exactly the same information as the original data of Table 4, even though the form of presentation is different. On the other hand equation (16) and Figure 2 do not provide the same information as Table 4. One cannot see from them for what values of t the quantity c(t) was measured (information lost), but one can see many values of t with the corresponding values of c(t) that were not observed experimentally (information added). A transformation of the data without any change in their information content is called *coding the data*. Equation (15) and Figure 1 are examples of coding the data of Table 4.

A transformation involving a change in the information content is called *modulating the data*. Examples of modulations are equation (16) and Figure 2.

Another example of modulation is given by equation

$$c(t) = t^3 - 6{\cdot}t^2 + 14{\cdot}t - 8. \tag{17}$$

All the above examples of modulation are consistent with the original data, i.e. the values of c(t) which can be obtained using Figure 2 and equations (16) and (17), are exactly the same as given by Table 4. But consistency with experimental data by itself is not sufficient to prove the validity of the hypotheses embedded in the equa-

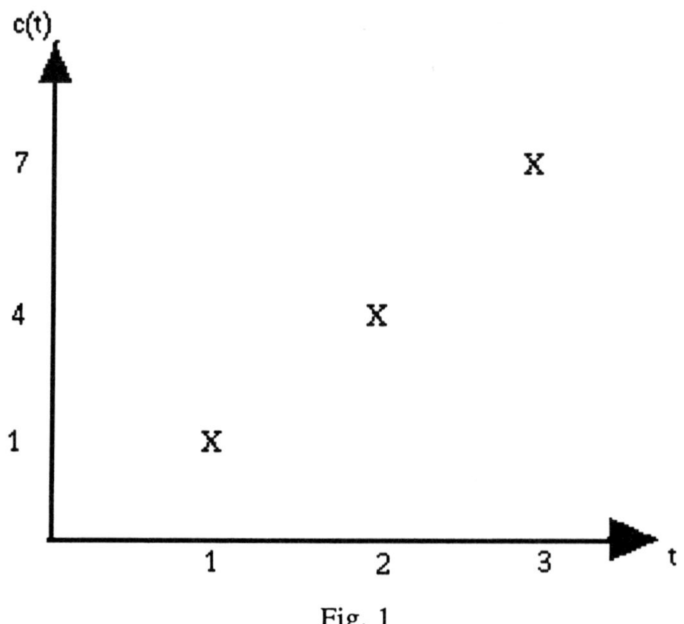

Fig. 1

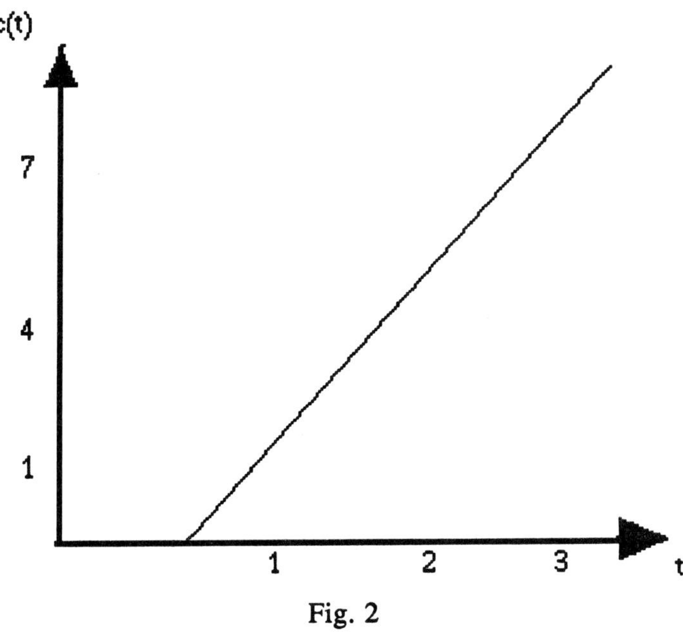

Fig. 2

tions used. As Popper [1935] repeatedly admonished, we can prove that a hypothesis is false, never that it is true.

Simulation and modeling

A special case of modulation is the *statistical regression*, namely the determination of the parameters of an equation of a given type, such that the divergence between computed and measured data is minimal, according to a specified criterion.

The selection of the equation to use for the statistical regression has one of these two aims:

1) To represent the experimental data in a form more concise or more evident.
2) To verify whether the hypotheses incorporated in that equation are consistent with the observations and to plan further observations that could reinforce these hypotheses or lead to their rejection.

In the first case, we speak of *simulation*, in the second of *model* [Rescigno and Beck, 1987]. This distinction is of a fundamental importance; simulation is only a description of the experimental observations, and is neutral as to the causes involved; model implies specific hypotheses that can be accepted or rejected.

The fitting of experimental data with a sum of exponential functions can easily be done with many computer programs; but it is essential to know the rationale behind such fitting. In the first case, the result of the computation is an efficient representation of the data, useful for the description of the experimental results and for interpolation, but deprived of any physical or physiological meaning. In the second case, we proceed because we have reasons to believe we are in the presence of compartments as defined above, and the parameters resulting from the fitting have a very precise physiological, as opposed to mathematical, meaning.

The microparameters

What we have said above may sound pessimistic, but it is not really so. Compartmental analysis is not always justified by the existing experimental conditions, but sometimes it is; and when it is not, the compartmental equations may still be useful as a heuristic tool. By observing more closely the meaning of the parameters involved in the compartmental equations, we shall try to gain a better understanding of the physiological parameters of the system under observation.

Let us consider the equation of radioactive decay seen previously,

$$\frac{dX}{dt} = -K \cdot X, \tag{18}$$

with the initial condition

$$X(0) = X_0. \tag{19}$$

Equation (18) and condition (19) describe the amount $X(t)$ of a drug present in a compartment where it was introduced at time 0 in a single dose X_0.

Suppose now that the drug is fed to the same compartment in a continuous infusion with rate r; equation (18) becomes

$$\frac{dX}{dt} = -K \cdot X + r, \tag{20}$$

and upon reaching steady state, i.e. when $X(t)$ becomes constant,

$$r = K \cdot X$$

or

$$\frac{1}{K} = \frac{X}{r} \; ; \tag{21}$$

but under this condition X is the amount of drug present in the compartment, while r is its rate of elimination; therefore 1/K is the average time the drug spends in that compartment before leaving it. This is called the *turnover time* or *transit time*.

With two connected compartments the pertinent equations are

$$\begin{aligned}
\frac{dX_1}{dt} &= -K_1 X_1 + k_{21} X_2 \\
\frac{dX_2}{dt} &= -K_2 X_2 + k_{12} X_1
\end{aligned} \tag{22}$$

with an obvious meaning for the coefficients K_1, K_2, k_{12}, k_{21}, called *microparameters* ; let the initial conditions be

$$X_1(0) = X_0, \; X_2(0) = 0.$$

Integrate all terms of equations (22) from 0 to t,

$$\int_0^t \frac{dX_1}{dt} dt = -K_1 \int_0^t X_1(t)dt + k_{21} \int_0^t X_2(t)dt$$

$$\int_0^t \frac{dX_2}{dt} dt = -K_2 \int_0^t X_2(t)dt + k_{12} \int_0^t X_1(t)dt$$

or

$$X_1(t) - X_0 = -K_1 \int_0^t X_1(t)dt + k_{21} \int_0^t X_2(t)dt$$

$$X_2(t) = -K_2 \int_0^t X_2(t)dt + k_{12} \int_0^t X_1(t)dt$$

If the two compartments are not closed, i.e. if the drug does not stay there forever, for $t \to \infty$ both $X_1(t)$ and $X_2(t)$ vanish, therefore

$$-X_0 = -K_1 \int_0^\infty X_1(t)dt + k_{21} \int_0^\infty X_2(t)dt$$

$$0 = -K_2 \int_0^\infty X_2(t)dt + k_{12} \int_0^\infty X_1(t)dt$$

thence

$$\frac{1}{X_0} \cdot \int_0^\infty X_1(t)dt = \frac{1}{K_1} \cdot \frac{1}{1 - \dfrac{k_{12}}{K_1} \cdot \dfrac{k_{21}}{K_2}} \tag{23}$$

$$\frac{1}{X_0} \cdot \int_0^\infty X_2(t)dt = \frac{1}{K_2} \cdot \frac{\dfrac{k_{12}}{K_1}}{1 - \dfrac{k_{12}}{K_1} \cdot \dfrac{k_{21}}{K_2}} \tag{24}$$

In identity (23) the factor $1/K_1$ is the turnover time of the first compartment. The following factor can be written as

$$v = \frac{1}{1 - \dfrac{k_{12}}{K_1} \cdot \dfrac{k_{21}}{K_2}} = 1 + \gamma + \gamma^2 + \gamma^3 + \cdots$$

where

$$\gamma = \frac{k_{12}}{K_1} \cdot \frac{k_{21}}{K_2}$$

is the fraction of drug that returns to the compartment after leaving it; therefore v is the average number of passages of the drug through that compartment. The product of this number of passages by the time spent in each passage is the average total time spent by the drug in all its passages through that compartment. This time is called *permanence time* [Rescigno, Thakur, Brill and Mariani, 1990].

Identity (24) can be analyzed in a similar way. The factor $1/K_2$ is the turnover time of the second compartment, the factor k_{12}/K_1 is the fraction of drug transferred from the first to the second compartment; the product of these two factors with v is the average time spent in the second compartment by the drug introduced into the first one; this time is called *residence time.*

With a simple extension of the above definitions we shall later be able to extend the concepts of turnover time, permanence time, and residence time to non-compartmental systems.

Matrix equations

Equations analogous to the (22) can be written for three or more compartments, and from those we can obtain expressions analogous to (23) and (24); the number of microparameters though increases very rapidly with the number of compartments, and their interpretation becomes laborious. It is simpler [Rescigno, Lambrecht and Duncan, 1983] to write those equations in matrix form,

$$\frac{d\mathbf{X}}{dt} = -\mathbf{X} \cdot \mathbf{K}, \tag{25}$$

where

$$\mathbf{X} = [X_1 \ X_2 \ ... \ X_n]$$

is the vector formed by the variables $X_1(t)$, $X_2(t)$, $X_3(t)$, ... of the different compartments, while

$$\mathbf{K} = \begin{bmatrix} K_1 & -k_{12} & -k_{13} & \cdots & -k_{1n} \\ -k_{21} & K_2 & -k_{23} & \cdots & -k_{2n} \\ -k_{31} & -k_{32} & K_3 & \cdots & -k_{3n} \\ \cdots & \cdots & \cdots & \cdots & \cdots \\ -k_{n1} & -k_{n2} & -k_{n3} & \cdots & K_n \end{bmatrix}$$

is the matrix of the microparameters.

It is easy to prove [Rescigno, Lambrecht and Duncan, 1983] that if the system is open, the matrix $\mathbf{K}$ is non singular, therefore it has an inverse; putting

$$\mathbf{K}^{-1} = \mathbf{T}$$

equation (25) becomes

$$-\frac{d\mathbf{X}}{dt} \cdot \mathbf{T} = \mathbf{X}, \tag{26}$$

and by integration

$$-\left(\mathbf{X}(t) - \mathbf{X}_0\right) \cdot \mathbf{T} = \int_0^t \mathbf{X}(t)dt ,$$

where $\mathbf{X}_0$ is the value of the vector $\mathbf{X}$ at time $t = 0$. For $t \to \infty$ we have

$$\int_0^\infty X(t)dt = X_0 \cdot T. \tag{27}$$

Let us now try to interpret physically the meaning of equation (27). Call t_{ij} the element of row i and column j of matrix T. Think of an experiment where the drug is introduced into compartment i at time t=0 as a dose $X_i(0)$. In this case all elements of vector X_0 are zero except the one in position i equal to $X_i(0)$. The product $X_0 \cdot T$ therefore is equal to $X_i(0)$ times row i of matrix T, i.e.

$$X_0 \cdot T = X_{io} \cdot [t_{i1} \; t_{i2} \; t_{i3} \; ... \; t_{in}];$$

calling $X_{ij}(t)$ the amount of drug introduced into compartment i at time t=0 and present in compartment j at time t, then

$$[\int_0^\infty X_{i1}(t)dt \; \int_0^\infty X_{i2}(t)dt \; ... \; \int_0^\infty X_{in}(t)dt \;] = X_i(0) \cdot [t_{i1} \; t_{i2} \; t_{i3} \; ... \; t_{in}] \;,$$

or

$$\frac{1}{X_i(0)} \cdot \int_0^\infty X_{ij}(t)dt = t_{ij} \;. \tag{28}$$

If $X_{ij}(t)$ is the amount of drug present in compartment j at time t, the amount of drug leaving that compartment in the interval of time from t to t+dt is $K_j X_{ij}(t)dt$ and

$$\int_0^\infty K_j X_{ij}(t)dt$$

is the total amount of drug introduced into compartment i that leaves from compartment j. But the amount of drug introduced into compartment i is $X_i(0)$, therefore

$$v_{ij} = \frac{1}{X_i(0)} \cdot \int_0^\infty K_j X_{ij}(t)dt = K_j t_{ij}$$

is the average number of passages of the drug through compartment j.

Remembering that each passage corresponds to the transit time $1/K_j$, we conclude that

$$v_{ij} \cdot \frac{1}{K_j} = t_{ij} = \int_0^\infty \frac{X_{ij}(t)}{X_i(0)} dt$$

is the total time the drug introduced into compartment i spends in compartment j.

For $i = j$, this is the permanence time; for $i \neq j$, it is the residence time. The ratio t_{ij}/t_{jj} is equal to the fraction of drug introduced into compartment i that reaches compartment j. This quantity, when related to the appropriate compartments, is called *bioavailability*.

17

Other moments

If we multiply both sides of equation (26) by t, and integrate from 0 to ∞,

$$-\int_0^t t \cdot \frac{d\mathbf{X}}{dt}\, dt \cdot \mathbf{T} = \int_0^t t \cdot \mathbf{X}(t)dt$$

$$\left(-t \cdot \mathbf{X}(t) + \int_0^t \mathbf{X}(t)dt\right) \cdot \mathbf{T} = \int_0^t t \cdot \mathbf{X}(t)dt;$$

with the usual hypothesis that the system is open, for $t \to \infty$ we have

$$\int_0^\infty \mathbf{X}(t)dt \cdot \mathbf{T} = \int_0^\infty t \cdot \mathbf{X}(t)dt$$

and remembering equation (27),

$$\int_0^\infty t \cdot \mathbf{X}(t)dt = \mathbf{X}_0 \mathbf{T}^2. \qquad (29)$$

By induction we get the general equation

$$\int_0^\infty \frac{t^p}{p!} \mathbf{X}(t)dt = \mathbf{X}_0 \mathbf{T}^{p+1} \qquad p=0,1,2,\ldots$$

and from it we can interpret the successive moments of the curves representing the amount of drug in all compartments accessible to measurements.

Let us now consider only equation (29); proceeding as with equation (28), we obtain

$$\frac{1}{X_i(0)} \cdot \int_0^\infty t \cdot X_{ij}(t)dt = (t^2)_{ij},$$

where $(t^2)_{ij}$ means the element of row i and column j of matrix $\mathbf{T}^2$.

Observing that

> $X_{ij}(t)$ = the amount of drug introduced into compartment i which is present in compartment j at time t,
>
> $K_j \cdot X_{ij}(t)$ = the rate of exit from compartment j at time t of the drug introduced into compartment i,
>
> $K_j \cdot X_{ij}(t) \cdot dt$ = the amount of drug introduced into compartment i that leaves compartment j in the interval of time t, t+dt;

the mean exit time from compartment j of the drug introduced into compartment i is therefore given by the fraction

$$\frac{\displaystyle\int_0^\infty t \cdot K_j X_{ij}(t)dt}{\displaystyle\int_0^\infty K_j X_{ij}(t)dt} \; ;$$

but K_j is constant, therefore

$$\frac{\displaystyle\int_0^\infty t \cdot K_j X_{ij}(t)dt}{\displaystyle\int_0^\infty K_j X_{ij}(t)dt} = \frac{\displaystyle\int_0^\infty t \cdot X_{ij}(t)dt}{\displaystyle\int_0^\infty X_{ij}(t)dt} = \frac{(t^2)_{ij}}{t_{ij}} \; .$$

This is the same ratio that appeared in the definition of transfer time, and is called the *moment of order one* of function $X_{ij}(t)$.

We discussed the properties of this ratio with reference to the drug introduced into compartment i; but the hypothesis of homogeneity of a compartment implies that the properties of the drug it contains are not dependent upon its origin. Therefore nothing changes if the drug originates from a different compartment. It follows that the moment of order one of $X_{ij}(t)$ is identical to the moment of order one of $X_j(t)$.

Nothing changes if in the expressions above to the amount of drug we substitute its concentration, because that is equivalent to dividing both numerator and denominator by the volume of compartment j.

In short we can write

$$\frac{\displaystyle\int_0^\infty t \cdot X_{ij}(t)dt}{\displaystyle\int_0^\infty X_{ij}(t)dt} = \frac{\displaystyle\int_0^\infty t \cdot c_j(t)dt}{\displaystyle\int_0^\infty c_j(t)dt} \; .$$

Other properties of the moments can be found in Rescigno, Lambrecht, and Duncan [1983a, b].

NON-COMPARTMENTAL MODELS

Moments of order zero

When the compartmental hypotheses are not justified, most of the above conclusions must be modified. If equation (18) is not valid, we cannot use definition (21); moreover we cannot any longer rigorously define transit time.

Nevertheless, we can write an equation analogous to equation (18):

$$\frac{dc}{dt} = - K \cdot c$$

where K is not constant; by integration from 0 to ∞ we get

$$c(0) = \int_0^\infty K \cdot c(t)\,dt;$$

we cannot export the coefficient K from the integral, but we can write

$$\frac{\displaystyle\int_0^\infty c(t)\,dt}{\displaystyle\int_0^\infty K \cdot c(t)\,dt} = \frac{\displaystyle\int_0^\infty c(t)\,dt}{c(0)}\,. \tag{30}$$

The fraction at the left hand side cannot be reduced to the form $1/K$, but it is analogous to it, and can be assumed as a definition of average turnover time of the drug at the point where $c(t)$ was measured. The fraction at the right hand side can be computed if $c(t)$ has been sampled for a sufficiently long time, and if both extrapolations for $t \to 0$ and $t \to \infty$ are acceptable.

Consider now equation (3) and rewrite it as

$$\frac{\displaystyle\int_0^\infty c(t)\,dt}{\displaystyle\int_0^\infty Cl \cdot c(t)\,dt} = \frac{AUC}{F \cdot D}\,; \tag{31}$$

the fraction at the right hand side can often be determined experimentally; it is equal to the inverse of the clearance only if this is constant.

A comparison of equation (30) with equation (31) shows the relationship between clearance and turnover time.

Consider a special case where the clearance is not constant. Suppose that a drug is eliminated with a process of order zero at concentrations above a critical value c_x and with a process of order one at lower concentrations. These hypotheses lead to the differential equations

$$\frac{dc(t)}{dt} = -k_0 \qquad \text{for } c(t) > c_x$$

$$\frac{dc(t)}{dt} = -k_1 c(t) \quad \text{for } c(t) \le c_x$$

with the initial condition

$$c(0) = c_0 > c_x\,;$$

for the continuity of the elimination process, it is also required that

$$k_0 = k_1 c_0.$$

The solution of these equations is

$$c(t) = c_o - k_1 c_x \cdot t \qquad \text{for } t < t_x$$

$$t_x = \frac{c_o - c_x}{k_1 c_x} \qquad\qquad c_x = c(t_x)$$

$$c(t) = c_x \cdot e^{-k_1(t - t_x)} \qquad \text{for } t \geq t_x$$

and by integration,

$$\int_0^{t_x} c(t)dt = \frac{c_o^2 - c_x^2}{2k_1 c_x}$$

$$\int_{t_x}^{\infty} c(t)dt = \frac{c_x}{k_1}$$

Therefore

$$\text{AUC} = \frac{c_x}{k_1} + \frac{c_o^2 - c_x^2}{2k_1 c_x} \; .$$

On the other hand, if the clearance had been constant, the solution would have been

$$\text{AUC} = c_o/k_1.$$

The difference between those two values might be very large; using equation (4) when Cl is not constant may lead to wrong values of the clearance.

Moments of order one

We have seen before that the moment of order one of function $X_i(t)$ measuring the amount of drug in a compartment, or of $c_i(t)$ measuring its concentration, is equal to the mean exit time of the drug from that compartment:

$$\frac{\displaystyle\int_0^{\infty} t \cdot X_i(t)dt}{\displaystyle\int_0^{\infty} X_i(t)dt} = \frac{\displaystyle\int_0^{\infty} t \cdot c_i(t)dt}{\displaystyle\int_0^{\infty} c_i(t)dt} \; .$$

If the compartmental hypotheses are not valid, we can still compute the ratio

$$\frac{\displaystyle\int_0^{\infty} t \cdot c_i(t)dt}{\displaystyle\int_0^{\infty} c_i(t)dt} \; ,$$

but this is not any longer the time of exit of the drug, but rather the average age of the drug at the point where the concentration $c_i(t)$ was sampled. We can still define the transfer time as the difference between two moments of order one, as in (5), even for non-compartmental systems.

Model-independent parameters

The term "model-independent" has been used by some authors as a synonym of "non-compartmental"; this use must be discouraged. Strictly speaking, no parameters of physical or physiological interest can be measured without a model.

For instance, the AUC, or Area Under the Curve measuring the concentration of a drug in the plasma versus time, as we have already seen, has meaning only in the context of some very precise hypotheses. It is equal to the amount of drug leaving the circulation divided by the clearance, when this last quantity is constant. In other conditions it may have a very different value. For instance, if the elimination of the drug follows a second order process, we have

$$\frac{dc}{dt} = -k \cdot c^2;$$

the solution of this equation is

$$c(t) = \frac{c_0}{1+c_0 kt}$$

where

$$c_0 = c(0)$$

is the initial condition. The integral of this function is

$$\int_0^t c(t)dt = \frac{1}{k}\ln(1+c_0 kt);$$

therefore in this case

$$AUC = \infty.$$

There may be many cases when the hypothesis of a constant clearance is not justifiable. The living body contains excellent homeostatic devices which control, for instance, carbohydrate metabolism and keep the blood-sugar levels within a relatively narrow range despite wide variations in ingestion and body activity. In such a case, any increase of D due to external administration would cause an increase of Cl or a decrease of F to keep equation (3) valid. Any measurement of

$$\int_0^\infty c(t)dt$$

is insufficient to decide whether Cl increased or F decreased. Therefore, in the absence of a precise model, the AUC has no practical value.

Furthermore, the method frequently used to estimate the AUC may be deceptive. This method consists in measuring the function c(t) up to a certain time t_1; to the integral computed over the finite interval of time $(0, t_1)$, one adds the integral

$$\int_{t_1}^{\infty} c(t)dt$$

estimated as though c(t) were an exponential function over the interval (t_1, ∞). The value of AUC obtained by this method is correct if the following hypotheses are valid:

a) c(t) is a sum of exponential functions,
b) the eigenvalue used to estimate

$$\int_{t_1}^{\infty} c(t)dt$$

is much smaller in absolute value than all other eigenvalues.

Failing these conditions, the estimated AUC may be smaller, and possibly much smaller, than the true AUC.

THE FUTURE OF PHARMACOKINETICS

All sciences, including Pharmacokinetics, may progress in three different areas, namely the solution of new problems, the elaboration of new methods, and the development of new symbolisms.

Each new molecule of pharmacological interest needs to be investigated for its pharmacokinetic properties; this is the routine job of the pharmacokineticist, and will never end as long as new substances will be added to the list of useful or potentially useful drugs.

The investigation of pharmacokinetic properties needs the help of several other sciences, for instance Analytical Chemistry, Physical Biochemistry, Numerical Analysis, etc. The methods of these sciences will progress concurrently making the solution of particular problems easier or more precise.

The third aspect in the progress of all sciences is often overlooked; it consists in the development of a new or an improved symbolism. Algebra never developed in a substantial way in Greece by reason of a lack of an appropriate notation, while the Chinese and Indians were able to solve algebraic equations several centuries before the Europeans did, just because they had efficient symbols for them. In this respect the science of Pharmacokinetics is in its state of infancy. Some efforts have been made to develop a unified nomenclature, notably the proposals by Rowland and Tucker [1980] and by Rescigno, Thakur, Brill and Mariani [1990], but besides the fact that they have not yet reached a general acceptance, those proposals are more on the line of abbreviations than symbols.

For instance AUC (Area Under the Curve) means the integral

$$\int_{0}^{\infty} c(t)dt,$$

where c(t) is the concentration of a drug in the plasma following a bolus injection at time 0; but it is not always clear whether the drug was injected IV or IM, and whether the dose injected was a unit dose or not. Furthermore, there are many properties connected to the above integral that could be used directly in the description of the fate of a drug in vivo. For instance [Rescigno, 1973; Rescigno and Michels, 1973a, b], the ratio of the AUC's measured in two different points of an organism, irrespective of the dose and the mode of injection of the drug, is another invariant quantity for a linear system. In addition to the advantage of being dimensionless, it has an interesting property, namely it can be used as the element of an algebra to describe the connectivity of the organism.

In these few pages it is impossible to review all concepts that constitute today the basis of Pharmacokinetics that could lead to important developments of this science. If a summary conclusion is possible from the concepts illustrated above, one could say that the present trend in Pharmacokinetics is mostly to move away from descriptive models and toward interpretative models. To this end it is necessary to pay more attention to the mathematical methods necessary to transform the hypotheses of a physical and physiological character into differential or integral equations.

The numerical solution of those equations is a separate problem; it is an important one but not crucial as it was a few years ago, thanks to the very efficient hardware and software available today.

Much more important is the logical approach to model building, namely to the problem of determining the minimum number of hypotheses necessary for the explanation of observed phenomena, and to planning the experiments in the most efficient way in order to verify the validity of the hypotheses postulated.

REFERENCES

Artom, C., G. Sarzana and E. Segré, 1938. Influence des grasses alimentaires sur la formation des phospholipides dans les tissues animaux (nouvelles recherches). Arch. internat. Physiol. **47**:245.

Beck, J. S. and A. Rescigno, 1964. Determination of Precursor Order and Particular Weighting Functions from Kinetic Data. J. Theoret. Biol. **6**:1.

Becquerel, H., 1896. Sur les radiations émises par phosphorescence. Comptes rendus Acad. Sci. **122**:420, 501.

Bergner, P.-E. E., 1962. The Significance of Certain Tracer Kinetical Methods, Especially with Respect to the Tracer Dynamic Definition of Metabolic Turnover. Acta Radiol. Suppl. **210**:1.

Biehler, W., 1925. Blutconzentration und Ausscheidung des Alkohols im Hochgebirge. Arch. exp. Path. Pharmacol. **107**:20.

Dost, F. H., 1953. "Der Blutspiegel. Kinetik der Konzentrationsverläufe in der Kreislaufflüssigkeit." Leipzig.

Gehlen, W., 1933. Wirkungsstärke intravenös verabreichter Arzneimittel als Zeitfunktion. Arch. exp. Ther. Pharmak. **171**:541.

Popper, K. R., 1935. "Logik der Forschung." Vienna.

Rescigno, A., 1960. Synthesis of a Multicompartmented Biological Model. Biochim. Biophys. Acta. **37**:463.

Rescigno, A., 1973. On Transfer Times in Tracer Experiments. J.Theor.Biol. **39**:9.

Rescigno, A. and J. S. Beck, 1972. Compartments. In: "Foundations of Mathematical Biology" (R. Rosen ed.), Vol. 2, page 255. Academic Press, New York.

Rescigno, A. and J. S. Beck, 1987. The Use and Abuse of Models. J. Pharmacokin. Biopharm. **15**:327.

Rescigno, A., R. M. Lambrecht and C. C. Duncan, 1983 a. Mathematical Methods in the Formulation of Pharmacokinetic Models. In: "Tracer Kinetics and Physiologic Modeling" (R. M. Lambrecht and A. Rescigno ed.), page 59. Springer-Verlag, Berlin.

Rescigno, A., R. M. Lambrecht and C. C. Duncan, 1983 b. Stochastic Modelling of Physiologic Processes with Radiotracers and Positron Emission Tomography. In: "Applications of Physics to Medicine and Biology" (G. M. Alberi, Z. Bajzer, and P. Baxa ed.), page 303. World Scientific Publishing Co., Singapore.

Rescigno, A. and L. D. Michels, 1973 a. On Dispersion in Tracer Experiments. J. Theor. Biol. **41**:451.

Rescigno, A. and L. D. Michels, 1973 b. Compartment Modeling from Tracer Experiments. Bull. Math. Biol. **35**:245.

Rescigno, A. and G. Segre, 1961 a. "La Cinetica dei Farmaci e dei Traccianti Radioattivi." Boringhieri, Torino.

Rescigno, A. and G. Segre, 1961 b. The Precursor-Product Relationship. J. Theor. Biol. **1**:498.

Rescigno, A. and G. Segre, 1965. On Some Metric Properties of the Systems of Compartments. Bull. Math. Biophys. **27**:315.

Rescigno, A., A. K. Thakur, A. B. Brill and G. Mariani, 1990. Tracer Kinetics: A Proposal for Unified Symbols and Nomenclature. Phys. Med. Biol. **35**:449.

Rowland, M. and G. Tucker, 1980. Symbols in Pharmacokinetics. J. Pharmacokin. Biopharm. **8**:497.

Rutherford, E., 1904. The succession of changes in radioactive bodies. Royal Soc. London Phil. Trans. **204**:169.

Rutherford, E. and B. A. Soddy, 1902. The cause and nature of radioactivity. Phil. Mag. **4**:370.

Teorell, T., 1937 a. Kinetics of Distribution of Substances Administered to the Body. I. The Extravascular Modes of Administration. Arch. Internat. Pharm. Thérapie. **57**:205.

Teorell, T., 1937 b. Kinetics of Distribution of Substances Administered to the Body. II. The Intravascular Modes of Administration. Arch. Intern. Pharm. Thérapie. **57**:226.

Widmark, E. M. P., 1919. Studies in the concentration of indifferent narcotics in blood and tissue. Acta Med. Scand. **52**:87.

Widmark, E. M. P., 1932. "Die Wissenschaftliche Grundlagen und die praktische Verwendbarkeit der Gerichtlich-Medizinschen Alkoholbestimmung." Urban & Schwarzenberg, Berlin.

Widmark, E. M. P. and J. Tandberg, 1924. Über die Bedingungen für die Akkumulation indifferenter Narkotiken. Theoretische Berechnungen. Biochem. Z. **147**:358.

EPISTEMOLOGY IN PHARMACOKINETICS

James S. Beck

The Bragg Creek Institute for Natural Philosophy
Calgary, Alberta, Canada

INTRODUCTION

There is a growing use of the term "model" and a growing recognition of the concept of model in pharmacokinetics. Current literature contains much evidence that there is considerable misuse of both the term and the idea. It is a costly misuse. This discourse is the result of an effort to clarify in a scientific context the nature of a model. Some limitations on what models can do and on what we can do to create them are indicated. A few examples serve to illustrate some uses and some pitfalls.

To put the issues in context and to clarify the terminology to be used, something must be said about statements. Any statement can be put into the form "A is a subset of B" ($B \supset A$), or "a is an element of the set A" ($a \in A$), where sets and elements can be primitive concepts. If **a, A, B** remain only symbols without representing things in the world of matter, energy and information, then these are analytic statements. If the relationships indicated by the statements are true or false by definition and/or by logic only — even if **a, A, B** are terms representing physical things — then again the statements are analytic. On the other hand, if they are identified as representative of things composed of matter and/or energy and their truth or falsity is a matter of experience, then the statements are synthetic [Pap, 1962]. The statement "bears are warm-blooded animals" ("the set of things which are bears is a subset of the things which are warm-blooded animals") is synthetic. The statement "bears are bears" ("the set of things which are bears is a subset of the set of things which are bears") is analytic.

Scientists, of course, are concerned professionally with synthetic statements. It is possible to put synthetic statements into three classes. The conceptually simplest may be called "statement of fact". A statement of fact is a representation of a specific event (the fact) such as "rat #23 died at 1407 h, September 3, 1990". Note that "the three coin-tosses resulted in tails/tails/heads" is also a statement of fact where the event has obviously separate components. It will cause no confusion to say "fact" in place of "statement of fact", as context will clarify any significant possibility of confusion.

A second class of synthetic statement may be called "law". A law is a representation of a relation which has occurred consistently in the past (and thus represents many facts) and is expected to recur consistently in the future. An example is "the volume of a fixed quantity of gas at constant pressure varies in direct

New Trends in Pharmacokinetics, Edited by A. Rescigno and A.K. Thakur
Plenum Press, New York, 1991

proportion to its temperature" or, in symbols, "V = KT where V represents volume, T temperature and K is a constant".

A third class of statement is a "theoretical statement" or "theory", where the theory is a set of theoretical statements. A theory is a body of statements of relations among words or symbols defined to represent in a general way components of the existing universe of energy, matter and/or information.

Whereas the distinctions between fact and law and between fact and theory are clear, the distinction between law and theory is less so. Theory includes non-observable constructs absent from simple statements of laws. Another distinction is that a law is not explanatory, whereas a theory — in a more or less limited context — is. To exemplify the difference between law and theory, consider the above statement (law): "The volume of a gas at constant pressure varies in direct proportion to its temperature". This tells us what has happened and what will happen to a defined component of the universe under defined conditions. But it does not tell us why. "Why" would tell us what the law is if we did not already know.

With this example, we may invoke the kinetic theory of gases to see the differences between law and theory. In the context of this theory we expect the components of matter to have greater kinetic energy at higher temperature and therefore to strike the container with a greater average energy so that at the same pressure there will be fewer collisions in a given time which implies lesser density and thus greater volume. With a little more detail we will have deduced the law from the theory. The kinetic theory applies to a very wide range of things and is based on very general relations and ideas about mechanisms. Clearly it differs from a law, even from a large set of laws. It involves understanding — a mysterious state of mind which gives us satisfaction, a feeling of security and a capacity to ask questions which promise to result in new laws, broader understanding and more questions. Another interesting feature of theories is that they force us to confront, much more than do laws, the question of scope — that is, to what part of the universe and in what conceptual context may it reasonably be applied. The gas law above requires a definition of "gas", "temperature", etc. — given the more general language which gives us "equal", "proportional", etc. But the kinetic theory and its use — its operational rules — requires much of logic, mathematics, concepts of probability, momentum, molecules and on and on. We have to be aware that, in this way, other theories include many of the same components of kinetic theory that should, in the interests of sanity and convenience, have the same meanings. To go a step further, we might want to regard kinetic theory as embedded in the theory of classical physics or of quantum mechanics or of some other theory more general than kinetic theory itself.

MODEL AND EXPERIMENT

These sorts of considerations apply to pharmacokinetics. The principal reason we are not forced to confront these problems consciously in pharmacokinetics (or in any other active, defined field of science) is that we who do pharmacokinetics are members of a community composed of individuals who — we assume — share common definitions and assumptions about pharmacokinetics. So it would be wasteful and annoying to start every paper we write or every talk we give with a lengthy, tortured exposition of definitions and premises based on primitive concepts which all of us accept.

What, then, is the investigator in pharmacokinetics after? Is it more facts? Is it refinement of accepted laws or new laws? Is it refinement or extension of a theory or a new theory?

In circumstances where we want to apply our knowledge to predict an outcome or, conversely, to choose conditions which will lead to a specified outcome, our goal may be very limited: perhaps we want to predict on the basis of a law or laws the concentration of a drug added to a known volume of its solvent. This may be very important, but it is not science. Like engineering or the practice of medicine, it is an application to build something or to manage a disease, or the like, but it is not a quest for new scientific law or theory. As a matter of fact, the victim of beriberi may be completely indifferent to the theoretical basis on which she or he is given thiamin, be that basis modern medical knowledge of vitamin deficiency or Amerindian spiritualism.

It has been argued that a mathematical representation of the behavior of a system which is designed to serve only as a tool to recall or to predict how the system has behaved, or will behave, in some particular context, might be called a *simulator,* and should be distinguished from a mathematical model, the latter being for the purpose of testing hypotheses about the system [Rescigno and Beck, 1987]. An example of a simulator is an artificial limb, which is intended to behave as a limb, but clearly is not intended to help gain insight into the nature of the natural limb. Another simulator is an equation which produces numbers closely following the numbers observed for a primary system over a narrow range of the independent variable. The equation might be a fourth-order polynomial and the primary system a quantity of gas observed for the values of its pressure at varying temperature.

In contrast to a simulation or an application, a scientific pharmacokinetic investigation has the purpose of explanation of phenomena of the material universe. New explanation requires new theory or extension of extant theory or new application of extant theory. Now, how, precisely, do we do that? Let us imagine ourselves investigating pharmacokinetically a system and watching philosophically what we do. Say we are interested in a suspension of cells identical with respect to transmembrane transport of the water-soluble substance A. We have observed that after two hours we find only a trace of A in the supernatant; essentially all of it is sedimented with the cells. In another laboratory, highly regarded by us, it has been observed that A is irreversibly degraded in the cytoplasm of these cells in a way consistent with a first-order reaction with reaction constant $K_e = 0.03$ min^{-1}.

The question we ask is whether the internalization of A into the cytoplasm $(A_o \rightarrow A_i)$ is first-order or second-order (autocotransport). We formulate two conceptual models:

<table>
<tr><td>Model 1</td><td>Model 2</td></tr>
<tr><td>$A_o \rightarrow A_i$</td><td>$2A_o \rightarrow 2A_i$</td></tr>
<tr><td>$A_i \rightarrow$</td><td>$A_i \rightarrow$</td></tr>
</table>

We see that we can add to the extracellular phase of the cell suspension a known amount of A and mix it so rapidly with respect to the rate of transmembrane transport that it is an effectively instantaneous addition. Furthermore, we can measure the intracellular A_i by cooling the suspension, centrifuging, breaking the cells, etc. very rapidly with respect to the rates of transport and degradation.

So we formulate mathematical models corresponding to the two conceptual models, observe their behavior (determine the general forms of the temporal variations in the variables to be measured, using educated guesses as the values of the unknown parameters), make corresponding measurements on the cell suspension, and compare the sets of data from the models to the set of data from the physical system. The mathematical models in the form of differential equations and initial conditions, where $a_o(t)$, $a_i(t)$ are $[A_o]$, $[A_i]$ at time t are:

Model 1	**Model 2**
$da_o/dt = -K_1 a_o$	$da_o/dt = -2K_2 a_o^2$
$da_i/dt = K_1 a_o - K_e a_i$	$da_i/dt = 2K_2 a_o^2 - K_e a_i$
$a_o(0) = 1$	$a_o(0) = 1$
$a_i(0) = 0$	$a_i(0) = 0$

The mathematical models give us some idea of what to expect so we can design an experiment with the expectation that it will either distinguish between the models or eliminate both of them as useful possibilities. Say we do our experiment and get the data represented by the X's in Fig. 1. The mathematical models can be solved analytically or numerically with one unknown parameter in either: K_1 of Model 1 and K_2 of Model 2. We can choose the respective parameters by fitting the curves to the data by some criterion such as minimum sum-of-squares-of-residuals. The two results might be like those shown in Fig. 1, solid curve for Model 1 and dotted curve for Model 2.

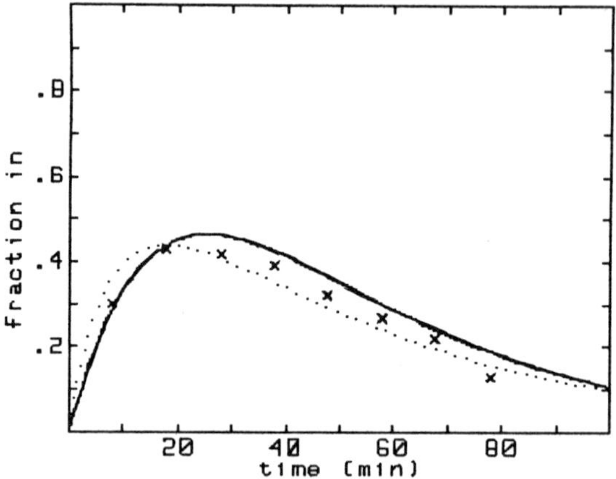

Figure 1. Observed data (x) and predictions of Model 1 (solid curve) and Model 2 (dotted curve) for fraction of injected amount of A in intracellular space.

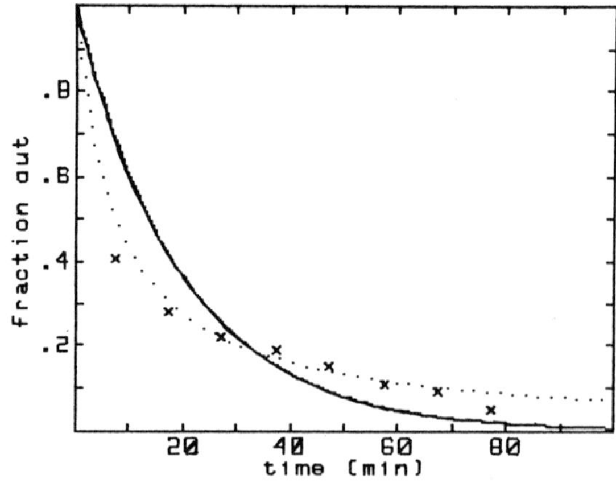

Figure 2. Observed data (x) and predictions of Model 1 (solid curve) and Model 2 (dotted curve) for fraction of injected amount of A in extracellular space.

It seems that this is not a good experiment to discriminate between the two hypotheses. So we must design a new experiment. We are equipped to measure $a_0(t)$, so we can try that. The mathematical models in integral form are:

Model 1 for A_0   **Model 2 for A_0**

$a_0(t) = e^{-K_1 t}$ $a_0(t) = 1/(2K_2 t + 1)$

$a_0(0) = 1$ $a_0(0) = 1$

(These are analytic solutions for the first dependent variable of the differential equations given above.) Choosing again the single unknown parameter for either model to fit the data, we see in Fig. 2 that there is a greater possibility for discrimination than we had with observation of $a_i(t)$. If the data we obtain are as shown by the X's in Fig. 2 and if they are sufficiently reliable, we may have succeeded in eliminating one of our models (Model 1 in this case) and retaining the other as a useful possibility.

MODEL AND EXPERIMENTAL DATA

To this point we have been concerned with an investigation by which we want to decide whether a particular system is better represented by one or another of two proposed mechanisms. We do this by comparing the behavior of the system itself with the behaviors of the two models of the system. But, of course, we excluded from consideration a lot of reasonable models. The structure of our question has been: of the two Models 1 and 2, which better represents the system? Actually we might have concluded that neither represents the system. In any case the assumption that the set of two models includes all possibilities is untrue in the context of the theory of transmembrane transport.

We are often tempted to take a next step and claim that the system is one of transport of two molecules of A at a time. Our experiment, of course, has not demonstrated that. Perhaps we can say that the system does *not* behave under these conditions as one-step, first-order transport of A from extracellular medium to cytoplasm. But we cannot say with any certainty that it does behave as a second-order translation of A. More precisely, we can say that it behaves as a second-order system would behave under these conditions, but we cannot say with certainty that the mechanism of the system is second-order. This enigma confronts us on all sides. From the physical nature of the system, we might expect alternative mechanisms which would exhibit the behavior we have observed. In the conceptual domain, we can imagine alternative mechanisms with similar behaviors under these conditions. And in the mathematical domain, there are infinities of functions within many classes of functions, all of which are consistent with the data to an arbitrary closeness [Beck and Rescigno, 1970].

Properly, we focussed on the big question, the mechanism of the system. But if we are to use our result to design more experiments or to anticipate for other purposes the behavior of such a system, we also need the value of the determined parameter K_2. The value of K_2 which gave the best fit to the data for $a_0(t)$ is 0.065 min^{-1}; for $a_i(t)$ it is 0.045 min^{-1}. While we did find the data for $a_0(t)$ better for distinguishing between the two models, we have no obvious reason for thinking them better for the estimation of K_2. While considering that question we will be encouraged that observations of two distinct aspects of the behavior of the system lead us to values so close to one another. On this issue the relative reliability of the two kinds of data would be significant.

Whether the data are sufficiently reliable to make it possible to reach a useful conclusion is a matter of judgement which may or may not be based on results of statistical inference. We have not yet addressed statistical matters. We have worked with mathematical models of the relations among certain variables, parameters and initial conditions. We have ignored the problems of accuracy and precision of measurement, except to say that these must be taken into account in evaluating the similarity or disparity of model output and experimental data — that is, the "fit" of the "theoretical curve" to the observations. Normally, we make multiple measurements on what is taken to be the same state of the system. And we usually observe that the results are not the same for a given state. To deal with this problem we may choose the methods of inferential statistics to get some sort of estimate of reliability of the averages of the data or of the estimated parameters, and so on. This requires a model of, a hypothesis about, the measurement — perhaps as outcomes of independent events with a normal distribution. Suffice it to say that this is another model having to do with another system — also involved in the investigation of the pharmacokinetic system, but different.

USES AND LIMITS OF THE MODEL

Recently van der Steen [1990] argued that "fact" and "meaning" should be distinguished and not mixed. In the context of our discussion of pharmacokinetic models, the data are clearly facts. The conceptual models introduce terms and concepts which have meaning. The discussion of van der Steen pertains to taxonomy principally and does not explicitly deal with the kinds of terms which result in symbols which are assigned values as in our case. But the distinction between fact and meaning is important for us.

Is a fractional turnover constant of a compartmental model a fact of nature waiting to be observed? Given that there are alternative models with constants having different meanings, we must conclude that they are not facts, except perhaps in a contingent sense within the context of an unambiguously specific model — fact of the mind rather than fact of nature. The facts associated with the system are the events of its behavior, not our mathematical models or conceptualizations of its structure or mechanism. In our imaginary example the value of K_2 exists within a law of nature; its meaning exists within a theory of nature.

As we watched ourselves doing this pharmacokinetic investigation, we might have noted some different uses of a model. The model incorporates a lot of history. In our present case both Model 1 and Model 2 include some specific past behavior such as the irreversibility of the transport of A and of the cytoplasmic degradation of A, the value of K_e, etc. The model is a mnemonic device which helps us remember these features. It also can tell us to some degree of accuracy what happened in a quantitative sense, what data were generated, when we observed the system in the past. This use of a model might be called retrodictive.

The retrodictive use of a model is the time-reverse of the predictive use. The predictive use within the context of the experiment is two-fold. It enables us to do rational experimental design; we can decide purposefully what to observe, where and when to observe it and often what accuracy and precision is required to make conclusions about particular hypotheses. Secondly, it produces a behavior resulting from our conjectures about the system which we can compare with our observations of the system. And this comparison serves as a basis for judgements about our hypotheses. In a broader context, if the model is confirmed positively — that is, if we conclude that it represents the pertinent characteristics of the system — then it serves generally to predict particular aspects of the behavior of this class of system under a certain range of conditions. And this predictive use may go beyond experimental design, beyond science, into applications of all sorts; the model may also be a simulator.

Practically, retrodiction and prediction are very different, but in form they are very similar. Hooke's law (a model well enough accepted to be called a law) relates force to extension of an elastic body by a single constant: $F = kL$. Retrodictively, it relieves us of the need to have access to a table for each material of interest (value of k) with virtually an infinite number of number-pairs (L, F) representing past observations. Predictively, it substitutes for countless observations yet to be made on materials for which k is known.

There is a third use of a model in the realm of conceptualization and understanding. Our hypothetical pharmacokinetic investigation began with our curiosity about a system defined by considerable history of investigation. From the beginning we had a model in mind. This model gave us a basis for our own conceptual models representing the alternative hypotheses about the process we were interested in. This use of the model is the last step of what went before and the first step of the investigation to come. At this point the model clarified our assumptions, defined the system of interest and enabled us to ask our questions clearly and to put them into usable form.

Subsequently the conceptual models allowed us to generate mathematical models. The mathematical models, in turn, generated data to be used to design rationally our experiment and data to compare with the results of the experiment on the physical system. Then the model made it possible to evaluate hypotheses on the basis of this comparison. And then the influence of the experiment appeared in the form of ideas about the model — its rejection or how it might be used or modified. And we come full circle, since at this point we are on the threshold of further inquiry and investigation.

Whether we recognize the fact or not, this process of using models is far older than science as we know it. The "new trends in pharmacokinetics" which have to do with models are two. One is the increasing explicit attention to models, and to methods involving the conscious use of models, that we see in pharmacokinetic — and other — literature. (Of the twelve titles given in the publicity about this advanced-study institute three include the word "models" and two the word "modeling".) The other trend is the writing and use of computer programs designed to extract information about models or relations between models from raw data or averages of raw data. These trends have led some investigators to claim that models can be generated from data alone, a claim which needs examination.

We cannot ask what a system is without having decided what it might be; that is, we must have models whose behavior can be compared with the behavior of the system. This seeming paradox is hard to accept. We want to approach a system of interest and have it tell us what it is. But is that possible? We consider an object in a room. What is it? Perhaps we identify it as a chair. This seems a straightforward and simple process. But on further examination of the process (not the chair) we recognize that "chair" is a model and our simple conclusion is unavoidably the end of a process of comparing the data we collect from the object — shape, size, feel as we sit — to the shape, size, feel of the model. In fact we may not be certain that it is a chair; it might be a modern sculpture which was a chair until the artist mounted it and gave it a title such as "Chair, 1990". On the other hand we see that it is not a blackboard, not a pharmacologist, etc.

Understandably, some of us don't want to accept this epistemological reality. We now see in the literature efforts to minimize the influence of our preconceptions in determining our choices of models, to rely entirely on data collected from a physical system. The extreme of this sort of effort has been accompanied by the term "model-free analysis". The idea of so-called model-free analysis is to start from a set of data taken from observation of a system and have the data tell what the mathematical model is, the inverse of predictive use of a model. There is no bias, there are no preconceptions. There is only pure empiricism and a conceptual *tabula rasa*. But we simply cannot do that.

THE LIMITED POWER OF EXPERIMENTAL DATA

To illustrate the impossibility, imagine that we have the set of data:

x	y
0	0
1	2
2	4

If we plot these on a rectilinear co-ordinate system, we can draw a straight line through the three points and then we might say $y = 2x$. We have a mathematical model. But that is not what happened. What happened is that we had the model "y is directly proportional to x" and then used the data to find the orientation and location of the straight line. Or, equivalently, we used a straight-edge to connect the points. That is, we said $y = ax + b$ and found that, when $a = 2$ and $b = 0$, the line passed through the data points.

As a matter of fact, the same three points lie exactly on the curves of many other classes of function. And on an infinity of cases of some of them. So, if the data could talk, they would only give us a lot of choices. Then the question arises: what model(s) do we choose to consider? Of course, there will be an answer to that question, but it will be answered on the basis of pharmacokinetics or physiology or physics or the like, not on the basis of the data alone. So why not do that thinking before producing the data? At least to some extent we must. The data tell us nothing except where they lie with respect to models we impose. Put another way, the mapping from model to data is one-to-one (that is, a particular deterministic mathematical model gives us one set of data for a particular set of conditions); the mapping from data to model is one-to-many (that is, a single set of data points may be represented by many models, possibly an infinite number) and therefore is indeterminate.

Perhaps the quintessential example of a methodological effort to extract a model from data is represented in a paper by Guardabasso, Munson and Rodbard [1988]. In this paper the authors address the problem of deciding on a model of a system, or a set of systems assumed to be meaningfully related, from which various sets of data have been collected. What the authors mean by "system" here requires particular attention, because they propose to alter curves derived from the sets of data so that they all coincide with a single curve they call the "template". So before any conclusion can be drawn, it is necessary that one specify whether this template is some sort of "average" of curves fit to different data sets or is drawn to minimize the "distance" of the other curves from itself or whether it represents the behavior of an ideal (in a sense to be designated) system or of a system under some standard conditions, whether the adjustments required are changes in conditions or changes of parameters or changes of relations among parameters or among variables or both. These questions cannot be answered without a specific mathematical model a priori except for a template determined from averages or minimum sum of "distances", in which cases no substantive question exists. However, some limited conclusions about the relationships among the data sets are possible.

The procedure of Guardabasso, Munson and Rodbard [1988] is to derive curves from the sets of data by forming spline functions representative of the data sets. (We are dealing with data pairs — that is, functions of a single variable — and therefore with two-dimensional plots.) A spline function is a continuous curve through a series of points, called "knots", composed of polynomials over the intervals between knots of degrees n_i whose first n_i-1 derivatives are continuous. The knots may or may not coincide with points to be approximated by the spline function. Fig. 3, taken from the above authors, shows three examples fit to the same

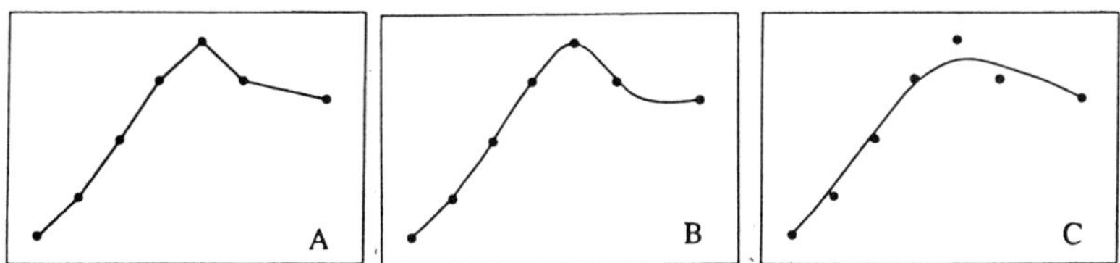

Figure 3. Illustration of spline functions fitted to seven
data points. See text for explanation. From Fig. 3 of
Guardabasso, Munson and Rodbard [1988].

set of seven data points. Panel A is the simple case where the knots are the data points and the polynomials are all of order 1; the continuity condition is simply that imposed on the function. Panel B shows the case where the knots again are the data points, the polynomials are all of order 3 and the function and its first and second derivatives are continuous. Panel C shows the result where the knots are at the horizontal values of the data points but not necessarily congruent and certain continuity and smoothness conditions are met at a minimum average distance from data points. Superficially, one might find the spline functions attractive for their relative lack of conceptual bias; they include only conditions of continuity, smoothness and acceptable measure of closeness to data points. And they may be good predictors of behavior of a system. But the lack of bias is unavoidably accompanied by a lack of scientific content.

These spline-function curves are then manipulated into coincidence with the template by changes in as many as four plotting factors. These correspond to shift and scaling of the two axes, individually for the plotting of each set of data. Another way to state this is that the component of either dimension of each point is multiplied by a constant and summed with another constant — the constants (possibly) being different for each set of data. The result is that all the curves are collapsed, more or less, onto a single curve. One also knows the shifts and scale changes required to do this for each curve. That might be useful to know — but only in the context of a model which can be interpreted in a physical (perhaps pharmacokinetic) sense.

Whether the canonical template curve is defined by a standard set of conditions or by some kind of averaging routine applied to the sets of data or to the spline functions, it is still nothing but a curve until it is compared to the behavior of a model

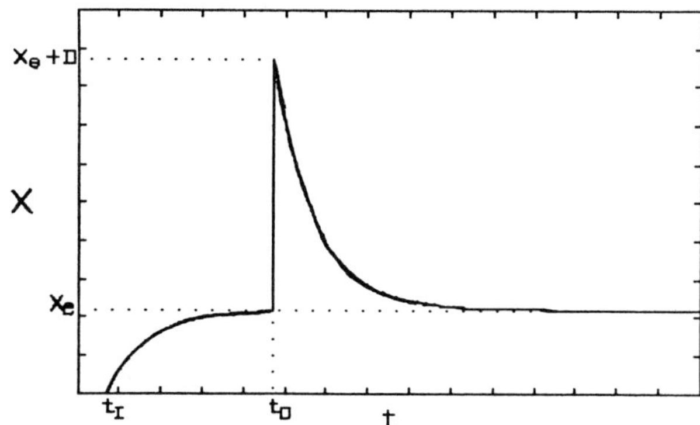

Figure 4. Hypothetical time course of amount of X
infused from time t_I at a constant rate and injected in
amount D at time t_D in a one-compartment system
with constant fractional washout.

— mathematical or physical. And then the usefulness depends as much on the model as on the canonical curve.

Skepticism about the data-driven transformation of data simply to find a canonical curve should not divert us from the purposeful application of scale changes based on realistic models. Shifting of axes can be convenient. In Fig. 4 we see what might be the result of measuring the amount of X in an open, well-mixed system at constant volume and constant flow as an infusion of X is begun at t_I and a bolus injected at t_D. If we are interested in estimating the fractional turnover constant which governs the washout of X, we may find it very convenient to make the scale shifts $t \rightarrow t-t_D$ and $x \rightarrow x-x_e$. If we do that, however, it is both based on a model and done for an a priori purpose.

An example of a useful scale change is the use of a dimensionless variable to eliminate the inconvenience of accounting for, say, different units in which an amount of a chemical is expressed. Another is normalization of a response by relating it to a maximal response. Again these are scale changes based on a model and for a specific purpose. And they may well have the same benefits as some of the maneuvers recommended by Guardabasso, Munson and Rodbard [1988], though in these cases the canonical form is a canonical space for display of the behavior of the system and of the model rather than a manufactured behavior.

A more complex problem of comparison which is important in pharmacology is that of making judgements across species lines or across age differences with respect to responses to drugs. Here dimensionless variables and normalizations can be very helpful. And here again, to the extent that they are helpful, they are model-based and specifically purposeful.

TWO EXAMPLES

Goresky and Rose [1977], preceding the paper by Guardabasso, Munson and Rodbard [1988] by a decade, give a nice example of purposefully shifting data. These investigators used a procedure which was based on a very specific physiological model and which was designed to answer a clear physiological question posed in advance. One of the things they wanted to know was the intralobular, extravascular, and extracellular volumes of the liver for a set of soluble substances. Goresky and Rose [1977] had reason to assume that the permeabilities of the substances they studied were such that equilibrium across the capillary wall was effectively instantaneous and that the concentrations in the extracellular space were uniform. They formulated a mathematical model of the concentration at an experimentally accessible outflow point with a single parameter which is the ratio of accessible volume to the volume in the sinusoids.

Then they made the measurement on erythrocytes and relied on the well-founded assumption that these were confined to the capillary/sinusoid space. Fig. 5 is taken from their paper; in the top and middle panels it shows the observed results for erythrocytes and various substances. The lowest panel shows the superposition onto the canonical (erythrocyte) curve which results from multiplying the measured concentrations by the factor which results from choosing the parameter appropriately. This superposition confirms a common mechanism for the behavior of these various substances in this system and provides a means of estimating, for each, the parameter of interest. The specificity and a priori purposefulness of the model contrast sharply with scale changes applied solely to produce fits to a curve chosen without clear rationale.

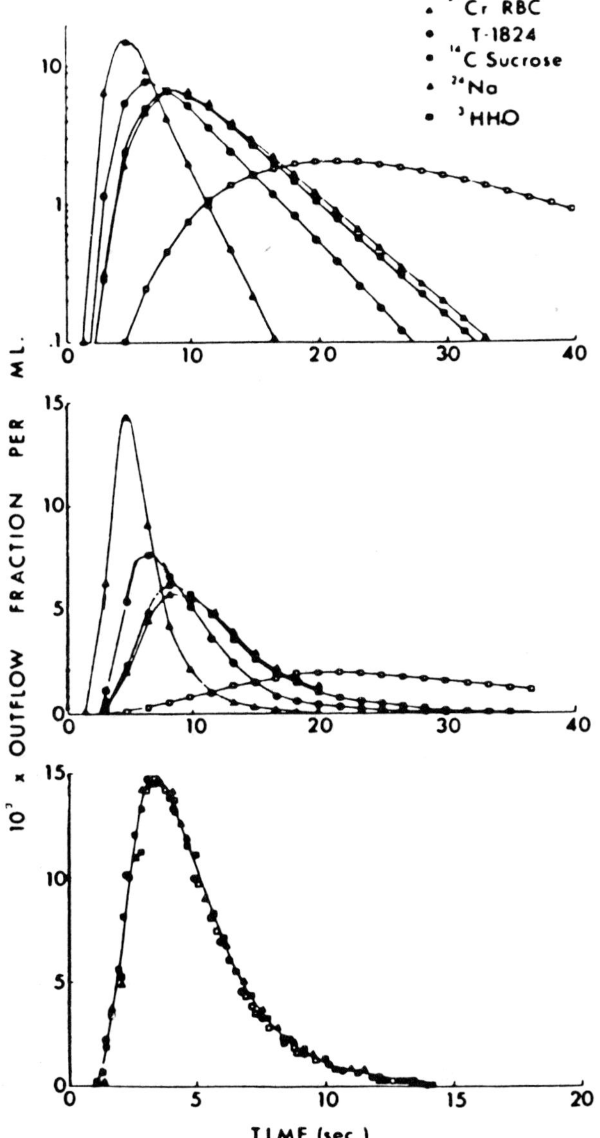

Figure 5. From Fig. 1 of Goresky and Rose [1977]. Top two panels are linear/linear and log/linear plots of observations on substances indicated. Bottom panel is superposition resulting from adjustment of the single parameter representing accessible volume.

One example of the use of models has taken on the form of a specific dispute extending over some 25 years. It is perhaps of particular interest to pharmacologists, as it involves the distribution of substances in systems, such as the human body, and has included some criticism of compartmental models. We should remind ourselves that the compartmental model has a general, reasonable, physical basis. Perhaps it is true that some of us have used it arbitrarily without consideration of its origins, perhaps out of habit or because we have computer programs to fit it to data. But its basis is the conceptual model of stochastic transitions between states (chemical, spatial, etc.) of entities (e.g., molecules, atoms) behaving independently and with constant transition probability densities (e.g., probability per unit time).

Recently, Wise and Borsboom [1990] have added to this long-standing discussion of the choice of compartmental models or power-law models in an analysis of what they consider exceptionally good sets of data. One of the issues they raise is that of parsimony; of two acceptable models the authors much prefer the one with fewer parameters. In comparing the compartmental model with the power-law model they count each coefficient of an exponential term and each coefficient of time in the exponent as an independent parameter. But if we are considering a compartmental model for the variations in amounts in certain states, in which the mathematical model results from a conceptual model, then the only parameters are those of the conceptual model, which should have physical meanings. In the compartmental model the coefficients are composed of sums, products, ratios, square roots of the parameters in the case of the exponents and are similar expressions involving parameters and boundary conditions in the case of the coefficients of the terms. So, for example, in a two-compartment model where there is only one exit from the system and the substance represented is initially added to only one of the compartments, there are only three parameters and one initial condition, though the expressions do have two exponential terms.

The alternative to the compartmental model in this dispute is a mathematical model consisting of two power laws [Anderson et al., 1963]. Up to time $T > 0$ the system behaves as $A_1 t^{a_1}$ and for $t > T$ as $A_2 t^{a_2}$. (In the 1990 paper Wise and Borsboom connect the two curves smoothly with a gamma function in which case there are eight parameters when the parameters of the gamma function and the additional transition time are added.) The four parameters A_1, a_1, A_2, a_2 and the fifth parameter T are chosen to minimize the sum-of-square-residuals. The mechanism behind such a representation is not discussed in the 1990 paper, though at least one of the authors has dealt with the matter before [Wise, 1971] and Norwich and Siu [1982] have discussed it in detail. The parameters in the power-law expressions do have physical meanings; the exponents are related to the geometry of the system represented and to the paths through which the substance of interest moves. The transition to a second power law is unexplained and apparently is used only to have a satisfactory fit to the data. Given that at least in some cases the sums of residuals are smaller for the compartmental model, one might reject the power-law model *because* of that unexplained and perhaps unrealistic transition.

DISCUSSION

Having looked at some examples of how models in pharmacokinetics could be and have been used, we may review some truths and caveats about models. First of all, if we are doing scientific investigation we are using models, whether we admit it or not. It is a fantasy to think that the data we gather from a physically real system can alone tell us the nature or mechanism of the system observed. It follows that manipulating data and finding representations of the data in mathematical expressions are useless exercises in the absence of meaningful correlates, in a scientific context, for the terms and relations in the mathematical expressions. Without such correlates in the observation language [Carnap, 1966, p. 258] and relations among them — which are our hypotheses — there is nothing scientific to be learned from an experiment. Nor is there any question or idea for a future experiment.

In contrast to a search for a model that behaves so as to duplicate the data observed in the experiment, the desirable and productive strategy is to design a model that is as likely as possible to be shown to be wrong — where its being wrong *is a consequence of an interesting hypothesis from which it was derived*. We may be fond of our models, the products of our knowledge and imagination, but we must put them at maximum risk. The experiment must be one with the best chance of showing the model to be a representation of a system different from the one we are trying to understand. If the critical hypothesis is wrong we want to know that it is wrong with as much certainty as possible. If the model does seem to represent the system well, then the greater the risk it survived, the more confidence we can have in it as a fruitful starting point for the rest of our investigations, predictions and understanding.

Finally — given the evidence in the literature of elaborate and expensive pursuits of curves which "fit" particular data and given the claim of "model-free analysis" — it bears repeating that we know a priori that when we are not constrained by hypotheses and models we can always find curves from data. Indeed, we can find a limitless number of curves passing as close to the data as we wish. Since we already know that, why waste our resources demonstrating it?

ACKNOWLEDGEMENT

This work was done when the author was in the Division of Medical Biophysics, University of Calgary.

REFERENCES

Anderson, J., S.B. Osborn, R.W.S. Tomlinson and I. Weinbren, 1963. Some applications of power law analysis to radioisotope studies in man. Phys.Med.Biol. **8**:287.

Beck, J.S. and A. Rescigno, 1970. Calcium kinetics: the philosophy and practice of science. Phys. Med. Biol. **15**:566.

Carnap, R., 1966. "Philosophical Foundations of Physics". Basic Books, New York.

Goresky, C.A. and C.P. Rose, 1977. Blood-tissue exchange in liver and heart: the influence of heterogeneity of capillary transit times. Fed. Proc. **36**:2629.

Guardabasso, V., P.J. Munson and D. Rodbard, 1988. A versatile method for simultaneous analysis of families of curves. FASEB J. **2**:209.

Norwich, K.H. and S. Siu, 1982. Power functions in physiology and pharmacology. J. Theoret. Biol. **95**:387.

Pap, A., 1962. "An Introduction to the Philosophy of Science". The Free Press of Glencoe, New York.

Rescigno, A. and J.S. Beck, 1987. The use and abuse of models. J. Pharmacokin. Biopharm.. **15**:327.

van der Steen, W.J., 1990. Concepts in biology: a survey of practical methodological principles. J. Theoret. Biol. **143**:383.

Wise, M.E., 1971. Skew probability curves with negative powers of time and related to random walks in series. Statistica Neerlandica **25**:159.

Wise, M.E. and G.J.J.M. Borsboom, 1990. Two exceptional sets of physiological clearance curves and their mathematical form: test cases?. Bull. Math. Biol. **51**:575.

MODEL: MECHANISTIC vs EMPIRICAL

Ajit K. Thakur

Hazleton Washington, Inc.
Vienna, VA 22182

INTRODUCTION

According to the Oxford English Dictionary, the word Model (French *modèle*, Italian *modello*, Latin *modulus*) has many meanings. In our context, the word could mean 'something that accurately resembles something else', 'an object of imitation', or 'a perfect exemplar of some excellence'. The Sanskrit equivalent of Model is 'Pratirupa', i.e. a perfect copy or imitation. From sociology to science, models have been used for centuries. National heroes like George Washington, Mahatma Gandhi and many others are models that many parents wish their children to follow (obviously there are hundreds of counter examples as well!). In science, the ancient Hindu mathematicians used picture models to calculate astronomical parameters. And of course, we have the model of gravity in Isaac Newton's Apple! There is no branch of science today that does not employ models to understand the system under study. Unfortunately, even with such a long tradition, the word model still brings out a lot of confusion and disagreement. This is particularly true when the statisticians are asked to make certain statements regarding observations made in various fields. The purpose of this discussion is not to settle any controversies, but to open the door for understanding such a broad concept.

From a systems analytic standpoint, we need some terms to be defined before we can discuss the concept of a model. These definitions are elegantly expressed in Rescigno and Beck [1987]. Accordingly:

(1) A system under study is the primary system.
(2) Any aspect of this primary system that an investigator uses to study the system is a secondary system. The data or a graph representing the primary system will then be secondary system.
(3) A model is a secondary system used to verify any hypotheses on the primary system.

Obviously, if one knew everything about the primary system, one would not need a model. Unfortunately, many of the real systems one wishes to study may not be wholly accessible. As a result, one has to use models to explain or verify certain aspects of such systems. Legal, social, religious, and ethical practices of the civilized world do not always allow us to pry into living subjects (some times not even dead subjects) to understand all intricate biological functions. So there must be needs for

New Trends in Pharmacokinetics, Edited by A. Rescigno and A.K. Thakur
Plenum Press, New York, 1991

models. Besides, modeling is like magic. It is so challenging to predict what is happening inside a black box, and then find out that many or all of the predictions were true!

MODELS: MECHANISTIC VS EMPIRICAL

For the purpose of our discussion, we will design two types of models: systems analytic or mechanistic and empirical or often designated as statistical (although the present author has some problem in using 'statistical' as a synonym for 'empirical'). A mechanistic model, as the name implies, should have as many features of the primary system built into it as observations or data will allow. Such a model should be consistent with the observed behavior of the system — retrodiction — [Rescigno and Beck, 1987]; it should further be predictive of the system's future behavior or behavior under perturbation — prediction — [Rescigno and Beck, 1987]. One must have some knowledge of the primary system in terms of structural connectivity and functional mechanisms. Some prefer to call this type of models realistic, intrinsic, and various other names. Many great discoveries in biology, medicine, and other branches of science have been made using such models. In this context one must remember that such models do not necessarily have to have an explicit mathematical expressions; they could be just conceptualizations.

On the other hand, when the system under study is complex and hardly anything is known about its structural connectivity and functional mechanisms, yet one has to produce hypotheses about it based on some external characteristics such as a dose-response (secondary system), one often relies on mathematical functional forms for such a system. These mathematical functions are empirical models. They may incorporate some mechanistic assumptions so that they may look realistic. Numerically, these models are generally easier to handle as opposed to many mechanistic models. Most normal theory based statistical hypothesis testing and confidence interval procedures are based on such models. One should not get the wrong impression that mechanistic models are not useful for such statistical techniques; they may be more difficult to handle numerically from estimation standpoints. Some people would call empirical models extrinsic because they are based purely on the external behavior of the system. Some call them statistical models. As mentioned earlier, it is unfair to assume that statisticians always like to use empirical models for their purposes. The reasons why there are abundance of this type of models in literature are obvious. Our knowledge about the primary system may be inadequate-to-none to allow us the formulation of a mechanistic model or one may not be interested in understanding the inherent structure of the system. In the present author's mind, the phrase statistical model includes both types of models. One must remember that an empirical model may be 'retroactive' (explaining what happened from a secondary system) and even locally 'predictive' (i.e. interpolation may be performed within the range of observations), but it is, in general, not globally 'predictive' (indicating outcome of future experiments). In fact, empirical models should never be used with any authority for extrapolative purposes.

According to Fisher [1925], K.F. Gauss in the early 1800's may have been instrumental in developing empirical modeling concept with his work on maximum likelihood and least squares theories. 'Gauss, further, perfected the systematic fitting of regression formulae, simple and multiple, by the method of least squares, which, in the cases to which it is appropriate, is a particular example of the method of maximum likelihood' [Fisher, 1925]. A slight variation of empirical modeling is defined by Ashby [1958]. In this form, one takes the system and examines its individual components. One makes hypotheses on these individual components with models and finally one tries to draw a global conclusion about the system. According to Ashby, the systems theoretic approaches by Bertalanffy [1950] fall in this category of empirical models. Most empirical models are generalized exponential or polynomial functions. The better ones of these also are mathematically well behaved.

The importance of empirical modeling is evident in all types of literatures. Huxley [1932] in his classical work used a simple empirical model known as the Allometric Equation:

$$y = b\, x^a \tag{1}$$

where

    y = size (linear or volumetric) of an organ
    b = relative size of the organ at inception
    x = (size of the total body − size of the organ) at the same stage of development
    a = rate of growth of the organ

to examine the problems of relative growth. Modified forms of the above equation are still widely used by investigators in different fields. Simple inter-species scaling uses variations of Equation (1).

Verhulst [1839], Pearl and Reed [1920], Lotka [1925], Volterra [1926], Gause [1934], Haldane [1936], and Rescigno and Richardson [1965, 1967], Rescigno [1968], May [1973], and many others did extensive work with the so-called predator-prey equations:

$$dN_1/dt = b_1 N_1 - f_1(N_1 N_2)$$
$$dN_2/dt = F(N_1 N_2) \tag{2}$$

where

    N_1 and N_2 = number of preys and predators respectively
    f_1 = a function characterizing the killing of the preys by the predators per unit time
    F = a function characterizing simultaneously the natality and mortality of the predators
    b_1 = natural increase rate of the preys = (birth rate − death rate)

to explain the struggle for existence and many other aspects of population dynamics. Similar equations are also used to explain many interesting nonlinear aspects of chemical reactions [Lotka, 1920; Nicholis and Prigogine, 1977; Turner et al, 1981]. It is interesting to note that functional forms of f_1 and F in Equations (2) incorporating physical and biological phenomena give rise to mechanistic models. In fact, many mechanistic models in various fields originated from preliminary empirical forms. In that respect, there is a continuum between these two types of models.

In the next section we will discuss some examples to illustrate the usefulness of both mechanistic and empirical models. The preference will be obvious in some cases from their discussions.

Examples

First example

First let us consider a simple first order chemical hydrolysis of phenylacetate [Jencks, 1969]. The reaction is represented in chemical equation form as follows:

$$A \xrightarrow{k} P$$

We can write the process in terms of an ordinary linear differential equation:

$$dA/dt = -kA, \quad A(0) = A_0 \tag{3}$$

whose solution is

$$A = A_0 \exp(-kt) \tag{4}$$

The data for the reaction appear in columns 1 and 2 of Table 1.

Table 1

Results of Analysis of Phenylacetate Hydrolysis Data from Jencks [1969]

Time (minute)	Observed Amount (arbitrary units)	Expected Amount from Equation (3)	from Equation (5)
0.00	0.550	0.556	0.551
0.25	0.420	0.413	0.418
0.50	0.310	0.306	0.312
0.75	0.230	0.227	0.230
1.00	0.170	0.169	0.168
1.25	0.120	0.125	0.121
1.50	0.085	0.093	0.085
0.80	-	0.214	0.217
2.00	-	0.051	0.029
10.00	-	$3.68 \cdot 10^{-6}$	-24.304
SS		$2.03 \cdot 10^{-4}$	$1.61 \cdot 10^{-5}$
Number of Runs		3	5

Equation (3) is a constant coefficient ordinary first order linear differential equation representative of a simple one-compartment system. The general practice among compartmental analysts is to express a system of n compartments as a sum of n exponentials. It will be shown later that such models are, strictly speaking, empirical. Equation (4) is an example of such systems, except, in this case, it is also a mechanistic model. Of course, Equations (3) and (4) provide exactly the same information regarding this phenylacetate hydrolysis case. Let us now see what other model could be appropriate for fitting the data in Table 1. As it turns out, a cubic polynomial of the form

$$A = a_0 + a_1 t + a_2 t^2 + a_3 t^3 \tag{5}$$

fits the data well with

$$a_0 = 0.5507, \ a_1 = -0.5934, \ a_2 = 0.2552, \text{ and } a_3 = -0.0444.$$

For comparative purpose, the data were also fitted with Equation (3) using the PAR programme (BMDP Statistical Software, Los Angeles, California, 1990). This nonlinear estimation program provides estimates of $A_0 = 0.5562$ and $k = 1.1926$ at convergence. The expected values from fitting the data with Equations (3) and (5) are shown in columns 3 and 4 respectively of Table 1. Figure 1 shows the observed and the two fitted curves. As can be seen, they both provide good visual agreements. In fact, both from Figure 1 and Table 1, the cubic seems to provide better fit than Equation (3). The residual sum squares for the cubic is smaller than the one obtained using Equation (3), which may further favor the cubic. Finally, from Table 1, the runs in the residuals (i.e., changes in signs in the observed - expected values) are 5

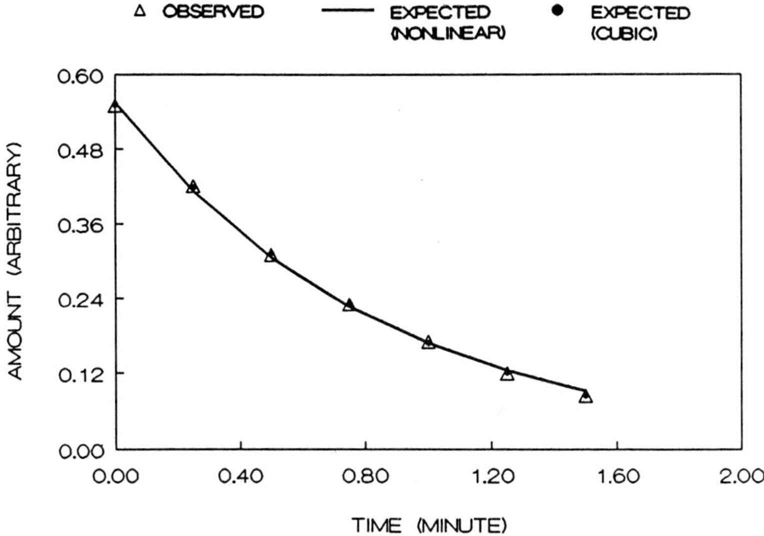

Fig. 1
Phenyl Acetate Hydrolysis Data
Experimental and Fitted Values

for the cubic and 3 for the model given by Equation (3). This fact would seem to add more weight in favor of the cubic. Also, when interpolating within the range of observation, as in the case for t = 0.80 minute, both models provide comparable values as can be seen in Table 1. Yet, as a general model for this chemical reaction, the cubic polynomial (or any other polynomial for that matter) is a poor model. This is clearly demonstrated by the expected values from these two models beyond the data range at time = 2 and 10 minutes. As a matter of fact, at 10 minutes, the cubic predicts a nonsensical negative value of −24.304 for A, whereas Equation (3) predicts a value of 3.68×10^{-6}, which is virtually zero for the same. Since the system is simple first order without any continuous feeding, it is approaching complete washout. Even at 2 minutes, the cubic is not a good predictor for this system. Even though A_0 is very well estimated by the cubic, it does not throw any light on the first order reaction constant k.

Lesson from this example is very simple. If the purpose of the model is to interpolate or to obtain a good smooth curve within the data range, an empirical function (the cubic in this case) may be a good model; but if the purpose is to understand any underlying mechanism of a system or to extrapolate beyond observable range, a mechanistic model must be the choice.

Second Example

Another classic example is estimation of median lethal dose (LD_{50}) for a chemical. This is an important tool to design toxicological experiments. The data are quantal, generally death due to treatment with a chemical. Table 2 is a typical example of such an experiment.

We know the cause of death of these animals, the chemical in question; but we do not know the mechanism behind it. As a consequence, we do not have any choice but to use empirical models to estimate the LD_{50}. Several chemicals can be compared statistically as well as toxicologically knowing their LD_{50}'s. Two popular models with different assumptions are generally used for this purpose.

(a) The logistic or the logit model: If one assumes that the tolerance of these animals to the chemical insult is binomially distributed, one can show that the probability of a response is given by a logistic function [Cox, 1977]:

$$P(x) = \frac{\exp(\alpha + \beta \cdot x)}{1 + \exp(\alpha + \beta \cdot x)} \tag{6}$$

where $\quad$ $P(x)$ = Probability of death

$\quad\quad$ α, β = Two parameters to be estimated.

Then

$$\frac{P(x)}{1 - P(x)} = \exp(\alpha + \beta \cdot x) \tag{7}$$

and

$$\ln\left(\frac{P(x)}{1 - P(x)}\right) = \alpha + \beta \cdot x \tag{8}$$

The logarithmitized quantity on the left hand side is called a logit transformation of proportions. Obviously, since the probability varies from 0 to 1, its logit must lie between $-\infty$ and $+\infty$. Equation (8) is a simple linear model whose parameters α and β can be estimated by least squares or maximum likelihood algorithms easily. The maximum likelihood estimates of LD_{50} and its standard error (SE) are shown in Table 2.

(b) The probit model: If one assumes that the tolerance is, instead, log-normally distributed (i.e., the log-tolerance is normally distributed), one can use the normal equivalent deviate to describe a probit transformation of the proportion $P(x)$ [Bliss, 1934a,b]. The probit of $P(x)$ is the abscissa corresponding to a probability P in a normal distribution with mean 5 and variance 1; in other words, the probit of $P(x)$ is Y where

$$P(x) = \frac{1}{\sqrt{2\pi}} \int_{-\infty}^{Y-5} \exp(-u^2/2)\, du \tag{9}$$

Effectively, both the logit and the probit transformations reduce a sigmoidal curve in semilogarithmic coordinates into a straight line. Another alternative transformation for such quantal data may be hyperbolic tangent.

The maximum likelihood estimates of LD_{50} and its SE using the probit model of the same data are shown in Table 2. For this particular example, both empirical models provide excellent fit to the data ($P_2 = 0.6569$ for the logit and $= 0.3392$ for the probit models). Although the goodness of fit P_2 for the logit is larger than that of the probit, the latter still cannot be rejected as having significantly worse fit. As Table 2 shows, the predicted LD_{50}'s from the two models are slightly different.

Table 2
Results of an LD_{50} Experiment

Dose (mg/kg BW)	Total Number	Number Dead
0.625	5	1
1.250	5	0
2.500	5	3
5.000	5	4
10.000	5	5
LD_{50} by Logit	2.8933	
SE	0.6993	
$P\chi^2$	0.6569	
LD_{50} by Probit	2.2627	
SE	0.5984	
$P\chi^2$	0.3392	

In any case, here we have an example where practically nothing is known about the mechanism of death, yet we have to model the mortality curve for some important decisions. This is where empirical models find their strength.

Third Example

For the third example, let us consider the case of radioimmunoassay. A generalized mechanistic model can be given [Thakur and DeLisi, 1978; Thakur and Rodbard, 1979] for ligand-receptor binding at equilibrium as follows:

$$\frac{B}{F} = \sum_{j=1}^{n} \left(R_j K_j \frac{1 + \beta K_j F}{1 + 2 K_j F + \beta K_j^2 F^2} \right) \tag{10}$$

where B = [Bound ligand], F = [Free ligand], R_j = [Receptor site$_j$]

$$R_0 = \sum_{j=1}^{n} R_j$$

K_j = Affinity constant of the jth receptor site

β = Cooperativity factor (a factor determining the change in affinity constant due to nearest neighbor interaction)

$\beta = 1$: No cooperativity

$\beta > 1$: Positive cooperativity

$\beta < 1$: Negative cooperativity

When n = 1, there is a single homogeneous class of receptor sites with affinity K. In this case, Equation (10) is rewritten in the Scatchard coordinates (B vs B/F) [Scatchard, 1949] as follows:

$$B/F = K\{(R_0 - 2B) + [(R_0 - 2B)^2 + 4(R_0 - B)]^{1/2}\}/2 \tag{11}$$

when $\beta = 1$, i.e., there is no cooperativity in the system, we get

$$B/F = K(R_0 - B) \tag{12}$$

which is the well known Scatchard Equation [Scatchard, 1949] for single homogeneous sites. In actual experimental situations, on top of sampling and alliquoting errors, one also has what is known as nonspecific binding in the system. Generally such nonspecific binding appears with very small affinity. If one can ignore or eliminate such nonspecific binding, one can plot B/F vs B in rectangular coordinates to obtain a Scatchard plot [Scatchard, 1949]. One can then estimate the affinity and the concentration of receptor sites from the slope and intercept. However, there are some serious problems with such reciprocal plots [Rodbard, et al, 1980, Finney, 1983, Thakur, 1990]:

(a) B is not a true independent variable because one needs to know F to calculate B from Total (T) ligand concentration.

(b) Both B and B/F are subject to error and the errors are highly correlated. As a consequence, ordinary least squares algorithms are not applicable for curve fitting.

(c) There is serious nonuniformity of variances which make the least squares algorithms inapplicable.

It has been shown previously [Rodbard et al., 1980, Thakur, 1990] that one should perform the curve fitting of Equation (12) and similar ones in B or B/T vs T or log (T) coordinates which are well behaved. To that goal, one can transform Equation (12) to

$$B/T = \{(KT + R_0K+1) - [(KT+R_0K+1)^2 - 4R_0K^2T]^{1/2}\}/(2KT) + NS \qquad (13)$$

where NS = Extent of nonspecific binding.

If one is interested in estimating the binding parameters of the system, one uses least squares algorithms. Equation (13) is nonlinear in parameters and would require good initial estimates which can be obtained graphically [Thakur et al, 1980]. On the other hand, there are situations where the primary interest in such experiments is not the physical chemical behavior of the system. An investigator may want to use the experiment for dose interpolation purposes, comparing several such curves, or simple quality control of assays [Thakur et al, 1984, Thakur, 1990]. Under those circumstances, one can actually automate the whole process using some empirical models. One such example is the four-parameter logistic model [Rodbard et al, 1980] given by

$$Y = (a - d)/[1 + (X/C)^b] + d \qquad (14)$$

where
$\quad$ a = Expected response at $X = 0$
$\quad$ b = Slope of the logit-log plot
$\quad$ c = ED_{50}, the dose at which the response is (a+d)/2
$\quad\quad$ (similar to LD_{50} in Example 2)
$\quad$ d = Nonspecific binding

Since many of the radioimmunoassay (RIA) response curves are sigmoidal, this empirical model has become a very popular tool in this field. Furthermore, for such simple cases, it is possible to provide physical correlation of these four parameters with the actual binding parameters. From this standpoint, one may want to think of the four-parameter logistic model as pseudo-mechanistic for simple homogeneous binding.

On the other hand, if the binding is a little more complex, any attempt of correlating the parameters of these two types of models becomes futile. The four-parameter logistic model in those cases will be purely empirical, to be used for the purposes of dose interpolation, assay validation, and other such statistical procedures only.

Fourth Example

As the fourth example, let us examine the case of a two-compartment system [Thakur, 1983] as in Figure 2.

The exchange of materials between the two compartments can be described by the following equation:

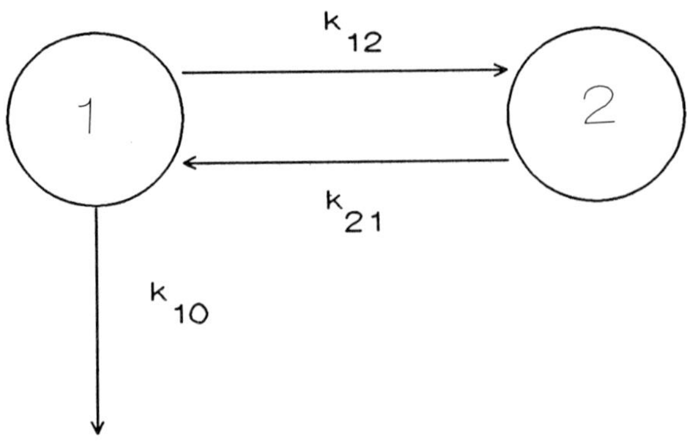

Fig.2
A Two-compartment Model

48

$$dx_1/dt = -(k_{10} + k_{12})x_1 + k_{12}\,x_2, \quad x_1(0) = x_{10}$$

$$dx_2/dt = k_{12}\,x_1 - k_{21}x_2, \quad x_2(0) = 0$$

(15)

where $x_1(t)$ and $x_2(t)$ are the amounts of a material in compartments 1 and 2 and k_{ij} are the rates of transfer from compartments i to j and outside. If 1 is the central compartment, its solution, after adjustment for its volume v_1, in terms of concentration becomes

$$c_1 = \frac{c_{10}}{\lambda_1 - \lambda_2}\left[(\lambda_1 - k_{21})\exp(-\lambda_1 t) + (k_{21} - \lambda_2)\exp(-\lambda_2 t)\right] \qquad (16)$$

where

$$\lambda_1, \lambda_2 = \tfrac{1}{2}\left(k_{10} + k_{12} + k_{21} \pm \sqrt{(k_{10} + k_{12} + k_{21})^2 - 4{\cdot}k_{10}k_{21}}\,\right)$$

As mentioned in Example 1, a common practice among many pharmacokineticists is to fit the experimental data from such a system as a sum of exponentials of the form

$$c_1(t) = \alpha_1\exp(-\beta_1 t) + \alpha_2\exp(-\beta_2 t) \qquad (17)$$

and estimate its parameters α_i and β_i and report them as the parameters of the system. Let us compare Equations (16) and (17) more carefully. The explicit forms of α_i and β_i are not simple. They have all the intrinsic parameters of the system lumped in a complex fashion. In physical terms, α_i and β_i do not have any simple meaning. Moreover, their error structures are also more complex then those of the actual parameters. Even though a model such as Equation (17) may provide "retrodiction" and "prediction" equally well as Equation (16), with λ_i substituted with their corresponding expressions, it is still empirical in the strict sense. At best one should call such a model pseudo-mechanistic or "lumped parameter" type since information about the individual intrinsic parameters as in Equation (16) is lost. Many nonlinear least squares programs have such pre-defined "models" in terms of sums of exponentials. One should be careful about interpreting the parameters from such programs.

DISCUSSION

We have seen several examples of mechanistic and empirical models in the foregoing discussion. As we saw, there are many instances where empirical models, because of their numerical simplicity, may be preferable or even necessary for certain types of analysis. Modeling is a technique used to understand and explain both the "knowns" and the "unknowns" of a system. It is especially important in biological sciences where often one is unable to probe into a living system at ones own will. Even where one may be able to physically isolate part of a system to study, its behavior or characteristics may not be exact or even appropriate when the part is intact and undisturbed. Cellular and biochemical integrity of the subsystem may have been totally lost when it is isolated. A classic example is the famous paucimolecular membrane model [Davson and Danielli, 1952]. The model was proposed to study the mechanisms of movements of molecules across the cell membranes. Even though cell membranes conformed to such a model *in vitro* in many instances, many present day critics take this model as a minimal one, if not inappropriate for many living systems. Their argument is based on the fact that once isolated, the cell may go through conformational and other changes which make the paucimolecular model unrealistic for the living system.

From a modeling standpoint, it does not necessarily mean that the model is a bad one. Modeling is an ongoing process. One builds it slowly and improves upon it as more and more information and technology become available. Given the techniques of probing available to them during the forties through the sixties, Davson, Danielli and many others gathered extensive knowledge from such a system that made more recent investigations possible. One can find an analogy in a child's quest for a model airplane. When the child first starts with it his/her main objective is to have a toy airplane which looks like a real one. But as the child grows with it, the model plane must be able to move on the ground on wheels and then it must simulate the flight pattern (although, as a child, I was never interested in the noise a flying plane makes, it was too scary!).

In many ways, R.A. Fisher [1925] said it all:
"A hypothesis is conceived and defined with all necessary exactitude; its logical consequences are ascertained by a deductive argument; these consequences are compared with the available observations; if these are completely in accord with the deductions, the hypothesis is justified at least until fresh and more stringent observations are available."

Unfortunately the zeal behind modeling is so overwhelming that some investigators refuse to heed the above lesson. They forget, irrespective of whether the model is mechanistic or empirical, that it is only as good as available information content from a system. When data are gathered under different experimental conditions, they should be willing to either modify or change their model if there is need for that. Examples of such zealous endeavors are abundant in literature. We will have some brief discussions of them in several later chapters under carcinogenic risk assessment.

The present author does not intend to be fatalistic, but there is a lesson to be learned from life, which itself may be considered a model. One gets what one deserves, or as the sometimes mathematician, sometimes statistician, sometimes singer, and sometimes pianist, Tom Lehrer [1959] once said in his sarcastic way:
"Life is like a sewer; what you get out of it, depends on what you put into it."

REFERENCES

Ashby W.R., 1958. General systems theory as a new discipline, General Systems, 3:1.

Bliss, C.I., 1934 a. The method of probits, Science, **79**:38.

Bliss, C.I., 1934 b. The method of probits — a correction, Science, **79**:409.

Cox, D.R., 1977. Analysis of Binary Data, Chapman and Hall, London (Reprinted).

Finney, D.J., 1983. Response curves for radioimmunoassay, Clin. Chem., **29**:1762.

Fisher, R.A., 1925. Statistical Methods for Research Workers, Oliver and Boyd, Edinburgh.

Gause, G.F., 1934. The Struggle for Existence, Williams and Wilkins, Baltimore, Maryland.

Haldane, J.B.S., 1932. The Causes of Evolution, Longmans, Green and Co., London.

Huxley, J.S., 1932. Problems of Relative Growth, Methuen, London.

Jencks, W.P., 1969. Catalysis in Chemistry and Enzymology, McGraw-Hill, New York, page 559.

Lehrer, T., 1959. An Evening Wasted with Tom Lehrer, Lehrer Records TL202.

Lotka, A.J., 1925. Elements of Physical Biology, Williams and Wilkins, Baltimore, Maryland.

Lotka, A.J., 1920. Undamped oscillations derived from the Law of Mass Action, J. Am. Chem. Soc., **42**:1595.

May, R.M., 1973. Stability and Complexity in Model Ecosystems, Monographs in Population Biology No. 6, Princeton University Press, Princeton, New Jersey.

Nichols, G. and Prigogine, I., 1977. Self-organization in Nonequilibrium — from Dissipative Structures to Order through Fluctuations, Wiley, New York.

Pearl, R. and Reed, L.J., 1920. On the rates of growth of the population of the United States since 1790 and its mathematical representation, Proc. Nat. Acad. Sci., **6**:275.

Rescigno, A. and Richardson, I.W., 1965. On the competitive exclusion principle, Bull. Math. Biophys., **17**:85.

Rescigno, A. and Richardson, I.W., 1967. The struggle for life: I. Two species, Bull. Math. Biophys., **29**:377.

Rescigno, A., 1968. The struggle for life: II. Three competitors, Bull. Math. Biophys., **30**:291.

Rescigno, A., Beck, J.S. (Comments: Thakur, A.K.), 1987. The use and abuse of models, J. Pharmacokin. Biopharm., **15**:327.

Rodbard, D., Munson, P.J., and Thakur, A.K., 1980. Quantitative characterization of hormone receptors, Cancer, **46**:2907.

Scatchard, G., 1949. The attractions of proteins for small molecules and ions, Ann. N.Y. Acad. Sci., **51**:660.

Thakur, A.K. and DeLisi, C., 1978. Theory of ligand binding to heterogeneous receptor populations: Characterization of the free-energy distribution function, Biopolymers, **17**:1075.

Thakur, A.K., and Rodbard, D., 1979. Graphical aids to interpretation of Scatchard plots and dose-response curves, J. Theoret. Biol., **80**:383.

Thakur, A.K., Jaffe, M.L., and Rodbard, D., 1980., Graphical analysis of ligand-binding systems: Evaluation by Monte-Carlo studies, Analyt. Biochem., **107**:270.

Thakur, A.K., 1983. Some statistical principles in compartmental analysis. In , "Compartmental Distribution of Radiotracers" (J.S. Robertson ed.), Chapter 6. CRC Press, Boca Raton, Florida.

Thakur, A.K., Listwak, S.J. and Rodbard, D., 1985. Quality Control of Radioimmunoassay, IAE-CN-45/108. In "Proceedings of Conference on Radiopharmaceuticals and Labelled Compounds". Vienna.

Thakur, A.K., 1990. Statistical Methods for Serum Hormone Assays. In "Handbook of the Laboratory Diagnosis and Treatment of Infertility" (Keel, B.A. and Webster, B.W. eds.), Chapter 15. CRC Press, Boca Raton, Florida.

Turner, J.S., Roux, J.C., McCormick, W.D., and Swinney, H.L., 1981. Alternating periodic and chaotic regimes in a chemical reaction — Experiment and theory. Physics Letters, **85A**:9.

Verhulst, P.F., 1839. Notice sur la loi que la population suit dans son accroissement. Corr. math. et phys. publ. par A, **10**:113.

Volterra, V., 1926. Variazioni e fluttuazioni del numero d'individui in specie animali conviventi. Mem. Accad. Lincei, **2**:31.

Von Bertalanffy, L., 1950. An outline of general system theory. Brit. J. Phil. Sci., **1**:13

STATISTICAL FOUNDATIONS OF PHARMACOKINETIC MODELING

D. Krewski, R.T. Burnett and W. Ross

Health Protection Branch
Health and Welfare Canada
Ottawa, Ontario, Canada

INTRODUCTION

Compartmental models have a long history of application in describing the pharmacokinetic properties of pharmaceutical agents [Wagner, 1971], and have also found application in the study of toxic chemicals such as methylmercury [Rice et al., 1989] and styrene [Withey and Collins, 1979]. Such models are useful in describing the absorption, uptake, distribution, metabolism and elimination of xenobiotic agents, particularly with respect to temporal levels of the compound of interest in blood and other tissues.

The mathematical formulation and solution of compartmental models are discussed in detail in texts by Gibaldi and Perrier [1975], O'Flaherty [1981], and Godfrey [1983]. Recently, Matis et al. [1989] have proposed generalized stochastic compartmental models that permit nonexponential retention times within compartments.

Classical statistical methods for estimating the pharmacokinetic parameters in compartmental models are based on nonlinear regression methods [Rustagi and Singh, 1977; Metzler, 1981]. Kodell and Matis [1976] use stochastic compartmental models to determine the autocorrelation of the observer data prior to estimation using nonlinear least squares. Minder and McMillian [1977] adopt a marginal likelihood approach to parameter estimation. Robust estimation procedures have been proposed by Rodda et al. [1975], Frome and Yakatan [1980], and Atkins and Nimmo [1981]. Rupert et al. [1989] discuss the use of weighted least squares to estimate the parameters in Michaels-Menten models used to describe saturable kinetic phemonema. Murdoch [1991a] considers transformations of the kinetic parameters to achieve more stable estimates. Nonparametric estimation of kinetic absorption data using kernel smoothing is discussed by Hougaard et al. [1987].

In this chapter, we propose a general statistical approach to estimating the parameters of compartmental models based on Gaussian estimation [Crowder, 1985]. This is a generalization of maximum likelihood estimation in which the distribution of the data does need to be fully specified. We begin with an exposition of the mathematical characterization of some simple compartmental

New Trends in Pharmacokinetics, Edited by A. Rescigno and A.K Thakur
Plenum Press, New York, 1991

models. The application of Gaussian estimation in the case of general multi-exponential models used to describe pharmacokinetic systems is then outlined along with diagnostic plots of residuals used to detect heteroscedastic errors. We conclude with an illustrative application of these procedures in describing the elimination of pyrene from the blood of rats following intravenous injection.

COMPARTMENTAL MODELS

Compartmental models for describing the fate of xenobiotics upon entering the body have recently been discussed by Collins [1990] and Murdoch [1991b]. In this section, we describe simple one and two compartmental models which have found application in the analysis of pharmacokinetic data. We also consider a general multi-exponential model which can be useful in the statistical analysis of pharmacokinetic data.

One Compartment Model

Consider first the administration of a single intravenous dose D in the simple one compartment model shown in Figure 1. Under first order kinetics, the amount $X_1(t)$ of chemical in the central compartment satisfies the differential equation

$$\frac{dX_1(t)}{dt} = -k_e X_1(t), \tag{1}$$

Since $X_1(0) = D$, the solution to (1) is

$$X_1(t) = D\, e^{-k_e t}. \tag{2}$$

Assuming that the chemical is uniformly dispersed throughout the apparent volume of distribution V_1, (2) may be divided by V_1 to obtain

$$C_1(t) = Ae^{-\alpha t}, \tag{3}$$

where $C_1(t)$ denotes the concentration of the substance of interest in the body compartment, $A = D/V_1$, and $\alpha = k_e$. Since

$$\ln C_1(t) = \ln A - \alpha t,$$

a semi-logarithmic plot of concentration $C_1(t)$ versus time t will be linear.

Suppose now that the compound of interest is administered orally. The amount of chemical $X_0(t)$ present in the gastrointestinal tract satisfies the differential equation

$$\frac{dX_0(t)}{dt} = -k_a X_0(t), \tag{4}$$

where k_a denotes the kinetic rate coefficient for absorption. With oral dosing, $X_1(t)$ satisfies the equation

$$\frac{dX_1(t)}{dt} = k_a X_0(t) - k_e X_1(t). \tag{5}$$

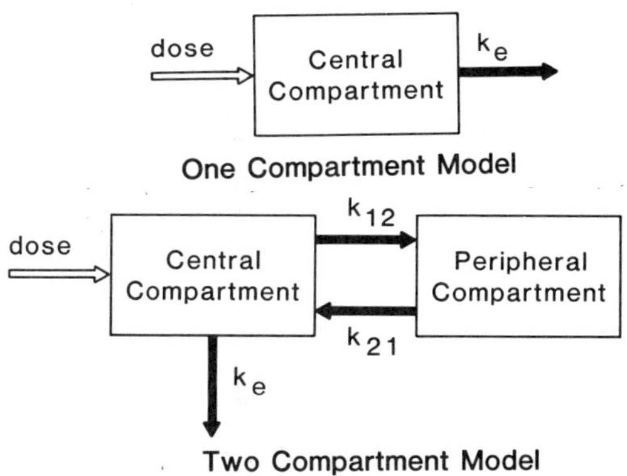

Figure 1.
One and Two Compartment Models

Equations (4) and (5) may be solved simultaneously using Laplace transforms to obtain

$$X_1(t) = \frac{Dk_a}{(k_a - k_e)} \left(e^{-k_e t} - e^{-k_a t}\right), \tag{6}$$

provided $k_a \neq k_e$. (A different solution obtains when $k_a = k_e$.) Dividing (6) by the apparent volume of distribution V_1 yields

$$C_1(t) = A \left(e^{-\alpha t} - e^{-\beta t}\right),$$

where $A = Dk_a [(k_a - k_e)V_1)]^{-1}$, $\alpha = k_e$, and $\beta = k_a$. A semi-logarithmic plot of $C_1(t)$ versus t has a terminal linear component with slope α.

In studies of the pharmacokinetics of volatile organic compounds such as styrene, exposure may occur by inhalation [Withey and Collins, 1979]. In this case, pulmonary uptake of the test chemical occurs at an essentially constant rate k_0, with

$$\frac{dX_1(t)}{dt} = k_0 - k_e X_1(t). \tag{7}$$

Integration of (7) and division by the apparent volume of distribution leads to

$$C_1(t) = A(1 - e^{-k_e t}),$$

where $A = k_0/[k_e V_1]$.

Two Compartment Model

If the one compartment model fails to provide an adequate fit to the available data, models comprised of more than one compartment may be entertained. Consider, for example, the two-compartment model in Figure 1 and the application of a single intravenous dose.

The amounts $X_1(t)$ and $X_2(t)$ of chemical present in the central and peripheral compartments respectively satisfy the differential equation

$$\frac{dX_1(t)}{dt} = -(k_e + k_{12})X_1(t) + k_{21}X_2(t) \tag{8}$$

and

$$\frac{dX_2(t)}{dt} = k_{12}X_1(t) - k_{21}X_2(t). \tag{9}$$

Here, k_e denotes the kinetic rate coefficient for elimination, and k_{12} and k_{21} denote the rate coefficients governing transfer from the central to peripheral compartment and from the peripheral to central compartment respectively. The solutions to (8) and (9) are given by

$$X_1(t) = \frac{D}{(\beta - \alpha)} \left[(k_{21} - \alpha)e^{-\alpha t} - (k_{21} - \beta)e^{-\beta t} \right] \tag{10}$$

and

$$X_2(t) = \frac{Dk_{21}}{(\beta - \alpha)} \left(e^{-\alpha t} - e^{-\beta t} \right),$$

where α and β satisfy the equation

$$\alpha + \beta = k_{21} + k_{12} + k_e \tag{11}$$

and

$$\alpha\beta = k_{21}k_e. \tag{12}$$

Dividing (10) by the apparent volume of distribution V_1 yields

$$C_1(t) = Ae^{-\alpha t} + Be^{-\beta t}, \tag{13}$$

where $A = D(k_{21} - \alpha)/[(\beta - \alpha)V_1]$ and $B = D(k_{21} - \beta)/[(\alpha - \beta)V_1]$. The kinetic constants in the two compartment model can be recovered from the parameters in (13) using the inverse relation

$$k_{21} = \frac{A\beta + B\alpha}{A + B},$$

$$k_e = \frac{\beta\alpha}{k_{21}},$$

and

$$k_{12} = \alpha + \beta - k_e - k_{21}.$$

The case of oral dosing in the two compartment model in Figure 1 can be solved in a similar fashion. The concentration-time profile in the central compartment can be shown to be

$$C_1(t) = Ae^{-\alpha t} + Be^{-\beta t} - (A + B)e^{-\gamma t} \tag{14}$$

[Collins, 1990, pp. 361-365], where the parameters $\{A, B, \alpha, \beta, \gamma\}$ depend on the kinetic constants $\{k_a, k_e, k_{12}, k_{21}\}$ and the ratio D/V_1.

For inhalation studies, with constant infusion at rate k_0, the blood concentration profile under the two compartment model can be shown to be

$$C_1(t) = A(1 - e^{-\alpha t}) - B(1 - e^{-\beta t}),$$

where α and β are hybrid rate coefficients satisfying (11) and (12),

$$A = \frac{k_0(\alpha - k_{21})}{V_1\alpha(\alpha - \beta)},$$

and

$$B = \frac{k_0(\beta - k_{21})}{V_1\beta(\beta - \alpha)},$$

[Collins, 1990, pp. 365-367]. The pharmacokinetic rate coefficients can be recovered from $\{A, B, \alpha, \beta\}$ using the relations

$$k_e = (A\alpha + B\beta)/(A + B),$$
$$k_{21} = \alpha\beta/k_e,$$

and

$$k_{12} = \alpha + \beta - k_e - k_{21}.$$

Multi-Exponential Models

The results provided to this point for the one and two compartment models can be generalized to more complex compartmental models [Gibaldi and Perrier, 1975; O'Flahery, 1981]. Models with three or more compartments may be configured in different ways (Figure 2), leading to different possible solutions, not all of which may be identifiable [Griffiths, 1979; Collins, 1990, p. 371]. The biological interpretation of such models is also unclear since the compartments themselves may have no clear physiological interpretation.

For these reasons, pharmocokinetic models involving three or more exponential terms may be motivated more on a statistical than biological basis. A natural generalization of the single and double exponential models in (3) and (13) is the multi-exponential model

$$C(t) = \sum_{k=1}^{K} A_k e^{-\alpha_k t}. \tag{15}$$

[Withey, 1990, p. 309]. This provides a convenient model for statistical analysis, but will be difficult to interpret in terms of simple compartmental models when K is moderately large. Multi-exponential models of the form (15) arising as solutions to compartmental systems may also be subject to certain constraints. For example, the two compartment oral dosing model in (14) may be modeled using (15) with $K = 3$, with the constraint $A_3 = -(A_1 + A_2)$.

FITTING PHARMACOKINETIC MODELS

In this section, we describe a general statistical approach to fitting compartmental pharmacokinetic models of the type described previously to experimental data in which measurements $y_1, \ldots, y_n$ of the concentration of the compound of interest in the central (blood) compartment are taken at n distinct times $t_1, \ldots, t_n$ following exposure to a single dose of the test substance. For purpose of statistical inference, such data may be represented by the nonlinear regression model

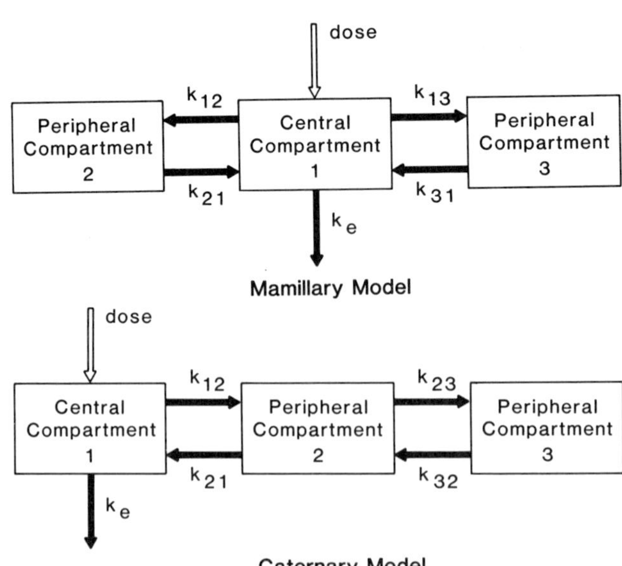

Figure 2.
Three Compartment Mamilliary and Catenary Models

$$y_i = f(t_i; \theta) + \epsilon_i,$$

where f is the underlying pharmacokinetic model involving a vector $\theta = (\theta_1, \ldots, \theta_p)$ of p unknown parameters, and $\epsilon_1, \ldots, \epsilon_n$ are mutually independent stochastic errors representing random deviations from the structural model f.

In many applications, the variance of y_i depends on the value of $f(t_i; \theta)$ [Beal and Sheiner, 1988]. A useful representation of this dependency is given by

$$\mathrm{Var}(y_i) = \sigma^2 f(t_i; \theta)^\xi \equiv \sigma_i^2. \tag{16}$$

If $\xi = 0$, then $\sigma_i^2 \equiv \sigma^2$ $(i = 1, \ldots, n)$, and the errors are said to be homoscedastic. A value of $\xi = 2$ corresponds to a constant coefficient of variation $[\mathrm{Var}(y_i)^{1/2}/f(t_i; \theta)] = \sigma$; $\xi = 1$ implies the $\mathrm{Var}(y_i)$ is proportional to $E(y_i) = f(t_i; \theta)$.

If the distribution of the observations is known, maximum likelihood methods may be used to estimate the unknown parameters $\omega = (\theta_1, \ldots, \theta_p, \sigma^2, \xi)$. If the distribution is unknown, it is possible to estimate α by specifying the mean and variance as in (16) using Gaussian estimation [Crowder, 1985; Beal and Sheiner, 1988]. This will be done here.

The objective function, $\mathbf{0}$, used in Gaussian estimation has the form

$$\mathbf{0} \propto -\frac{1}{2} \sum_{i=1}^{n} \left\{ \ln \sigma_i^2 + [\sigma_i^2]^{-1}(y_i - f_i)^2 \right\}, \tag{17}$$

where $f_i = f(t_i; \theta)$. A Gauss-Newton iterative procedure may be used to determine the value $\hat{\omega}$ that maximizes (17). Given $\hat{\omega}^{(h)}$, the current value of ω at the hth iteration, the updated value is given by

$$\hat{\omega}^{(h+1)} = \hat{\omega}^{(h)} + \left(E\left\{ -\frac{\partial^2 \mathbf{0}}{\partial \omega \partial \omega^{\mathrm{T}}} \right\} \right)^{-1} \left\{ \frac{\partial \mathbf{0}}{\partial \omega} \right\}, \tag{18}$$

where

$$\frac{\partial 0}{\partial \omega_\ell} = \sum_{i=1}^{n} [\sigma_i^2]^{-1} \left[\left(\frac{\partial f_i}{\partial \omega_\ell} \right) (y_i - f_i) \right.$$
$$\left. + \frac{1}{2} \left(\frac{\partial \sigma_i^2}{\partial \omega_\ell} \right) \left\{ [\sigma_i^2]^{-1} (y_i - f_i)^2 - 1 \right\} \right], \tag{19}$$

$(\ell = 1, \ldots, p+2)$ and

$$E\left\{ -\frac{\partial^2 0}{\partial \omega_\ell \partial \omega_m} \right\} = \sum_{i=1}^{n} [\sigma_i^2]^{-1} \left[\left(\frac{\partial f_i}{\partial \omega_\ell} \right) \left(\frac{\partial f_i}{\partial \omega_m} \right) \right.$$
$$\left. + \frac{1}{2} \left(\frac{\partial \sigma_i^2}{\partial \omega_\ell} \right) \left(\frac{\partial \sigma_i^2}{\partial \omega_m} \right) [\sigma_i^2]^{-1} \right]$$

$(\ell, m = 1, \ldots, p+2)$. Note that $\partial f_i / \partial \omega_\ell = 0$ for $\ell = p+1$ and $p+2$, since the expectation f_i is not a function of either σ^2 or ξ. For $\ell = 1, \ldots, p$, we also have

$$\frac{\partial \sigma_i^2}{\partial \theta_\ell} = \sigma^2 \xi f_i^{\xi-1} \left(\frac{\partial f_i}{\partial \theta_\ell} \right),$$

with

$$\frac{\partial \sigma_i^2}{\partial \sigma^2} = f_i^\xi,$$

and

$$\frac{\partial \sigma_i^2}{\partial \xi} = \sigma^2 f_i^\xi \ln f_i.$$

Under mild regularity conditions [Inagaki, 1973; Crowder, 1985; Thall and Vail, 1990], $n^{1/2}(\hat{\omega} - \omega)$ converges in distribution as $n \to \infty$ to a multivariate normal variate with zero mean and covariance matrix

$$n \left(E\left\{ -\frac{\partial^2 0}{\partial \omega \partial \omega^T} \right\} \right)^{-1} \text{Cov} \left(\frac{\partial 0}{\partial \omega} \right) \left(E\left\{ -\frac{\partial^2 0}{\partial \omega \partial \omega^T} \right\} \right)^{-1}, \tag{20}$$

where

$$\text{Cov} \left(\frac{\partial 0}{\partial \omega_\ell}, \frac{\partial 0}{\partial \omega_m} \right) = \sum_{i=1}^{n} [\sigma_i^2]^{-2} \left[\left(\frac{\partial f_i}{\partial \omega_\ell} \right)^T \left(\frac{\partial f_i}{\partial \omega_m} \right) \sigma_i^2 \right.$$
$$+ \frac{1}{4} \left(\frac{\partial \sigma_i^2}{\partial \omega_\ell} \right)^T \left(\frac{\partial \sigma_i^2}{\partial \omega_m} \right) [\sigma_i^2]^{-2} \text{Var}\{(y_i - f_i)^2\}$$
$$+ \frac{1}{2} \left(\left(\frac{\partial f_i}{\partial \omega_\ell} \right) \left(\frac{\partial \sigma_i^2}{\partial \omega_m} \right) + \left(\frac{\partial f_i}{\partial \omega_m} \right) \left(\frac{\partial \sigma_i^2}{\partial \omega_\ell} \right) \right)$$
$$\left. \times [\sigma_i^2]^{-1} \text{Cov}\{(y_i - f_i), (y_i - f_i)^2\} \right]. \tag{21}$$

Note that if the data are normally distributed then

$$\mathrm{Var}\{(y_i - f_i)^2\} = 2\left[\sigma_i^2\right]^2 \tag{22}$$

and

$$\mathrm{Cov}\{(y_i - f_i),\ (y_i - f_i)^2\} = 0 \tag{23}$$

$(i = 1, \ldots, n)$. Thus

$$\mathrm{Cov}\ \left(\frac{\partial 0}{\partial \omega}\right) = \mathrm{E}\left\{-\frac{\partial^2 0}{\partial \omega \partial \omega^{\mathrm{T}}}\right\}$$

and

$$\mathrm{Cov}(\hat{\omega}) = \left(\mathrm{E}\left\{-\frac{\partial^2 0}{\partial \omega \partial \omega^{\mathrm{T}}}\right\}\right)^{-1}.$$

If the observations are not assumed to be normally distributed, the third and fourth moments of the data are required in (21) to obtain the asymptotic dispersion matrix of the parameter estimates. If these higher order moments are unknown, they may be estimated empirically by

$$\widehat{\mathrm{Var}}\{(y_i - f_i)^2\} = (y_i - \hat{f}_i)^4 - [\hat{\sigma}_i^2]^2 \tag{24}$$

and

$$\widehat{\mathrm{Cov}}\{(y_i - f_i),\ (y_i - f_i)^2\} = (y_i - \hat{f}_i)^3 \tag{25}$$

[Thall and Vail, 1990], where $\hat{f}_i = f(t_i; \hat{\theta})$ and $\hat{\sigma}_i^3 = \hat{\sigma}^2 f(t_i; \hat{\theta})^{\hat{\xi}}$.

With this approach, the unknown parameters θ, σ^2, and ξ are estimated simultaneously. Beal and Sheiner [1988] estimate θ, σ^2 and ξ separately employing iterative reweighed nonlinear least squares. Separate estimation of θ and (σ^2, ξ) employing Gaussian estimation may be accomplished by defining two sets of estimating equations, one for $\theta = (\theta_1, \ldots, \theta_p)$ and the other for the pair (σ^2, ξ). The Gauss-Newton equations have the same form as the joint estimation procedure (18), with the score function for θ given by the first term in (19) and the corresponding function for the variance parameters given by the second term in (19). Separate estimation in this fashion corresponds to iteratively reweighed nonlinear least squares using Gaussian weights. If desired, three sets estimating equations may be formed for θ, σ^2 and ξ separately.

Some loss in statistical efficiency in estimating θ is expected with the separate estimation approach since only the residuals $(y_i - f_i)$ are employed in estimating θ. There is, however, information on θ in the variance terms σ_i^2 since they are assumed to be a function of the structural model f_i. On the other hand, Giltinan and Ruppert [1989] have developed a convenient method of implementing the separate estimation approach using the statistical analysis package SAS, thus making it attractive in terms of utilizing existing computer software.

DIAGNOSTIC PLOTS FOR HETEROSCEDASTICITY

Diagnostic techniques are used to assess the adequacy of the assumptions underlying the regression model used, and to identify unusual characteristics of the data which may unduly influence the statistical analysis. A large number of such techniques are available for linear regression models. They include graphical techniques, such as residual plots, and numerical indicators, such as Cook's statistic [Cook, 1977]. A useful survey of these methods is provided in Cook and Weisberg [1982]. Many of the diagnostic methods for linear regression can be extended to nonlinear regression models [Cook and Weisberg, 1982, 1987; Dzieciolowski and Ross, 1990].

The possibility of heteroscedasticity distinguishes pharmacokinetic models from other nonlinear regression problems. Diagnostic methods are easily constructed to examine this model assumption. The most direct approach is to fit a homoscedastic model with $\xi = 0$ in equation (16). In this case, Gaussian estimation corresponds to ordinary least squares (OLS). Then, given the fitted values and the corresponding residuals, plots or numerical summaries can be constructed based on techniques analogous to those used in linear regression.

Let $\epsilon = (\epsilon_1, \ldots, \epsilon_n)^{\mathrm{T}}$ denote the $n \times 1$ vector of errors in the regression model with $\epsilon_i = y_i - f(t_i; \theta)$. For notational convenience, let $W = \mathrm{diag}\{w_i\}$ denote the $n \times n$ diagonal matrix with i^{th} diagonal entry $w_i = f(t_i; \theta)^\xi$ and let $V = ((\partial f(t_i; \theta)/\partial \theta_j))$ be the $n \times p$ matrix of derivatives of f with respect to the pharmacokinetic parameters θ. Then $H = V(V^t V)^{-1} V^t$ is the projection onto the column space of V and $M = I_n - H$ is the projection orthogonal to the column space of V, where I_n is the $n \times n$ identity matrix. Note that the elements of each of V, H, M and W depend on the parameter vector θ. In addition, the elements of W also depend on the heteroscedasticity parameter ξ. With this notation, $\mathrm{Var}(\epsilon) = \sigma^2 W$. In the case of the homoscedastic regression model, $W = I_n$. Finally, let $\eta(\theta) = ((f(t_i; \theta)))$ denote the $n \times 1$ vector of expectations. The set of all possible values of $\eta(\theta)$ in $\mathbb{R}^n$ is called the solution locus of the regression model.

Let $\hat{\theta}_0$ denote the estimates of the parameter vector θ obtained using ordinary least squares. Under quite general regularity conditions, $\hat{\theta}_0$ is a consistent, although possibly inefficient, estimate of θ [Beal and Sheiner, 1988].

Several authors, including Bates and Watts [1980], have examined the geometry of the solution locus. These studies indicate that for nonlinear regression models of the type considered in pharmacokinetics, the tangent plane approximation

$$(\eta(\hat{\theta}_0) - \eta(\theta)) \simeq H\epsilon \tag{26}$$

is adequate, at least for diagnostic purposes. The vector of OLS residuals,

$$\hat{\epsilon}_0 = ((y_i - \hat{f}_i)),$$

can be expressed as

$$\hat{\epsilon}_0 = \epsilon - (\eta(\hat{\theta}_0) - \eta(\theta)). \tag{27}$$

Substituting approximation (26) into (27), the OLS vector of residuals can be approximated by

$$\hat{\epsilon}_0 \simeq M\epsilon. \tag{28}$$

Based on the approximation given in (28), the squared OLS residuals $\hat{\epsilon}_{0i}^2$ are approximated by

$$\hat{\epsilon}_{0i}^2 \simeq \sum_{j=1}^{n} \sum_{k=1}^{n} m_{ij} m_{ik} \epsilon_j \epsilon_k$$

$(i = 1, \ldots, n)$, where the m_{ij} are the elements of the matrix M. As a result,

$$E\{\hat{\epsilon}_{0i}^2\} \simeq \sigma^2 \sum_{j=1}^{n} w_j m_{ij}^2. \tag{29}$$

When the errors are in fact homoscedastic, (29) reduces to

$$E\{\hat{\epsilon}_{0i}^2\} \simeq \sigma^2 m_{ii}. \tag{30}$$

Thus, even for this simple error structure, the resulting residuals are not homoscedastic.

When $\xi = 0$, the degree of heteroscedasticity in the OLS residuals depends on the range of values of the diagonal elements of M. Using the fact that M is symmetric and idempotent, it is easy to show that $0 \leq m_{ii} \leq 1$. Thus, for those cases where m_{ii} is close to zero, equation (30) suggests that the resulting residual will be deceptively small. Therefore, it is important in such cases to adjust the residuals accordingly. This result parallels that of linear regression models, where (30) is exact. An important difference for nonlinear regression models, however, is that the right hand side of (30) is a function of expected value of the response variable.

A first order Taylor series approximation of w_i about $\xi = 0$, has the form

$$w_i \simeq 1 + \xi \ln(f(t_i; \theta)).$$

Substituting this approximation into (29) yields

$$E\{\hat{\epsilon}_{0i}^2\} \simeq \sigma^2 m_{ii} \{1 + \sigma_i \sum_{j=1}^{n} \ln(f(t_j; \theta)) m_{ij}^2 / m_{ii}\}. \tag{31}$$

Define $z_i = m_{ii} \ln(f(t_i; \theta))$. Then (31) can be expressed as

$$E\{\hat{\epsilon}_{0i}^2 / \sigma^2 m_{ii}\} \simeq 1 + \xi z_i + \xi \sum_{\substack{j=1\omega \\ (j \neq i)}}^{n} z_j (m_{ij}/m_{ii}). \tag{32}$$

For many regression problems the final term on the right hand side of (32) is small and can be dropped from the approximation [Cook and Weisberg, 1983]. The resulting expression gives the expectation of a properly standardized squared OLS residual as a monotonic function of the parameter ξ. This suggests a possible diagnostic plot for heteroscedasticity.

Fixing $\xi = 0$, the Gaussian estimation procedure estimates σ^2 as

$$\hat{\sigma}_0^2 = \sum_{i=1}^{n} \hat{\epsilon}_{i0}^2 / n.$$

Let $\hat{m}_{0ii}$ denote the diagonal elements of M evaluated at $\eta(\hat{\theta}_0)$, and define

$$r_i^2 = \hat{\epsilon}_{0i}^2 / \hat{\sigma}_0^2 \hat{m}_{0ii}.$$

Finally, let $\hat{z}_i$ denote z_i evaluated at $\eta(\hat{\theta}_0)$, and consider the plot of r_i^2 vs $\hat{z}_i$. For positive ξ, r_i^2 will generally increase with $\hat{z}_i$. Similarly, negative ξ will result in generally decreasing r_i^2 with increasing $\hat{z}_i$.

APPLICATIONS

The estimation and diagnostic methods given in this paper are illustrated with an example of the pharmacokinetics of pyrene in the rat [Withey et al., 1991]. A single intravenous dose of ^{14}C-labelled pyrene (9mg/kg) was given in an emulphor/water solvent vehicle. The concentration of unchanged pyrene was measured at 13 distinct points in time from 15-480 minutes post dosing (Table 1).

Preliminary analysis of the data suggested that the two-exponential model with $K = 2$ in (15), provides an reasonable fit to the data. Parameter estimates, and their corresponding standard errors obtained under the assumption of constant variance ($\epsilon = 0$) using the SAS procedure NLIN [SAS, 1988], are given in Table 2. A plot of the raw data and fitted two-exponential model is given in Figure 3. The starting values of the parameters A, α, B and β are based on the method of feathering [Collins, 1990, p. 364].

The quantities $\hat{m}_{0ii}, \hat{z}_i, \hat{\epsilon}_i$ and r_i^2 are all easily determined from standard output of the OLS fit to the data set. One notable feature, common to nonlinear regression models of this form, is the extremely small value of $\hat{m}_{0ii}$, corresponding to the leading design value having largest expectation. In this case $\hat{m}_{001} = 0.084$, whereas $0.642 \leq \hat{m}_{0ii} \leq 0.810$ for $i > 1$. Thus, it is expected that the ordinary residual corresponding to the initial measurement will be quite small. In other words, for these types of nonlinear regression models, the first data point will always be well fit, making it an influential point in the fitting process.

A plot of $\hat{r}_i$ versus $\hat{z}_i$ is given in Figure 4. There are two prominent features of this plot. First, the adjusted squared OLS residual, r_1^2, is comparatively large and distinct from the rest of the plotted values. This is consistent with the characteristically small value of $\hat{m}_{011}$. Second, the plot exhibits a distinctive upward curve with increasing $\hat{z}_i$. This suggests a positive value for ξ, and the plausibility of a heteroscedastic regression model.

Based on the diagnostic produces discussed previously, the data were re-analysed assuming the power function model for the variance-mean relationship as given in (16). Here, Gaussian estimation is used to estimate the pharmocokinetic parameters θ along with the dispersion parameters ξ and σ^2. Standard errors of the parameter estimates based on (20), computed first assuming the data follow a normal distribution as in (22)-(23) and then using empirical estimates of the third and fourth moments as in (24)-(25), are also given in Table 2.

As shown in Table 2, the fit to this data set of the more general model results in $\hat{\xi} = 1.8$. In view of the standard error of this parameter, this is not inconsistent with the constant coefficient of variation model ($\xi = 2$), common in pharmacokinetics [Beal and Sheiner, 1988]. One notable aspect of these later results is the substantial change in precision of the estimates. In particular, the standard errors of $\hat{\alpha}$ and $\hat{\beta}$ are remarkably reduced in comparison with those based on OLS. However, the standard error of the estimate of A is increased. This may be due to the fact that the initial observation, which strongly influences the estimate of A, has the largest predicted variance under the power function variance model. There is little difference between standard errors and based on the power function variance model obtained using normal and empirical third and fourth moments.

Table 1. Observed and Predicted Levels of Pyrene in Rat Blood Following Intravenous Injection

Observations	Time t_i Post Dosing (min)	Blood Level ($\mu g/mg$) Observed y_i	Predicted f_i $\xi = 0$	$\xi = 1.8$
1	15	37.10	36.79	35.48
2	30	22.90	23.54	23.57
3	45	15.30	15.66	16.05
4	60	11.40	10.95	11.29
5	70	9.46	8.90	9.12
6	90	6.68	6.37	6.33
7	120	4.28	4.60	4.26
8	180	3.28	3.40	2.88
9	240	2.50	2.87	2.42
10	330	2.09	2.30	2.00
11	420	1.85	1.85	1.67
12	570	1.58	1.29	1.23
13	720	1.13	0.90	0.91

Table 2. Estimates of Pharmacokinetic Parameters of Two-Compartmental Model Describing the Elimination of Pyrene in the Blood Compartment Following Intravenous Dosing in Rats (standard error in parentheses)

Parameter (units)	Method of Estimation Starting Values	$\xi = 0$	$\xi \neq 0$
$A(\mu g/mg)$	52.59	54.05 (1.20)	50.52 ($2.15^a; 2.19^b$)
$\alpha(\min^{-1})$	0.0322	0.0352 (0.0016)	0.0309 (0.0010; 0.0011)
$B(\mu g/mg)$	4.04	5.09 (0.71)	3.86 (0.19; 0.20)
$\sigma^2(\mu g/mg)^2$	na	0.1865 (na)	0.0026 (0.0019; 0.0012)
ξ	na	na	1.80 (0.37; 0.21)

[a] Standard error based on normal third and fourth moments.
[b] Standard error based on empirical third and fourth moments.
na: not applicable

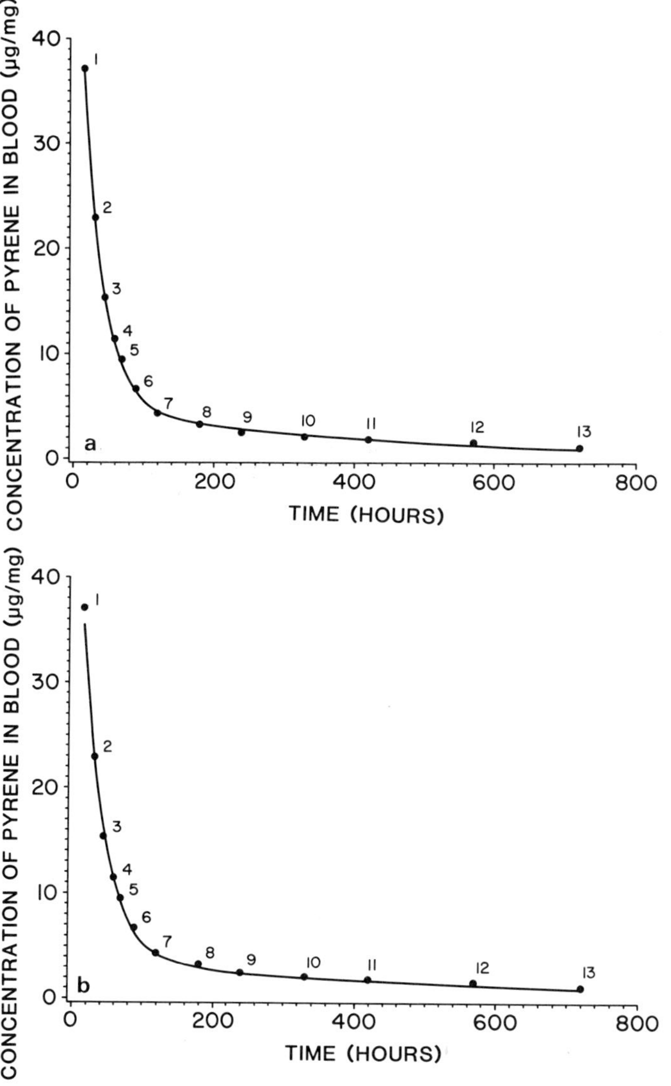

Figure 3.
Two Compartment Model Fitted to Levels of Pyrene in Rat Blood
Following Intravenous Injection
($\xi = 0$, Top Panel; $\xi = 1.8$, Bottom Panel)

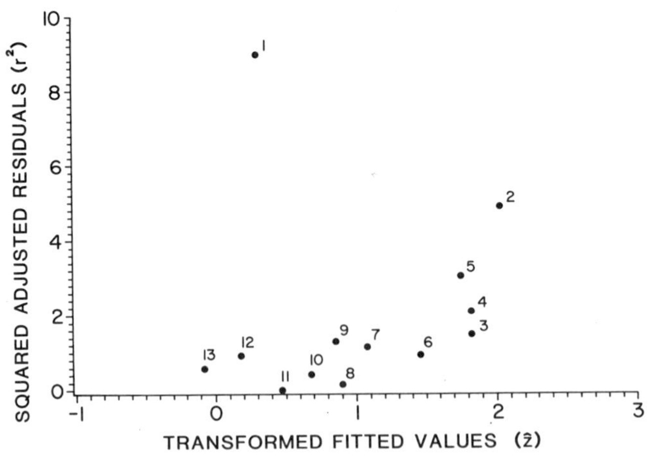

Figure 4.
Squared Adjusted Residuals versus Transformed Fitted Values
Based on Fitted Two Compartment Model ($\xi = 0$)

ACKNOWLEDGEMENTS

We are grateful to Dr. Mietek Szyszkowicz for developing the computer
program used to implement the Gaussian estimation procedure used here, and
to Dr. James R. Withey for providing the pharmacokinetic data on pyrene.

REFERENCES

Atkins, G.L. and I.A. Nimmo, 1981. Robust alternatives to least squares fitting. In: "Kinetic Data Analysis: Design and Analysis of Pharmacokinetic
Experiments" (L. Endryeni, ed.), page 121. Plenum Press, New York.

Bates, D.M. and D.G. Watts, 1980. Relative curvature measures of nonlinar-
ity. J. Roy. Statist. Soc. **B42**:1.

Beal, S.L. and L.B. Sheiner, 1988. Heteroscedastic nonlinear regression. Technometrics **30**:327.

Collins, B.T., 1990. Pharmacokinetic models. In: "Handbook of *In Vivo*
Toxicity Testing" (D.L. Arnold, H.C. Grice and D.R. Krewski, eds.), page
339. Academic Press, San Diego.

Cook, R.D., 1977. Detection of influential observations in linear regression.
Technometrics **19**:15.

Cook, R.D. and S. Weisberg, 1982. "Residuals and Influence in Regression."
Chapman and Hall, London.

Crowder, M., 1985. Gaussian estimation for correlated binomial data. J.R.
Statist. Soc. **B47**:229.

Dzieciolowsky, K. and W.H. Ross, 1990. Assessing case influence on confidence intervals in nonlinear regression. Canad. J. Statistics **18**:127.

Frome, E.L. and G.J. Yakatan, 1980. Statistical estimation of the pharmacokinetic parameters in the one-compartment open model. Communications in Statistics **B9**:202.

Gibaldi, M. and D. Perrier, 1975. "Pharmacokinetics". Marcel Dekker, New York.

Giltinan, D.M. and O. Ruppert, 1989. Fitting heteroscedastic regression models to individual pharmacokinetic data using standard statistical software. J. Pharmacokin. and Biopharm. **17**:601.

Griffiths, D., 1979. Structural identifiability for compartmental models. Technometrics **21**:257.

Godfrey, K., 1983. "Compartmental Models and Their Application." Academic Press, New York.

Hougaard, P., A. Plum and U. Ribel, 1989. Kernel function smoothing of insulin absorption kinetics. Biometrics **45**:1041.

Inagaki, N., 1973. Asymptotic relations between the likelihood estimating function and the maximum likelihood estimator. Annal. Inst. Statist. Math. **25**:1.

Kodell, R.L. and J.H. Matis, 1976. Estimating the rate constants in a two-compartment stochastic model. Biometrics **32**:377.

Matis, J.H., T.E. Wehrly and W.C. Ellis, 1989. Some generalized stochastic compartment models for digesta flow. Biometrics **45**:703.

Metzler, C.M., 1981. Statistical properties of kinetic estimates. In: "Kinetic Data Analysis: Design and Analysis of Pharmacokinetic Experiments" (L. Endryenyi, ed.), page 25. Plenum Press, New York.

Murdoch, D.J., 1991a. A note on a new parameterization of linear compartmental models. Biometrics. To appear.

Murdoch, D.J., 1991b. Compartmental pharmacokinetic models. In: "Statistics in Toxicology" (D. Krewski and C.A. Franklin, eds.), page 143. Gordon and Breach Science Publishers, New York.

O'Flaherty, E., 1981. "Toxicants and Drugs: Kinetics and Dynamics". Wiley, New York.

Rice, D.C., D. Krewski, B.T. Collins and R.F. Willes, 1989. Pharmacokinetics of methylmercury in the blood of monkeys (*Macca fascicularis*). Fundam. Appl. Toxicol. **12**:23.

Rudd, B.E., C.B. Sampson and D.W. Smith, 1975. The one-compartment open model: some statistical aspects of parameter estimation. Appl. Statist. **24**: 309.

Ruppert, D., N. Cressie and R.J. Carroll, 1989. A transformation/weighting model for estimating Michaelis-Menten parameters. Biometrics **45**:637.

Rustagi, J.S. and U. Singh, 1977. Statistical analysis of compartmental models with applications to pharmacokinetics and bioavailability. In: "Applciations of Statistics" (P.R. Krishnaiah, ed.), page 461. North-Holland, New York.

SAS , 1988. "SAS/STAT User's Guide Release 6.03". SAS Institute, Carey, North Carolina.

Thall, P.F. and S.C. Vail, 1990. Some covariance models for longitudinal count data with overdispersion. Biometrics **46**:657.

Wagner, J.G., 1971. "Biopharmaceutics and Relevant Pharmacokinetics". Drug Intelligence Publications, Hamilton, Illinois.

Withey, J.R. and P.G. Collins, 1979. The distribution of styrene monomer in rats by the pulmonary route. J. Environ. Path. Toxicol. **2**:1329.

Withey, J.R., 1990. Pharmacokinetics: principles, mechanisms and methods. In: "Handbook of *In Vivo* Toxicity Testing" (D.L. Arnold, H.C. Grice and D.R. Krewski, eds.), page 303. Academic Press, New York.

Withey, J.R., F.C.P. Law and L. Endrenyi, 1991. Pharmacokinetics and bioavailability of pyrene in the rat. J. Toxicol. Environ. Health **32**:429.

THE USEFULNESS OF PHARMACOKINETICS IN THE DEVELOPMENT OF ANTINEOPLASTIC AND ANTI-AIDS AGENTS

Julie L. Eiseman

University of Maryland Cancer Center
Division of Developmental Therapeutics and Department of Pathology
University of Maryland School of Medicine
Baltimore, MD 21201

INTRODUCTION

There are many difficulties to be faced, both ethical and scientific, in the introduction of a new agent into clinical practice. In the development of cytotoxic agents such as antineoplastic drugs and antiviral agents directed against HIV (human immunodeficiency virus) the path is similar. Clinicians deliver and patients take experimental treatments with the hope of therapeutic benefit in the face of a life threatening illness. In most cases, the chance of therapeutic benefit from the experimental treatment is low; in 187 clinical trials on 54 experimental agents the response rate was only 4% [Estey et al., 1986]. Since the likelihood of toxicity is much greater, investigators involved in the development of these cytotoxic agents must plan their experiments well and gain the most amount of information regarding the efficacy and toxicity of the agent with the fewest number of patients. It is the goal of this chapter to illustrate the successful use of pharmacokinetics in the development of two such therapeutic agents which have recently been introduced to the clinic as approved therapies: Carboplatin as an antineoplastic agent and Zidovudine as an anti-AIDS agent. Through these examples, the importance of pharmacokinetics during the Phase I testing of cytotoxic agents will be established and the use of pharmacokinetics in the development of these two agents will provide guidelines for future phase I testing strategies.

The development of cytotoxic drugs such as antineoplastic agents and anti-AIDS therapeutics is a long process involving several phases of study and raising a number of questions of both scientific and ethical nature. The first stage in the development of cytotoxic agents is preclinical testing. These studies are conducted both in vitro and in vivo, generally in small animal models, to establish efficacy and toxicity of the agent under investigation. During the preclinical studies, analytical methods are developed to measure the potential therapeutic agent in biological fluids and the pharmacokinetics of the agent are investigated in two species, generally in rodents and dogs. If a drug makes it through the preclinical testing with documented efficacy and limited toxicity, it enters clinical trials.

New Trends in Pharmacokinetics, Edited by A. Rescigno and A.K. Thakur
Plenum Press, New York, 1991

Clinical trials are divided into Phases I, II, and III. The first clinical trials are Phase I trials and the goals of these studies are to detail the adverse effects of the agent, to establish the human pharmacokinetics, and determine the maximum tolerated dose (MTD) of the drug. In retrospective analyses of 187 Phase I trials conducted using 54 potential antineoplastic drugs, the response rate was only 4.2% [Estey et al., 1986] underscoring the difficulties of determining proper treatment regimens. Phase I studies generally involve dose escalation using a modified Fibonacci scheme to define the MTD. Generally the starting dose is one tenth of the LD_{10} in mice. The reason for a starting dose at one tenth the LD_{10} in mice is that the ratio of the MTD in humans and the LD_{10} in mice generally falls between 0.1 and 10 [Freireich et al, 1966]. If patients are treated with a low dose in the dose escalation scheme, they are very unlikely to receive any benefit from the treatment since most cytotoxic drugs are therapeutic only near their MTD. The number of patients receiving ineffective therapy during a Phase I trial should be limited.

Collins et al. [1986,1990] examined whether pharmacokinetic or pharmacodynamic factors contributed to the disparity between MTD and LD_{10}. They compared the areas under the plasma concentration time curves (AUC's) in mice and man for a number of drugs when administered to man at the MTD and to mice at their LD_{10}. These investigators found that AUC ratios were closer to one than were the ratios of the doses suggesting that the discrepancy between the doses was largely due to differences in pharmacokinetics between the species. The exceptions to this finding were the anti metabolite class of antineoplastic agents and the anti-AIDS drug, dideoxycytidine. Three of four exceptional drugs were compounds that require activation by the nucleoside kinases. Based on their observations, Collins et al. [1986] recommended two methods to use preclinical pharmacokinetic data to reduce the number of dose escalations in a Phase I trial. The prerequisite was that the preclinical pharmacokinetics was linear with dose. The central assumption was that the AUCs were more closely related to pharmacological effect (i.e. toxicity) than were dose. Hence by measuring the AUC in a Phase I trial starting dose, one could compare the AUC to that obtained at the LD_{10} in mice. If the AUC was close to the mouse AUC, then escalation should be conservative. If the AUC was much less than that of the mouse, escalation could be higher. In the first method, one doubled the dose until the AUC in man reached 40% of the mouse AUC, then completed the study using a modified Fibonacci scheme. In the second method, the first dose escalation was the square root of the ratio of the mouse AUC at LD_{10} to the human starting dose AUC. The rest of the Phase I study would then follow a modified Fibonacci scheme as well. The major problems encountered in using their methods are analytical assay sensitivity, species differences handling the drug, such as metabolism and protein binding, and patient variability. Although their recent studies have shown that there is no standard design, pharmacologically guided dose escalation based on application of pharmacokinetic principles can reduce the length of time required to conduct a Phase I trial and reduce the number of patients exposed to ineffective therapeutic levels. This assumes that one can identify the causes of patient variability and correct for them in calculating doses. With the conscientious application of pharmacokinetic principles, Phase I studies can be conducted with better utilization of time and resources.

Using pharmacokinetic/pharmacodynamic principles to obtain accurate human pharmacokinetic information is both useful and necessary for the further development of the drug through Phase II and Phase III trials and is important to the clinical use after the agent has been approved. To illustrate the importance of pharmacokinetics to the development of both antineoplastic and anti-AIDS agents, key studies with an antineoplastic drug, Carboplatin, and with Zidovudine, an anti-AIDS drug, will be presented.

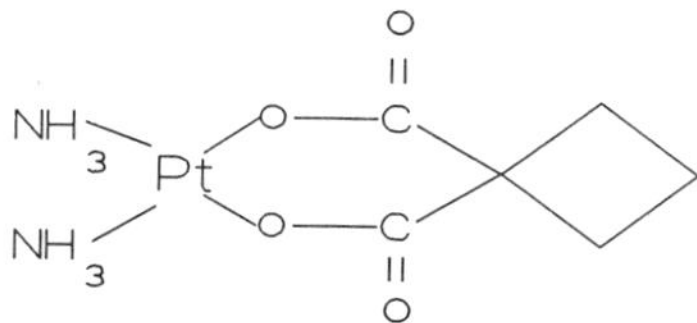

Fig.1. Carboplatin structure

CARBOPLATIN

Carboplatin is an analog of cisplatin that was developed because although cisplatin is effective, its usefulness is limited due to renal toxicity and severe nausea and vomiting. The structure of Carboplatin is shown in Figure 1.

Carboplatin interacts in vitro with DNA in a manner similar to cisplatin to produce intrastrand breaks. The preclinical studies of carboplatin were summarized by Schurig et al. [1990]. Carboplatin was active against several NCI tumor panel malignancies including B16 melanoma, subcutaneous CD8F1 mouse mammary carcinoma, murine colon 26 adenocarcinoma, and human breast carcinoma xenografts implanted under the renal capsule of athymic mice. Carboplatin was more active against murine P388 leukemia when given by the i.p. route every six hours on a day 1,5,9 schedule (Table 1).

Table 1
Antitumor Activity of Carboplatin

Tumor Type	Activity of Carboplatin			
	Inactive	Mild	Moderate	Marked
Murine Tumors Responsive to Cisplatin				
B16		X		
CD8F1			X	
C26		X		
C38		X		
M109	X			
M5076	X			
Murine Tumors Resistant to Cisplatin				
L1210			X	
P388		X		
Human Tumor Xenografts				
Bladder SW780		X		
Broncho P246			X	
Mammary MX-1				X
Ovarian MRI-H-207				X

Preclinical toxicity indicated that carboplatin was less emetigenic and nephrotoxic than cisplatin. In animals, the dose limiting toxicity was myelosuppression (Table 2).

Table 2
Toxicity of Carboplatin Relative to Cisplatic

Toxicity	Species	Carboplatin	Cisplatin
Nephrotoxicity(measured as Increased BUN or by Histology			
	Mouse	+	+++
	Rat	0	+++
	Dog	0	+++
Emetic Effects			
	Ferret	+	+++
	Dog	+/-	++++
Myelosuppressive Effects			
	Mouse	++	0
	Rat	+++	++
	Dog	++	+

Carboplatin and ultrafilterable (free) platinum exhibited linear pharmacokinetics in dogs. The maximum plasma concentrations and the AUCs increased linearly with dose. The elimination half lives for carboplatin and free ultrafilterable platinum were about one hour in the dog. In the rat, they were about 0.5 hour. More than 90% of the free, ultrafilterable platinum in dog plasma was carboplatin. The same was true for rat plasma (Table 3).

Table 3
Pharmacokinetic Values of Free, Ultrafilterable Platinum after IV
Administration to Animals

Species	Dose (mg/kg)	Half-life (hr)	CL(L/hr/kg)	%Pt in Urine
Mouse	5	---	---	92% in 72 hr
	80	---	---	91% in 4 hr
Rat	20	0.43	0.62	89% in 4 hr
	20	0.62	0.42	62% in 24 hr
Dog	3	0.7	0.25	74% in 96 hr
	6	0.9	0.22	78% in 96 hr
	12	0.9	0.22	61% in 96 hr
	24	1.2	0.2	67% in 96 hr

Cisplatin was excreted by glomerular filtration whereas carboplatin was excreted by active tubular secretion as well as glomerular filtration.

Based on this preclinical information, a number of clinical Phase 1 trials were initiated. A summary of these trials is presented (Table 4).

The data from the Phase I clinical trial at the University of Maryland [Van Echo et al.,1984 and Egorin et al.,1985] and its extension into another Phase I trial will be elaborated on here because these studies elegantly outline use of pharmacokinetics and pharmacodynamic principles to improve clinical regimens.

The first study was undertaken to determine the MTD (maximum tolerated dose) of Carboplatin when administered as a single i.v. bolus daily for five days. The courses were repeated every 4 5 weeks and pharmacokinetic studies were performed on all patients entered at doses of 11-77 mg/m^2.

To be eligible for this protocol, patients had to have histologic proof of a malignant disease which had failed conventional therapy and for which no conventional therapy existed. Furthermore, they must have recovered from the toxic effects of prior treatments and had not received either chemotherapy or radiation therapy for at least four weeks. All patients had a minimum life expectancy of 12 weeks and performance status of at least 60% on the Karnofsky scale (a scale specifying the pretreatment performance status of a patient) [Karnofsky and

Table 4
Phase 1 Clinical Trials of Carboplatin

Institution	Administration	Side Effects
Royal Marsden Hospital	Bolus	Vomiting, Peripheral Neuropathy
Japan Carboplatin Study Group	Bolus	Nausea and Vomiting, Peripheral Neuropathy
Hospital San Giovanni	Bolus	Vomiting
Mt. Sinai	Bolus	Nausea and Vomiting
Wisconsin	Bolus	Vomiting, Peripheral Neuropathy
CI Tokyo	Bolus	Nausea and Vomiting
USA NCI	CI	Nausea and Vomiting
Mt. Sinai	CI	Nausea and Vomiting
Institut Jules Bordet	d x 5	Nausea and Vomiting, Peripheral Neuropathy
University of MD	d x 5	Nausea and Vomiting, Peripheral Neuropathy

Berchenal, 1949]. All patients had evidence of adequate bone marrow function and platelet counts, adequate liver and kidney functions.

The starting dose of Carboplatin was based on animal toxicity. In this case, one tenth the LD_{10} in dogs. A cohort of three patients were studied at the starting dose, 11 mg/m^2. An interval of 1 week passed between the entry of each of the patients at this treatment level. Subsequent patients were entered at higher dosages if no dose limiting toxicity was seen at the previous treatment level. New doses were chosen with a modified Fibonacci search scheme. A minimum of two patients were entered at each dose level. Dosages were escalated in the same patient after the patient had received two courses of Carboplatin with mild reversible toxicity. Six or more patients were studied at doses which produced acceptable, reversible toxicity. Patients did not receive antiemetics during their first course, so this adverse effect could be evaluated.

On each treatment day and weekly thereafter, serum tests including Na, K, Cl, Mg, P, BUN (blood urea nitrogen), and creatinine were measured. Twenty four-hour urine creatinine clearances were performed prior to Carboplatin and every two weeks thereafter in all patients. Serum liver function tests, electrocardiograms and audiograms were measured prior to treatment and at intervals during treatment. All tests were repeated when the patient was removed from study.

At specified times after treatment, blood samples were collected in heparinized tubes and centrifuged to separate the plasma. Aliquots of plasma were stored frozen until analysis. Protein free ultrafiltrates of plasma were prepared by centrifuging the remaining plasma from each sample over protein exclusion membranes for 20 minutes at 4°C. Urine samples were collected at the time of voiding, measured and stored frozen as pooled four hour collections. Plasma, ultrafiltrates, and urine samples were analyzed for platinum content by flameless atomic absorption spectrophotometry.

Plasma pharmacokinetic analysis was performed using MLAB (Division of Computer Research and Technology, NIH, Bethesda, MD) using nonlinear regression analysis with and without weighting with $1/SE^2$. AUC's for ultrafilterable platinum were estimated MLAB's integration routine. Total body clearance and the steady state volume of distribution were also calculated. Data from this study are summarized in Table 5.

Table 5
Pharmacokinetic Parameters of Total and Ultrafilterable Platinum in Plasma of
Patients Treated with Carboplatin (Data Taken from Van Echo et al., 1984)

Dose (mg/m^2)	α-half-life (min^{-1})	β-half-life (min^{-1})	AUC (μg/min/ml)	CL (ml/min/m^2)	Vd$_{ss}$ (L/m^2)
Total Platinum					
11	28	576	394	15	22.6
55	26	492	745	39	43.2
77	49	1188	1358	30	69.6
Ultrafilterable Platinum					
11	12.2	236	145	40	23.7
22	7.6	146	235	49	17.7
55	21.4	133	290	100	24.0
77	10.2	102	709	57	16.1

Dosages evaluated were 11, 22, 55, 77, and 99 mg/m^2. Nonhematologic toxicity was confined to emesis which was dose related. Lower extremity myralgias were noted at the highest dose levels beginning 12 hours after dosing and persisting for 8 12 hours. Hematologic toxicities were noted and the dose limiting toxicity was thrombocytopenia. For patients who had received prior chemotherapy, the median nadir (the lowest point) was 89000/μ at 77 mg/m^2 while for those without chemotherapy the platelet nadir occurred at 99 mg/m^2 and was 100000/μl.

The plasma concentration of total platinum declined in a biexponential fashion with α and β half lives of 26-49 and 492-1088 minutes respectively. As expected, the administration of higher doses of carboplatin produced proportionally higher peak plasma concentrations and AUCs for total platinum. There were no consistent effects of dosage on plasma half life or total body clearance of platinum. The total body clearance of platinum approximated creatinine clearance. Very little platinum remained in the plasma after 24 hours and in 14 courses there was no accumulation of platinum in the plasma.

The ultrafilterable plasma also declined in biexponential function with α and β half lives of 7.6-21.4 and 102-236 minutes respectively. These values were three to five fold shorter than those for total platinum. The administration of higher doses of carboplatin produced proportionally higher peak plasma concentrations and AUCs of ultrafilterable platinum, but did not effect the total body clearance, half lives or volume of distribution at steady state. The total body clearance for ultrafilterable platinum was 3 4 fold greater than that for total platinum. Ultrafilterable platinum accounted for greater than 70% of total platinum during the first two hours after dosing, and declined slowly to 50% in 6 hours. The kidney was the major route of excretion and by 24 hours, 53% of injected platinum had been excreted in the urine.

This study demonstrated that carboplatin is an effective antineoplastic agent with thrombocytopenia as the dose limiting toxicity. A dose of 77 mg/m^2 was recommended as a starting dose for Phase II studies. The pharmacokinetics of Carboplatin was linear in the dose ranges examined. At all doses examined, the total body clearance of ultrafilterable platinum was 3 4 fold higher than creatinine clearance and the volume of distribution at steady state approximated total body water.

Egorin et al. [1985] went on to further examine two relationships observed in this Phase I study. They examined Carboplatin treatment in patients with renal impairment since the clearance of total platinum approximated creatinine clearance. They also examined more closely the relationship of previous pretreatment on the dose limiting toxicity of Carboplatin, thrombocytopenia. The goal of the second study was to examine the pharmacokinetics and toxicities of Carboplatin in patients with reduced renal function and to define a dosage reduction scheme for such patients.

Egorin et al. [1985] examined the relationship between glomerular filtration as measured by creatinine clearance and the amount of platinum excreted in the urine. As was expected, reduced creatinine clearance correlated with decreased urinary excretion of Carboplatin platinum. Further, the relationship of creatinine clearance to total body clearance was also linear with a correlation coefficient of 0.82. This implied that other means of drug elimination, such as increased protein binding or biliary excretion, were not activated to compensate for the decreased renal clearance.

To examine whether a relationship existed between the pharmacokinetics of Carboplatin and its dose limiting toxicity, comparisons of peak plasma concentrations and AUCs were made to nadir. Comparisons of the peak concentration of total platinum to ultrafilterable platinum bore no relationship to the myelosuppression produced by the course of therapy nor did the AUC. Since the patients platelet counts, at the time of Carboplatin administration, varied from 150000 to 760000/μl and a platelet nadir of 100000/μl produced in the first case might represent more myelosuppression than in the second case, the percentage of platelet reduction was examined in relationship to the pharmacokinetic parameters of Carboplatin. The percent reduction of platelet count was calculated as:

$$\% \text{ Reduction} = (\text{Pretrt. count-nadir})/(\text{Pretrt. count}) \times 100.$$

When the AUCs were compared to the percent platelet reduction, two parallel empirical linear relationships were found:

$$\% \text{ Reduction in Platelets} = 0.72 \text{ AUC} + 48.5 \text{ and}$$
$$\% \text{ Reduction in Platelets} = 0.72 \text{ AUC} + 31.2. \tag{1}$$

The constants 31.2, 48.5, and 0.72 in the above expressions are least squares estimates with correlation coefficients of 0.73 and 0.89, respectively, and not overlapping. Closer inspection of the data revealed that the first relationship represented patients who had been heavily pretreated with chemotherapy while the latter equation represented patients who had not received prior chemotherapy. Not only did a relationship exist between easily determined pharmacokinetic parameter, AUC, and a pharmacodynamic event, reduction in platelets, but for every AUC, heavily pretreated patients experienced a 17% greater reduction in platelet count than did patients not receiving prior therapy [Egorin et al. 1985]. These workers used basic pharmacological principles and the relationship established between creatinine clearance and total body clearance and the relationship between AUC and percent reduction in platelet count to define a dosage calculation equation based on the desired reduction in platelet count. Since total body clearance was directly related to creatinine clearance in the case of Carboplatin and total body clearance is equal to the dose/AUC, the dose of Carboplatin could be determined for each individual patient based on the patient's creatinine clearance, body surface area and desired percentage in platelet count.

The usefulness of these observations was examined in a prospective study. The doses of Carboplatin were calculated using the following empirical linear relationships:

For previously untreated patients:

$$\text{Dosage (mg/m}^2) = 0.091 \ (C_{cr}/\text{Body Surface Area}) \times 100 + 86$$

For previously treated patients:

$$\text{Dosage (mg/m}^2) = 0.091 \left((C_{cr}/\text{Body Surface Area}) \times 100 - 17 \right) + 86 \qquad (2)$$

$$\text{where } X = \frac{\text{Pretrt. Platelet Count} - \text{Desired Platelet Count}}{\text{Pretrt. Platelet Count}} .$$

In Equation (2), 0.091 and 86 are least squares estimates and 17 is the rounded difference between the intercepts 48.5 and 31.2 in Equation (1).

The applicability of these dosing relationships was validated in the prospective study. The population with the poorest fit was the previously pretreated population, and this would be expected as this population is heterogeneous with differences in both the type and actual degree of pretreatment.

These studies by Egorin et al. [1984,1985] document the relationship between the pharmacokinetics and pharmacodynamics of Carboplatin and demonstrate the ability to use this relationship to benefit the patient. The patient can now receive an effective course of therapy with acceptable toxicity. This work is being extended and the use of this dosage adjustment scheme for Carboplatin should be applicable when Carboplatin is used in combination trials with other antineoplastic agents [Balani et al. 1990].

For pharmacologists and pharmacokineticists, this work represents the thoughtful use of pharmacokinetics and its relationship to pharmacodynamics to improve antineoplastic therapy. These techniques are not foolproof and their application requires the determination of an acceptable percent reduction in platelet count which varies for each patient. Although these studies only examined the relationships for one agent, Carboplatin, careful examination of the relationship between pharmacokinetics and toxicity (pharmacodynamic effect) is important at all levels in the development of any therapeutic agent.

ZIDOVUDINE

Zidovudine or AZT (3'-Azido-3'-deoxythymidine) is a nucleoside analog of thymidine (Figure 2) and is the first therapeutic agent approved for the treatment of human Acquired Immunodeficiency Syndrome (AIDS).

AZT readily enters cells and is converted to nucleotide forms by the intracellular kinases which activate thymidine. The selectivity of its action is based on the preferential utilization of AZT triphosphate by the retroviral reverse transcriptase of HIV (human immunodeficiency virus) compared to cellular DNA polymerases. In vitro studies [Mitsuya et al. 1985] indicated that concentrations of AZT at or above 1.0 μM could selectively inhibit the virus with no detrimental effects on cells. Bilello et al. [1988] examined the inhibitory effects of AZT on both de novo and chronic viral infection of cells using murine retroviruses in tissue culture and found similar results. In animal models of retrovirus infection, AZT was effective in preventing infection if given prophylactically and delayed disease progression if given after infection [Ruprecht et al., 1987, Bilello et al., 1988]. The major dose limiting toxicity in the animal studies was hematopoietic toxicity with hematocrits dropping rapidly and the development of aplastic anemia if treatment was not terminated.

Analytical methods to measure AZT in biological tissues have relied on high performance liquid chromatography and the peak area versus concentration was linear with a lower detection limit of 27 ng/ml [Yarchoan et al., 1986, Klecker et al., 1987, Eiseman et al., 1988]. Pharmacokinetics of AZT in mice were linear with dose, the drug was bioavailable by the oral route, and no accumulation was seen

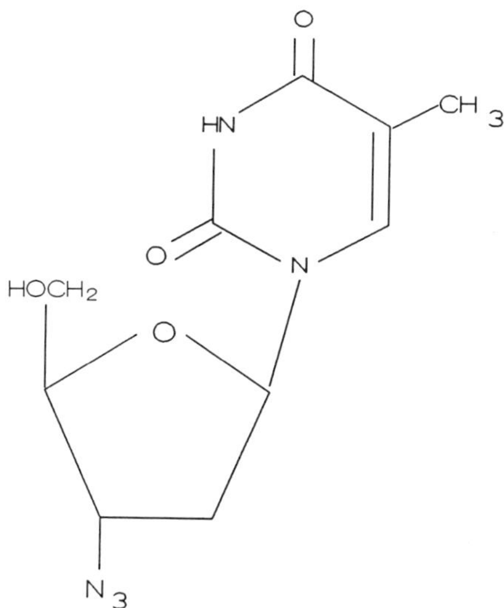

Fig. 2. Structure of Zidovudine (AZT)

with repeated dosing [Eiseman et al., 1988]. The terminal half-life was approximately 20 minutes in the study by Eiseman et al. [1988] and 1 hour in the studies of Doshi et al. [1989].

Most of these animal efficacy studies were not completed until after the introduction of AZT into clinical trials, so the initial human trials were conducted to determine if the effective in vitro concentrations could be attained in plasma. Well designed pharmacokinetic studies by a number of investigators have played a central role in the rapid clinical development of this agent. Information obtained by Yarchoan et al. [1986] showed that one could achieve plasma concentrations in man similar to those required to inhibit the retroviral reverse transcriptase in vitro. This information helped in the selection of a dosing interval. Studies by Klecker et al. [1987] and Blum et al. [1988] encouraged the use of oral administration and documented cerebrospinal fluid penetration of AZT as well as identified the elimination pathways. Klecker et al. [1987] compared the oral and intravenous administration of AZT to patients. AZT clinical pharmacokinetics was linear with dose, the terminal half life was about 1 hour, and CL (total body clearance) was about 1.3 L/kg/hr. The volume of distribution at steady state ranged between 1.3 and 1.6 L/kg. Bioavailability studies indicated that about 60% of the agent was available by the oral route. The time to reach peak occurred at about 0.6 hour after dosing. Most endogenous nucleosides and analogs distribute readily throughout extracellular space with the exception of the central nervous system, however AZT is more lipophilic than most and was able to directly cross cellular membranes. The data from Klecker et al. [1987] demonstrated the entrance of AZT into cerebrospinal fluid and the levels attained were about one fifth of those in the plasma. This finding is very important since the central nervous system is thought to be a reservoir for the AIDS virus.

The data of Klecker et al. [1987] are in close agreement with the results of a Phase I/II study conducted by Blum et al. [1988]. These workers also examined the metabolism of AZT to its major inactive metabolite GAZT (5'-O-glucuronide of AZT). Most of the AZT is excreted in urine as GAZT. After an intravenous dose only about 17% of the dose is recovered as unchanged AZT while greater than 60% is recovered as GAZT in the urine. After an oral dose, about 75% of the drug is recovered in the urine as GAZT while only about 14% of the parent drug is excreted unchanged. The short half life of AZT in man influenced the dosing schedule. Most protocols require round the clock dosing every four hours and the demonstration of bioavailability of AZT by the oral route greatly improved the continuous use of this agent. Generally doses of 200-250 mg are administered every four hours, but even with this schedule, peak to trough ratios exceed 10.

The data from Yarchoan et al. [1986] clearly demonstrate that AZT administration on this schedule is able to reverse or slow the AIDS symptoms. The number of T-helper cells increased as did the ratio of T-helper to T-suppressor cells. Further, the number of incidents of opportunistic infection were reduced.

Balis et al. [1989] extended the studies of AZT to children with AIDS and found that the pharmacokinetic parameters of AZT in children greater than 1 year of age were similar to those in adults. In children, as in adults, the fraction bioavailable was 0.68 ± 0.25. The terminal half life of AZT in the plasma was approximately 90 minutes and the volume of distribution at steady state was about 45 L/m^2. Because of compliance problems with children on a four hour round the clock dosing schedule, Pizzo et al. [1990] made use of portable constant infusion pumps and Hickman or Broviac catheter to deliver the AZT intravenously to children.

Morse et al. [1990] in their studies in HIV infected hemophilia patients demonstrated that the kinetics of AZT did not change during continuous administration. Parameters were not significantly different between weeks 1, 6, and 12 of treatment. Since AZT is primarily metabolized to GAZT by glucuronidation and both are excreted in the urine, de Miranda et al. [1989] examined the effect of probenecid, a drug extensively metabolized in humans by glucuronidation and which utilizes renal tubular organic acid secretion pathway, on the elimination of AZT and GAZT. As expected, the mean plasma levels, AUCs, and half lives of both AZT and GAZT were elevated after probenecid treatment. This study pointed out a serious problem that may exist if AZT is administered concurrently with other agents that are glucuronidated and excreted by the organic acid secretory mechanism of the kidney. Because AZT has a very small therapeutic window, coadministration of such agents could result in toxicity without proper clinical management. The pharmacokinetic studies of Singlas et al. [1989] pointed out the difficulties in administering AZT to uremic patients and showed the extensive accumulation of AZT and GAZT in such patients.

The final determination of the usefulness of an agent such as AZT requires years of careful investigation. However, the studies outlined here have shown the contribution of pharmacokinetics to the development of this agent. Already this nucleoside has become the cornerstone of treatment for a disease that has high priority. Unfortunately, the period of useful therapy is often limited. In the case of AZT, the occurrence of hematologic toxicity often becomes dose-limiting. Furthermore, AZT resistant strains of HIV have been isolated in vitro from samples of patients who have been on AZT for prolonged periods. Whether this phenomenon has clinical significance is not known at this time. Nonetheless, the trend is now to extend the time of treatment by combining AZT with other antiviral agents such as DDC (Dideoxycytidine), another nucleoside analog with a markedly different spectrum of toxicity. Such a regimen was proposed by Bozzette et al. [1990] with alternating and intermittent AZT and DDC and is still under investigation. It is the thorough understanding of pharmacokinetics and pharmacodynamics that forms the logical basis for combination trials with these drugs.

CONCLUSIONS

The studies presented here have shown the importance of pharmacokinetics in the development of two therapeutic agents, Carboplatin, an antineoplastic agent and Zidovudine, an anti-AIDS agent. The principles utilized so elegantly in the studies of the many investigators referenced here can be applied to the development of any therapeutic agent. The goal and the challenge for us is to use the principles of pharmacokinetics and the relationship with pharmacodynamics to improve therapeutic outcome.

REFERENCES

Balis, F.M., Pizzo, P.A., Eddy, J., Wilfert, C., McKinney, R., Scott, G., Murphy, R.F., Jarosinski, P.F., Falloon, J., and Poplack, D.G., 1989. Pharmacokinetics of Zidovudine Administered Intravenously and Orally in Children with Humna Immunodeficiency Virus Infection. J. Pediatrics 114:880.

Balis, F.M., McCully, C., Gough, L., Pizzo, P.A., and Poplack,D.G., 1989. Pharmacokinetics of Subcutaneous Azidothymidine in Rhesus Monkeys. Antimicrob. Agts. Chemother. 33:810.

Belani, C.P., Egorin, M.J., Abrams, J.S., Hiponia, D., Eisenberger, M., Aisner, J., and Van Echo, D.A., 1990. A Novel Pharmacodynamically Based Approach to Dose Optimization of Carboplatin when Used in Combination with Etoposide. In: "Carboplatin (JM 8) Current Perspectives and Future Directions" (P.A. Bunn, Jr., R.F. Canetta, R.F. Ozols, and M. Rozencweig, M., ed.), page 39. W.B. Saunders Co., Philadelphia.

Bilello, J.A., Eiseman, J.L., MacAuley, C., Bell, M.M., and Yetter, R.A., 1988. 3'Azidothymidine Prevents the Dissemination of Retrovirus in LP-BM5 MuLV-Infected Mice. Int. Conf. AIDS Res., Stockholm, p. 170.

Blum, M.R., Liao, S.H.T., Good, S.S., and deMiranda, P., 1988. Pharmacokinetics and Bioavailability of Zidovudine in Humans. Am. J. Med. 85: Suppl. 2A, p. 189.

Bozzette, S.A. and Richman D.D., 1990 Salvage Therapy for Zidovudine Intolerant HIV™Infected Patients with Alternating and Intermittent Regimens of Zidovudine and Dideoxycytidine. Am. J. Med. 88:Suppl. 5B, p. 24S.

Collins, J.M., Zaharko, D.S., Dedrick, R.I., and Chabner, B.A., 1986. Potential Roles for Preclinical Pharmacology in Phase I Clinical Trials. Cancer Treat. Rep. 70:73.

Collins, J.M. and Unadkat, J.D., 1989. Clinical Pharmacokinetics of Zidovudine, An Overview of Current Data. Clin. Pharmacokinet. 17:1.

Collins, J.M., Grieshabar, C.K., and Chabner, B., 1990. Pharmacologically Guided Phase I Clinical Trials Based Upon Preclinical Drug Development. J. Natl. Cancer Institute. 82:1321.

de Miranda, P., Good, S.S., Yarchaon, R., Thomas, R.V., Blum, M.R., Myers C.E., and Broder, S., 1989. Alteration of Zidovudine Pharmacokinetics in Patients with AIDS or AIDS Related Complex. Clin. Pharm. Ther. 46:494

Doshi, K.J., Gallo, J.M., Boudinot, F.D., Schinazi, R.F., and Chu, C.K., 1989. Comparative Pharmacokinetics of 3' Azido 3'Deoxythymidine (AZT) and 3' Azido 2',3' dideoxyuridine (AZddU) in Mice. Drug Met. Disp. 17:590.

Egorin, M.J., Van Echo, D.A., Tipping, S.J., Olman, E.A., Whitacre, M., Thompson, B.W., and Aisner, J., 1984. Pharmacokinetics and Dosage Reduction of cis Diammine(1,1™cyclobutanecarboxylato)platinum in Patients with Impaired Renal Function. Cancer Res. 44:5432.

Egorin, M.J., Van Echo, D.A., Olman, E.A., Whitacre, M.Y., Forrest, A., and Aisner J., 1985. Prospective validation of a pharmacologically based dosing scheme for the cis-Diamminedichloroplatinum (II) analog diamminecyclobutanedicarboxylatoplatinum, Cancer Res. 45:6502.

Eiseman, J.L., Bell, M.M., Bilello, J.A., Wetherall, D.L., MacAuley, C., and Yetter, R.A., 1988. Pharmacokinetics of 3'Azidothymidine (AZT) in Retrovirus Infected Mice. Fed. Proc. 2:3596.

Estey, E., Hoth, D., Simon, R., Marsoni, S., Leyland Jones, B., and Wittes, R., 1986. Therapeutic Response in Phase I Trials of Antineoplastic Agents. Cancer Treat. Rep. 70:1105.

Freireich, E.J., Gehan, E.A., Rall, D.P., Schmidt, L.H., and Skipper, H.E., 1966. Quantitative Comparison of Toxicity of Anticancer Agents in Mouse, Rat, Hamster, Dog, Monkey and Man. Cancer Chemother. Rep. 50:219.

Garraffo, R., Cassuto Viguier, E., Barillon, J., Lapalus, P., and Duplay, H., 1989. Influence of Hemodialysis on Zidovudine (AZT) and its Glucuronide (GAZT) Pharmacokinetics: Two Case Reports. Int. J. Clin. Pharmaco. Ther. Toxicol. 27: 535.

Hollander, H., Lifson, A., Maha, M., Blum, M.R., Rutherford, G.W., and Nusinoff-Lehrman, S., 1989. Phase I Study of Low Dose Zidovudine and Acyclovir in Asymptomatic Human Immunodeficiency Virus Seropositive Individuals. Am. J. Med. 87:628.

Karnofsky, D.A. and Berchenal, J.H., 1949. The clinical evaluation of chemotherapeutic agents in cancer. In: "Evaluation of Chemotherapeutic Agents" (C.M. Macleod ed.), page 191. Columbia University Press, New York.

Klecker, R.W., Collins, J.M., Yarchoan, R., Thomas, R., Jenkins, J.F., Broder, S., and Myers, C.E., 1987. Plasma and Cerebrospinal Fluid Pharmacokinetics of 3'-Azido-3'-deoxythymidine: A Novel Pyrimidine Analog with Potential Application for the Treatment of AIDS and Related Diseases. Clin. Pharm. Ther. 41:407.

Meng, T C., Fischl, M.A., and Richman, D.D., 1990. AIDS Clinical Trials Group: Phase I/II Study of Combination 2',3' Dideoxycytidine and Zidovudine in Patients with Acquired Immunodeficiency Syndrome (AIDS) and Advanced AIDS Related Complex. Am J. Med. 88:Suppl. 5B, p. 27S.

Mitsuya, H., Weinhold, K.L., Furman, P.A., St. Clair, M.H., Nusinoff-Lehrman, S., Gallo, R.C., Bolognesi, D., Barry, D.W., and Broder, S., 1985. 3'-azido-3'-deoxythymidine (BW A509U): A New Antiviral Agent that Inhibits the Infectivity and Cytopathic Effect of Human T-lymphotrophic virus type III/lymphadenopathy-associated virus in vitro. Proc. Natl. Acad. Sci.(USA) 82:7096.

Morse, G.D., Portmore, A., Olson, J., Taylor, C., Plank, C., and Reichman, R.C., 1990. Multiple Dose Pharmacokinetics of Oral Zidovudine in Hemophilia Patients with Human Immunodeficiency Virus Infection. Antimicro. Agts. Chemother. 34:394.

Pizzo, P., 1990. Treatment of Human Immunodeficiency Virus Infected Infants and Young Children with Dideoxynucleosides. Am. J. Med. 88: Suppl. 5B, p. 16S.

Ruprecht, R.M., O'Brien, L.G., Rossoni, L.D., and Nussinoff-Lehrman, S., 1987. Suppression of mouse viremia and retroviral disease by 3'-azido-3'-deoxythymidine. Nature 323:467.

Schurig, J.E., Rose W.C., Catino, J.J., Gaver R.C., Long B.H., Madissoo, H., and Canetta, R., 1990. The Pharmacologic Characteristics of Carboplatin: Preclinical Experience. In: "Carboplatin (JM 8) Current Prospectives and Future Directions" (P.A. Bunn, Jr., R.F. Canetta, R.F. Ozols, and M. Rozencweig, M. eds.), page 39. W.B. Saunders Co., Philadelphia.

Singlas, E., Pioger, J.C., Taburet, A.M., Colin, J.N., and Fillastre, J.P., 1989. Zidovudine Disposition in Patients with Severe Renal Impairment: Influence of Hemodialysis. Clin.Pharm. Ther. **46**: 190.

Skowron, G., and Merigan T.C., 1990. Alternating and Intermittent Regimens of Zidovudine (3'-azido-deoxythymidine) and Deoxycytidine (2',3'-deoxycytidine) in the Treatment of Patients with Acquired Immunodeficiency Syndrome (AIDS) and AIDS Related Complex. Am. J. Med. **88**:Suppl. 5B, p. 20S.

Terasaki, T. and Pardridge, W.P., 1988. Restricted Transport of 3' Azido 3' Deoxythymidine and Deoxynucloesides Through the Blood Brain Barrier. J. Infect. Dis. **158**:630.

Van Echo, D.A., Egorin, M.J., Whitacre, M.Y., Olman, E.A., and Aisner, J., 1984. Phase I clinical and Pharmacologic Trial of Carboplatin Daily for 5 Days. Cancer Treat. Rep. **44**:1103.

Van Echo, D.A., Egorin, M.J., and Aisner, J., 1989. The Pharmacology of Carboplatin. Seminars in Oncology **16**:Suppl. 5, p. 1.

Yarchoan, R., Weinhold, K.J., Lyerly, H.K., Blum, M.R., Shearer, G.M., Mitsuya, H., Collins, J.M., Myers, C.E., Klecker, R.W., Markham, P.D., Durack, D.T., Nusinoff-Lehrman, S., Barry, D.W., Fischl, M.A., Bolognesi, D.P., and Broder, S., 1986. Administration of 3'-Azido-3'-deoxythymidine, an Inhibitor of HTLV III/LAV Replication, to Patients with AIDS or AIDS Related Complex. Lancet **1**:575.

PHYSIOLOGIC MODELS OF HEPATIC DRUG ELIMINATION

Malcolm Rowland and Allan M. Evans

Department of Pharmacy
University of Manchester
Manchester, U.K.

INTRODUCTION

Pharmacokinetic models are used primarily to describe the time course of drugs and metabolites in the body following various routes of administration. Such models take a variety of forms. Some are simply descriptive, comprising mathematical equations which make no reference to underlying physiology. The ability to use such descriptive models to interpret pharmacokinetic data and to predict outcome under a variety of conditions is extremely limited. Pharmacokinetic models which are physiologically based have greater application and have enjoyed wide usage, particularly those applied to the elimination of drugs by the liver and, to a lesser extent, by the kidneys [Rowland & Tozer, 1989]. The present chapter reviews the physiologic models that have been applied to hepatic clearance, focusing on recent advances, and comments on some problems and outstanding issues. Mention is also made of the usefulness of the isolated perfused liver for investigating drug distribution and elimination kinetics.

REVIEW OF EXISTING MODELS

Any model of hepatic drug elimination must be able to relate output to input, generally with reference to drug concentration. Physiologic factors that need to be considered include organ perfusion, solute binding to blood components, membrane permeability, enzymatic and cellular activity, and the architecture of organ microvasculature. An assortment of models, varying immensely in their mathematical complexity, have been used to relate the hepatic extraction of eliminated substrates to some, or all, of these physiologic factors (Figure 1). Such models have most frequently been applied to data generated using an isolated perfused liver preparation (usually of the rat) in which perfusate flow rate, solute binding within the perfusate, and the input concentration-time profile of solute can be controlled.

New Trends in Pharmacokinetics, Edited by A. Rescigno and A.K Thakur
Plenum Press, New York, 1991

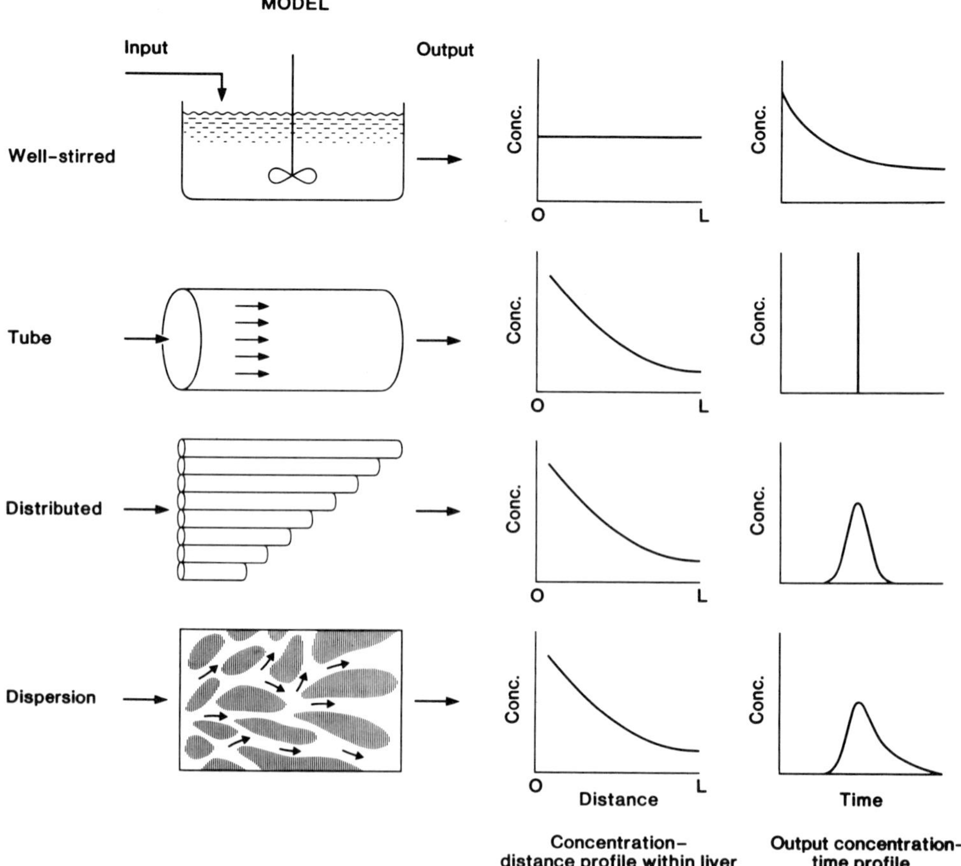

Figure 1. Schematic representation of microcirculatory events assumed by various models of hepatic drug clearance, together with the concentration-distance profile within the liver during constant drug input, and the output profile predicted after a unit impulse of drug into the hepatic portal vein [Roberts & Rowland, 1986b]. The distributed and dispersion models, which attempt to account for organ heterogeneity, predict solute residence time distributions which are most compatible with experimental observations.

Ideal Models

One of the fundamental differences between the various models of hepatic drug elimination is the relationship which is assumed to exist between the unbound solute concentration at the site of metabolism (i.e. within the hepatocyte) to that entering and leaving the organ. On this point, the two simplest and most frequently applied models differ sharply. Thus, the *well-stirred* (venous equilibration) model assumes that the liver comprises a single, "well-stirred" homogenous compartment, in which the emergent venous unbound drug is in equilibrium with that in the liver. In contrast, the *undistributed parallel-tube* (plug-flow) model views the liver as a set of identical, unconnected parallel tubes (sinusoids) in which, at steady state, the concentration of solute declines exponentially in the direction of blood flow. The average unbound drug concentration in blood within the liver is assumed to be the geometric mean of that entering and leaving the organ. Because the undistributed parallel-tube model predicts intrahepatic concentration-distance gradients, such as those which have been observed experimentally [Gumucio, 1983], it is perhaps more physiologically realistic than the well-stirred model, which fails to define where and how arterio-venous concentration differences develop [Forker & Luxon, 1985b]. In both models, enzyme activity is assumed to be uniformly distributed in cells along the length of the liver and the distribution of solute within the liver is usually taken to be perfusion-rate limited. Although enzyme kinetics are normally assumed to be first-order, both models can be extended to incorporate Michalis-Menten kinetics [Pang & Rowland, 1977a].

Theoretical analysis shows that the well-stirred and undistributed parallel-tube models (and the steady state solution of other models described below) reduce to a common form of a simple point source filter for drugs of low extraction ratio. Both models predict that the clearance of such drugs will be determined primarily by solute binding to blood components and the activity of the relevant enzyme system(s). Even for highly extracted drugs, the two models are similar inasmuch as they both predict that drug clearance will be relatively insensitive to a change in any factor other than organ blood flow. Choice of model only becomes important when considering the availability of highly extracted drugs [Morgan & Raymond, 1982; Pang & Rowland, 1977a]. This is because hepatic arterio-venous concentration differences for such drugs are large and model assumptions regarding sinusoidal concentration gradients become important. When considering the effects of altered hepatic blood flow or changes in drug binding to blood components on the fraction of drug escaping elimination during a single pass through the liver (or the systemic availability of orally administered drugs which are extensively cleared during hepatic first-pass) divergence between model predictions are extreme. Such divergence has formed the basis of experimental protocols designed to ascertain the most appropriate physiologic model. However, when the well-stirred and undistributed parallel-tube models have been so evaluated, the results have been ambivalent. For example, whereas the well-stirred model better predicts the influence of altered perfusion on the hepatic handling of lidocaine [Pang & Rowland, 1977b] and meperidine [Ahmad et al., 1983] and of altered protein binding on the hepatic handling of propranolol [Jones et al., 1984], the undistributed parallel-tube model better predicts the influence of altered perfusion on the elimination of galactose [Keiding & Chiarantini, 1978] and of altered protein binding on the hepatic clearance of diazepam [Rowland et al., 1984].

In addition to the disparity between the well-stirred and undistributed parallel-tube models in terms of the predicted change in availability in the event of an alteration in the physiologic determinants of elimination, they differ markedly in their ability to extrapolate in vitro microsomal drug metabolism data to the in vivo condition [Roberts & Rowland, 1986a]. This is because, for a given hepatic extraction ratio, the undistributed parallel-tube model predicts a higher intrahepatic substrate concentration, and therefore a lower organ metabolic capacity, than does the well-stirred model. Conversely, for a given organ metabolic capacity estimated from in vitro studies, the undistributed parallel-tube model predicts more efficient organ extraction in vivo.

Because the fundamental difference between the well-stirred and undistributed parallel-tube models lies in their contrasting assumptions regarding hepatic microvasculatory events, a logical progression towards a more realistic and perhaps more unifying physiologic model of hepatic drug clearance is to consider more closely intrahepatic solute distribution patterns. Thus, it is the intent of stochastic models of hepatic drug elimination to more accurately represent organ physiology, albeit at the expense of greater mathematical complexity.

Stochastic Models

Perhaps the most crucial evidence that the well-stirred and undistributed parallel-tube models apply unrealistic views of hepatic microcirculatory events is that they fail to describe the residence time distribution (RTD) of blood elements within the organ. This distribution, which is most readily seen from the venous outflow concentration-time course following an impulse input of tracer into the hepatic portal vein (Figure 1), is caused by the known heterogeneity of the liver, the complexity of which defies detailed description. In attempts to account for this heterogeneity, a number of stochastic models have been evoked. One of these, the *distributed sinusoidal model*, views the sinusoids in the liver as a collection of segregated tubes with differing magnitudes of such properties as blood flow and enzyme activity. The degree of organ heterogeneity is dictated by the variance of the statistical distribution chosen to represent each property [Bass et al., 1978; Forker & Luxon, 1978].

Another stochastic model, proposed by Roberts and Rowland [1986b,c], is the *axial dispersion* (convection-dispersion) model, widely used to represent non-ideal flow behaviour in chemical reactors [Wen & Fan, 1985]. The axial dispersion model is characterised by a dispersion number, D_N, a dimensionless, stochastic parameter which quantifies the relative spreading of solute elements within the liver. Unlike the distributed sinusoidal model, the axial dispersion model incorporates the known intermixing of blood elements within the liver, but is more complicated than the distributed models in that it is described by a second-order partial differential equation, for which boundary conditions must be imposed. The well-stirred and undistributed parallel-tube models are asymptotic limits of the dispersion model, representing infinite dispersion and zero dispersion, respectively [Roberts & Rowland 1986b,c]. In other words, the fundamental assumptions of the ideal models regarding microcirculatory events are incorporated into the dispersion model as a stochastic parameter. The ability of the axial dispersion model (and other stochastic models) to accomodate hepatic extraction data obtained under conditions of altered flow or solute binding is therefore hardly surprising, given the additional flexibility. A key issue, however, is the magnitude of D_N and the extent to which it may vary among substrates.

Bass et al. [1987] have extended the properties of the distributed sinusoidal and axial dispersion models, with the view to enable their direct comparison. For this purpose, the distributed model was extended to incorporate intra-hepatic mixing sites, and the dispersion model was modified to include distribution of enzyme activity along the liver length. Roberts et al. [1988] have shown that of the stochastic models, the dispersion model and the distributed model with either inverse Gaussian or log normal statistical distribution better describe the hepatic RTD of radiolabelled erythrocytes, albumin, sucrose and water (frequently used non-eliminated indicators of hepatic physiology), than do normal or gamma distributed models. However, when operating under linear conditions, all stochastic models that accurately describe the hepatic RTD of blood elements are equivalent with respect to predictions of availability for eliminated solutes, irrespective of enzyme heterogeneity or distribution kinetics of drug between blood and hepatocytes. Differences in the predictions of these stochastic models only become apparent when saturable elimination kinetics are operable [Roberts et al., 1989, 1988].

The tank-in-series (series-compartment) model is a stochastic model which views the liver as a number (N) of identical, well-mixed tanks connected in series [Weisiger, 1985]. When a single tank is considered (N=1), the model is identical to the well-stirred model, while for large values of N its behaviour is identical to that of the undistributed parallel-tube model. For intermediate values of N, the model is a hybrid of the these two simpler systems.

Over recent years, as data on hepatic drug elimination has become available, there has been a progressive increase in the mathematical complexity of the models required to adequately describe the data. It is important to emphasise, however, that in many cases, the drug extraction data used to test the models has, through necessity, been obtained under conditions which depart from physiological reality — the alterations in perfusate protein content (and therefore solute binding), and perfusate flow rate, are substantially greater than those normally encountered in vivo. Nevertheless, the choice of physiologic model in the clinical setting is important, and consideration of both the well-stirred and undistributed parallel-tube extremes is recommended when considering factors which may influence the systemic availability of highly extracted drugs which are administered orally [Morgan & Smallwood, 1990].

The application of these various models, and being aware of their advantages and limitations, has assisted immensely in providing insight into the factors governing drug distribution and elimination at a physiologic level. Notwithstanding the significant advances that have been made in recent times, there are a great many problems and issues which still need consideration. Some of these issues are considered below, mostly within the context of the axial dispersion model, although many of the conclusions drawn are not so limited.

ESTIMATION OF HEPATIC DISPERSION

An essential requirement of a stochastic model of hepatic drug clearance is that it must accurately represent the RTD within the eliminating tissue (the hepatic sinusoidal bed). As mentioned above, most detailed knowledge of hepatic elimination is gained from studies of the isolated perfused liver. Here the liver is positioned between the input and collection systems, both of which may cause appreciable dispersion of moving solute. Consequently, the RTD observed experimentally, following an impulse input of tracer, is that in the liver distorted by that in the experimental devices [Goresky & Silverman, 1964; Luxon & Forker, 1982]. Transit through the non-exchanging hepatic vessels leading to, and from, the sinusoidal bed must also be considered. It is particularly important to consider apparatus dispersion when modelling the outflow profiles of substances which have short hepatic mean residence times, including those that are extensively bound within perfusate and have small hepatic distribution volumes.

One common method of correcting for transit through the non-hepatic region of the perfused liver system is to subtract, from each experimental time point, the mean residence time associated with the non-hepatic system (MRT_{NH}). Thus

$$\text{Corrected time} = \text{Observed time} - MRT_{NH} \qquad (1)$$

An estimate of MRT_{NH} can be made in the absence of a liver by connecting the inflow and outflow cannulas. An alternative approach is to use the multiple-indicator dilution method, whereby the mean residence time within all non-exchanging vessels (including those within the liver) can be estimated by transforming the hepatic effluent profile of a non-eliminated tracer which distributes into the extravascular space, such as radiolabelled albumin, so that it superimposes upon that of labelled erythrocytes, which are confined to the vascular space [Goresky, 1983].

Another method of correction uses a linear systems approach in which the experimentally observed RTD is corrected not only for the transitional delay but also for the dispersion of tracer within non-hepatic regions. Thus, for example, if $g_H(t)$ and $g_{NH}(t)$ are the unit impulse responses for the hepatic and non-hepatic regions of

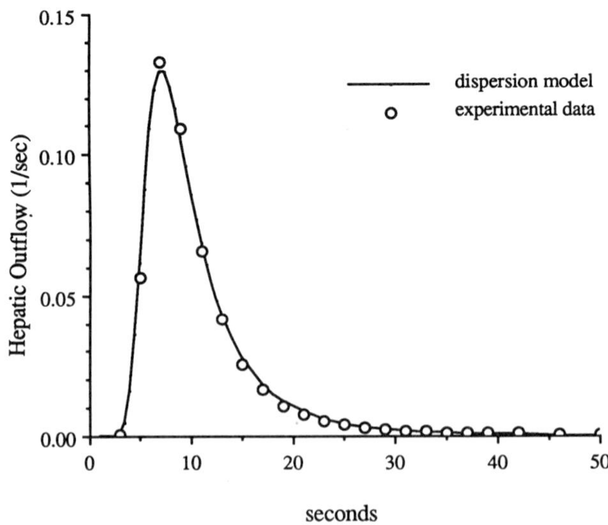

Figure 2. A representative outflow versus time profile for iodinated human serum albumin in an isolated perfused rat liver preparation, and the relationship predicted by Eq. (5), after incorporating the transfer parameters of the non-hepatic system. Data are expressed as frequency outflow (fraction of the injected dose eluting per second) versus mid-point sampling time. The estimated D_N and V_H were 0.41 and 1.52 ml, respectively. The perfusate, delivered to the liver at a constant rate of 15 ml/min, contained 5 g/L of unlabelled human serum albumin. The derived dispersion number for HSA signifies a relatively high degree of dispersion within the liver [Evans et al., 1991].

the experimental system, respectively, then the impulse response for the entire system $g_T(t)$ is given by the convolution integral

$$g_T(t) = g_H(t) * g_{NH}(t) \tag{2}$$

or, taking Laplace transform

$$w_T(s) = w_H(s) \cdot w_{NH}(s) \tag{3}$$

where $w_T(s)$ is the transfer function.

If the RTD of a solute in the experimental apparatus in the absence and presence of a liver are known, two approaches can be taken to estimate $g_H(t)$. One is numerical deconvolution of $g_T(t)$ by $g_{NH}(t)$. However, this approach while having the advantage of making no assumption regarding the form of $g_T(t)$ and $g_{NH}(t)$, can be beset by numerical problems. We chose another approach, through modelling of $g_T(t)$ and $g_{NH}(t)$ using the axial dispersion model. For erythrocytes, albumin and water, a one-compartment (flow-limited) dispersion model with mixed boundary conditions adequately describes the hepatic output-time profiles [Roberts & Rowland, 1986b; Roberts et al., 1988]. For non-eliminated solutes such as these, the hepatic transfer function, $w_H(s)$, is given by

$$w_H(s) = \exp[(1-a)/(2D_N)] \tag{4}$$

where D_N, the hepatic dispersion number, characterises the degree of dispersion within the liver. The term $a = (1 + 4D_N \cdot MRT_H \cdot s)^{1/2}$, where MRT_H is the mean residence time of the solute within the liver, and s is the Laplace operator.

Dispersion in the non-hepatic (NH) regions of the liver perfusion system is due almost entirely to mixing, and we have observed that the outflow profile of a tracer injected as a bolus into an experimental system in which the hepatic inflow and outflow cannulas are connected, can also be described by a one-compartment dispersion model, with characteristic transfer parameters $D_{N,NH}$ and MRT_{NH}. Inserting appropriately into Eq. (3) yields

$$w_T(s) = \exp[(1-a)/(2D_N)]_H \cdot \exp[(1-a)/(2D_N)]_{NH} \tag{5}$$

When analyzing outflow profiles from liver perfusion experiments, we used the following approach. Firstly, the inverse of Eq. (4) was fitted to the outflow data for the non-hepatic system to determine $D_{N,NH}$ and MRT_{NH}. A rapid numerical inversion technique followed by non-linear regression was used [Yano et al., 1989]. The so-derived parameters were inserted into Eq. (5) and the inverse of this modified equation was fitted to the outflow profile obtained using the entire system (apparatus plus liver) to generate estimates of D_N and MRT_H. For non-eliminated tracers, the hepatic distribution volume (V_H) was estimated using the relationship,

$$V_H = MRT_H \cdot Q \tag{6}$$

where Q is the perfusate flow rate. This method of correcting for the transitional delay and dispersion within the apparatus is particularly useful for analyzing the hepatic RTD of human serum albumin (HSA), which has a small hepatic distribution volume (about 0.15 ml/g of liver) and therefore a short hepatic mean residence time (Figure 2).

Another poorly appreciated issue is the influence of discrete sampling on estimation of RTDs [Levenspiel, 1972]. Equation (4) predicts fractional outflow at an instant of time, t, yet in many experimental systems hepatic effluent is collected continuously over finite times, yielding average estimates of output at the midpoint of the collection interval. The difference between instantaneous and average output is greatest at earliest times and will be of importance for rapidly eluting solutes. Thus, even if the experimental apparatus is designed to minimize non-hepatic dispersion, sampling times must be very short to avoid appreciable errors in the analysis of organ RTDs.

PERMEABILITY

In the simpler models of hepatic elimination, permeability of the hepatocyte to eliminated solute is assumed to be so high that distribution is perfusion rate-limited. This is certainly true for the entry of drugs into the space of Disse and although it may apply to the movement of most drugs across the hepatocyte membrane, permeability still needs to be considered. Thus, for the well-stirred model, the steady state availability, F, is usually given by

$$F = 1/[1 + (fu_b \cdot CL_{int})/Q] \tag{7}$$

and for the undistributed parallel-tube model, by

$$F = \exp[-(fu_b CL_{int})/Q] \tag{8}$$

where fu_b is the fraction of drug in blood unbound, CL_{int} is the intrinsic clearance of the drug (defined as the rate of elimination divided by the unbound drug concentration in the hepatocyte) and Q is organ blood flow. Yet, the correct and complete equations are, for the well-stirred model,

$$F = 1/(1 + R_N) \qquad (9)$$

and for the tube model

$$F = \exp(-R_N) \qquad (10)$$

where R_N, termed the efficiency number, is given by

$$R_N = (fu_b \cdot P \cdot CL_{int})/[Q \cdot (P+CL_{int})] \qquad (11)$$

and P is the hepatocyte permeability to the drug.

The corresponding equation for the axial dispersion model, assuming mixed boundary conditions, is

$$F = \exp[(1-a)/(2D_N)] \qquad (12)$$

where $a = (1 + 4\,D_N R_N)^{1/2}$.

The ability of drugs to diffuse into the hepatocyte is extremely variable. At one extreme are small lipophilic drugs of low extraction ratio, for which the condition $P \gg CL_{int}$ is likely to prevail, so that $R_N \approx fu_b \cdot CL_{int}/Q$. However, for highly extracted drugs the reduction of Eqs. (9) and (10) to Eqs. (7) and (8) respectively, is more problematic. For some solutes, distribution into hepatocytes may still be perfusion rate-limited but permeability may be less than CL_{int}, so that $R_N \approx fu_b \cdot P/Q$. Clearly, in such instances, to equate R_N with enzymatic activity is inappropriate.

Although hepatic clearance is often claimed to be capacity limited ($CL \approx fu_b \cdot CL_{int}$) or perfusion limited ($CL \approx Q$), the influence of diffusional barriers are rarely considered. Even if permeability exceeds intrinsic clearance, as is likely in most instances, solute movement across the hepatocyte may influence intrahepatic drug distribution patterns, and therefore the RTD of solute within the liver. For this reason, insight into the relative influences of perfusate flow, solute permeability and intrinsic clearance on hepatic distribution and elimination can be gained from dynamic experiments such as the impulse-response technique. Here, the liver must be represented, at a minimum, as a two compartment system : the vascular and extracellular space as one kinetic compartment and the hepatic cell as the other. A barrier-limited distributed model was used to describe the kinetics of warfarin [Tsao et al., 1986, 1988] and 4-methylumbelliferone [Miyauchi et al., 1987] after bolus input into the perfused rat liver. A barrier-limited (or two-compartment) form of the axial dispersion model can also be applied to describe such transient hepatic data. Although the time varying form of the two-compartment axial dispersion model is complex [Roberts et al., 1988] it can readily be solved numerically [Yano et al., 1989].

Permeability is an important factor in the hepatic elimination of drug metabolites. Many metabolites are polar molecules and for them, the hepatocyte can represent a diffusional barrier to elimination [de Lannoy & Pang, 1987]. If the parent drug readily passes into the hepatocyte to liberate a relatively impermeable metabolite, then the metabolite will accumulate within the cell and may be eliminated by the liver at a faster rate than occurs with circulating metabolite, the hepatic clearance of which may be permeability limited. This situation has been observed

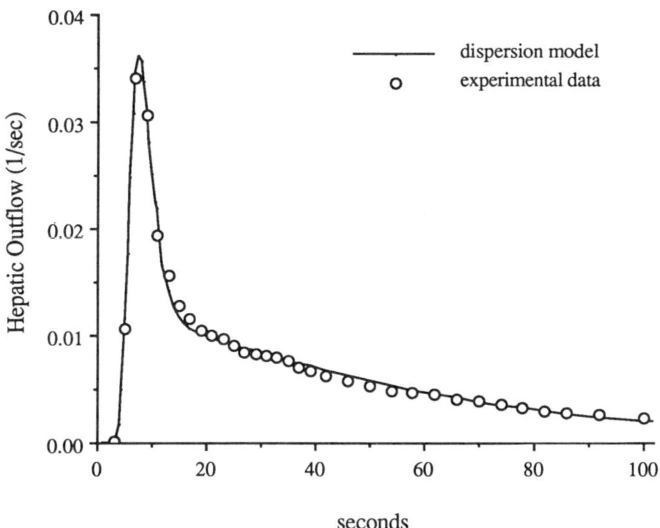

Figure 3. A two-compartment dispersion model, which allows for non-instantaneous distribution of substrate into the hepatocyte, is needed to describe the observed hepatic outflow of diclofenac in the presence of its plasma binding protein, HSA, following an impulse input of drug. The one-compartment dispersion model failed to adequately describe the observations. The perfusate, containing 0.5% HSA, was delivered to a rat liver at a constant rate of 15ml/min. Data are expressed as frequency output versus mid-point sampling time.

with enalaprilat, the polar dicarboxylic acid metabolite of enalapril [de Lannoy & Pang, 1987; Schwab et al., 1990], and is probably quite a common phenomenon. Under these circumstances, systemically administered metabolite cannot be used to reliably predict the quantitative fate of hepatically formed metabolite.

The role of enzyme heterogeneity and localization (zonation) is also an important determinant of the quantitative fate of drugs and their metabolites [Pang, 1983; Pang & Mulder, 1990; Pang & Stillwell, 1983]. By considering sinusoidal distribution patterns of sulfation, glucuronidation and hydroxylation activities in the rat liver, Xu & Pang [1989] were able to account for the dissimilar fate of gentisamide as the generated metabolite of salicylamide as compared to that of pre-formed gentisamide. It was suggested that sub-compartmentalization of enzyme systems may need to be considered in hepatic modelling of drugs and metabolites. Factors such as hepatocyte permeability and enzyme zonation may be of particular importance if formed metabolites elicit pharmacological or toxicological effects.

Another potential barrier to hepatic distribution and elimination is the erythrocyte membrane. Lee & Chiou [1989a,b] found that erythrocytes had a marked effect on the availability of doxorubicin and propranolol in the isolated perfused rat liver, and previously, Goresky and colleagues [1975], using multiple-indicator dilution methods, found that pre-equilibration of thiourea with red blood cells resulted in appreciable "carriage" of the compound through the liver. Given the short transit time of blood between the gastrointestinal tract and the liver, orally administered drug may have insufficient time to equilibrate completely within blood before reaching the hepatic sinusoidal bed. Consequently, the extraction ratio of absorbed drug during its initial transit through the liver may be greater than that of systemically circulating drug, the mean residence time of which generally ensures that distribution equilibrium in blood has been achieved [Lee & Chiou 1989a,b].

INFLUENCE OF PROTEIN BINDING

Examination of Eqs. (9) and (10) indicates that availability increases as fu_b decreases. The same conclusion is drawn from consideration of the more complex dispersion and distributed models. Not so apparent, however, is the influence of protein binding within blood or perfusate on the temporal pattern of hepatic drug outflow such as that which is seen after bolus input into the hepatic portal vein. The one-compartment dispersion model predicts a unimodal hepatic outflow profile, with a shortening of the time of the maximum outflow as drug binding to plasma protein is increased [Roberts & Rowland, 1986b]. This model is inconsistent with the hepatic RTD of the acidic anti-inflammatory drug diclofenac in the presence of its binding protein, HSA. For this drug, the RTD appears as two distinctive regions — a rapidly eluting peak followed by a sustained tail of slowly eluting material [Evans et al., 1991]. A two-compartment dispersion model was needed to describe the data, providing a D_N of 0.34 for the profile depicted in Figure 3. Interestingly, this dispersion number is similar to that found for various non-eliminated substances [Roberts et al., 1988], including HSA (Figure 2).

Albumin rapidly exchanges between the vascular space and the space of Disse but is excluded from the hepatocytes. Accordingly, the driving force behind the hepatocellular uptake of albumin-bound drugs must be the availability of unbound species at the cell surface. If a tracer of drug and its binding protein are simultaneously injected into the perfused liver, a higher ratio of protein to drug in the initial venous efflux, compared to that injected into the organ, signifies hepatocellular drug uptake, with the magnitude of the difference being a measure of cellular permeability [Goresky & Rose, 1977]. If the rate of cellular uptake is dependent only on unbound drug in the vascular space, then the initial slope of ln(protein/drug), when corrected for the binding of drug within the perfusate, should be constant. However, when this approach was used to investigate the hepatic uptake of warfarin, in the presence of varying concentrations of its binding protein (albumin) in the perfusion medium, the corrected slope rose as protein binding increased. This phenomenon was interpreted as an increase in apparent unbound permeability of warfarin as binding increases [Tsao et al., 1988], and a kinetic model incorporating albumin-mediated uptake of warfarin was evoked.

The concept of albumin mediating the uptake of its bound ligand is not new. Similar interpretations have been given for the influence of albumin on the uptake of a number of organic anions, including oleate [Weisiger et al., 1981], tauracholate [Forker & Luxon, 1981] and rose bengal [Forker et al., 1982] by the perfused rat liver. For example, it was observed that hepatic uptake of oleate correlated more closely to the perfusate concentration of total rather than unbound ligand, and uptake appeared to saturate with increasing perfusate protein concentration [Weisiger et al., 1981]. It was originally envisaged that albumin delivered its bound ligand directly to the cell surface via a transient interaction with the hepatocyte membrane, possibly via a specific "albumin receptor". The formulation of the albumin receptor model stimulated a great deal of debate regarding the physiologic model most suitable for interpreting hepatic extraction data [Colburn, 1982, 1983; Forker & Luxon, 1983a, 1985b, 1986; Morgan et al., 1985]. The central issue is that albumin-mediated uptake need only be evoked if the parallel-tube model is used to interpret the data. Although analysis using the well-stirred model conformed to the preconception that intrinsic clearance is independent of perfusate protein concentration, physiologists have been understandably reluctant to accept the well-stirred model because of its physiological irrelevance [Forker & Luxon, 1983a]. However, the fact that the data for tauracholate, for example, may be adequately explained by any physiologic model incorporating a large degree of intrahepatic heterogeneity must not be overlooked [Ching et al., 1989; Smallwood et al., 1988]. Subsequently, "albumin-mediated uptake" was observed in preparations in which the confounding influence of the intact hepatic microvasculature (and its bearing on model selection) is removed, including well-stirred suspensions of hepatocytes [Barnhart et al., 1983; Mizuma et al., 1985; Nunes et al., 1988], and hepatocyte monolayers [Burczynski et al., 1989; Fleischer et al., 1986].

There remains a great deal of controversy surrounding the existence and physiologic role of a specific albumin receptor which mediates ligand uptake. Although Stremmel et al. [1983] failed to isolate or detect a specific albumin binding site on the plasma membrane of rat hepatocytes, the possible existence of non-specific sites to which albumin binds in a transient manner with very low affinity cannot be precluded [Weisiger, 1986; Weisiger et al., 1981]. However, the ability of other proteins, (including beta-lactoglobulin) to elicit mediatory effects similar to albumin [Nunes et al., 1988] and the fact that "albumin-mediated" uptake has been observed in organs other than the liver [Jones et al., 1988; Pardridge, 1986] and in the perfused liver of marine species for which albumin is not an endogenous protein [Smith et al., 1987; Weisiger et al., 1984], suggests the involvement of factors other than a specific albumin receptor on the hepatocyte surface. One consideration is that the ligand binding properties of albumin (and other proteins) may be modified by non-specific interactions with cell surfaces or by the microenvironment at the surface of biological membranes [Forker & Luxon, 1981; Horie et al., 1988; Nunes et al., 1988; Weisiger, 1986].

When evaluating the evidence for albumin-mediated uptake, it must be considered that a number of factors may increase the apparent intrinsic clearance of a ligand in the presence of increasing concentrations of its binding protein. One hypothesis concerns the well known phenomenon of unstirred-layers bordering physiologic membranes. In the absence of binding protein, the "static-region" adjacent to the cell surface may be rapidly depleted of substrate, upon which uptake will be limited by diffusion of substrate from the bulk phase into the unstirred layer. In the presence of binding protein, however, co-diffusion of the protein-ligand complex into the unstirred layer may enhance the rate of cellular influx of drug, above that expected from bulk considerations, by providing a continual source of unbound drug as it partitions into the cell. This mechanism, previously considered by Forker & Luxon [1981, 1985a], can explain the increase in the unbound clearance of some highly bound drugs as plasma protein binding increases [Bass and Pond, 1988]. However, the extent to which co-diffusion in a static diffusion layer explains the effects of protein on ligand uptake and elimination is open to debate [Fleischer et al., 1986; Forker & Luxon, 1985a], and recent experiments, comparing the uptake of palmitate by synthetic (polyethylene) membranes and hepatocyte monolayers, suggest that albumin may have an effect on uptake over and above that which is due to co-diffusion into a stagnant layer [Burczynski et al., 1989].

The co-diffusion idea raises the issue of the kinetics of protein binding, in that the rate at which ligand dissociates from its binding protein (determined by the concentration of the protein-ligand complex and the "off-rate" constant) in particular is a major consideration in the rate of replenishment of unbound drug transferring across the cell membrane. If the removal of drug is more rapid than the rate at which it dissociates from its binding protein, non-equilibrium binding conditions will arise and uptake will become, at the extreme, dissociation-rate limited [Jansen, 1981; van der Sluijs et al., 1987; Weisiger, 1985, 1986; Weisiger et al., 1984]. In most models of hepatic elimination, however, the kinetics of protein binding (which can be extremely rapid) are ignored; bound and unbound drug are assumed to be in equilibrium at all times. Whether this assumption is valid needs to be reconsidered. Although the kinetics of protein binding within the bulk phase, together with mass-balance considerations, are sufficient to explain the nature of most observations attributed to albumin-mediated uptake [Bardsley, W.G., personal communication; Weisiger, 1985, 1986] there is a body of data suggesting that the rate of uptake by the liver (and other organs) greatly exceeds the rate of dissociation of the protein-ligand complex, predicted from *in vitro* binding studies [Pardridge, 1986; Weisiger, 1986; Wolkoff, 1987]. A central issue, however, is the magnitude of the on- and off-rate constants of albumin-ligand interactions under physiologically realistic conditions, about which there is currently limited information.

The common thread in these assortment of findings is that an increase in the concentration of binding protein in the perfusion medium (or in the bathing medium of hepatocyte suspensions or monolayers) leads, in an absolute manner, to a decrease in ligand uptake, but when expressed in terms of unbound species, uptake appears to increase. As a matter of interest, Øie & Fiori [1985] found that the clearance of antipyrine by the perfused rat liver increased upon addition of albumin to the perfusion medium. The conflict here is that antipyrine does not bind to albumin. Clearly, the influence of plasma proteins on drug distribution and elimination may not be as simple as originally thought, and the relative importance of unstirred-layer phenomenon, non-equilibrium binding and surface-mediated uptake or dissociation must be established before firm conclusions can be drawn. Such conclusions may lead to a reconsideration of the physiologic models applied to drug clearance.

CONTRIBUTIONS TO HEPATIC DISPERSION

As with any statistical distribution, the RTD of non-eliminated substances within the liver can be characterised using moment theory — providing model-independent estimates of total recovery and the mean and variance of the distribution [Roberts et al., 1988]. The coefficient of variation, CV (standard deviation/mean) of the distribution offers a useful dimensionless, non-parametric descriptor of variability. Any function which is normally distributed can be described completely by its mean and CV. Similarly, in the case of the dispersion model, MRT and D_N (the parametric counterpart of CV^2) are sufficient to completely describe the outflow profile of a non-eliminated solute (Eq. 4). However, when elimination of solute occurs within the liver, moment analysis alone is of limited use, and outflow data cannot be used to interpret hepatic events without reference to a structural model.

Roberts & Rowland [1986b] found that the dispersion model adequately describes the hepatic output profiles of a variety of non-eliminated substances. They argued that D_N is a characteristic of the liver, and found that the value of D_N, approximately 0.2 to 0.5, [Roberts et al., 1988] was of similar magnitude for a number of non-eliminated solutes. They left open the question of whether the magnitude of D_N is dependent on physiologic factors, such as perfusate flow rate. Using the isolated perfused rat liver, we have studied urea, whose hepatic uptake is identical to that of water, and found that the perfusate flow rate (15 to 30 ml/min) had no effect on D_N, which varied between 0.3 and 0.4 [unpublished findings]. A similar lack of effect of perfusion on D_N has been established by others on theoretical [Bass et al., 1987] and experimental [Roberts et al., 1990] grounds. We have also studied the hepatic effluent profiles of a homologous series of barbiturates whose lipophilicity and intrinsic clearance both increase with the addition of each methylene group. Despite the observed differences in output-time profiles (Figure 4), preliminary investigations suggest that the magnitude of D_N is similar for all compounds (0.3 to 0.4). These findings support the concept that D_N is a characteristic of the liver, predominantly reflecting the intricate architecture of the microvasculature.

The hepatic dispersion number can be estimated not only by modelling transient outflow data after bolus input, but also by applying the steady state form of the dispersion model (Eq. 12) to availability data in much the same way as steady state experiments have been used to discriminate between the well-stirred and

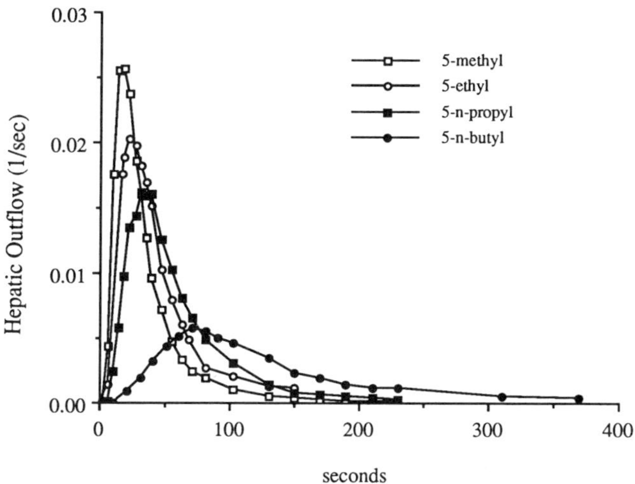

Figure 4. Frequency output versus mid-point time profiles for a homologous series of barbiturates (5-alkyl-5-ethylbarbituric acids), following an impulse of each into a rat liver perfused under single-pass conditions (15 ml/min) with protein-free media. The value of D_N was similar (approximately 0.35) for all four compounds [Fornasini, G.F., unpublished findings].

undistributed parallel-tube models. In theory, D_N may be estimated by considering the effect of altering any of the physiological determinants (fu_b, CL_{int} or Q), although changes in intrinsic clearance, which may be brought about by enzyme induction or inhibition, are difficult to quantify and provide minimal discriminatory capacity under linear conditions. Roberts & Rowland [1986c] found that the dispersion model with a D_N of 0.25 best described the effect of altered protein binding on the availability of diazepam. Under conditions of altered perfusate flow rate, the changes in the availability of colloidal chromic phosphate followed the predictions of the dispersion model with a D_N of 0.3 to 0.7 (depending on the relationship between D_N and Q; Roberts et al., 1988], and the effect of flow on tauracholate availability conformed to the predictions of the dispersion model with a D_N of 0.24 [Roberts et al., 1990]. While these data add support for a dispersion number which is in the range of 0.2 to 0.5, much higher values have been reported from steady state experiments examining the relationship between solute binding to albumin and availability for tauracholate [Ching et al., 1989; Smallwood et al., 1988]. An important consideration here, however, is that the dispersion number estimated for tauracholate may be artefactually high due to an unidentified effect of albumin on its uptake mechanism, as discussed previously. Failure of the dispersion model to account for such protein-dependent effects will confound attempts to characterise the magnitude of the dispersion number.

The concept of a solute-independent dispersion number in the order of 0.2 to 0.5 may not explain fully the lack of effect of changes in perfusate flow rate on the steady state reservoir concentration of lidocaine and meperidine upon direct infusion into the hepatic portal vein [Ahmad et al., 1983; Pang & Rowland, 1977b]. Such observations are compatible with the well-stirred model. The reasons for these inconsistencies have not been resolved. One factor which must be considered is that application of the dispersion model to steady state availability data will provide an estimate of dispersion in regions of elimination only. Conversely, in impulse experiments, a composite measure of dispersion within all regions of the liver is made. Consequently, greater dispersion in regions of elimination would manifest as higher values of D_N determined from the steady state approach than derived using the impulse-response technique. Such is likely to be the case if elimination occurs primarily within the periportal region of the hepatic acinus, where the degree of anastomosing is greatest [Gumucio, 1983]. The behaviour of a compound eliminated in regions of high dispersion will be closer to that predicted by the well-stirred model.

METHODS OF ASSESSING PHARMACOKINETIC PARAMETERS

In the isolated perfused liver system, several methods are commonly used to assess pharmacokinetic parameters: single-pass perfusion with either impulse or continuous drug administration, and recirculation with impulse or continuous administration directly into the reservoir or into the hepatic portal vein. Retrograde perfusion, in which the direction of flow is reversed, may be useful for providing insight into such factors as intrahepatic enzyme patterns. The recirculatory mode corresponds closest to events *in vivo*; direct administration into the reservoir and hepatic portal vein being analogous to intravenous and oral modes of administration, respectively. Each method provides either different information or estimates a particular pharmacokinetic parameter with a different degree of precision. Although selection of a modality may be influenced by economic and biological factors, choice is commonly dictated by the kinetic properties of the drug and the pharmacokinetic parameters of primary interest, as discussed briefly below.

Single-Pass, Impulse Input

The popularity of this modality stems from its simplicity and the richness of information provided for both non-eliminated and eliminated solutes. For a non-eliminated solute, given the observed hepatic outflow versus time profile, one can calculate MRT (and hence V_H; Eq. 6), and the variance (hence CV^2) of the distribution. Furthermore, if the dispersion model (or any other stochastic model) adequately describes the hepatic output data, values can be calculated for D_N and distribution rate constants if a permeability barrier exists. For an eliminated solute, although extraction ratio (and hence clearance), MRT and CV^2 can be calculated from the effluent data using moment analysis, further interpretation cannot be made without recourse to a parametric model. The reason here is that although the amount of solute entering and leaving the liver at any time is known, the amount of unchanged drug within the liver is not — it must be calculated. Conceivably, experimental measurement of hepatic content with time could be used to distinguish between various models of hepatic elimination. However, there are the practical problems of instantaneously stopping elimination, of determining the distribution of solute between the intrahepatic vasculature, space of Disse and hepatic cells, and of inherent variability due to the destructive nature of the experiment, with the corresponding need to define a composite picture based on data from many animals.

Of further note, the volume of distribution of injected material is calculated with reference to drug in hepatic effluent, which is in contrast to the conventional *in vivo* experiment, where the reference is a peripheral venous site in which the concentration of drug is closer to that entering the liver than that in the hepatic vein. As the hepatic venous concentration-time profile is a better measure of events within the liver than is the input profile, the *in vitro* estimate is the more realistic of the two values in terms of organ physiology, and will be less than the *in vivo* estimate by a factor of F, the steady state availability.

Experimentally, a problem exists in accurately estimating AUC (and hence availability) and particularly MRT and CV, for drugs of large volume of distribution and high hepatic clearance. For such drugs, the effluent concentrations are very low and may fall below the detection limit of the analytical method before much of the available drug has left the liver, thereby providing only truncated data with the well known problems of extrapolation. Furthermore, limited solute solubility, and saturation of binding or elimination processes, with the introduction of many additional complexities in modelling and interpretation, may severely limit the use of higher drug input concentrations. Clearance estimates may also be unreliable for substrates which have a very low hepatic extraction ratio, although simultaneous injection of a non-eliminated tracer may help to overcome this difficulty.

Single-Pass, Continuous Input

This modality is generally chosen to provide data at steady state, and is particularly useful for accurately measuring the availability of highly extracted solutes, with copious volumes of hepatic effluent being an additional benefit if analytical sensitivity is limited. Data both on the approach to steady state and following the cessation of drug input can also be used to provide kinetic information, although the so-derived parameters tend to be estimated with less precision than obtained following bolus input. The sole use of steady state data is to provide an accurate estimate of the extraction ratio (and hence clearance and availability) of an eliminated solute; in contrast to the impulse input modality, no information can be provided for a non-eliminated solute. Experimentally, the problem is the balance between perfusing drug long enough to ensure the attainment of steady state and the viability of the preparation, which tends to decrease with time. Here, the problem is greatest for drugs which have a large volume of distribution within the liver.

Recirculatory Mode

Usually sampling is from the reservoir, which mimics the systemic circulation and, as such, the parameters derived correspond most closely to those obtained *in vivo*. When the drug is placed into the reservoir, the area under the reservoir concentration versus time curve can be used to estimate clearance and therefore extraction ratio. However, as extraction ratio increases, the ability to accurately estimate availability decreases. The recirculatory approach has been shown to be suitable for measuring the first-order hepatic transfer rate-constants, being more suitable than the impulse-response technique in some circumstances [Forker & Luxon, 1983b; Luxon et al., 1982]. However, because drug within the reservoir may have undergone many circulations through the liver, details of rapid kinetic events such as distribution are more poorly defined than those derived from single-pass impulse experiments. In addition, as the reservoir volume is often quite large, estimates of the volume of distribution of the drug within the liver (which must be determined by difference between the estimated volume of the entire system and the volume of the reservoir) will tend to be poor for drugs which are not extensively distributed. An additional complexity which may arise with the recirculatory mode is the accumulation of metabolites which may compete with the parent drug for binding and metabolic sites.

Bolus input with fractional sampling from venous outflow can be used in conjunction with the recirculatory mode if the rate of removal of venous outflow is much less than total perfusate flow rate, and the total volume removed does not significantly influence that of the reservoir. However, the contribution of material which undergoes more than one circulation will confound attempts to interpret the outflow-concentration versus time profile. For highly extracted drugs, infusion directly into the hepatic portal vein to attain steady state with respect to reservoir concentrations, is a useful alternative to the single-pass, continuous input mode for accurately estimating availability, particularly if perfusate contains expensive additives — the single pass modality may require up to 2 to 4 litres of perfusate over a 2 hour period, whilst recirculation can be performed with a reservoir volume less than 50 ml.

WHOLE BODY PHARMACOKINETICS

Drug in tissues and organs of the body reach distribution equilibrium with that in blood at different times. Among the determinants of the kinetics of distribution are organ perfusion, membrane permeability and relative affinity of drug for tissue and blood components. Distribution equilibrium is reached earlier as perfusion rate and permeability increase, and as the affinity of the tissue or organ for drug decrease [Rowland & Tozer, 1989].

The liver, like most organs of elimination such as the kidney and lungs, is highly perfused. As such, drug in liver will reach distribution equilibrium with that in systemic blood much faster than many other organs and tissues of the body. Accordingly, for much of the period of drug distribution within, and elimination from, the body, the liver can be assumed to be at virtual steady state, with the steady-state parameters, clearance and extraction ratio (and the volume of distribution of drug in the liver, if it is a major organ of distribution) being of greatest value. Nonetheless, controlled, detailed impulse-response studies can provide considerable insight into the physiologic processes controlling the kinetics of hepatic drug elimination.

ACKNOWLEDGEMENTS

This work was supported by MRC grant G8516637-SA, with partial support for one of us (AME) by a Merck Sharp & Dohme Research Fellowship.

REFERENCES

Ahmad, A.B., Bennett, P.N., Rowland, M., 1983. Models of hepatic drug clearance: discrimination between the "well-stirred" and "parallel-tube" models. J. Pharm. Pharmacol. **35**:219.

Barnhart, J.L., Witt, B.L., Hardison, W.G., Berk, R.N., 1983. Uptake of iopanoic acid by isolated rat hepatocytes in primary culture, Am. J. Physiol. **244**:G630.

Bass, L., Pond, S.M., 1988. The puzzle of rates of cellular uptake of protein bound ligands. In: "Pharmacokinetics: Mathematical and Statistical Approaches to Metabolism and Distribution of Chemicals and Drugs" (A. Pecile and A. Rescigno, eds), page 245. Plenum Press, New York.

Bass, L., Roberts, M.S., Robinson, P.J., 1987. On the relation between extended forms of the sinusoidal perfusion and of the convection-dispersion models of hepatic elimination. J. Theor. Biol. **126**:457.

Bass, L., Robinson, P., Bracken, A.J., 1978. Hepatic elimination of flowing substrates: The distributed model. Theor. Biol. **72**:161.

Burczynski, F.J., Cai, Z.-S., Moran, J.B., Forker, E.L., 1989. Palmitate uptake by cultured hepatocytes: albumin binding and stagnant layer phenomenon. Am. J. Physiol. **257**:G584.

Ching, M.S., Morgan, D.J., Smallwood, R.A., 1989. Models of hepatic elimination: implications from studies of the simultaneous elimination of tauracholate and diazepam by isolated rat liver under varying conditions of binding. J. Pharmacol. Exp. Ther. **250**:1048.

Colburn, W.A., 1982. Albumin does not mediate the removal of tauracholate by the rat liver. J. Pharm. Sci. **71**:373.

Colburn, W.A., 1983. Albumin binding and hepatic uptake: the importance of model selection — a response. J. Pharm. Sci. **72**:1233.

de Lannoy, I.A.M., Pang, K.S., 1987. Diffusional barriers on drug and metabolite kinetics. Drug Metab. Dispos. **15**:51.

Evans, A.M., Hussein, Z., Rowland, M., 1991. A two-compartment dispersion model describes the hepatic outflow profile of diclofenac in the presence of its binding protein. J. Pharm. Pharmacol. *in press.*

Fleischer, A.B., Shurmantine, W.O., Luxon, B.A., Forker, E.L., 1986. Palmitate uptake by hepatocyte monolayers. Effect of albumin binding. J. Clin. Invest. **77**:964.

Forker, E.L., Luxon, B., 1978. Hepatic transport kinetics and plasma disappearance curves: distributed modeling versus conventional approach. Am. J. Physiol. **235**:E648

Forker, E.L., Luxon, B.A., 1981. Albumin helps mediate removal of tauracholate by rat liver. J. Clin. Invest. **67**:1517.

Forker E.L., Luxon, B.A., 1983a. Albumin binding and hepatic uptake: the importance of model selection. J. Pharm. Sci. **72** :1232.

Forker, E.L., Luxon, B.A., 1983b. Analyzing tracer disappearance curves to study hepatic transport kinetics. Am. J. Physiol. **244**:G573.

Forker, E.L., Luxon, B.A., 1985a. Effects of unstirred Disse fluid, nonequilibrium binding, and surface-mediated dissociation on hepatic removal of albumin-bound organic anions. Am. J. Physiol. **248**:G709.

Forker, E.L., Luxon, B.A., 1985b. Lumpers vs. distributers. Hepatology **5**:1236.

Forker, E.L., Luxon, B.A., 1986. Models of hepatic elimination: a critical commentary. Hepatology **6**:340.

Forker, E.L., Luxon, B.A., Snell, M., Shurmantine, W.O., 1982. Effect of albumin binding on the hepatic transport of rose bengal: surface mediated dissociation of limited capacity. J. Pharmacol. Exp. Ther. **233**:342.

Goresky, C.A., 1983. Kinetic interpretation of hepatic multiple-indicator dilution studies. Am. J. Physiol. **245**:G1.

Goresky, C.A., Bach, G.G., Nadeau, B.E., 1975. Red cell carriage of label: its limiting effect on the exchange of materials in the liver. Circ. Res. **36**:328.

Goresky, C.A., Rose, C.P., 1977. Blood-tissue exchange in liver and heart: the influence of heterogeneity of capillary transit times. Fed. Proc. **36**:2629.

Goresky, C.A., Silverman, M., 1964. Effect of correction of catheter distortion on calculated liver sinusoidal volumes. Am. J. Physiol. **207**:883.

Gumucio, J.J., 1983. Functional and anatomic heterogeneity in the liver acinus: impact on transport. Am. J. Physiol. **244**:G578.

Horie, T., Mizuma, T., Kasai, S., Awazu, S., 1988. Conformational change in plasma albumin due to interaction with isolated rat hepatocyte. Am. J. Physiol. **254**:G465.

Jansen, J.A., 1981. Influence of plasma protein binding kinetics on hepatic clearance assessed from a "tube" model and a "well-stirred" model. J. Pharmacokin. Biopharm. **9**:15.

Jones, D.B., Morgan, D.J., Mihaly, G.W., Webster, L.K., Smallwood, R.A., 1984. Discrimination between the venous equilibrium and sinusoidal models of hepatic drug elimination in the isolated perfused rat liver by perturbation of propranolol protein binding. J. Pharmacol. Exp. Ther. **229**:522.

Jones, D.R., Hall, S.D., Jackson, E.K., Branch, R.A., Wilkinson, G.R., 1988. Brain uptake of benzodiazepines: effect of lipophilicity and plasma protein binding. J. Pharmacol. Exp. Ther. **245**:816.

Keiding, S., Chiarantini, E., 1978. Effect of sinusoidal perfusion on galactose elimination kinetics in perfused rat liver. J. Pharmacol. Exp. Ther. **204**:465.

Lee, H.-J., Chiou, W.L., 1989a. Erythrocytes as barriers for drug elimination in the isolated rat liver. I. Doxorubicin. Pharm. Res. **6**:833.

Lee, H.-J., Chiou, W.L., 1989b. Erythrocytes as barriers for drug elimination in the isolated rat liver. II. Propranolol. Pharm. Res. **6**:840.

Levenspiel, O., 1972. "Chemical Reaction Engineering", pages 253-315. Wiley, New York.

Luxon, B.A., Forker, E.L., 1982. Simulation and analysis of hepatic indicator dilution curves. Am. J. Physiol. **243**:G76.

Luxon, B.A., King, P.D., Forker, E.L., 1982. How to measure first-order hepatic transfer coefficients by distributed modeling of a recirculating rat liver perfusion system. Am. J. Physiol. **243**:G518.

Miyauchi, S., Sugiyama, Y., Sawada, Y., Morita, K., Iga, T., Hanano, M., 1987. Kinetics of hepatic transport of 4-methylumbelliferone in rats. Analysis by multiple indicator dilution method. J. Pharmacokin. Biopharm. **15**:25.

Mizuma, T., Horie, T., Awazu, S., 1985. The effect of albumin on the uptake of bromosulfophthalein by isolated rat hepatocytes,. J. Pharmacobio-Dyn. **8**:90.

Morgan, D.J., Jones, D.B., Smallwood, R.A., 1985. Modeling of substrate elimination by the liver: has the albumin receptor model superseded the well-stirred model ? Hepatology **5**:1231.

Morgan, D.J., Raymond, K., 1982. Use of unbound drug concentration in blood to discriminate between two models of hepatic drug elimination. J. Pharm. Sci. **71**:600.

Morgan, D.J., Smallwood, R.A., 1990. Clinical significance of pharmacokinetic models of hepatic elimination. Clin. Pharmacokin. **18**:61.

Nunes, R., Kiang, C.-L., Sorrentino, D., Berk, P.D., 1988. "Albumin-receptor" uptake kinetics do not require an intact lobular architecture and are not specific for albumin. J. Hepatology **7**:293.

Øie, S., Fiori, F., 1985. Effect of albumin and alpha-1-acid glycoprotein on elimination of prazocin and antipyrine in the isolated perfused rat liver. J. Pharmacol. Exp. Ther. **234**: 636.

Pang, K.S., 1983. The effects of intercellular distribution of drug metabolizing enzymes on the kinetics of stable metabolite formation and elimination by liver: first-pass effects. Drug Metab. Rev. **14**:61.

Pang, K.S., Mulder, G.J., 1990. The effect of hepatic blood flow on formation of metabolites. Drug Metab. Dispos. **18**:270.

Pang, K.S., Rowland, M., 1977a. Hepatic clearance of drugs. 1. Theoretical considerations of a "well-stirred" model and a "parallel-tube" model. Influence of hepatic blood flow, plasma and blood cell binding and hepatocellular enzymatic activity on hepatic drug clearance. J. Pharmacokin. Biopharm. **5**:625.

Pang, K.S., Rowland, M., 1977b. Hepatic clearance of drugs. II. Experimental evidence for acceptance of the "well-stirred" model over the "parallel-tube" model using lidocaine in the perfused rat liver *in situ* preparation. J. Pharmacokin. Biopharm. **5**:655.

Pang, K.S., Stillwell, R.N., 1983. An understanding of the role of enzymic localization of the liver on metabolite kinetics: a computer simulation. J. Pharmacokin. Biopharm. **11**:451.

Pardridge, W.M., 1986. Transport of plasma protein-bound drugs into tissues in vivo. In: "Symposia Medica Hoechst, Volume 20: Protein binding and drug transport" (J.-P. Tillement and E. Lindenlaub, eds.). Schattauer Verlag, New York.

Roberts, M.S., Donaldson, J.D., Jackett, D., 1989. Availability predictions by hepatic elimination models for Michaelis-Menten kinetics. J. Pharmacokin. Biopharm. **17**:687.

Roberts, M.S., Donaldson, J.D., Rowland, M., 1988. Models of hepatic elimination: Comparison of stochastic models to describe residence time distributions and to predict the influence of drug distribution, enzyme heterogeneity and systemic recycling on hepatic elimination. J. Pharmacokin. Biopharm. **16**:41.

Roberts, M.S., Fraser, S., Wagner, A., McLeod, L., 1990. Residence time distributions of solutes in the perfused rat liver using a dispersion model of hepatic elimination. 1. Effect of changes in perfusate flow and albumin concentration on sucrose and tauracholate. J. Pharmacokin. Biopharm. **18**:209.

Roberts, M.S., Rowland, M., 1986a. Correlation between in-vitro microsomal enzyme activity and whole organ hepatic elimination kinetics: analysis with a dispersion model. J. Pharm. Pharmacol. **38**:177.

Roberts, M.S., Rowland, M., 1986b. A dispersion model of hepatic elimination. 1. Formulation of the model and bolus considerations. J. Pharmacokin. Biopharm. **14**:227.

Roberts, M.S., Rowland, M., 1986c. A dispersion model of hepatic elimination. 2. Steady-state considerations — influence of hepatic blood flow, binding within blood, and hepatocellular enzyme activity. J. Pharmacokin. Biopharm. **14**:261.

Rowland, M., Leitch, D., Fleming, G., Smith, B., 1984. Protein binding and hepatic clearance: Discrimination between models of hepatic clearance with diazepam, a drug of high intrinsic clearance, in the isolated perfused rat liver preparation. J. Pharmacokin. Biopharm. **12**:129.

Rowland, M., Tozer, T.N., 1989. "Clinical Pharmacokinetics: concepts and applications", Second edition. Lea & Febiger, Philadelphia.

Schwab, A.J., Barker, F., Goresky, C.A., Pang, K.S., 1990. Transfer of enalaprilat across rat liver cell membranes is barrier limited. Am. J. Physiol. **258**:G461.

Smallwood, R.H., Morgan, D.J., Mihaly, G.W., Jones, D.B., Smallwood, R.A., 1988. Effect of plasma protein binding on elimination of tauracholate by isolated perfused rat liver: comparison of venous equilibrium, undistributed and distributed sinusoidal, and dispersion models. J. Pharmacokin. Biopharm. **16**:377.

Smith, D.J., Grossbard, M., Gordon, E.R., Boyer, J.L., 1987. Tauracholate uptake by isolated skate hepatocytes: effect of albumin. Am. J. Physiol. **252**:G479.

Stremmel, W., Potter, B.J., Berk, P.D., 1983. Studies of albumin binding to rat liver plasma membranes. Implications for the albumin receptor hypothesis. Biochim. Biophys. Acta. **756**:20.

Tsao, S.C., Sugiyama, Y., Sawada, Y., Iga, T., Hanano, M., 1988. Kinetic analysis of albumin-mediated uptake of warfarin by perfused rat liver. J. Pharmacokin. Biopharm. **16**:165.

Tsao, S.C., Sugiyama, Y., Sawada, Y., Nagase, S., Iga, T., Hanano, M., 1986. Effect of albumin on hepatic uptake of warfarin in normal and analbuminemic mutant rats: analysis by multiple indicator dilution method. J. Pharmacokin. Biopharm. **14**:51.

van der Sluijs, P., Postema, B., Meijer, D.K.F., 1987. Lactosylation of albumin reduces uptake rates of dibromosulfophthalein in perfused rat liver and dissociation rate from albumin in vitro. Hepatology 7:688.

Weisiger, R., Gollan, J., Ockner, R., 1981. Receptor for albumin on the liver cell surface may mediate uptake of fatty acids and other albumin bound substances. Science **211**:1048.

Weisiger, R.A., 1985. Dissociation from albumin: A potentially rate-limiting step in the clearance of substances by the liver. Proc. Natl. Acad. Sci (U.S.A) **82**:1563.

Weisiger, R.A., 1986. Non-equilibrium drug binding and hepatic drug removal. In: "Symposia Medica Hoechst, Volume 20: Protein binding and drug transport" (J.-P. Tillement and E. Lindenlaub, eds.). Schattauer Verlag, New York.

Weisiger, R.A., Zacks, C.M., Smith, N.D., Boyer, J.L., 1984. Effect of albumin binding on extraction of sulfobromophthalein by perfused elasmobranch liver: evidence for dissociation-limited uptake. Hepatology 4:492.

Wen, C.Y., Fan, L.T., 1985. "Models for Flow Systems and Chemical Reactors". Marcel Dekker, New York.

Wolkoff, A.W., 1987. The role of an albumin receptor in hepatic organic anion uptake: the controversy continues. Hepatology 7:777.

Xu, X., Pang, K.S., 1989. Hepatic modeling of metabolite kinetics in sequential and parallel pathways: salicylamide and gentisamide metabolism in perfused rat liver. J. Pharmacokin. Biopharm. 17:645.

Yano, Y., Yamaoka, K., Aoyama, Y. Tanaka, H.,1989. Two-compartment dispersion model for analysis of organ perfusion system of drugs by Fast Inverse Laplace Transform (FILT). J. Pharmacokin. Biopharm. 17:179.

TRENDS IN THE PHARMACOKINETICS OF DRUG-RECEPTOR INTERACTIONS

David E. Schafer

Department of Veterans Affairs Medical Center
West Haven, CT 06516

HISTORICAL PERSPECTIVE

The opportunity to speak about receptor mathematics on the ancient and beautiful island of Sicily is both a challenge and an inspiration. Sicily, after all, was the home both of the mad philosopher Empedocles of Agrigento (c. 495-35 BCE), who might be said to have first dimly envisioned something like receptors [Kirk, Raven, and Schofield, 1983], and of the greatest mathematician up to the time of Isaac Newton, Archimedes of Syracusa (d. 212 BCE), who, unlike his predecessors, did not disdain to use approximate methods to achieve a practical result [Struik, 1987]

Empedocles, whose ideas antedated the notion of atoms, vaguely suggested that a form of recognition might occur when "effluents" having exactly the right shape and size would fit perfectly into corresponding "pores". For this and other ingenious arguments Empedocles was deified by his neighbors, and having the courage of his convictions, he put his deity to experimental test, it is said, by leaping into the crater of Mount Etna. As a result, perhaps, little further progress was made in the matter of receptors until the end of the 19th century, when Paul Ehrlich rather less flamboyantly enunciated the doctrine that *corpora non agunt nisi fixata* — to have an effect, particles must be attached to something [Ehrlich, 1909].

As we know, Ehrlich was right. The vast majority of drugs exert their effects by binding to specific receptor molecules on or in their target cells. Over the past century, mere speculation about receptors has given way to experimental confirmation in a long series of notable instances, like the discovery of morphine receptors. The volume of quantitative descriptions has also grown, mainly based on the model of the bimolecular reaction A + B <=> AB. After around 1940, when it became possible to label ligands with radioisotopes [Artom et al., 1937; Hevesy and Hahn, 1941], notebooks began to fill up with dissociation constants. The ensuing years were a classic period of positivistic data collection.

If most drugs are ligands, then most pharmacokinetics is ligand pharmacokinetics. The part having to do with drug-receptor interactions is the peripheral aspect, the quantitative description of the way in which the binding of a drug by its receptor affects the movements and local concentrations of each. The past decade has seen spectacular advances in methodology and a flood of new information about details of ligand-receptor interactions that might only have been

New Trends in Pharmacokinetics, Edited by A. Rescigno and A.K. Thakur
Plenum Press, New York, 1991

suspected before. A fundamental change has resulted, not just in the ways we study these interactions but in the kinds of questions we ask about them, heralding new and uncharted directions in pharmacokinetics. The kinetics of interaction between ligand and receptor is no longer concerned merely with a black box — "compartment" if you like — entered by association and exited by dissociation of the complex — perhaps with a bit of speculation about what, if anything, might go on inside the box. A picture is gradually emerging of an intricate network of processes, in series and parallel, before, during, and after the association and dissociation of ligand and receptor.

We are now starting to get a true representation of how Empedocles' "effluents" really interact with their "pores." And over this enterprise of increasing complexity hovers the spirit of Archimedes of Syracusa, urging us to seek appropriate simplifications that will permit practical quantitative results without the sacrifice of really essential details.

SCOPE OF THIS CHAPTER

In this short survey we will focus primarily on a handful of publications during the years 1988-90 which seem particularly representative of the new directions mentioned above. This presentation is divided into three sections, dealing with: "neo-classic" pharmacokinetics of ligand-receptor association and dissociation; the new kinetics of ligand and receptor reactions in the cell membrane and cell interior; and the new kinetics of ligand-receptor interactions in intact organs and living subjects. In an effort to capture some of the distinctive flavor of current research in each area, in each section we will focus on one or more illustrative examples of research that has already produced exciting results, and appears to hold promise for the future.

The choice of topics reflects the present author's judgment of their interest, and many subjects that others would consider essential are omitted. No attempt will be made to explain "classic" ligand-receptor kinetics, which has been described hundreds of times, usually rather well. Many useful references in the field are contained in an earlier publication [Schafer, 1983]. Among the available comprehensive treatments are two excellent books which have appeared within the past four years [Limbird, 1986; Williams, Glennon, and Timmermans, 1989].

TRENDS IN "CLASSIC" LIGAND-RECEPTOR KINETICS

Notation

Although "classic" ligand-receptor kinetics has sometimes assumed complex, even esoteric forms, in its essence it has always been based on the elementary reactions of single ligand and ideal receptor molecules, as represented by the equation:

$$x \text{ (ligand)} + y \text{ (free receptor)} \Leftrightarrow z \text{ (ligand-receptor complex)},$$

and the unidirectional reactions:

$$\text{rate of ligand binding} = k_a x y \text{ (association) and}$$
$$\text{rate of ligand liberation} = k_d z \text{ (dissociation)},$$

where k_a and k_d are, respectively, second- and first-order rate constants. Sometimes it is helpful to eliminate the variable y from consideration, and write the first of these reactions in the form

$$\text{rate of ligand binding} = k_a x (R - z),$$

where R is the constant total receptor concentration, $R = y + z$.

In pharmacokinetics these reactions are often represented as those of entry into and exit from a compartment, "bound x," since they formally resemble those of a compartment in which

$$\text{rate of entry of } x = k_{in} x_{out}, \text{ and}$$
$$\text{rate of exit of } x = k_{out} x_{in}.$$

Here the pseudo-first-order rate constant k_{in} represents $k_a y$ and x_{in} represents "bound x," in the form of the complex z. The resemblance to a linear compartment is stronger if receptor is in great excess, and therefore y and k_{in} are essentially constants. In general, though, the binding rate is proportional both to y and to x_{out}, so that as free receptor is used up the reaction "saturates" — slows down. By contrast, dissociation of the ligand-receptor complex is exactly like the decay of a radioisotope or exit from a compartment, in that its rate depends only on the concentration of bound ligand, or of radioisotope, or x in a compartment.

At equilibrium, when association and dissociation are balanced,

$$k_a xy = k_d z,$$
$$\frac{xy}{z} = \frac{k_d}{k_a} = K_d,$$

the dissociation constant. I shall refer to K_d as K and use it in preference to its reciprocal, the affinity constant, so that x,y,z, and K can all be expressed in the same unit of concentration, M. A well-known and particularly simple expression describes the relation between free and bound ligand concentrations, x and z:

$$\frac{z}{R} = \frac{x}{K + x},$$

or, even more simply,

$$z' = \frac{x'}{1 + x'},$$

where z' = z/R and x' = x/K are dimensionless quantities.

Observations

Dissociation constants K for ligand-receptor interactions vary over many orders of magnitude, but are usually less than 10^{-6} M and may be as small as 10^{-12} M or smaller. Receptors have evolved in parallel with the systems which they serve — neurocrine, paracrine, endocrine. Receptors involved in fast neurotransmission generally do not bind their naturally occurring ligands tightly; for example, the muscarinic acetylcholine receptor has K values around 10^{-4} M. Inhibitors for the same receptors, whether naturally occurring or synthetic, may have much lower K's, of course.

The following approximate values will give a rough idea of the rate constants k_d and k_a in the numerator and denominator of K:

$$k_a = 10^7\text{-}10^9 \ M^{-1} \ s^{-1}, \text{ or less}$$
$$k_d = 0.1\text{-}1.0 \ (\text{or more}) \ s^{-1} \ (\text{unstable complexes})$$
$$< 10^{-7}\text{-}10^{-6} \ s^{-1} \ (\text{highly stable complexes}).$$

The values given for k_a are close to those for "diffusion-limited association" of molecules of medium size, i.e., the fastest collision rates possible for these molecules, given their finite diffusion rate. As illustrative examples, here are the association rate constants for two well-known reactions in aqueous solution:

association of H^+ and OH^- (an extremely fast reaction)
$$k_a = 1.5 \times 10^{11} \ M^{-1} \ s^{-1} \ [\text{Eigen, 1954}];$$
binding of hydrogen peroxide to peroxidase
$$k_a = 0.9 \times 10^7 \ M^{-1} \ s^{-1} \ [\text{Chance, 1943}].$$

The dissociation rate constant k_d varies several orders of magnitude more than the association rate constant k_a. In physical terms, factors that primarily govern association, like the frequencies of collision and rotation of molecules in aqueous solution, are relatively uniform from one ligand-receptor pair to another, while factors that govern dissociation, the balance of attractive and repulsive forces between ligand and receptor in the complex, vary enormously. One practical consequence of the wide variation of k_d is that separation methods for measuring binding equilibrium can be used only with a few ligands whose complexes dissociate within minutes or hours; if the rate of dissociation is more rapid, the complex cannot be isolated, and if it is less rapid, equilibrium may not be achieved within a reasonable time.

"Classic" descriptive kinetics of ligand-receptor interactions was concerned overwhelmingly with the behavior of such ideal systems, singly and in combination — or with unexplained deviations from such ideal behavior. The former category includes the binding of a single ligand species to two or more receptors or non-specific binding sites, or that of two or more ligands competing for a single receptor. In the latter category are concentration-dependent deviations (positive or negative cooperativity), or time-dependent deviations (desensitization, or up- or down-regulation). In "neo-classic" studies these familiar phenomena are beginning to yield to precise molecular interpretations through the critical application of techniques such as those illustrated below.

The Current Status of Competitive Interaction Kinetics

The recent literature on the kinetics of ligand competition reveals that some relatively simple practical problems are slow to go away. For example, papers continue to appear regularly [e.g., Horovitz and Levitzki, 1987; Munson and Rodbard, 1988; Epps et al., 1989] offering simpler or more accurate alternatives to the most common method [Cheng and Prusoff, 1973] of determining K for the binding of a ligand to its receptor by measuring its competitive inhibition of the binding of another ligand to the same receptor. These papers are intended for experimental characterization of ligands in the pharmacology laboratory, but the underlying question is essential to pharmacokinetics in intact tissues, where drugs stimulate or inhibit in the presence of endogenous ligands. An important current example will appear in the last section of this survey.

If we understand the system, the problem is really much simpler than it might seem at first. Since this topic is still an active one in the current literature, I should like to describe again a general approach I suggested in the chapter cited above [Schafer, 1983]. Using the symbols introduced above with the subscripts 1 and 2 to denote the two ligands, we can describe the system with eight variables, x_1, x_2, z_1, z_2, K_1, K_2, y, and R, and three equations (y links all the other variables),

$$K_1 = x_1 y/z_1, \tag{1}$$

$$K_2 = x_2 y/z_2, \tag{2}$$

and $$R = y + z_1 + z_2. \tag{3}$$

Typically we know the total ligand concentrations L_1 and L_2 rather than x_1 and x_2. So we have two more equations,

$$L_1 = x_1 + z_1 \tag{4}$$

and $$L_2 = x_2 + z_2 \tag{5}$$

for a total of ten variables and five equations. Usually we know three more of the variables — R, K_1, and the measured value of z_1 in the presence of ligand 2 — so we end up with five equations in five unknowns. We can calculate the complete solution (or program a desktop computer to give it to us) in just *five* steps: calculate x_1 (Eq. 4), then y (Eq. 1), then z_2 (Eq. 3), then x_2 (Eq. 5), and finally K_2 (Eq. 2). Some variant of this method will usually be found to work in other situations.

The physics of competitive interaction resides in the competition of the two ligands for free receptor, expressed in Equations 1-3, which we can combine in a single equation showing how the total receptor R is divided up among the various forms — free and bound to ligands 1 and 2:

$$R = y(1 + x_1/K_1 + x_2/K_2). \tag{3a}$$

There is no reason to obtain a single explicit algebraic expression for K_2, and I do not think it is even a good idea to do so, since such an expression carries much less information than the individual equations and also has little intuitive meaning for most of us.

The Current Status of Receptor Cooperativity

The way in which drug movements depend on binding and release, and ultimately on drug concentrations, lies at the heart of drug-receptor pharmacokinetics. "Classic" ligand-receptor kinetics tells us that if a ligand binds to an ideal receptor, the ratio K is independent of x, the concentration of free ligand. It follows from this that since $y = R - z$ and $z/x = y/K = R/K - z/K$, a plot of z/x vs. z ought to give a straight line for such a receptor, with y intercept at R/K, x intercept at R, and a slope of $-1/K$.

Frequently this line is not straight, however. Sometimes the plot is concave upward, suggesting that K increases as z and x increase. Sometimes the plot is convex upward, suggesting the opposite change in K. It is as though there were some kind of interaction between free ligand/receptor molecules and the receptor-ligand complex: in other words, they "cooperate." When K increases with binding, the cooperation is negative; when K decreases, positive. Unfortunately, though, other mechanisms can give similar curves; for example, the typical "negative cooperativity" curve also results from the presence of two binding site populations with different dissociation constants for the same drug. (In any case, the ligand is not bound to a single species of ideal receptor.)

One of the most thoroughly studied receptors, the insulin receptor [Cuatrecasas and Jacobs, 1990], was at one time thought to exist in pig liver in two forms — a "high affinity, low capacity" receptor, with an apparent K of 1.7×10^{-10} M, and a "low affinity, high capacity" receptor, with an apparent K of 7×10^{-9} M. An alternative hypothesis was suggested, that there were two negatively cooperating equivalent sites in a single receptor [De Meyts et al., 1978]. It is now accepted that experimental evidence of (1) binding heterogeneity at steady state and (2) ligand-induced dissociation of bound insulin supports the second hypothesis [Gammeltoft, 1984]. It has been shown that the subunit structure of the insulin receptor consists of a tetramer with two alpha (binding) subunits at the outer cell surface in contact with each other and one transmembrane beta subunit at each end, carrying tyrosine kinase activity [Roth, 1990]; when the two alpha-beta dimers are separated from each other, they bind insulin noncooperatively, lending further credence to the hypothesis of negative cooperation between sites [Gu et al., 1988].

Positive cooperativity, observed in oxygen binding to hemoglobin, is not very well documented among ligand-receptor interactions [Titeler, 1989]. It was recently shown that receptors for melanocyte-stimulating hormone (MSH) in Cloudman melanoma cell cultures synchronized in phase G_2 of the cell cycle display strong positive cooperativity [McLane and Pawelek, 1988], whereas MSH receptors of random cultures do not. Simple dilution of G_2 cells by other cells in random cultures could not account for this difference, and it was suggested that synchronized cultures may release autocrine factors that regulate the number and characteristics of MSH receptors. The same paper also cited earlier reports of positive cooperativity with other peptide hormone receptors.

Both positive and negative cooperativity, then, are accepted facts in drug-receptor kinetics. But, granted that cooperativity is a fact in certain instances, what does it mean? What are its causes and its kinetic and functional consequences? The classic case of hemoglobin and oxygen emphasizes the problem: here positive cooperativity is due to allosteric interactions among four subunits, with an increase of almost 300-fold in binding affinity between the first and last subunits, and results in a dramatic enhancement of the oxygen transport function of hemoglobin [Roughton, Otis, and Lyster, 1955]. No story like this has so far been pieced together for any receptor. What we can say at present is that purely descriptive information about cooperativity can be useful in calculating the distributions of various drugs. An example is again the insulin receptor, where at typical serum levels around 10^{-10} M the effects of negative cooperativity on kinetics are significant.

Trends in Molecular Analysis of Drug-Receptor Interaction

In spite of the admittedly bleak tone of the preceding paragraph, it is clear that the "classic" descriptive phase of drug-receptor kinetics, concerned mainly with documenting binding-rate and equilibrium constants and anomalies like cooperativity, is fast drawing to a close. Within the next few years a more quantitative molecular understanding not only of cooperativity, but indeed, of the whole "classic" catalog of events in drug-receptor association and dissociation, may well become possible. Recently, knowledge of 3-dimensional relationships between ligands and receptors has entered a strikingly rapid phase of growth as newer techniques in applied genetics (cloning and DNA sequencing), protein physics and chemistry (separation and spectroscopy), and the application of computers in molecular imaging, dynamics, and design have come to be exploited alone and in combination. Even at the present time, the structures of several cloned receptors are sufficiently well known to permit serious attempts to explain the kinetics of drug-receptor interactions at the molecular level, and also to design improved ligands and modify the receptors themselves for specific purposes.

Ultimately, the ambitious goals of this phase of investigation are nothing less than to explain and predict the paths, in space and time, of ligand-receptor association and dissociation, at spatial and temporal resolutions commensurate with atomic motions in both molecules. For the purposes of this survey, a brief selection of representative references will have to suffice [Marshall, 1987; Marshall and Cramer, 1988; Mercier, Osman, and Weinstein, 1988; Tilton et al., 1988; Hruby and Gehrig, 1989; Lefkowitz, Kobilka, and Caron, 1989; Martin and Danaher, 1989; Mercier, Osman, and Weinstein, 1989; Perun and Propst, 1989; Cohen et al., 1990; and Kawasaki et al., 1990].

The Nicotinic Acetylcholine Receptor

Among the most important and powerful drugs are those which act directly on ionic channels in the plasma membrane. The pharmacology of these drugs has been totally transformed in the years since 1976, when a technique for recording electrical activity from a single channel was first described [Neher and Sakmann, 1976]. The technique, known as patch-clamping, has been in use in major laboratories around the world since 1981, when it was described in detail [Hamill et. al., 1981] — and especially since 1982, when a course on the technique was held at the Ettore Majorana Centre for Scientific Culture. To illustrate how single-channel recording may help to elucidate the pharmacokinetics of ion channels, a representative recent article is chosen here from a rather extensive literature on nicotinic acetylcholine (ACh) receptors from the electric organ of the electric ray *Torpedo californica* [Sine, Claudio, and Sigworth, 1990].

A moment's reflection will remind us that the "classic" description of association and dissociation yields the time-average behavior of two populations of molecules — ligand and receptor. Single-channel recording, on the other hand, can tell us something about the stochastic behavior of single ligand (here ACh) molecules in relation to one or more single receptor molecules tightly coupled to the ion channel. Whereas "classic" wet biochemical separation techniques commonly have a temporal resolution of the order of minutes or a few seconds, at best, single-channel recording has a resolution of the order of microseconds. It is thus well suited to analysis of the interactions between a single ACh molecule and the nicotinic ACh receptor, which typically responds to ACh levels around 10^{-6}-10^{-3} M, and thus has a k_d in the range of 10-10^4 s^{-1} or greater — much too fast for separation analysis.

The patch-clamp instrumentation requires a very low noise broad-band current amplifier, capable of measuring ionic currents in the range of 1 pA (1 pA = 6.24 elementary charges per microsecond). The recording microelectrode forms a very high-impedance seal ("gigaseal") with the plasma membrane around the receptor/channel, electrically isolating the channel from its surroundings. The average noise level recorded in the absence of ACh is typically around 0.5 pA. A single open ACh receptor (AChR) channel passes a current of around 8.3 pA when the cell membrane potential is "clamped" at −70 mV. Depending on the medium, the AChR channel can admit Na$^+$, K$^+$, or various other cations.

The AChR from *Torpedo*, used in these experiments, is a nicotinic AChR similar to that at the skeletal neuromuscular junction. Because it can be purified in milligram quantities, it has been the AChR of choice for structural studies of all kinds, including those having to do with the functional roles of individual amino acid residues [e.g., Imoto et al., 1988]. On the other hand, the cells of the *Torpedo* electric organ are so small and tightly packed that they are not well suited to single-channel measurements. This problem has previously been attacked by the use of purified receptor incorporated into lipid bilayers [Tank et al, 1983], for example. The work cited here takes advantage of the recent development of an expression system in which All-15 cells, a clonal isolate of mouse NIH3T3 fibroblasts, have been stably transfected with the four genes coding for *Torpedo* AChR [Claudio et al., 1987]. The AChRs in these cells have been investigated by patch-clamping techniques.

The electrical activity of single neuromuscular junctions was first studied with microelectrodes almost 40 years ago [Fatt and Katz, 1952], and has been intensively investigated ever since. From the start it was recognized that the binding of ligand appeared to be intimately coupled to the electrical effects in this system, so that it was a meeting place of pharmacokinetics and pharmacodynamics. It soon proved necessary to incorporate two important features of the behavior of the AChR into mathematical models: (1) variations in efficacy [Stephenson, 1956] between drugs (e.g., full and partial agonists), and (2) cooperative interaction of two AChR subunits. The first of these was introduced by Katz and co-workers (Castillo and Katz, 1957; Katz and Miledi, 1972], in the so-called "KM theory," postulating two distinct states of receptor bound to ligand: inactive (closed), AR, and active (open), AR*:

$$
\mathrm{R} + \mathrm{A} \underset{k_{-1}}{\overset{k_{+1}}{<===>}} \mathrm{AR} \underset{a}{\overset{b}{<===>}} \mathrm{AR}^{*}
$$

By adding a step separating binding from effect, this equation could account for variations in drug efficacy, but not for cooperativity.

The basis for much of the present-day theory of channels was laid down in two papers [Colquhoun and Hawkes, 1977,1981], which explored the general probabilistic principles of channel operation in detail, by treating a channel as a set of transition probabilities. Among many possibilities, these papers described the behavior of a "KM-type" system that also includes cooperativity between two receptor sites:

$$
\mathrm{R} + 2\mathrm{A} \underset{k_{-1}}{\overset{k_{+1}}{<===>}} \mathrm{AR} + \mathrm{A} \underset{k_{-2}}{\overset{k_{+2}}{<===>}} \mathrm{A_2R} \underset{a}{\overset{b}{<===>}} \mathrm{A_2R}^{*}
$$

A recording of current from a single AChR/ion channel may be described, somewhat over-simplified, as a series of rectangular pulses that differ in width (open duration) and separation (closed duration). The task of analysis is to discriminate, measure, and interpret these two classes of intervals. The data are represented as open-duration and closed-duration histograms and fitted by continuous theoretical curves. When the channel is closed, the receptor may be in any one of the three states R, AR, or A_2R, but it can open only from the doubly liganded state A_2R. As the concentration of ACh is increased, all the reactions shift to the right.

The complete solution for this system of reactions involves all six rate constants. It could be extremely difficult to determine them all. Fortunately, however, at low ACh concentrations (10^{-6} M) the data show two clearly resolved populations of closed durations, one around 10^{-1} s and the other around 10^{-5} s. In simple physical terms this says that at such low ACh concentrations the mean time required for a single A_2R complex to lose a single ACh, recombine with a single ACh, and then open is some 10^4 times longer than the mean time required for the same complex simply to open. It follows that several short-duration closures should occur together in "bursts" separated by a long closure, and they do. The theory [Colquhoun and Hawkes, 1981] predicts that for these brief closures the time constant is $1/(b + k_{-2})$, the number of closures per burst is b/k_{-2}, and the time constant for the open duration is $1/a$. From low-concentration data, then, the authors estimate that a, b, and k_{-2} are around 15, 26, and 74 ms^{-1}, respectively (at 22 C inCa-free medium, used to obtain large enough currents to be resolved by the method); they use these values in interpreting experiments at higher ACh concentrations, to determine the three remaining constants.

At increasing ACh concentrations, up to 10^{-4} M, the temporal distribution of brief closures remains practically unchanged, but that of longer closures shifts to shorter times and reveals several components. According to the basic theory [Colquhoun and Hawkes, 1981], the closed durations are described by the sum of exponentials that are functions of the rate constants. At higher concentrations it also becomes necessary to use a more complex model, incorporating an anesthetic-type blockade of the open channel by ACh [Ogden and Colquhoun, 1985],

$$R + 2A \overset{k_{+1}}{\underset{k_{-1}}{\rightleftharpoons}} AR + A \overset{k_{+2}}{\underset{k_{-2}}{\rightleftharpoons}} A_2R \overset{b}{\underset{a}{\rightleftharpoons}} A_2R^* + A \overset{k_{+b}}{\underset{k_{-b}}{\rightleftharpoons}} A_2R_b$$

which contributes another exponential term. At high ACh concentrations, also, a smaller additional correction must be made for very long closures attributed to receptor desensitization. In fitting the data, the initial assumption was made that the dissociation constant for the first binding site, k_{-1}, is $k_{-2}/2$, as would be predicted if the two binding sites are equivalent. This gave a very poor fit, together with an unreasonable value for k_{+1}; in the end it was necessary to reduce k_{-1} all the way to 500 s^{-1}, roughly 100 times less than k_{-2}, to fit the data. In other words, the data appear to show that the dissociation constant for the first ACh binding site is on the order of 100-fold lower than that for the second site, indicating marked negative cooperativity. The other two constants, k_{+1} and k_{+2}, were estimated to be in the expected range of 10^7 M^{-1} s^{-1}. Accordingly, the calculated dissociation constants for the first and second binding sites were 8.3 x 10^{-6} M and 7 x 10^{-4} M, respectively (still in Ca-free solution).

This paper is representative of recent studies employing the patch clamp method to characterize the binding behavior of individual receptor molecules. The paper also reports other experiments on the effects of low calcium concentrations and of temperature variation. The latter is of interest in part because the experimental model incorporates a receptor from a poikilotherm into a homeothermic cell membrane.

ALTERATIONS OF RECEPTOR FUNCTION IN MEMBRANES AND CELLS

Membrane effects on association and dissociation rates

Around 15 years ago particular interest began to develop in possible ways that receptors in cell membranes might behave differently from receptors in solution [Berg and Purcell, 1977; DeLisi, 1980]. Theoretically several factors could contribute to a difference in function: receptors are usually oriented similarly in the plane of the membrane; they may be able to diffuse laterally within the membrane;

their local concentration in the membrane may be much higher than the average concentration in free solution; they become even more concentrated ("clustered") in particular regions of the membrane if they and their ligands are multivalent.

The theory distinguishes two processes in series: (1) ligand diffusion, both translational and rotational, from the bulk phase to a proper position and orientation for binding in the immediate vicinity of the surface, and (2) the binding reaction itself. Each of these processes is reversible, with forward and reverse rate constants. Diffusion in solution is such that a drug molecule which collides with the membrane is likely to recollide with it several times, with potentially interesting consequences. In the last decade a considerable literature has grown up around this subject; I shall cite two representative recent examples.

A recent article makes the case for the practical importance of the "collisional limit" [Abbott and Nelsestuen, 1988]. If the receptor density in the membrane is sufficiently high, a drug molecule in the vicinity of the membrane is very much more likely to bind than to escape into the bulk phase. In this limiting case, every collision results in binding, and the overall association rate depends not on the receptor concentration, but on the cell concentration instead. The dissociation rate is similarly altered. The article provides simulated curves of association and dissociation rates for activated platelets of 500 nm radius containing receptors of 1 nm radius for factor Va, the blood clotting protein, at densities up to 10,000 receptors/platelet (1% of membrane area). It is apparent that the collisional limit is a realistic possibility for this system, and several other examples are suggested. While such a mechanism, thermodynamically, cannot affect equilibrium, it can profoundly change behavior in the steady state. The authors argue that the collision limit might offer a function for "spare receptors" far in excess of the number required to elicit a maximum cell response. They also point out that receptors reconstituted in membranes very different in geometry from the original cells may exhibit misleading behavior.

The kinetics are further altered when receptors in solution can compete with the cell surface receptors. Recently a combined theoretical and experimental investigation of competition between membrane and solution receptors has been described [Goldstein et al., 1989]. The method employs as a receptor a monoclonal immunoglobulin E (IgE) against dinitrophenol (DNP). Briefly, the anti-DNP IgE reversibly binds a monovalent DNP derivative, 2,4-dinitrophenol aminocaproyl-L-tyrosine (DCT) in solution; when the IgE is tightly bound to its high-affinity Fc_ε receptor on rat basophilic leukemia (RBL) cells, it becomes also a surface receptor for DCT. To quantitate binding, IgE is labeled with fluorescein-5-isothiocyanate (FITC), whose fluorescence is quenched by binding of DCT. The surface receptor concentration is controlled by saturating the RBL Fc_ε binding sites with labeled IgE mixed in various ratios with unlabeled IgE which does not bind DNP.

In this study the authors have developed and tested equations for dissociation of the DCT-IgE complex at the cell surface in the presence of solution IgE in varying amounts. In a typical experiment, DCT was added to a labeled cell suspension, causing immediate partial quenching of the FITC-IgE fluorescence. Unlabeled IgE was then added, initiating gradual fluorescence recovery as the DCT was released from membrane IgE and taken up by solution IgE. According to the theory, for the system under investigation the concentration of solution IgE would have to be much greater than 2,400 nM to prevent DCT rebinding to surface IgE. At the solution IgE concentrations used (200-1,700 nM), the net rate of DCT dissociation from surface IgE was slower than from solution IgE, as the theory predicts. Also, the release of DCT increased the number of free sites and the likelihood of rebinding; consequently, the rate of net dissociation gradually slowed by 40-50% during the experiment — in agreement, within experimental error, with quantitative predictions.

G proteins and the ternary complex

The "classic" static model of drug-receptor binding treated the total number of receptors as fixed; receptors were either free or complexed with ligand (R = y + z, in the symbols used above). Today it is clear that, while this model is still a useful approximation for many purposes, it describes only one of many elements in the life of a receptor. One particularly active field of current research has to do with interactions between drug-receptor complexes and G proteins, a family of GTP-binding transducer proteins that reside in the membrane and link the drug-receptor complex with an effector molecule at the inner membrane surface [Freissmuth et al., 1989; Nathanson and Harden, 1990]. The linkage may be either stimulatory or inhibitory; the effector molecule is commonly adenylate cyclase. Among receptors which interact with G proteins, those whose primary structure is known (β-adrenergic, α_2-adrenergic, muscarinic) are homologous proteins with seven membrane-spanning regions [Weiss et al., 1988].

In systems with G proteins the function of the drug-receptor complex is to catalyze the conversion of a G-GDP complex to G-GTP-Mg via the formation of a ternary intermediate complex, which in turn causes the drug-receptor complex to dissociate. Thus in addition to the reaction x + y <==> z we have at least four more reactions forming a cycle:

$$z + G\text{-}GDP <==> z\text{-}G\text{-}GDP$$
$$z\text{-}G\text{-}GDP <==> z\text{-}G + GDP$$
$$z\text{-}G + GTP\text{-}Mg <==> z\text{-}G\text{-}GTP\text{-}Mg$$
$$z\text{-}G\text{-}GTP\text{-}Mg <==> x + y + G\text{-}GTP\text{-}Mg \text{ or } x + y\text{-}G\text{-}GTP\text{-}Mg, \text{ etc.}$$

The result is that GDP is released, z dissociates to x and y, allowing the receptor to recycle, and most importantly, the G protein complexes to GTP-Mg. The complex G-GTP-Mg undergoes further reactions leading to inhibition or (in the following case) activation of the enzyme:

$$G\text{-}GTP\text{-}Mg <==> G_\alpha\text{-}GTP\text{-}Mg + G_{\beta\gamma}$$

$$G_\alpha\text{-}GTP\text{-}Mg + E <==> G_\alpha\text{-}GTP\text{-}Mg\text{-}E^*$$

$$G_\alpha\text{-}GTP\text{-}Mg\text{-}E^* + HOH <==> G_\alpha\text{-}GDP\text{-}E + P_i + Mg$$
$$G_\alpha\text{-}GDP\text{-}E <==> G_\alpha\text{-}GDP + E$$
$$G_\alpha\text{-}GDP + G_{\beta\gamma} <==> G\text{-}GDP$$

If we take ternary complex formation into account, the total receptor concentration R is now divided among the forms y, z, z-G-GDP, z-G, and z-G-GTP-Mg. These reactions provide mechanisms by which the kinetics of the drug-receptor interaction might be altered. As was mentioned above, dissociation of the complex z when it is bound to G is dependent on binding of z-G to GTP-Mg. Also, if appreciable amounts of any of the three z-G forms were to accumulate, the concentration of free receptor would be reduced, slowing association. Such accumulation does not appear to be significant normally, since z-G-GDP and z-G-GTP-Mg are unstable and z-G, while relatively stable, reacts almost immediately with GTP-Mg, which is usually present in sufficient amounts. Theoretically, z-G might become quantitatively important under abnormal circumstances. A possibility of interaction among receptors arises from the fact that different G proteins have functionally interchangeable $G_{\beta\gamma}$ subunits. The effector specificity of these proteins, including the difference between stimulatory and inhibitory functions, arises from G_α.

Desensitization of β-adrenergic receptors

Current drug-receptor pharmacokinetics views the receptor and its ligand as a dynamic system, in which the number, state, and disposition of receptors vary constantly in response to changing conditions. Morphological and biochemical studies track the "receptor itinerary" from its synthesis and insertion into the membrane, to its binding to ligand, to its migration and binding to specialized membrane structures, to its internalization and subsequent fate — degradation, recycling, or retention within the cell. Sometimes the ligand is taken into the cell with the receptor ("receptor-mediated internalization") and is then handled separately.

Like many other receptors, the β-adrenergic receptor undergoes "desensitization" — a rapid decrease in the cellular responses to a sustained level of agonist. A recent review [Hausdorff, Caron, and Lefkowitz, 1990] distinguishes three major processes of β-adrenergic receptor desensitization: (1) *phosphorylation* of the plasma membrane receptor by cAMP-dependent protein kinase and by βARK, a novel cAMP-independent β-adrenergic receptor kinase; (2) *sequestration*, a rapid translocation, also induced by the agonist, of receptor from the plasma membrane to a distinct vesicular compartment; and (3) *down-regulation*, a decrease in receptor number, with degradation of the receptor, presumably via a lysosomal pathway, upon prolonged exposure to agonist. Arguments are given for the view that sequestration and down-regulation can account for long-term densitization, due to exposure to agonists for hours or days, but that rapid desensitization, within a few minutes, is probably due to more immediate effects of receptor phosphorylation. The review points out that cAMP-dependent phosphorylation of the receptor occurs at amino acid residues in the cytoplasm near the sixth and seventh membrane-spanning regions, while βARK phosphorylation sites are distinct from these, and still nearer to the C-terminal of the receptor. The cAMP sites are near receptor regions which site-directed mutagenesis has shown are involved in coupling to stimulatory G protein. A mechanism is proposed whereby cAMP-dependent phosphorylation alone elicits weak desensitization at nanomolar levels of agonist, and the "concerted action" of both kinases gives rise to the profound loss of responsiveness seen at micromolar levels of agonist. From the viewpoint of pharmacokinetics, this scheme extends previous discussions of the contrasting implications of short-term and long-term desensitization for ligand-binding activity at the cell surface [Brodde, 1989].

Cellular processing of insulin and its receptor

The "insulin receptor itinerary" has been extensively studied. The following summary is taken from a recent review [Levy and Olefsky, 1990]. The initial association of insulin to its receptor and the subsequent dissociation of the complex are both much faster processes than the endocytotic and later steps. At pH 3.5 and 4°C, surface-bound insulin is released from cells; internalized insulin is not. When cultured hepatocytes are mixed with insulin, the insulin that quickly becomes associated with the cells is almost all acid-extractable for 1 min; the non-acid-extractable part first becomes appreciable after 2 min and thereafter increases steadily. Microscopically, the insulin-receptor complex is endocytosed — by coated pits in some cell types but not in others — into "endosomal" vesicles, where a pH <6.0 causes the insulin to be released into the fluid, leaving the receptor in the membrane. The insulin is degraded by lysosomes to intermediates and then small molecular weight peptides, while the receptor is degraded, or recycled to the plasma membrane, or stored. In an earlier review [Sonne, 1988] the description is similar, except that insulin is degraded by insulin-specific peptidases in the "endosomes."

Theory of receptor internalization and turnover

An important quantitative approach to receptor internalization was introduced a decade ago [Wiley and Cunningham, 1981]. To the variables x, y_s, and z_s (the subscript s here refers to the cell surface) were added several new variables to account for insertion of free receptor into the surface, turnover (internalization of free receptor), endocytosis (internalization of ligand-receptor complex), and hydrolysis of internalized ligand:

V_r — rate of insertion of receptor into plasma membrane
k_t — turnover rate constant (min^{-1})
$k_t y_s$ — rate of free receptor turnover (internalization)
k_e — endocytotic rate constant (min^{-1})
$k_e z_s$ — rate of endocytosis of ligand-receptor complex
z_i — concentration of ligand-receptor complex inside the cell
k_h — rate constant for ligand hydrolysis in the cell (min^{-1})
$k_h z_i$ — rate of hydrolysis of internalized ligand.

Initially, the idea that these processes are first-order was only a simplifying assumption, but experience has shown it to be a useful one. A further major assumption was also made that could only be justified by proper choice of experimental conditions: that the relevant variables of the system (flows and concentrations) for the particular question we wish to answer can be brought into a steady state. When this assumption is shown experimentally to be valid, an otherwise awkward set of equations suddenly becomes a small number of linear relationships.

The authors tested their theory with a variety of experiments in which human fibroblasts were treated with labeled epidermal growth factor (EGF). The values obtained in this initial study were:

$V_r = 320$ receptors/cell
$k_t = 4 \times 10^{-3}$ min^{-1}
$k_e = 6 \times 10^{-2}$ min^{-1}
$k_h = 9 \times 10^{-3}$ min^{-1}
$R_s = 78,000$/cell (initial number of surface receptors/cell).

The fact that $k_e/k_t = 15$ means that occupied receptors are cleared from the surface of the cell 15 times faster than unoccupied receptors. It was suggested that this might explain "down-regulation." In a second paper the following year, the authors employed an improved experimental method with an acid-stripping technique to remove surface-bound ligand. With k_e was found to be 16×10^{-2} min^{-1}, and a lag time of 15 min after internalization was observed before the start of EGF degradation [Wiley and Cunningham, 1982].

In the ensuing years, the theory has been extensively refined as additional data have been obtained with a variety of other systems [Wiley, 1985, 1988]. The density of surface receptors can be increased dramatically by using cells expressing transfected receptor genes [Chen et al., 1989] or as a result of gene amplification, as in the A431 cell line, which has over 3×10^6 receptors per cell. It has recently been found [Lund et al., 1990] that whereas transferrin receptor endocytosis is a simple first-order process, EGF endocytosis has at least two paths, both saturable — one a low-capacity path with high affinity for occupied receptors and the other a high-capacity path with low receptor affinity. The former was associated with cells having receptors with tyrosine kinase activity. Mutant cells without such activity had only the low affinity pathway. These observations have led to a more elaborate model [Lund et al., 1990; Starbuck, Wiley, and Lauffenburger, 1990] for EGF endocytosis, with two parallel mechanisms, constitutive and inducible. The former path seems to involve smooth pits, and the latter coated pits. The coated pits contain specialized proteins which are thought to form a ternary complex with ligand and receptor. The model entails 13 equations — five for the surface pools (ligand, receptor, ligand-receptor complex, coated pit protein, ternary complex), a corresponding five for the intracellular pools in the induced pathway, and three (ligand, receptor, and ligand-receptor complex) for the intracellular pools in the simpler constitutive pathway. Despite the complexity of this system, it does not appear unreasonable to think that every component of it is susceptible to experimental test, e.g., in currently or potentially available mutants.

RECEPTOR KINETICS IN WHOLE ORGANS AND INTACT SUBJECTS

The challenge

At a still higher level of complexity, we are confronted by the immensely practical question of how to predict and interpret interactions of drugs and receptors in "real" situations in vivo — in tissues, organs, and intact subjects. Here, in general, it may be necessary to consider a long list of effects that can often be avoided in vitro, such as blood perfusion, transcapillary exchange, extracellular diffusion, nonspecific binding to tissue elements, cellular heterogeneity with multiple receptor populations, endogenous ligands, and so on. Moreover, in the ultimate application to human subjects, it is desirable and usually necessary to infer drug distribution from evidence obtained noninvasively or at most from blood or biopsy analysis. It is hardly surprising that methods for such studies have been slow to appear. In this section let us consider an approach to this problem.

Models

At the outset, it is probably better not to borrow trouble needlessly. Even the most general "comprehensive" model of the distribution of labeled receptor-binding ligands in tissue [Huang and Phelps, 1986] takes into account just five "compartments" — free ligand in plasma, tissue, and synapses, and ligand bound nonspecifically in tissue and specifically in synapses. A partially reduced model of ligand distribution in brain [Young, Frey, and Agranoff, 1986] combines free ligand in tissue and synapses into a single compartment. A still further reduction [Wong et al., 1986] leads to a three-compartment model in common use for current brain studies:

$$C_p \overset{k_1}{\underset{k_2}{<====>}} C_e \overset{k_3}{\underset{k_4}{<====>}} C_r.$$

(p=plasma, e=extracellular, and r=receptor). Notice that in this notation k_3 is a pseudo-first-order rate constant which is equal to $k_a y$ in our terminology, so that if k_a is constant, k_3 is directly proportional to y, the concentration of free receptors.

The dopamine D$_2$ receptor in human brain

One method that seems to offer unusual promise for noninvasive study of drug-receptor kinetics in vivo is positron emission tomography, or PET [Phelps, Mazziotta, and Schelbert, 1986]. Ligand labeled with positron-emitting isotopes is administered, and its two-dimensional distribution in a cross-section of a particular organ is inferred from analysis of positron detection in a 360° array of detectors around the subject. Repeated images are obtained at frequent intervals, permitting estimates of the time course of distribution of the ligand in a variety of different structures. The method is restricted to short-term studies, since the positron emitters have short half-lives — 20 min for ^{11}C, for example. It is necessary to produce the isotope locally and to label and purify the ligand very quickly.

A fascinating challenge to applied receptor kinetics in vivo began to develop around 1986 [Andreasen et al., 1988]. There was widespread interest in quantitation of brain dopamine D$_2$ receptors, especially because of the "dopamine hypothesis" of the origin of schizophrenia, which postulates that patients suffering from schizophrenia have an apparent overactivity of dopaminergic mechanisms in particular brain regions. Some anatomical studies on post-mortem brains suggested that the abnormality could be an excessive number of dopamine D$_2$ receptors in the neostriatum region. Since most or all of the patients from which these brains were obtained had been treated with neuroleptics, which block D$_2$ receptors, it was necessary to rule out the possibility that these increased numbers might result from proliferation induced by chronic blockade. So it was desirable to study schizophrenics who were drug-naive, preferably by using a technique such as PET. As it happened, two laboratories independently tackled the problem of measuring dopamine D$_2$ receptors in the brain neostriatum, using different methods, and reached opposite conclusions!

It is essential to understand the most important differences in the two methods used. The first group used (3-N-[^{11}C]methyl)-spiperone ([^{11}C]NMSP) as the radioligand [Wong et al., 1986a]. It has one of the lowest K's of any known ligand for the D$_2$ receptor, around 0.25 x 10^{-9} M in vitro, i.e., it dissociates very slowly, with a k$_d$ of about 0.022 min^{-1}. Therefore it essentially does not dissociate from the receptor during the time of the study — it binds and does not come off. For such a ligand it is appropriate to use the steady-state rate of binding, k$_3$, from the constant slope of the uptake curve, as a measure of y, the concentration of free receptor. The concentration of label in the cerebellum, which has no D$_2$ receptors, was used for the nonspecific binding in this study, and also in the other one.

Each study was performed twice. Four hours before the second study, another binding agent, unlabeled haloperidol, was administered orally in order to block the D$_2$ binding sites, in contrast with the first study. According to the theory developed [Wong et al., 1986b], the two values of k$_3$ measured in the two studies are related to the concentration of haloperidol in the brain by a straight line:

$$1/k_3 = a\,[H] + b,$$

where a is inversely proportional to R, the total receptor concentration, and 1/b is the apparent K for haloperidol binding to the receptor. So the two values of 1/k$_3$ are plotted against [H], and the slope and intercept of the straight line connecting them is used to obtain R. The method used by the second group was an equilibrium, not a steady-state, method [Farde et al., 1987]. It employed ^{11}C-raclopride, a substituted 6-methoxysalicylamide labeled at the methoxy carbon, as a low-affinity ligand, with a K of 7.3 x 10^{-9} M in vivo. Since this ligand comes to equilibrium quickly, conventional equilibrium kinetics was used. Two studies were performed, at high and low specific activity, giving two values of z at different values of x. The z/x ratio was plotted against time to verify that equilibrium was reached. It was reported that z/x was approximately constant from 36 min on. Since z/x plotted against z gives a straight line (Section C.4, above), the two points representing the two studies were connected by a straight line, from which R and $-1/K$ were calculated.

It was, and is, widely agreed that in principle these two methods should have given equivalent results, but they did not. The first group found that 11 control subjects had a total receptor count of 17 ± 3 pmol g^{-1}, while 10 drug-naive schizophrenics had counts of 42 ± 5 pmol g^{-1} [Wong et al., 1986a]. The second group found that 14 control subjects and 15 drug-naive patients had counts of 25 ± 6 and 25 ± 7 pmol g^{-1}, respectively [Farde et al., 1987]. What accounts for the discrepancy?

The two experiments were critically examined by a panel including representatives of both groups [Andreasen et al., 1988]. The panel found no major errors in either investigation that might help to explain the differences in results. Though there does not appear to be any consensus in this matter as yet, recent data [Walters, Chapman, and Howard, 1990], demonstrating that haloperidol induces a prompt and striking release of dopamine in the striatum, suggest that endogenous dopamine might play an important role in these results. Most recently, Gjedde and Wong et al. have proposed that at least three differences between normals and drug-naive schizophrenics may come into play in the PET studies: it is likely that schizophrenic subjects have higher basal dopamine levels, as well as a greater number of D_2 receptors, and also show a greater release of dopamine or some other endogenous competitor upon the administration of (unlabeled) raclopride or haloperidol [Gjedde and Wong, 1991; Wong et al., 1991]

DRUG-RECEPTOR PHARMACOKINETICS IN THE NINETIES AND BEYOND

The prospect for near-term future development in receptor kinetics is as brilliant in broad outlines as it is opaque in regard to specific details. If the rapid pace of recent years continues or accelerates, as we have good reason to expect, when we enter the twenty-first century the field will bear little resemblance to what we have known in the past. It seems clear, however, that the two tools likely to contribute most to progress during this decade are the computer and the genetic code.

With respect to the combination of these two powerful forces, Walter Gilbert has recently described a coming "paradigm shift in biology," amounting almost to a reversal of the familiar way of doing things [Gilbert, 1991]. He predicts that "by the end of the decade," for most purposes, the starting point of a biological investigation will no longer be experimental, but theoretical, in that the structure of all the "genes" will be resident in electronic databases, and experimentation will be employed not to generate, but almost exclusively to test, hypotheses. Increasingly, the ability to retrieve new data continually from the world information bank will become an essential skill for biologists. Underlying this thesis is the defensible conviction that in some sense the keys to all questions pertaining to life will lie hidden somewhere in this reservoir of information.

Nowhere will Gilbert's "paradigm shift" be more profound than in the world of drug-receptor kinetics, where the rate of discovery of new hormones and growth factors, and new functions for old ones, still shows no sign of diminishing. A genetic database that will not only, in effect, enumerate all human proteins, including all receptors and peptide messengers, but also imply their primary structures will fundamentally redefine the creative process in receptor kinetics within the foreseeable future. Continuing rapid growth, promoted by computer-assisted analysis and representation, in understanding of protein folding and the internal and external dynamics of receptors, ligands, and their complexes, in membranes and in solution, will make the consequences of this genetic information even more formidable. Already, computer simulation of protein-ligand interactions has begun to include kinesthetic as well as visual imaging [Stix, 1991]. Coming developments in parallel processing, a technique that seems made to order for the description of large molecules, will further hasten our better quantitative understanding of ligand-receptor interactions. Such quantitative understanding will inevitably be reflected in a significantly higher level of ability to control and modify these interactions, even within the decade.

So far, knowledge of drug and receptor cycles in the cell membrane and cell interior has been focused on different fragments of the total story. As the pieces of this story gradually come together over the next few years, drug-receptor kinetics will increasingly become concerned with the total "life history" of the receptor and its complexes with ligands, both endogenous and exogenous, and with other cell constituents.

At a still higher level of organization, the elusive physiological anatomy of organ permeation via the capillary bed, long the domain of elegant analytical approaches involving differential and integral equations, appears likely to yield, albeit gradually, to productive numerical analysis — another task for which parallel processing seems particularly well suited — in which an entire organ is viewed as a multidimensional vector field, characterized by the movements of prototypic marker substances. These movements, in turn, will be tracked by imaging techniques of steadily improved temporal and spatial resolution. Without too much effort or risk of exaggeration, one can foresee a world of drug-receptor kinetics in which compartments and exponentials will be superseded in practice, if not in spirit, by fast-moving multicolored images — with a printout containing more numbers than anyone could possibly use, for anyone who may want them anyhow.

ACKNOWLEDGMENTS

I am grateful to Victor Bloomfield, Albert Gjedde, Morley Hollenberg, David Levitt, Terry Lybrand, John Pawelek, Elliot Ross, Steve Sine, and Steve Wiley for helpful conversations. Gjedde, Pawelek, Sine, and Wiley provided reprints, preprints, or drafts of submitted abstracts. I especially thank Jean Logan for her very helpful discussion of the PET D_2 studies in the light of her experiences at Brookhaven.

REFERENCES

Abbott, A.J., and G.L. Nelsestuen, 1988. The collisional limit: an important consideration for membrane-associated enzymes and receptors. FASEB J. 2:2858.

Andreasen, N.C., R. Carson, M. Diksic, A. Evans, L. Farde, A. Gjedde., A. Hakim., S. Lal, N. Nair, G. Sedvall, L. Tune, and D. Wong, 1988. Workshop on schizophrenia, PET, and dopamine D_2 receptors in the human neostriatum. Schizophrenia Bull. 14:471.

Artom, C., G. Sarzana, C. Perrier, M. Santangelo, and E. Segré, 1937. Arch. Internat. Physiol. 45:32. Cited in Kamen, M.D., 1957, "Isotopic Tracers in Biology," Academic Press, New York.

Berg, H.C. and E.M. Purcell, 1977. Physics of chemoreception. Biophys. J. 20:193.

Brodde, O.-E., 1989. β-Adrenoceptors. In "Receptor Pharmacology and Function" (Williams, M., Glennon, R.A., and Timmermans, P.B.M.W.M., eds.), Chapter 8. Marcel Dekker, New York.

Castillo, J. del, and B. Katz, 1957. Interaction at end-plate receptors between different choline derivatives. Proc. Roy. Soc. (London) **B 146**:369.

Chance, B. 1943. The kinetics of the enzyme-substrate compound of peroxidase. J. Biol. Chem. **151**:553.

Chen, W.S., C. S. Lazar, K. A. Lund, J. B. Welsh, C.-P. Chang, G. M. Walton, C. J. Der, H. S. Wiley, G. N. Gill, and M.G. Rosenfeld, 1989. Functional independence of the epidermal growth factor receptor from a domain required for ligand-induced internalization and calcium regulation. Cell **59**:33.

Cheng, Y.-C., and W.H. Prusoff, 1973. Relationship between the inhibition constant (K_I) and the concentration of inhibitor which causes 50 percent inhibition (I_{50}) of an enzymatic reaction. Biochem. Pharmacol. **22**:3309.

Claudio, T., W. N. Green, D. S. Hartmann, D. Hayden, H. L. Paulson, F. J. Sigworth, S. M. Sine, and A. Swedlund, 1987. Genetic reconstruction of functional acetylcholine receptor channels in mouse fibroblasts. Science **238**:1688.

Cohen, N.C., J. M. Blaney, C. Humblet, P. Gund, and D.C. Barry, 1990. Molecular modeling software and methods for medicinal chemistry. J. Med. Chem. **33**:883.

Colquhoun, D. and A.G. Hawkes, 1977. Relaxations and fluctuations of membrane currents that flow through drug operated channels. Proc. Roy. Soc. (London) **B199**:231.

Colquhoun, D. and A.G. Hawkes, 1981. On the stochastic properties of single ion channels. Proc. Roy. Soc. (London) **B211**:205.

Cuatrecasas, P., and S. Jacobs ed., 1990. "Insulin", Vol. 92, Handbook of Experimental Pharmacology, Springer-Verlag, Berlin.

DeLisi, C., 1980. The biophysics of ligand-receptor interaction. Q. Rev. Biophys. **103**:32.

De Meyts, P., E. Van Obberghen, and J. Roth, 1978. Mapping of the residues responsible for the negative cooperativity of the receptor-binding region of insulin. Nature **273**:504.

Ehrlich, P., 1909. Ueber den jetzigen Stand der Chemotherapie. Ber. Dtsch. Chem. Ges. **42**:17.

Eigen, M., 1954. Methods for investigation of ionic reactions in aqueous solutions with half-times as short as 10^{-9} sec. Discuss. Faraday Soc. **17**:194.

Epps, D.E., H. Schostarez, C. V. Argoudelis, R. Poorman, J. Hinzmann, T. K. Sawyer, and F. Mandel, 1989. An experimental method for the determination of enzyme-competitive inhibitor dissociation constants from displacement curves: application to human renin using fluorescence energy transfer to a systhetic dansylated inhibitor peptide. Anal. Biochem. **181**:172.

Farde, L., F. A. Wiesel, H. Hall, C. Halldin, S. Stone-Elander, and G. Sedvall, 1987. No D_2 receptor increase in PET study of schizophrenia. Arch. Gen. Psychiat. **44**:671.

Fatt, P., and B. Katz, 1952. Spontaneous subthreshold activity at motor nerve endings. J. Physiol. (London) **117**:109.

Freissmuth, M., P. J. Casey, and A.G. Gilman, 1989. G proteins control diverse pathways of transmembrane signaling. FASEB J. **3**:2125.

Gammeltoft, S., 1984. Insulin receptors: binding kinetics and structure-function relationship of insulin. Physiol. Rev. **64**:1321.

Gilbert, W., 1991. Towards a paradigm shift in biology. Nature **349**:99.

Gjedde, A., and D.F. Wong, 1991. Dopamine D_2 receptors in schizophrenia: hypothetical reconciliation of PET results. Abstr. Proc. "Brain 91" (Miami FL, June 1-6, 1991). J. Cereb. Blood Flow Metab. 31 (Supp. 1).

Goldstein, G., R. G. Posner, D. C. Torney, J. Erickson, D. Holowka, and B. Baird, 1989. Competition between solution and cell surface receptors for ligand: dissociation of hapten bound to surface antibody in the presence of solution antibody. Biophys. J. 56:955.

Gu, J.-L., I. D. Goldfine, J. R. Forsayeth, and P. De Meyts, 1988. Reversal of insulin-induced negative cooperativity by monoclonal antibodies that stabilize the slowly dissociating state of the insulin receptor. Biochem. Biophys. Res. Commun. 150:694.

Hamill, O.P., A. Marty, E. Neher, B. Sakmann, and F.J. Sigworth, 1981. Improved patch-clamp techniques for high-resolution current recording from cells and cell-free membrance patches. Pfluegers Arch. 391:85.

Hausdorff, W.P., M. G. Caron, and R.J. Lefkowitz, 1990. Turning off the signal: desensitization of β-adrenergic receptor function. FASEB J. 4:2281.

Hevesy, G., and L. Hahn, 1941. Kgl. Danske Videnskab. Selskab, Biol. Medd. 16:1. Cited in Kamen, M.D., 1957. "Isotopic Tracers in Biology." Academic Press, New York.

Horovitz, A., and A. Levitzky, 1987. An accurate method for determination of receptor-ligand and enzyme-inhibitor dissociation constants from displacement curves. Proc. Natl. Acad. Sci. USA 84:6654.

Hruby, V.J., and C.A. Gehrig, 1989. Recent developments in the design of receptor specific opioid peptides. Med. Res. Rev. 9:343.

Huang, S.-C., and M.E. Phelps, 1986. Principles of tracer kinetic modeling in position emission tomography and autoradiography. In: "Positron Emission Tomography and Autoradiography: Principles and Applications" (Phelps, M.E., Mazziotta, J.C., and Schelbert, H.R., eds.), Chapter 7. Raven Press, New York.

Imoto, K., C. Busch, B. Sakmann, M. Mishina, T. Konno, J. Nakai, H. Bujo, Y. Mori, K. Fukuda, and S. Numa, 1988. Rings of negatively charged amino acids determine the acetylcholine receptor conductance. Nature 335:645.

Katz, B., and R. Miledi, 1972. The statistical nature of the acetylcholine potential and its molecular components. J. Physiol. (London) 224:665.

Kawasaki, A.M., R. J. Knapp, T. H. Kramer, W. S. Wire, O. S. Vasquez, H. I. Yamamura, T. F. Burks, and F.J. Hruby, 1990. Design and synthesis of highly potent and selective cyclic dynorphin A analogues. J. Med. Chem. 33:1874.

Kirk, G.S., J. E. Raven, and M. Schofield, 1983. "The Presocratic Philosophers." Cambridge University Press, Cambridge.

Lefkowitz, R.J., B. K. Kobilka, and M.G. Caron, 1989. The new biology of drug receptors. Biochem. Pharmacol. 38:2941.

Levy, J.R., and J.M. Olefsky, 1990. Receptor-mediated internalization and turnover. In: "Insulin" (Cuatrecasas, P., and S. Jacobs, eds.), Vol. 92, Chapter 12. Handbook of Experimental Pharmacology, Springer-Verlag, Berlin.

Limbird, L.E., 1986. "Cell Surface Receptors: A Short Course on Theory and Methods." Martinus Nijhoff, Boston.

Lund, K.A., L. K. Opresko, C. Starbuck, B. J. Walsh, and H.S. Wiley, 1990. Quantitative analysis of the endocytic system involved in hormone-induced receptor internalization. J. Biol. Chem 265:15713.

Marshall, G.R. 1987. Computer-aided drug design. Ann. Rev. Pharmacol. Toxicol. 27:193.

Marshall, G.R., and R.D. Cramer, 3rd, 1988. Three-dimensional structure-activity relationships. Trends Pharmacol. Sci. 9:285.

Martin, Y.C., and E. Danaher, 1989. Molecular modeling of receptor-ligand interactions. In: "Receptor Pharmacology and Function" (Williams, M.,

Glennon, R.A., and Timmermans, P.B.M.W.M. eds.), Chapter 6. Marcel Dekker, New York.

McLane, J.A., and J.M. Pawelek, 1988. Receptors for B-melanocyte-stimulating hormone exhibit positive cooperativity in synchronized melanoma cells. Biochem. 27:3743.

Mercier, G.A., Jr., R. Osman, and H. Weinstein, 1988. Role of primary and secondary protein sturcture in neurotransmitter receptor activation mechanisms. Protein Eng. 2:261.

Mercier, G.A., Jr., R. Osman, R., and H. Weinstein, 1989. A molecular theoretical model of recognition and activation at a 5-HT receptor. Prog. Clin. Biol. Res. 289:399.

Munson, P.J., and D. Rodbard, 1988. An exact correction to the "Cheng-Prusoff" correction. J. Recept. Res. 8:533.

Nathanson, N.M., and T.K. Harden ed., 1990. "G Proteins and Signal Transduction". 43rd Ann. Symp., Soc. Gen. Physiol. Rockefeller Univ. Press, New York.

Neher, E., and B. Sakmann, 1976. Single channel currents recorded from membrane of denervated frog muscle fibres. Nature 260:799.

Ogden, D.C. and D. Colquhoun, 1965. Ion channel block by acetylcholine, carbachol, and suberyldicholine at the frog neuromuscular junction. Proc. Roy. Soc. (London) B225:329.

Perun, T.J., and C.L. Probst, 1989. "Computer-Aided Drug Design: Methods and Applications". Dekker, New York.

Phelps, M.E., J. C. Mazziotta, and H.R. Schelbert, 1986. "Positron Emission Tomography and Autoradiography: Principles and Applications". Raven Press, New York.

Roth, R.A., 1990. Insulin receptor structure. In: "Insulin" (Cuatrecasas, P. and Jacobs, S. eds.). Handbook of Experimental Pharmacology, Vol. 92, Chapter 9. Springer-Verlag, Berlin.

Roughton, F.J.W., A. B. Otis, and R.L.J. Lyster, 1955. The determination of the individual equilibrium constants of the four intermediate reactions between oxygen and sheep haemoglobin. Proc. Roy. Soc. (London) B144:29.

Schafer, D.E., 1983. Measurement of Ligand-Receptor Binding: Theory and Practice. In: "Tracer Kinetics and Physiologic Modeling" (Lambrecht, R.M. and Rescigno, A., eds.), page 445. Springer-Verlag, Berlin.

Sine, S.M., T. Claudio, and F.J. Sigworth, 1990. Activation of *Torpedo* acetylcholine receptors expressed in mouse fibroblasts: Single channel current kinetics reveal distinct agonist binding affinities. J. Gen. Physiol., 96:395.

Sonne, O., 1988. Receptor-mediated endocytosis and degradation of insulin. Physiol. Rev. 68:1129.

Starbuck, C., H. S. Wiley, and D.A. Lauffenburger, 1990. Epidermal growth factor binding and trafficking dynamics in fibroblasts: Relationship to cell proliferation. Chem. Eng. Sci. 45:2367.

Stephenson, R.P. 1956. A modification of receptor theory. Br. J. Pharm. Chemother. 11:379.

Stix, G., 1991. Reach out: Touch is added to virtual reality simulations. Sci. Amer. 264:134.

Struik, D.J., 1987. "A Concise History of Mathematics", Dover, New York.

Tank, D.W., R. L. Huganir, P. Greengard, and W.W. Webb, 1983. Patch-recorded single-channel currents of the purified and reconstituted *Torpedo* acetylcholine receptor. Proc. Nat. Acad. Sci. (USA) 80:5129.

Tilton, R.F., Jr., U. C. Singh, I. D. Kuntz, Jr., and P.A. Kollman, 1988. Protein-ligand dynamics. A 96 picosecond simulation of a myoglobin-xenon complex. J. Mol. Biol. 199:195.

Titeler, M., 1989. Receptor binding theory and methodology. In: "Receptor Pharmacology and Function" (Williams, M., Glennon, R.A., and P.B.M.W.M. Timmermans eds.), Chapter 2. Marcel Dekker, New York.

Weiss, E.R., D. J. Kelleher, C. W. Woon, S. Soparkar, S. Osawa, L. E. Heasley, and G.L. Johnson, 1988. Receptor activation of G proteins. FASEB J. 2:2841.

Wiley, H.S. 1985. Receptors as models for the mechanisms of membrane protein turnover and dynamics. Curr. Top. Membr. Trans. 24:369.

Wiley, H.S. 1988. Anomalous binding of epidermal growth factor to A431 cells is due to the effect of high receptor densities and a saturable endocytic system. J. Cell. Biol. **107**:801.

Wiley, H.S. and D.D. Cunningham, 1981. A steady state model for analyzing the cellular binding, internalization and degradation of polypeptide ligands. Cell **25**:433.

Wiley, H.S. and D.D. Cunningham, 1982. The endocytic rate constant: A cellular parameter for quantitating receptor-mediated endocytosis. J. Biol. Chem. **257**:4222.

Williams, M., R. A. Glennon, and P.B.M.W.M. Timmermans, 1989. "Receptor Pharmacology and Function". Marcel Dekker, New York.

Wong, D.F., H. N. Wagner, Jr., L. E. Tune, R. F. Dannals, G. D. Pearlson, J. M. Links, D. A. Tamminga, E. P. Broussole, H. T. Ravert, A. A. Wilson, J. K. T. Toung, J. Malat, J. A. Williams, L. A. O'Tuama, S. H. Snyder, J. J. Kuhar, and A. Gjedde, 1986a. Positron emission tomography reveals elevated D_2 dopamine receptors in drug-naive schizophrenics. Science **234**:1558.

Wong, D.F., A. Gjedde, H. N. Wagner, Jr., R. F. Dannals, K. H. Douglass, J. M. Link, and M.J. Kuhar, 1986b. Quantification of neuroreceptors in the living human brain: II. Inhibition studies of receptor density and affinity. J. Cereb. Blood Flow Metab. **6**:147.

Wong, D.F., A. Gjedde, D. Young, T. Young, L. Tune, E. Shaya, G. Pearlson, B. Chan, D. Burkhardt, P. D. Wilson, R. F. Dannals, A. A. Wilson, H. T. Ravert, T. K. Natarajan, and H.N. Wagner, Jr., 1991. Elevated endogenous dopamine concentrations in striatum of psychotic patients in vivo. Abstr. Proc. "Brain 91" (Miami, Florida, June 1-6, 1991). J. Cereb. Blood Flow Metab. **11** (Supp. 1).

Young, A.B., K. Frey, and B.W. Agranoff, 1986. Receptor assays: *in vitro* and *in vivo*. In: "Positron Emission Tomography and Autoradiography: Principles and Applications" (M. E. Phelps, J. C. Mazziotta, and H.R. Schelbert, eds.). Raven Press, New York.

IN VITRO IMAGING

A. Bertrand Brill and Andrew Karellas

Departments of Nuclear Medicine and Radiology
University of Massachusetts Medical Center
Worcester, MA 01655

INTRODUCTION

In vitro imaging is a broad topic, with books written on each subtopic discussed
in summary form in this chapter. Thus, the presentation cannot be comprehensive,
but presents principles and examples in areas of current interest in pharmacokinetics
in which we have had some experience. The discussion starts with front end devices
for data acquisition, and then considers the different elements in the imaging chain.
Autoradiography (ARG) systems are described in detail to indicate some of the
techniques common to film and digital image analysis systems, as well as to illustrate
selected biological applications. The discussion concludes with selected examples
from oncology and cardiology research.

IMAGING TECHNOLOGY

The fundamental tool in the imaging chain is the light source and sensor used to
image the object. Typically, optics (such as microscopes of various kinds), pass
photons to position sensitive image receptors, which in turn pass the collected information to either an analog viewing system (such as film or a cathode-ray tube), or as
digital signals to a computer. Many different portions of the electromagnetic spectrum (X- and gamma-ray, UV, Visible, IR, and microwave), and many particles
(alpha, beta, electron, proton, neutron, and triton), have been used (Fig.1).
Traditional systems involve cone beam light sources (ideally emanations from a point
source), which may be tuned to different wave lengths. This ordinarily involves use
of optical filters, or diffraction gratings but high flux tunable light sources are becoming increasingly available. These may originate from dye lasers, or from high
energy electron accelerators, i.e. synchrotron light sources. These large devices typically provide high intensity tunable radiation beams ranging from the UV through the
visible range [Pounds et al., 1988]. Microscopes with UV illuminators and digital
read out systems now are readily available and widely used for quantitative imaging
of different wave length radiations. Typically, systems view the sample as a two-
dimensional projection. However, three-dimensional imaging systems are in common use for in vivo imaging in medicine, as well as in vitro research applications.

New Trends in Pharmacokinetics, Edited by A. Rescigno and A.K. Thakur
Plenum Press, New York, 1991

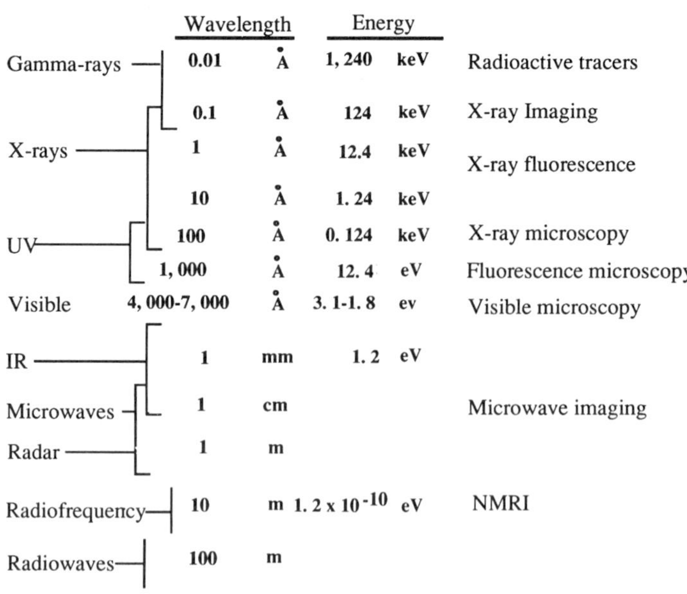

Figure 1. Overview of the electromagnetic spectrum showing the energy and wavelength relevant to biomedical imaging applications.

Tomographic images view the system at one point in time, and it is important that the system be stationary for a reconstruction free of motion blurring.

In vitro radiography has been used for many years, primarily for imaging of gross biological specimens. Typical applications include post-mortem angiographic studies for determining the integrity of blood vessels and for identification of certain bone lesions in excised specimens before microscopic examination. It can also be used as an adjunct to autoradiography for imaging very fine anatomic detail. Most in vitro X-ray radiography is performed at a spatial resolution not exceeding 30 line pairs (lp)/mm with an ultimate resolution of 15-30 μm. The most widely used equipment for this purpose are cabinet-type radiographic units employing an X-ray tube with a tungsten target. The tube potential typically has a range from 15-70 Kilovolts peak (kVp) with a tube current of about 0.5 mA. Unfortunately, much of in vitro X-ray imaging in many clinical and research environments is performed by trial and error without an in-depth understanding of the important physical parameters involved in this process. Understanding of the basic principles is extremely important for effective utilization of this technology. The spatial resolution depends on both the characteristics of the X-ray tube and those of the image receptor. Radiographic film is the most commonly used image receptor. The maximum attainable spatial resolution (R) (lp/mm) can be computed by:

$$R = \frac{M}{f\,(M-1)} \tag{1}$$

where f is the actual size of the X-ray focus and M is the magnification. The magnification is defined as S_i/S_o where S_i and So are the sizes of image and object respectively. The magnification can be determined easily by:

$$M = \frac{a+b}{a} \tag{2}$$

where a and b are the distances from the X-ray focus-to-object and from object to the image receptor respectively. Equations 1 and 2 have a profound effect and practical importance in specimen radiography. The limiting spatial resolution as defined by equation 1 is a result of the geometric blurring produced by an X-ray focus which has a finite size. For this reason, the size of the X-ray focus, (focal spot) must be chosen to satisfy the user's spatial resolution requirements according to equations 1 and 2. It is clear that for imaging a very thin slice of tissue, b is very small, and the magnitude of M becomes marginally greater than one resulting in high resolution. Therefore, a given focal spot may produce resolution with a range from a few microns to about 1 mm, depending on the magnification. If high magnification is required, (M= 1.5 to M=5), at relatively high resolution (15-30 lp/mm), the focal spot requirement becomes extremely critical. Specimen radiography equipment typically employ focal spots in the range of 0.05 mm to 1 mm. Some manufacturers claim focal spot sizes smaller than 10 microns. The user must be aware that a small focal spot size will require a reduction in the current in order to prevent overheating of the target in the X-ray tube. A lower tube current will require longer imaging time. This can be a problem in the presence of motion or vibrations. Laboratory X-ray units employ focal spots in the range of 0.050-1.0 mm. Some manufacturers claim focal spots smaller than 10 microns. In small units the focal spot size is not adjustable but it may be selectable to two options, large or small. It is important to note that the resolution capability of the focal spot should never be inferred from its nominal size. Often, the focal spot is asymmetric with rectangular shape and often, only its smaller dimension is close to the nominal value. This will result in a non-isotropic modulation transfer function (MTF) and consequently a loss in spatial resolution. The use of magnification relaxes the resolution requirement of the image receptor and it enhances contrast due to the suppression of scattered X-rays from the specimen. Scatter is a lesser concern with specimens of thickness less than 1.0 mm.

Most microscopic scenes are viewed as static snap shots, as they are literally frozen at a single point in time. However, there are several ways in which it is possible to view cell processes as a function of time. For this purpose, in vitro procedures most commonly employ reflection and transmission techniques using time lapse photography of cells taken from the body, and maintained in culture. Using ARG techniques, it is possible to image the temporal behavior of systems with respect to interventions by the appropriate choice of labeled tracers introduced into the system at appropriately selected times. This can be looked upon as a 3-D image, where the third dimension is time.

It is possible to image the third dimension spatially in various ways. Cross sectional imaging can be done based on longitudinal or cross-sectional (rotational) tomography. Rotational tomography views the object from all sides and is accomplished in living, as well as inanimate objects, using many different types of transmitted and/or emitted radiations [Ter-Pogosian et al., 1977]. With tissue materials, three-dimensional high resolution replicas of anatomy (and/or function) can be created by stacking spatially registered sequential sections by the use of computer-aided graphics tools. Good spatial resolution in the plane, but lower depth resolution, can be obtained by the use of longitudinal tomography. Longitudinal tomography images can be obtained from standard microscopes by varying the depth of focus. Higher Z axis resolution can be obtained by the use of the laser confocal microscope (LCM), a device which has sharp resolution in x,y, and z directions. New LCM devices can be attached to standard microscopes using laser illuminators, with potentially tunable wavelength, small focal spots and scanning motions. These devices provide good spatial resolution in three dimensions (typically, the Z axis depth resolution is of the order of 1.0 μm). In this application, the laser excites emissions from fluorescent dyes tagged to molecules of interest. When these are irradiated point-by-point by the scanning confocal beam, out-of-focus planes are also illuminated and two effects ensue. Image blurring occurs and needs to be corrected by appropriate deconvolution and some loss of fluorescence occurs, which needs to be corrected [Fay et al., 1989]. For this reason, it is important to use minimum excitation to preserve satisfactory signal levels. One of the ways in which fluorescence measurement systems are enhanced is through the use of image amplifiers, such as microchannel plates coupled to charge coupled devices (CCDs), originally developed for military and astronomical uses which are now widely used in low light measurement systems for biomedical, as well as for celestial observations [Hiraoka et al., 1987].

New dimensions in imaging research are being explored using synchrotron light sources at major facilities. At present, there are large facilities in the United States, France, Germany and Japan and additional large installations are being built in many countries, including China, Italy, and the USSR. They provide broad energy light sources in the visible and UV regions, as well as monoenergetic tunable energies from X-rays through to visible light bands. These systems permit multiple scientists to work on separate problems simultaneously at different specialized beam energies surrounding the central electron accelerator. There is much effort at present to develop smaller cheaper sources of synchrotron light, and smaller dedicated instruments are in production for industrial use.

The most commonly used image receptor is film. The analysis of film requires digitization and high quality digitizers providing as much as 4096x4096 data are now commercially available at relatively high cost. Radiographic and autoradiographic (ARG) studies typically use film emulsions to record the distribution of radioactivity emitted by objects placed in close contact with the film. Following film development, recorded images can then be digitized by electro-optical devices. Alternate means now exist for direct image quantitation. These use high sensitivity electrooptical devices which have lower resolution (0.5-2 mm) than film. These devices can record and analyze each event using energy and positional information [Dilmanian et al., 1990], or record the data on storage plates (photostimulable phosphors) in medium resolution (0.1-1.0 mm) using laser stimulation and photomultiplier sensors to read out the images [Sonoda et al., 1983]. These systems substitute for film, and provide higher sensitivity than film (much shorter imaging times), wider dynamic range, and provide digitized data directly, thereby minimizing the need for special high resolution film digitizers.

The major developments in imaging instrumentation rely heavily upon the advent of high performance, low cost, small digital computers. Computer image processing (archiving, data analysis, and display) is essential for most advanced applications. This has now become routine, with data presented in 2-, and 3-D, black and white, or color displays. Image restoration techniques are used on the raw data to remove characteristic artifacts and known sources of distortion, and the success of these operations is of primary importance for the production of high quality images. Computers now permit the extraction of features from scenes for classification (diagnosis), and for the analysis of temporal behavior. Without the aid of computers, it would not be possible to reconstruct data from complex multidetector stationary and moving systems, and for the assembly of three-dimensional representations of objects from sequential slices of objects of interest. Much of this technology is now well known, based on the widespread dissemination of X-ray Computed Tomography (CT), and Nuclear Magnetic Resonance Imaging (MRI) devices [Harris et al., 1986].

Image Acquisition and Analysis Systems

The two types of systems we have discussed for image analysis use images acquired on analog media, such as film or video media, while the second acquires data directly in digital form. Radiographic film has been the most commonly used image receptor for both quantitative and qualitative in vitro X-ray imaging. Intensifying screens increase sensitivity at the expense of loss in spatial resolution (upper limit on the attainable spatial resolution is about 20 lp/mm). Industrial-grade film with very fine grain is often used when high resolution is required at the expense of sensitivity. Radiographic film is typically used without screens and the films are processed in rapid automatic processors which are widely available in hospitals. In our experience,however, most specimen radiography can be performed by using a high resolution intensifying screen in contact with single-sided emulsion film. Such a system is used widely for mammographic imaging with a limiting resolution of 18-20 lp/mm. The resolving capability of the screen can be doubled if a 2x magnification is used, provided that the MTF of the focal spot is not severely degraded at this magnification. Depending on the size of the focal spot, an optimum magnification usually exists where the product of the MTFs of the focal spot and screen is at a maximum and this optimal magnification must be determined experimentally.

The digitization of analog data requires the use of analog-to-digital converters (ADCs), and there are several options in current use. Film analysis requires the use of an electrooptical device to sense the different amounts of light transmitted through or reflected from the film to be digitized. The most common devices use TV pick-up tubes, the output of which is digitized and stored in a computer. These devices are limited in resolution by the pick-up tube to about 500 samples per line, but the length of the line, and hence the field of view is determined by the optical system used in front of the tube. Thus, for example, one can obtain 500 samples across a chest X-ray, which would be inadequate to represent the true scene, but the same number of samples across a 1 cm structure of interest would result in 20 micrometer (μm) bins, which would be more than needed for clinical radiography. Various kinds of flying spot scanners are also used, and these use a small beam size to interrogate and quantitate grey scale changes in sequential regions by the amount of light transmitted through sequential points in the film. Since the detector and light source have essentially constant characteristics, these systems are subject to less distortion than other types of systems used, but suffer from the fact that they are expensive and clumsy to use. The most recent systems use charge coupled devices (CCDs). These devices typically have 512 x 512 sampling elements and are used in many commercial applications, including camcorders. When high resolution is required, this can be achieved with optical magnification of a particular region of the scene, or by the use of higher resolution linear arrays (4096 x 1) to scan line-by-line over the specimen. New CCDs with 4096 x 4096 arrays have been manufactured, but these devices are not widely available. These devices store charge from each of the regions of the scene in different elements in the mosaic detector, and the charge is read out to an ADC converter and recorded in a computer. For the very large CCDs, the time for read out limits the frequency of temporal sampling to a few images per second at a maximum. These devices provide high resolution, wide dynamic range and good spatial linearity. However, they are subject to nonuniformities due to sensitivity variations in the individual CCD elements. Since these are at fixed locations, these distortions can be corrected effectively using computer post processing.

In many applications, other factors besides spatial resolution are of primary importance, and often, image contrast is of equal or more importance for optimal detectability. Film typically is used both as image receptor and as the display medium. While film provides excellent spatial resolution, its response to electromagnetic radiation is non-linear and its dynamic range is limited. The use of film becomes especially difficult in quantitative imaging where the density of a tissue must be measured from the recorded shades of gray. The detection of X-rays by solid state sensors,on the other hand, can retain very wide dynamic range with a linear response to input intensity. Charge coupled devices and photodiode arrays are adaptable for this task. In this approach, a scintillator is used as the primary detector of X-rays. The scintillator may be placed in contact with the solid state sensor or optically coupled

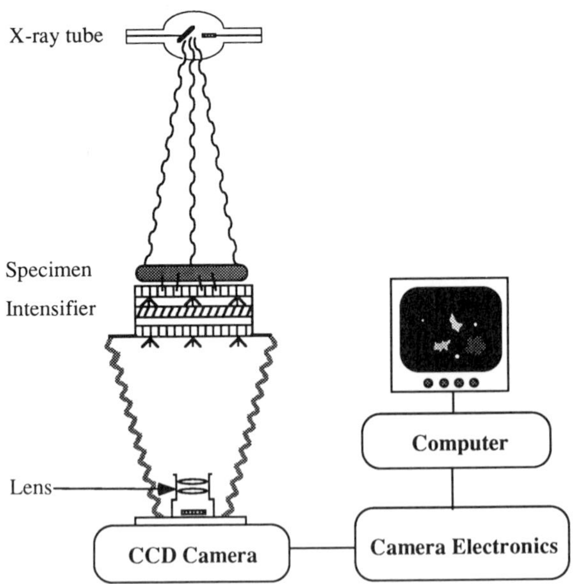

Figure 2. Schematic representation of the system under development
at the University of Massachusetts Medical Center, for specimen
X-ray transmission and radionuclide emission imaging.

via a lens or a fiberoptic device. The solid state sensor may be a stationary two-dimensional device, or one-dimensional linear devices can be scanned over the specimen. At the University of Massachusetts Medical Center, we have been using a 2,048 x 2,048 pixel CCD, optically coupled to an X-ray scintillator via a fast lens. In an alternative approach, we have used a microchannel plate image intensifier between the scintillator and the CCD (Fig. 2). A unique characteristic of the intensifier is that it incorporates an X-ray converting fiberoptic glass scintillating plate as the primary X-ray detector. This scintillator is an integral part of the intensifier and it is capable of producing spatial resolution of 20-25 lp /mm. This system allows for X-ray transmission and radionuclide emission imaging of the same specimen using the same detector. This approach enables the investigator to correlate the radioisotope distribution (physiologic) image with the X-ray transmission (anatomic) image. This system is currently at the prototype stage and testing is in progress.

A major advantage of the area devices is that they can be used for "on line" real time data acquisition, for example, when mounted on a fluorescent microscope. In this case, the high sensitivity to low light levels makes it possible to minimize the intensity of exciting light used. This makes it possible to decrease the amount of fading of the fluorescent signal during 3-D imaging of specimens using confocal light microscopy and thereby minimizing the extent of quantitative corrections needed [Fay et al., 1989].

In most routine applications, the computer requirements for microscopic image analysis can be well served by any of the modern personal computing systems and work stations available from many manufacturers. System requirements include a means of digitizing the input signal, and standard frame buffer interface cards for data acquisition and display tasks. An interactive graphics device (coordinate digitizer, or mouse) is needed for region of interest delineation, along with an appropriate means of generating and viewing black and white or color displays. The software for accomplishing these tasks is more expensive and difficult to generate, and applications requirements may dictate whether one buys a commercial system, or accomplishes the task using special purpose software generated by the user for one's particular needs.

The major tasks associated with image analysis include distortion correction, calibration, registration, quantitation, feature extraction, and display. Distortion correction is needed to compensate for nonuniformities in the light box used to transilluminate the film; spatial linearity and intensity response of the sensor; slice-to-slice variations (cutting anomalies, film artifacts); differences in blur distortion due to differences in energy of radiations (especially important in multi-tracer studies); and in the display system used. Correction procedures depend on the source of the problem, and these technical problems are discussed along with calibration techniques in the next section.

Image registration is a major task that needs to be accomplished in multitracer and multimodality studies. First, one must correct for the known sources of distortion, especially those that differentially affect the individual images to be registered . Different energy radiations blur their film representation as a function of the distance particles travel before being stopped. By imaging line sources of the different tracers, the degree of blurring can be established and suitable restitution filters designed to correct this distortion [Loats et al., 1988]. Once this has been accomplished, common fiducial markers need to be identified in each image and transformation accomplished to superimpose these points and intervening structures by interpolation. Usually spline functions are used to accomplish this task with minimum ripple artifacts. The more fiducials one can identify with accuracy, the better the registration of individual pairs of pixels [Venot et al., 1986].

Noise arises in digital- and film-based images for different reasons, and needs to be corrected. Systematic problems need to be corrected once the source is identified. Residual problems may require additional cosmetic corrections. In digital images, one can discard occasional outliers by the use of median filters or data bounding techniques. These techniques reject pixels that differ from the mean of the field in which they lie, and replace aberrant values by the mean itself. In film analysis systems, scratches due to handling or processing damage can be corrected by similar techniques applied to pixels which vary significantly from the pixel values that surround each point in space on that film , including registered locations on slices above and below the image being analyzed [H. Maeda, Unpublished BNL Technical Report, 1984]. Such corrections need only be applied to regions in which there are obvious defects, since some loss in spatial resolution occurs with application of these smoothing procedures.

Ratio, difference, or various parametric mappings can be applied to sets of registered data, depending on the particular comparisons required. For example, difference images reflect patterns of temporal or spatial change; ratio images relate two different processes, as in flow per unit volume or flow per unit of metabolic trapping; and normalized comparisons may be useful in making side-to-side comparisons in some collective function, such as glucose utilization in activated neurons.

Real-time video displays are used to guide the conduct of image analysis. They also make it possible to view scenes from different spatial or temporal perspectives. The individual can be coupled into the system by various tricks that make it possible to get a sense of being able to see the image from various perspectives, including removing outer structures in order to view internal features [Robb and Barillot, 1989]. Other display media, such as film or video can be viewed at remote locations, and communications channels make it possible to transmit scenes locally or to distant locations by telephone, cable, or satellite channels in black and white, or color as two- or three-dimensional displays. Some of the most impressive imaging applications have caught the attention of scientists and the lay public because of the great strides that have been made in high quality image display technologies.

The recording of ARG information requires the use of an adequate amount of radioactive tracer, and the choice of appropriate exposure circumstances. The sensitivity of film is low, which is the penalty one pays for high resolution (thicker films would be faster, but poorer in resolution). Low cost, high sensitivity 2-D imaging receptors can be used for image acquisition or for screening tissues or gels to determine the length of exposure needed for optimum film exposures. Wire chambers and related physics research tools fall into this category, and may help to bridge the gap between image acquisition and display of low activity images. The major advantage of such systems is their ability to decrease the time of image formation by 100- to 1000-fold. The problem is that it is difficult to achieve high resolution (i.e. better than 1-2 mm). Nonetheless, the field is new and there is promise that some combinations of different radiolabeled agents can be resolved more effectively by the use of these energy and position sensitive devices [Angelini et al., 1988].

Autoradiographic and radiographic methods have many features in common. Both involve film analysis, although as noted, modern systems may use electronic imaging receptors. The technical challenges common to these devices will be discussed as they serve as prototypes for the different imaging modalities. Each of the imaging devices have their own special requirements, but the basic aspects of calibration, distortion correction, image analysis and display have features in common, and these will be the subject of the following discussion.

Techniques for correcting system response variations

Uniformity of Response

Sources of uniformity distortion arise at many points of the imaging chain. In film analysis systems, these include the image receptor, the light source used to transilluminate or reflect light from the negative or positive film, as well as features that relate to the handling of the film. In addition, variables such as transient stability also affect the performance of the analysis systems. Correction for nonuniformities in intensity response is usually accomplished by acquiring and digitizing an image with the system in the absence of an attenuating object in the field of view. Correction factors are then established, pixel by pixel to restore the flat response that should have been achieved. The next problem is the stability of these factors, which may relate to intensity of the light source, imperfections in intervening lenses or filters and temporal variations. These may include such considerations as warm-up time before stable operation is achieved, fluctuations in power sources, or ambient lighting. Problems in the film developing or handling also influence system performance. Film developing problems are minimized by the use of simultaneous imaging standards.

Linearity Distortion

Beside uniformity (intensity distortion), the other major source of distortion arises from nonlinear spatial response. Various causes of distortion, including astigmatism can arise from irregular lenses, or poor quality digitizers. Linearity distortion is easily recognized when checkerboard-type patterns are imaged with the system. Correction factors can be established which adjust each of the fiducial points in the image so that they correspond with the true locus of points in the object [Y. J. C. Bizais, HOLPLT. Unpublished BNL Technical Report, 1983]. Images of graph paper can be imaged on the digitizer to establish the presence of linearity distortion. The imaging of a checkerboard hole pattern permits the generation of correction factors for real-time or post-processing correction. Since correction factors themselves may not be stable, and small errors propagate through the system, it is good practice to correct the causes of distortion rather than compensating for them by computer adjustments. When this is not feasible, computer methods are useful.

Spatial Resolution

Spatial resolution is limited by the information captured on the film which depends on the image formation and image analysis system used. Characteristics of the source of illumination, distance between the object and that source, and object-to-film distance all influence spatial resolution. In radiography, the size and shape of the focal spot, as noted earlier, determine the ultimate resolution achievable. In ARG systems, comparable factors that limit resolution include the energy of the radionuclide(s), specimen thickness, and proximity of film emulsion to the radioactive tissue, and film stopping power. These factors all limit the ultimate resolution which appears on film records generated from these studies. Knowledge of system response characteristics, however, make it possible to compensate in part for systematic known spatial distortions.

System resolution

System resolution is determined by the number of pixels into which the image can be digitized. Since one needs to have at least two samples across a feature in an object to reveal its presence, one attempts to oversample whenever possible. In the case of film analysis, it is possible to vary system resolution by choice of imaging parameters. Thus, using a 512 x 512 imaging matrix, one can digitize a 1 cm object (or portion of an object) at moderate resolution, or by changing lenses one can image a 1 mm object at 10 times higher resolution, or a 100 cm object at considerably coarser resolution. This is true for either a single element imaging receptor such as a vidicon camera, or a mosaic array detector, such as a CCD.

STAINING TECHNOLOGY

High quality medical imaging relies upon: 1) stains or tracers that delineate tissue and cell structures of interest with high specificity, and high contrast, and 2) imaging devices that preserve those outlined features. Recent advances in microscopic imaging rest heavily upon the special staining procedures recently developed using immunotargeted agents. Cells can now be classified with increasing confidence with respect to lineage, antigenic differentiation, the presence of oncogenes, or metabolic and other factors for which specific probes exist [Taylor, 1978]. Pharmacokinetic analyses require the use of labeled drug products to permit measurement of their passage and transformations as compounds move through different organs, tissues, and cells as a function of time. The use of radioactive and stable labels as tracers of specific substances and pathways makes it possible to determine how much of which substance goes where, and how long it stays. In addition, one needs and can obtain information on the integrity and metabolism of labeled compounds from animal studies, and have it validated in human subjects based on biopsy sample analyses.

During more than 100 years, pathologists developed a large number of special stains which they used with increasing effectiveness to identify and classify different cells, and intracellular organelles. At the present time there are many special stains, but the most standard stain used for light microscopy is Hematoxylin and Eosin (H&E). Hematoxylin stains cell nuclei blue, and eosin stains cytoplasm red. The major recent advance in staining technology uses antibodies against a rapidly growing number of cell markers, and this has ushered a new era in cytopathology. The primary image is formed by antibodies, labeled with different chromagens, fluorescent markers, enzymes that react with chromagens, or isotopes when they localize on specifically targeted sites.

The first of the powerful new microscopic techniques developed was the electron microscope. The short wave length radiations which it emits make it possible to obtain higher resolution spatial images than can be obtained with lower energy radiations. The ability to stain these ultra thin sectioned materials required the development of high electron density contrast agents (heavy metals) that could be shadow cast onto the cut specimens post mortem, or coupled to probes with a selective biodistribution [Hainfeld, 1987]. Variations in electron penetration of the sample and the image they cast depend upon the distribution of the heavy metal atoms in the specimen. In addition to the differential thickness images produced by transmission images, such systems can also reveal elemental composition by taking advantage of the fluorescent X-rays emitted when the electron beam is scanned over different portions of the specimen [Fiori and Newbuty, 1978]. Thus, one can quantitate and localize the distribution of elemental constituents in different cells, especially for elements with atomic numbers greater than or equal to 15 (phosphorous). Fluorescent X-ray spectra can be generated using a variety of incident beams. These include X-rays (including those generated by synchrotron light sources), and protons (proton induced X-ray emission-PIXE), both of which can provide accurate measurements of normally occurring levels of elemental content of individual cells, and sub-cellular structures [Jones et al., 1988; Duflow et al., 1989].

The laser confocal microscope is a new development that provides access to a low-priced effective tool for exploring the depth dimension in specimens. This device scans a finely focused spot of light over a specimen containing a fluorescent dye so that different depths are illuminated and interrogated successively. The optics interrogates the same spot that is illuminated, and in so doing provides significantly better resolution than can be obtained from computer-processed optical microscopes focused at successive depths in a specimen. The fluorescent dye marks the location of an anatomic structure or a physiological process, such as calcium channel activity. These systems permit 3-D mapping of the distribution of tracers in biological structures with sharp Z-axis focus and can be used to distinguish the location of markers on the cell surface vs. deeper locations in the cell [Fay et al., 1989].

The classification of normal and abnormal cells and tissues is based on morphology and staining characteristics. The use of immunospecific staining procedures now makes it possible to further specify cells in terms of their antigenic characteristics. Thus, in addition to identifying a cell as a lymphocyte, for example, one can identify it as a B, T, or Killer cell, as well as its stage of maturation [Braylan and Benson, 1989]. In the diagnosis of cancer, it is frequently difficult to determine the site of origin of metastases, and the delineation of antigenic markers on these cells is of great help in determining the cell of origin and its degree of malignancy [Nadji and Morales, 1984; Merkel, Dressler and McGuire, 1987]. Two ways of accomplishing this task involve the use of flow cytometry (for dispersed cell populations), or fluorescence microscopy (performed on thin tissue slices) [Shapiro, 1989; Tubbs and Bauer, 1989]. The nature of the cell line is inferred from light microscopy in which presence or absence of different antigenic markers is established in tissue sections prepared using selected members of a panel of immunoreactive stains. This is a tedious, time consuming task, and microscopic examination typically uses presence or absence of staining, rather than intensity as a criterion.

Alternatively, flow cytometry with multiparameter analysis can be used to screen for multiple factors simultaneously, using intensity as well as wavelength of the emitted fluorescence as a classification factor [Watson, Sikora and Evan, 1985]. Such systems use a single light source (typically a laser or Argon light source), and two (or more) fluorescent antibodies (which target different antigens and which emit different wave lengths of light following excitation by the same wave length) [Ploem-Zaaijer et al., 1979]. The advantages of the fluorescence-activated cell sorter (FACS) include the ability to use multiple emitted wave lengths plus intensity, as well as light scattering, which depends on molecular size and conformation. In addition, the FACS permits the collection of cells which share a combination of these features for additional special studies. The distribution of FACS markers can be displayed as a two-, or three-dimensional distribution which uses intensity information along the Z-axis plotted against the correlation between two associated antigens on the X- and Y-axes.

The immunologic methods upon which these tests are based involve the use of different markers (fluorescent and enzymatic) tied to antibody products. There are four types of methods in most common use. These involve the coupling of a marker, typically horse radish peroxidase (HRP) to an antibody directed at the antigen of interest. When a chromogen whose color is changed by HRP is applied to the preparation, the distribution of the "HRP-marked" antigen is revealed by a yellow color in the specimen. The direct method uses this approach. Thus, if there are two different antigens, two different antibodies are needed, each coupled to HRP applied to successive sections in the specimen, or each labeled with a different chromogen on the same slide. The second method is an indirect method. This uses an unconjugated antibody, derived from rabbit or murine cell lines which target the antigen of interest. To localize this antibody, a second antibody (coupled to HRP), is applied. This non-specific antibody is directed at rabbit or murine gamma globulin and is attached to HRP which is then visualized as in the direct method. The remaining methods, both of which are in more common use are the PAP and avidin-biotin techniques. The PAP method uses three reagents, primary and secondary antibodies, but in this case it is the third antibody directed against the second, or linking antibody that contains the HRP. The avidin-biotin method is based on the ability of avidin to chemically bind to four biotin (vitamin) molecules. As in the PAP method, the procedure involves three steps. First, the rabbit or murine antibody to the tissue antigen of interest is applied. Then the second antibody (to rabbit or murine protein) is added, coupled to biotin. The third reagent is a complex in which peroxidase is conjugated to biotin and avidin, and the unsaturated avidin binds to the biotin on the second antibody. Both the Avidin-Biotin, and PAP methods make it possible to use a single peroxidase coupled agent for all antigen determinations, thereby decreasing cost of reagents, and further, these methods have higher sensitivity than the first two methods. These immunoperoxidase (impox) methods can be used with fresh or fixed tissue. Microscopic evaluation of stained slides indicate whether the cells are positive or negative for the particular antigens based on yellow staining [Taylor, 1986].

FACS systems can be used with tissue minces from which cells are dispersed, or circulating blood cells can be analyzed and sorted. In this system, the intensity of staining is displayed as one-, two-, or three-dimensional histograms, based on two or more measurements which can be made on the same cell simultaneously. Thus, one can classify tumors based on the presence of particular tumor antigens, while at the same time quantitating, for example, the amount of DNA in individual cells. In this way it is possible to correlate S phase fraction, which provides an index of replication rate with oncogenes or tumor markers in a tumor specimen. The advantages of this method are:1) it is quantitative, and therefore has a high degree of objectivity, and 2) it is under automatic control (and thus requires little effort) and is able to coordinate function with antigenic presence in large numbers of cells. The disadvantage is that one does not know what fraction of the data comes from cells of interest, as opposed to cells that may be normal, necrotic, or otherwise uninteresting. Such distinctions are better appreciated from microscopic analysis of immunostained specimens [Seckinger, Sugarbaker and Frankfurt, 1989].

APPLICATIONS RELEVANT TO PHARMACOKINETICS

Introduction

The role of microscopic imaging in pharmacokinetic research includes the determination of the biodistribution and kinetics of new compounds and their metabolites. The radiotracer method has been key to the delineation of metabolic pathways based upon analysis of tissue minces. The analysis of the sub regions of organs and cells in which these processes take place can also be studied in vitro in tissue cultures but the in vivo behavior requires analyses in intact animals where the various systems interact to determine the absorption, fate and excretion (AFE) of new compounds. Initial studies are carried out in animals, and finally in man, before a new drug can be qualified for routine use.

In vitro Imaging

How it contributes to determining biodistribution of a new agent: the different pieces of information needed include:

1. Where does it go?
    Influence of route of administration
2. How long does it stay? (kinetics)
3. Stability of compound, and distribution of metabolites
4. Can one anticipate hazards?
    Unusual localization, Cofactors
5. Does it have a role in diagnosis?
    Adequate sensitivity/specificity
6. Might it have a role in therapy?
    Highly selective distribution
7. Influence of Co-factors on Therapy?
    Radiation, ChemoRx, ...

1. Where does it go?
Organ distribution studies can be performed by counting and weighing whole organs or tissue samples from animals sacrificed following injection or other intervention. Information on the temporal behavior of the tracer can be obtained from assays performed when times between drug administration, interventions, and tissue sampling are varied.

Given information on tracer turnover, one can fit the data to physiologically relevant models and extract parameters such as area under curve, residence times, and spaces of distribution. From these results, one can calculate various quantities of interest reflecting system behavior or drug related effects (i.e radiation dosimetry). To ensure that the label does not alter drug behavior these studies often use a compound labeled with C-14, or H-3 which substitute for an isotope of the same element in the animal. More detailed information on spatial and temporal distribution of labeled compounds can be obtained by ARG methods, and these will be discussed separately.

2. How long does it stay?
Quantitative *external imaging* techniques (discussed in a separate paper by Fishman in this conference report) have developed rapidly in recent years based on the same technology used for in vitro imaging. Temporal and spatial information on the biodistribution of new compounds is achieved in external imaging studies, with more detailed information available from in vitro imaging of biopsy samples. In addition to organ uptake and retention from organ imaging studies, whole body counting provides information on AFE, which can be obtained with less difficulty and greater accuracy than from intake/output studies. The technical, logistic, and ethical difficulties in conducting human metabolic research are significant, and hence most of the data upon which drug qualification is based rely upon data from animal studies. The data from human studies relate mostly to efficacy and safety, with less complete data regarding biokinetic behavior. Questions regarding stability of the labeled compound are resolved by detailed analyses of plasma, urine, and biopsy

samples. Data from animal studies are more easily obtained from many tissues, at many times, but the validity of extrapolating data from animal studies to man makes it important to obtain human data to validate the general trends established by the more detailed animal study data. How long a compound stays is assessed by counting and weighing gross tissue samples from animals sacrificed at different times following drug administration. Analysis of distribution at the cellular and sub-cellular level can best be established by sequential studies on the same or different animals at the microscopic level. The conduct of such studies in the same animal at two different time points requires dual tracer ARG methodology which involves methods to be discussed in the next section.

3. Compound stability and biodistribution

Compound stability and biodistribution ultimately need to be established in humans since the behavior of the compunds may differ in various species. Extrapolation of results of antibody studies between species is especially difficult since antigenic differences will complicate interpretation. Accurate biochemical analyses of tissue extracts are needed to characterize the kinetics of the originally labeled compound, which requires techniques such as High Performance Liquid Chromatography (HPLC). All too frequently this last requirement is overlooked. Imaging is needed to determine whether the compound is distributed uniformly through the organ or if it is localized in some special subregion. Does that distribution depend on route of administration? Clearly, gut and lung localization will be sensitive to whether an agent is ingested or inhaled. Likewise, uptake by an intraperitoneal (IP) tumor will be sensitive to whether an agent is administered IV or IP. The delineation of surface-to-depth distribution can be established best by microscopic assessment of the labeled compound, and the method used will depend on the available tracer and imaging methods, and the sensitivity and resolution required. Autoradiography and fluorescence imaging systems can and do play a major role in these studies.

4. Sources of hazard

The identification of possible sources of hazard from a new agent is a continuing problem. Concern by regulatory agencies licensing new drugs and physicians employing new drugs is shared with patients who want to have access to more effective agents, and who also want to be assured that the drugs are safe. The thalidomide experience made it clear that in utero data are important and that information on microdistribution in the developing fetus is needed. The development of whole body cryomicrotomes and their dissemination was spurred by that concern, and many studies have been carried out in animals varying in size from rodents to monkeys investigating these questions [Ullberg, 1977].

5. Role in diagnosis

The potential use of an agent for diagnosis relates to its biodistribution and kinetics in animals and ultimately in humans. Imaging studies play a role in the screening of new agents for desired organ uptake, retention and turnover. In the case of tumor diagnosis, the effectiveness of a new agent is determined by the sensitivity and specificity of that compound for the particular tumor. Clinical utility can often be anticipated from preclinical studies by inspection of experimental data on intensity of uptake and biodistribution of the labeled agent in tumor and normal organs in several animal species.

6. Role in therapy

The potential utility of an agent for therapy requires a higher degree of uptake in the tumor than is needed for diagnosis. In the case of radiation therapy using I-131 for thyroid cancer, fifty-fold higher uptake in the tumor than in the surrounding tissues is thought to be needed for successful tumor therapy, whereas a two- to five-fold increase is quite compatible with diagnostic utility. Concentration and duration of attachment at a radiosensitive site can be obtained from the study of radioisotope- or fluorescent-labeled tumor markers. If uniform distribution of tracer throughout the tumor is needed for therapeutic efficacy, imaging studies can provide the needed evidence. Chemotherapy and radiotherapy procedures employing low energy nuclides require highly uniform distributions within the tumor, and sometimes, within the cell nucleus per se.

7. Influence of cofactors

Influence of cofactors is especially important in cancer therapy, where combinations of different agents are used in the treatment of most cancers. Evidence of the enhanced uptake of internally administered compounds comes from workers at Johns Hopkins Hospital where they found a three-fold increase in the uptake of tumor directed antibodies following external radiation therapy to the tumor [Leichner et al., 1989]. The penetration of large molecules, such as antibodies into tumors, is limited by vascular and tumor pore sizes [Flessner et al., 1985]. Radiation damage to barriers to vascular egress and cell penetration may influence tumor uptake of radiolabeled antibodies.

Autoradiography:
Techniques and applications relevant to pharmacokinetics

Technical Factors

A. Standards: Quantitative ARG studies require simultaneous imaging of multiple dilutions of standards for each of the tracers used. These calibration data are fit by piecewise linear, higher order polynomials (or other transformations) which make it possible to translate activity per pixel in the image to microcuries or fraction of the injected dose per gram of tissue. Experiments using long-lived tracers, i.e. C-14 or H-3, can employ commercially available calibration step wedges. The C-14 step wedge for the brain is calibrated and corrected for attenuation in passage through a nominal 30 micron slice thickness (C-14 Standard: Microscales; Amersham Corp. Arlington Hts, IL.). The H-3 standard is set into a block which is cut along with the specimen, thereby achieving appropriate reference data (H-3 Standards: Dupont Corp., No. Billerica, MA.). When other tracers are employed, user developed standards are needed. Some experiments use filter paper standards which are applied to the films, and corrections for attenuation are based upon analytical considerations. More accurate standards are made from tissue paste emulsions into which various dilutions of the tracer are mixed. By embedding and cutting a set of these standards with the tissue samples, one minimizes uncertainties in the quantitative accuracy of results [Unnerstall et al., 1981; Ito and Brill, 1990]. For dual isotope studies, it is necessary to have two sets of standards, with graded dilutions for each tracer used. In addition, it is necessary to obtain several values for known mixtures to correct for interactions between the two tracers. The problem comes in at the low and high ends of the H and D curve. The influence of the second longer-lived tracer may influence the values for the short-half life tracer in the low end of the curve, where the curve rises from the low plateau region onto the ascending limb more rapidly due to the combined exposure from both tracers [Laskey, no date]. Variations in density of the different tissues also needs to be taken into account. This is especially true for bone and lung, at two ends of the density extrema [Ito and Brill, 1990].

B. System Characterization: Evaluation of system uniformity, linearity and resolution is a necessary prelude to qualifying a new system. Once these factors have been established, and distortion corrections developed, one chooses appropriate pixel size for the particular study, and adjusts the size of the imaging array and lens magnification appropriately. Imaging of linearly ruled graph paper with precise line spacing is a useful means of establishing pixel dimensions, as well as verifying system operational characteristics. Changes in performance when f-stop is altered also need to be monitored, as focus must also be altered, and it is important to avoid systematic variations in system performance. The use of a field limiting light enclosure may be useful in limiting temporal and spatial variations in ambient lighting. All systems have a warm-up time during which the system needs to equilibrate before use. This minimum time should be established so that comparable results can be achieved on successive days. Further, the dynamic range of the camera needs to be considered in analyzing a particular image. Accurate quantitation is best achieved by operating the system within the linear region of the camera response. When the image has wide variations in radioactive content, it is possible to increase system lattitude by digitizing the image together with standard calibration sources at several different camera lens f-stops. In this way, high intensity pixels are analyzed using a calibration curve generated from reference standards viewed through small lens openings, while low intensity regions are compared to curves generated from standards imaged with larger lens openings [Yonekura et al., 1983 a, 1983 b].

C. Types of tracer combinations that have been used.

1.Single tracer (single process at single time): The choice of sample thickness, film emulsion, exposure and development times are the major variables that affect the quality of single tracer studies. These are well understood and well described in standard texts [Rogers, 1979].

2. Dual isotope ARG studies are a valuable means of analyzing changes in the biodistribution of a single compound labeled with different tracers each of which is administered at a different time prior to sacrifice of the animal. Similarly, one can label two different compounds, one of which traces a physiological pathway, while the other delineates a body space (as a normalizing factor), or a metabolic process. Analytic separation of the two tracers is based on differences in radioactive half-life of the two tracers or on differential stopping power of their emitted ionizing radiations [Lear et al., 1984].

The first method is best accomplished when a long-lived tracer is given in small amount, and a short-lived tracer is given in much larger amount. Thus, in the first exposure of the tissue slice to film, one captures data almost exclusively from the short-lived tracer, with little contribution from the longer-lived tracer. When the short-lived tracer is gone (>10 half-lives), the remaining activity is only from the long-lived tracer. Use of the simultaneously imaged standards allows the distribution to be quantitated accurately. When the activities of the two tracers are comparable, the separation is more difficult, and standards, including a mixture of the two nuclides in graded amounts, are needed to correct for non-linear interactions between the two tracers and the film exposure process.

In the second method, separations are based on differences in stopping power of the tracers. This method is well-suited to combinations of low energy beta emitters, such as H-3 , C-14, or I-125 , and higher energy emitters, such as I-131, Y-90,. In this case, the sample is exposed to film with and without an intervening attenuator. The thickness of attenuator is chosen so that all of the low energy emissions are stopped, without appreciably attenuating the higher energy emissions. Reference to calibration standards imaged in both circumstances provides the necessary data for quantitative analysis. Typically, thin mylar foils are sufficient to stop the low energy beta particles without appreciably affecting the higher energy, more penetrating tracers.

3. Three or more tracers: (Radioactive & stable tracers, incl. fluors). As the number of tracers increases, the problems associated with cross talk increase when the additional tracers are radioactive. However, the addition of a histochemical or fluorescent dye marker permits the use of an additional tracer without such problems.There are also stable tracers that can be activated, thereby providing radioactive emissions which can be turned on after the radioactive labels have essentially disappeared.

4. Emission/Transmission: The multiple tracer studies referred to above all require image registration for quantitation and analytic comparisons. It is possible with sensitive imaging devices such as CCDs, or multi-wire chambers, to capture the

data from radioactive emissions directly into a machine readable device. These devices lend themselves to this application since they can store and read out data without altering the sample position on the detector. When the specimen is transilluminated with an appropriate energy X-ray or charged particle beam, it collects an image of the attenuating structures in the specimen based on differential transmissability of the sample.

5. Radiolabel Vs. Stable Tracer (H-3 Vs. H-2). There are several ways that have been used in which stable tracers can be imaged as if they were naturally radioactive. Both involve the use of nuclear activation techniques. The ability to image the distribution of deuterium in biological samples is based on the transmutation of H-2 into an alpha particle by bombarding the specimen, mounted against a film emulsion in a light sealed covering with energetic tritons from a Van derGraaf accelerator [Geisler et al., 1974]. These particles react with deuterium to form a compound nucleus which emits an alpha particle and is stopped in a short distance in the film emulsion. If the specimen also contains H-3, one could separate the mixture using differential stopping power method described above on the same sample or on sequential sections from the same specimen.

6. Other methods of separating different distributions have been used for special applications. One method based on differential volatility characterized the different chemical forms of the tracer from the sample in the following fashion. The frozen tissue section contained the native compound plus potentially volatile breakdown products. By imaging the frozen tissue section first, and then imaging it after it had warmed to room temperature, they were able to demonstrate the location of the volatile metabolite by regional losses in tracer content [Bergman and Tjalve, 1977].

Typically, analysis of the chemical form of the tracer in the tissue is established based on chromatographic separations of the labeled activity on 2-D gels. In this case one obtains physical samples, i.e. small tissue plugs taken from the ARG section for which radioactive content has been ascertained from the analyzed film record. The separated moieties detected on the gel are referenced against calibration compounds chosen because of their potential as breakdown products to estimate the metabolic fate of the material in different organs. Tissue sections usually are cut in 5-30 µm thickness for image analysis. Tissue plug analysis can be accomplished more easily on specimens taken from an adjacent section cut in 100 µm thickness. This sample is then registered with the adjacent ARG section in order to select samples from anatomical and physiologically interesting regions. In this way, one can count and then analyze the biochemical content of these materials, and also obtain accurate calibration data. The counting data reveal actual content in terms of counts/gm, thus avoiding questions concerning self absorption, or film development factors [d'Argy, 1977].

2. Examples of dual tracer applications

a. Blood Flow Correlations: The analysis of tracer uptake patterns reflects access of the tracer to localization sites, retention, and loss phenomena. The need to separate blood flow differentials from metabolic variations in the brain has focused attention on this problem. Several approaches have been taken, including the use of both physical and chemical microspheres. Physical microspheres may be composed of aggregates of albumen (often used in human studies as they are metabolized and cleared from the perfused capillaries), or ceramic materials (which are more uniform in size, but are not biodegraded, and hence are used mostly in animal studies). Either of these tracers can be imaged autoradiographically whether labeled with a stable or radioactive tracer, although the radioactive tracer studies are far simpler to conduct, and provide higher spatial resolution.

The separation of brain blood flow from glucose uptake as an index of brain metabolism has been approached by the use of F-18 2-deoxyglucose, followed several minutes prior to sacrifice (45 min. later) by I-125 antipyrine administration (I-131-Antipyrine/F-18 2DG: Unpublished BNL Studies, 1984). The analytic separation of these two tracers is easily achieved due to the great difference in half-lives of these two tracers. The difficulty is that relatively few places have easy access to the F-18 labeled compound, and the necessity to process samples very rapidly before the F-18 decays to low levels. The longer one waits, the greater is the contribution of the I-125 to the image, and the more computational effort is needed to correct for its presence. Alternatively, commercially available C-14-2fluoro-2deoxylglucose and I-131 antipyrine have been used [Lear at al., 1981]. Longer exposure times are needed, but good separations can be achieved using half-life or stopping power differences for separations.

The effect of various perturbations on blood flow can be delineated spatially using two or three radiolabeled microspheres, administered at different times, one before perturbation, and others at different times following experimental interventions. For example, one could administer a long-lived tracer as the control, and two tracers with signficantly shorter half life. These could be distinguished based on different pathlength, using a suitable thickness attenuator to block the radiations from the lower energy emitter, and by taking advantage of half-life differences. It is difficult but not impossible to find commercially available tracers that meet this requirement, but because of half-life considerations, they do have to be ordered for a particular experiment, which can be expensive. Ceramic microspheres are available which contain many different radioactive nuclides. However, the cross talk between different energy emissions ultimately limits the number of combinations that can be used simultaneously. In principle, several stable tracer labeled microspheres can be used in this application. Since the stable tracer does not undergo radioactive decay and doesn't leak from the microsphere, samples can be held for as long as required for study conduct and sample analysis. Imaging and counting of stable tracers requires that they be activated by neutron or charged particle exposure, or they can be mapped using X-ray fluorescence techniques. In some cases, it is possible to expose sections mounted on film emulsions in the beam (prompt reactions, such as used with B-10 imaging) [Green et al., 1978]. Films can be exposed or samples counted following radiation at times dictated by the half-life of the activated elements in the sample. In principle, one could expose successive adjacent tissue sections, each to an exciting beam chosen to optimize a particular desired radioactive transmutation. In so doing, one could minimize competing reactions and simplify analysis. Problems in experimental conduct require access to unusual nuclear research facilities and personnel. Physical microsphere studies define regional perfusion based on the the fraction of an injected dose of particles (in the 10-50 μm range), which lodge in the capillaries of perfused tissues of interest. Chemical microspheres, such as antipyrine, are diffusible substances which localize in the perfused region based on diffusion or metabolic localization, which reflect blood flow distribution.

b. Cell kinetic analyses. Analysis of cell turnover rates historically was based on the use of tritiated thymidine (H3-Thy) incorporation into dividing cells. The counting of grains over individual cells enables one to estimate the rate of division for particular cell populations.The use of H3-Thy is limited in patients due to radiation dose considerations. In order to avoid this problem, and to find a means of obtaining similar information, Slatkin developed and used H2-Thy and demonstrated the feasibility of using perdeuterated nucleotide precursors [D. Slatkin, Medical Dept., Brookhaven National Lab. Personal Communication 1/7/91]. In principle, one can chronicle cell turnover rates following ionizing radiation exposure, for example, by infusing one of these tracers prior to irradiation, and the second at selected times thereafter. Analysis of the H2-Thy data is complicated since access to a triton beam is required and the technique has not been used in human subjects. Better methods of assessing mitotic behavior have now been developed using FACS and DNA specific staining, and these techniques supplant the need for the more complex H-2 or H3-Thy studies, but other applications of the H-2 tracer method may justify its mention here.

c. Dual tracer ARG methods lend themselves to use in experimental tumor therapy research. Two examples that illustrate different applications include: 1) the distribution of radioactive therapeutics administered by intravenous (IV) vs. intra peritoneal (IP) routes, and 2) a delineation of the hypoxic component of a tumor.

Radiolabeled antibodies are used in the treatment of ovarian cancer, which commonly is limited to the IP space. Administration of large molecules, like antibodies, IP exposes the tumor cells to higher concentrations immediately following injection than can be achieved by IV infusion. (Where the tumor is large, surface encounters between antibody and antigen can result in coupling, with subsequent diffusion of material through the tumor.) In general, the center of the tumor receives less blood flow and tends to be hypoxic, and thus is poorly targeted by the antibody. Part of the IP infused tracer is absorbed into the bloodstream, and thus may reach more central well perfused tumor sites, as well as metastatic remote tumor foci. The choice of route of administration and properties of the radionuclide to be used for therapy determines the radiation dosimetry and probability of patient benefit. Dose depends strongly on the distribution of the tracer labeled compound in the tumor and normal radiosensitve tissues.

Principles governing the choice of route of administration have been assessed in animal studies using combinations of tracers administered by these different routes. [Ito et al., 1990]. These studies have used In-111 OC-125 infused IV, and Y-90 OC-125 introduced IP., and relatively greater depth dose but lesser surface dose has been reported from the use of IV administration. For tumors which are small with respect to the path length of the Y-90 beta particle, a lower energy beta emitter would be favored as a larger fraction of the dose would be delivered to the tumor, and less to adjacent normal structures. For larger tumors, it seems reasonable to expect that a mixture of IV and IP labeled agents would be desireable. One of the problems with the IV, IP combination indicated is that Y-90, if it elutes from OC-125, has a strong proclivity for bone uptake, whereas free In-111 does not. We have been investigating a similar question utilizing two different isotopes of Indium, namely -111, and -114m, which avoid the issue associated with different biodistribution factors [Aas et al., 1990]. The In-114m is separated from In-111 based on the differences in half-life (In-114m has a 50 day t 1/2 which is 18 times longer than In-111). The problem with the use of this combination is that In-111 has a small amount of In-114m as a natural contaminant (<0.15%), which must be taken into account.

The presence of hypoxic tissue in a tumor contributes significantly to radioresistance, which makes it more difficult to achieve successful tumor therapy. Thus, information on the extent and distribution of hypoxic tissue in tumors would be of significance to the radiotherapist. Chapman and his colleagues at the Univ. of Alberta in Edmonton have conducted a large number of studies documenting the importance of misonidazole as an indicator of hypoxic tissue. In collaboration with the Brookhaven National Laboratory Medical Dept. (BNLM), ARG studies were conducted in which C-14 labeled Misonidazole was administered along with F-18 2-DG, and a prominent central zone of increased misonidazole was seen to be surrounded by a ring of 2-DG, reflecting the growing periferal tumor and the central hypoxic slower growing necrotic appearing tumor [Garrecht and Chapman, 1983]. Subsequent studies in other laboratories have revealed the potential utility of misonidazole correlations with increased glucose uptake in myocardial and cerebral ischemia. It should be noted that in tumor patients receiving these in vivo studies, the opportunity for obtaining detailed ARG information on tracer distribution can be achieved by analysis of biopsy specimens, when these are available.

d. Cardiovascular disease processes have been studied using multitracer ARG studies in a variety of animal models of disease. A series of studies were performed at BNLM in Dahl strain inbred rats which develop severe hypertension following NaCl ingestion. Hypertensive animals were administered a number of experimental fatty acid analogs developed by Knapp and his coworkers at the Oak Ridge National

Laboratory. Dual tracer studies were conducted in which animals received Tl-201 as a myocardial flow tracer, along with a series of I-125 labeled fatty acid analogs. The findings clearly indicated a decrease in fatty acid uptake in the wall of the left ventricle in regions in which Tl-201 uptake was unchanged [Yonekura et al., 1985]. The comparison group used in these studies were Dahl strain rats which had not received salt additions to their diet, and were normotensive. Additional studies showed that F-18 2DG uptake was increased in the same region in which the fatty acid was decreased. The fact that there was no decrease in Tl-201 in these regions was not explained although the presence of an unusual metabolic defect in the Dahl strain rat was suggested. Current dual tracer investigations work at the Univ. of Massachusetts Medical Center (UMMC) are being performed to compare the biodistribution of Tl-201 with that observed with various new cardiac agents being proposed for clinical use [J. Leppo, Univ. of Massachusetts Medical Center. Nuclear Medicine Dept. Personal Communication. Jan 7, 1991].

3. Examples of Microscopic techniques in Cancer Diagnosis and Therapy

a. Cell Type: Immunohistochemistry, plus cytologic features are now able to provide strong objective support for what previously were subjective judgments based on staining and morphologic features [Stewart, 1989]. Tumor therapy depends on the cell of origin, its degree of malignancy, and extent of spread. The cell of origin can be classified based on its antigenic determinants, both in amount and kind. These features may be established by microscopic examination or by the use of FACS, where the coordination of several markers can be more easily achieved on a cell-by-cell basis, with good statistical confidence in the results. Information on antigens which are present in large amount in the tumor, and to a lesser extent elsewhere, may be amenable to diagnostic, as well as therapeutic uses of radiolabeled or toxin-coupled antibodies.

b. Degree of malignancy: Ploidy, S Phase Fraction. Ploidy and S-phase fraction can both be established based upon microscopic characterization of DNA content per cell, and grain counting of cells labeled with H-3 Thy. However, multiparameter FACS is faster, cheaper and more accurate, and promises to supercede the direct imaging methods for most applications. However, it is possible to attribute these measures to the cell population of interest with greater precision and reliability by microscopic methods than by FACS, where uninteresting cell populations could dominate the quantitative results.

c. Spatial distribution of therapeutic agent(s). The distribution of radiolabeled therapeutic drugs can be studied in human biopsy specimens. Studies are in progress in many institutions to evaluate the distribution of diagnostic and therapeutic agents. The distribution of radiolabeled antibodies is assessed with respect to percent of injected dose per gm to determine potential therapeutic utility. At the same time, these uptake values are correlated to antigen representation to establish the degree of correspondence and specificity of the distribution [Tuttle et al., 1988].

A new study is being established to evaluate the in vivo and in vitro distribution of Pt-195m cisplatin in cancer patients (Walter Wolf, IND 36-051 for Pt-195m-Cisplatin. Univ. of So. Calif. School of Medicine). ARG studies may assist in establishing the adequacy of the biodistribution obtained with this agent as it is used in tumor therapy. Is it in the cell nucleus, which is the site of action, or not? If not, where is it? Is it on the membrane or is it internalized? How long does it stay there? All of these questions can be addresssed using ARG and immunohistochemical techniques. Similarly, in patients receiving other forms of chemotherapy, it is not unreasonable to establish ARG evidence of the C-14 labeled drug distribution to determine the targeting and potential utility of one or more drugs singly, or in combination in tumor biopsy samples.

Some combinations of drugs have additive effects (which may be beneficial or adverse), while others are potentiated nonlinearly. In the case of chemotherapy, different drugs may supplement each others' antitumor activity by differential effectiveness on different cell lines. In the case of antibody uptake, a three-fold increase in labeled antibody uptake by tumor has been reported following external beam radiation therapy directed at the tumor.

d. Temporal distribution of agent. The utility of a diagnostic or therapeutic agent depends not only on where it goes, but how long it stays there. The residence time of a diagnostic agent has to equal or exceed the time at which imaging studies are anticipated. For a therapeutic agent to be effective, the agent has to stay at the desired site long enough to impart lethal damage, while sparing essential normal tissues. Given adequate data on temporal and spatial distribution, it is possible to compute radiation dose to the different body organs using well establish analytic methods. ARG studies on biopsy samples plus external imaging can provide useful data, although the biopsy data are limited at best to several time points. A novel method has recently been developed which results in an estimate of the integral dose to the tumor from an administered radioactive drug [Griffith et al., 1988]. The method requires insertion of a thermoluminescent (TLD) calcium sulphate (dysprosium) rod into an accessible tumor. The patient is then injected with the drug in question. After several radioactive half-lives have passed, the tumor is removed, fixed, and prepared for sectioning. The small piece of TLD cut on each tissue section is counted and an ARG is prepared from each slice. The TLD integrates the dose received at that point in the sample from elements on and adjacent to the section from which it was obtained. Using this method, Wessells et al computed the integral dose for different sections in the tumor from multiple TLDs in animal studies, and some human studies have now been performed [Wessels, Yorke and Lambiotte, 1990].

Potential utility of an agent for therapy depends on its access to critical sites within cells. In the case of chemotherapy, it is essential that the drug be internalized. In the case of radioactive drug therapy, spatial relationships are also important, but by appropriate choices, one's options are greater than with standard chemotherapy. If a tracer localizes in or immediately adjacent to the sensitive site, the radionuclide chosen should have a path length just longer than that needed to cover that target. If there is wider distance, say several cell diameters apart, successful tumor therapy may still be achieved by the use of longer range ionizing radiations. One of the imaging tools that has promise for assisting in obtaining detailed surface-to-depth distribution ratios is the laser confocal microscope. In this way, one can image fluorescent radiations from the bound antibody site with 1-2 μm Z axis resolution, and a better appreciation of intracellular tracer localization can be achieved than by other currently available microscopic techniques. Thus, one could use a fluorescent marker targeting a monoclonal antibody or a chemotherapeutic agent to reveal that agent's three-dimensional distribution.

e. Characteristics of tumors can be assessed at different times following therapy to determine whether the cells are from the same or a different clone, depending on changes in biochemical and antigenic markers. These can be achieved with immunofluorescent markers or with multiparameter FACS systems. Such information is of great importance in the clinical management of cancer patients.

SUMMARY

Major improvements in imaging technology coupled with high specificity staining methods have led to important advances in the ability to identify cells, and their intrinsic components, as well as opening the way to fundamental knowledge of how cells function in health and disease. In addition to characterizing biological system behavior, these new tools now permit tagging various compounds, drugs, toxins, and physiological substances and measuring their absorption, fate and excretion in greater detail than ever before was possible. This provides information on the anatomical and histological localization and Kinetics of drugs/drug products, and possible effects on metabolic processes at the cellular level. Information of great importance is obtained regarding

      1. Potential toxicity,
      2. Metabolic Investigations, and
      3. Diagnostic, and Therapeutic potential

ACKNOWLEDGMENT

Supported in part by P.H.S. grant R-29 CA45545 awarded by the National Cancer Institute, D.H.HY.S.

REFERENCES

Aas, M., Hnatowich, D.J., Mardirossian, G., Griffin T., Salimi A., Atcher R., Hines J., Brill A.B., 1990. Long-time Distribution of 114m-In Labeled Monoclonal Antibody (B72.3) in Mice Bearing Human Tumor Xenografts. Antibody, Immunoconj. and Radiopharm. **3:**57.

Angelini F., Bellazzini R., Brez A., Massai M.M., 1988. Digital Imaging with Advanced Gaseous Detectors Developed at Pisa University. In: "Nuclear Instruments and Methods in Physics Research" (K. Siegbahn editor), Volume A269, No.2, Proceedings of International Symposium on the Use of Wire Chambers in Medical Imaging, Belgium, June 28-30, 1987 (F. Deconinck, M. Defrise, and S. Tavernier, editors), page 430. North Holland, Amsterdam.

Bergman K., and Tjalve H., 1977. Three-step Autoradiography of Organic Solvents and Plastic Monomers to Register Total Radioactivity, Non Volatile Metabolites, and Non-extractable Metabolites. Acta Pharm et Toxic. **41:** Supp. 1, page 22.

Braylan R.C., and Benson, N.A., 1989. Flow Cytometric Analysis of Lymphomas. Arch Pathol Lab Med **113:** 627.

d'Argy R., 1977. Tissue Pieces from Whole Body Sections Used for Chromatographic Separation of Metabolites. Acta Pharm et Toxic. **41:** Supp. 1, page 16.

Dilmanian F.A., Weber D.A., Coderre J.A., Joel D.D., Shi K.-C., Meinken G.E., Som P., Tang Y.-N., Volkow N.D., Yee C., Brill A.B., Watanabe M., Inuzuka E., Oba K., Gerson R., Iida H., Hiruma A., 1990. A High-Resolution SPECT System Based on a Microchannel Plate Imager. IEEE Trans. Nucl. Sci. **37:**687.

Duflow H., Maenhaut W., and DeReuck J., 1989. Regional Distribution of Potassium, Calcium, and Six Trace Elements in Normal Human Brain. Neurochemical Res. **14:**1099.

Fay F.S., Carrington W., and Fogarty K.E., 1989. Three-Dimensional Molecular Distribution in Single Cells Analyzed Using the Digital Imaging Microscope. J. Microscopy. **153:**133.

Fiori C.E. and Newbury D.E., 1978. Artifacts Observed in Energy Dispersive X-Ray Spectrometry in Scanning Electron Microscopy. In "Scanning Electron Microscopy", Volume I, page 401. Scanning EM Inc., AMF O'Hare, IL.

Flessner M.F., Fenstermacher J.D., Blasberg R.G., and Dedrick R.L., 1985. Peritoneal Absorption of Macromolecules Studied by Quantitative Autoradiography. Am. J. Physiol **248:** H26-H32.

Garrecht B.M., and Chapman J.D., 1983. The labelling of EMT-6 Tumours in BALB/C Mice with C-14 Misonidazole. Brit. J. Radiol **56:**745.

Geisler F.H., Jones K.W., Fowler J.S., Ktaner H.W., Wolf A.P., Kronkite E.P., Slatkin D.N., 1974. Deuterium Micromapping of Biological Samples Using the D(T,n)He-4 Reaction and Plastic Track Detectors. Science **186:**361.

Green D., Howell G.R., Thorne M.C., and Watts R.H., 1978. Imaging of Tissue Sections on Lexan by Alpha-Particles and Thermal Neutrons: An aid in fissionable radionuclide distribution studies. Int. J. App. Rad. and Isotopes **29**:285.

Griffith M.H., Yorke E.D.,Wessels B.W., DeNardo G.L., Neacy W.P., 1988. Direct Dose Confirmation of Quantitative Autoradiography with Micro-TLD Measurements for Radioimmunotherapy. J. Nucl Med. **29**:1795.

Hainfeld J.F., 1987. A Small Gold-conjugated Antibody Label: Improved Resolution for Electron Microscopy. Science **236**: 450-453.

Harris L.D., Camp J.J., Ritman E.L. and Robb R.A., 1986. Three-dimensional Display analysis of Tomographic Volume Images Utilizing a Varifocal Mirror. IEEE Trans.Med.Imaging **5**:67.

Hiraoka Y., Sedat J.W. and Agard D.A., 1987. The use of Charged Coupled Device for quantitative optical microscopy of biological structures. Science **238**:36.

Ito T., and Brill A.B., 1990. Validity of Tissue Paste Standards for Quantitative Whole-Body Autoradiography Using Short-Lived Radionuclides. Appl. Radiat. Isot. **41**:661.

Ito T., Griffin T.W., Collins J.A., and Brill A.B., 1990. Immunotoxin Penetration into Tumor Foci Assessed by Quantitative Autoradiography. In: "Proceedings of Second International Symposium on Immunotoxins, June 14-16, 1990 Lake Buena Vista, FL."

Jones K.W., Gordon B.M., Hanson A.L., Kwiatek W.M., Pounds J.G., 1988. X-ray Fluorescence with Synchrotron Radiation. Ultramicroscopy **24**:313.

Laskey, R.A., no date. Radioisotope Detection by Fluorography and Intensifying Screens. Rev. 23. Amersham Corp. Arlington Hts, IL.

Lear J.L., Jones S.C., Greenberg J.H., Fedora T.J., Reivich M., 1981. Use of I-123 and C-14 in a Double Radionucle Autoradiography Technique for Simultaneous Measurement of LCBF and LCMRgl. Stroke **12**:589.

Lear J.L., Ackermann R., Kameyama M., Carson R., Phelps M., 1984. Multiple Radionuclide Autoradiography in Evaluation of Cerebral Function. J. Cer. Blood Flow and Metab. **4**:264.

Leichner P.K., Yang N.C., Frenkel T.L., Loudenslager D.M., Hawkins W.G., Klein J.L. and Order S.E., 1988. Dosimetry and Treatment Planning for Y-90 Labeled Antiferritin in Hepatoma. Int. J. Rad. Oncol, Biol, Physics. **14**:1033.

Loats H.L., Pittenger M., Unnerstall J.R., Loyd D.G. and Tucker R.W., 1988. Biomedical Image Analysis Applications. In: "Imaging Techniques in Biology and Medicine" (C. C Swenberg and J. J. Conklin, editors),.page 1. Academic Press, New York.

Merkel D.E., Dressler L.G., McGuire W.L., 1987. Flow Cytometry, Cellular DNA Content and Prognosis in Human Malignancy. J. Clin. Oncol. **5**:1690.

Nadji M. and Morales A.R., 1984. Immunoperoxidase II: Practical Applications. Lab Medicine **15**:343.

Ploem-Zaaijer J.J., Beyer-Boon M.E., Leyte-Veldstra L. and Ploem, J., 1979. Cytofluorometric and Cytophotometric DNA Measurements of Cervical Smears Stained using a new Bi-color Method. In: "The Automation of Cancer Cytology and Cell Image Analysis" (N.J. Pressman and G.L. Wied, editors), page 225. Tutorials of Cytology, International Academy of Cytology, Chicago, IL.

Pounds J.G., Long G.J., Kwiatek W.M., Jones K.W., Gordon B.M. and Hanson A.L., 1988. The Role of High Energy Synchrotron Radiation in Biomedical Trace Element Res. In: "X-Ray Microscopy" (D. Sayre, M. Howells, J. Kirz, and H. Rarback, editors), Volume 2, page 425. Springer-Verlag, N.Y.

Robb R.A. and Barillot C., 1989. Interactive Display and Analysis of 3-D. Images. IEEE Trans. Med. Imaging **8**:217.

Rogers A.W., 1979. "Techniques of Autoradiography". Elsevier/North Holland Biomedical Press. Amsterdam.

Seckinger D., Sugarbaker E., Frankfurt O., 1989. DNA Content in Human Cancer. Arch Pathol Lab Med **113**:619.

Shapiro H.M., 1989. Flow Cytometry of DNA Content and Other Indicators of Proliferative Activity. Arch Pathol Lab Med **113**:591.

Sonoda M., Takano M., Miyahara J. and Hisatoyo K., 1983. Computed Radiography utilizing Scanning Laser Stimulated Luminescence. Radiology **148**:833.

Stewart C.C., 1989. Flow Cytometric Analysis of Oncogene Expression in Human Neoplasia. Arch Pathol Lab Med **113**: 634.

Taylor C.R., 1978. Immunoperoxidase Technique. Practical and Theoretical Aspects. Arch. Pathol. Lab. Med. **102**:113.

Taylor C.R., ed., 1986. "Immunomicroscopy: A Diagnostic Tool for the Surgical Pathologist". Chapter 1, Principles of Immunomicroscopy. 1-22. W.B. Saunders Co., Philadelphia, PA.

Ter-Pogosian M., Phelps M., Brownell G., Davis D. and Evens R., editors, 1977. "Reconstruction Tomography in Diagnostic Radiology and Nuclear Medicine". University Park Press, Baltimore, MD.

Tubbs R.R. and Bauer T.W., 1989. Automation of Immunohistology. Arch Pathol Lab Med **113**:653.

Tuttle S.E., Jewell S.D., Mojzisik C.M., Hinkle G.H., Colcher D., Schlom J. and Martin E.W., 1988. Intraoperative Radioimmunolocalization of Colorectal Carcinoma with a Hand-held Gamma Probe and Mab B72.3: Comparison of In Vivo Gamma Probe Counts with In Vitro Mab Localization Int. J. Cancer **42**:352.

Ullberg A., 1977. The Technique of Whole Body Autoradiography Cryosectioning of Large Specimens. Science Tools, LKB Inst. Journal, page 2.

Unnerstall J.R.,Kuhar M.J., Niehoff D.L. and Palacios J.M., 1981. Benzodiazepine Receptors are Coupled to a Subpopulation of Gamma-aminbutyric Acid (GABA) Receptors: Evidence from a Quantitative Autoradiographic Study. J. Pharmacol.Exp. Ther. **218**:797.

Venot A., Liehn J.-C., Lebruchec J.-F. and Roucayrol J.-C., 1986. Automated Comparison of Images. J. Nucl Med. **27**:1337.

Watson J.V., Sikora K. and Evan, G.I., 1985. A Simultaneous Flow Cytometric Assay for c-myc Oncoprotein and DNA in Nuclei from Paraffin-Embedded Material. J. Immunol. Methods. **83**:179.

Wessels B.W., Yorke E.D. and Lambiotte, M., 1990. "Multicellular Level Dosimetry and Low Dose Rate Effects. Frontiers in Nuclear Medicine. Dosimetry of Administered Radionuclides." ACNP/SNM/DOE Publication, page 295.

Yonekura Y., Brill A.B. Som P., Bennett G.W. and Fand I., 1983 a. Quantitative Autoradiography with Radiopharmaceutical. Part I. Digital Film Analysis System by Videodensitometry: Concise Communication. J. Nucl Med **24**:231.

Yonekura Y. and Brill A.B., 1983 b. Film Analysis Systems and Applications. In: "Digital Imaging in Medicine" (R. C. Reba, D. J. Goodenough, and H. F. Davidson,.editors), page 347. NATO ASI Series E. Applied Sciences No. 61. Martinus Nijhoff, The Hague.

Yonekura Y., Brill A.B., Som P., Yamamoto K., Srivastava S. and Iwai J., 1985. Regional Myocardial Substrate Uptake in Hypertensive Rats: A Quantitative Autoradiographic Measurement. Science **227**:1494.

IMAGING TECHNIQUES IN PHARMACOLOGY

Alan J. Fischman, Robert H. Rubin
and H. William Strauss

Division of Nuclear Medicine of the Department of Radiology
Massachusetts General Hospital
and the Departments of Radiology and Medicine
Harvard Medical School
Boston, MA 02114

INTRODUCTION

The development of new drugs is increasingly based on a knowledge of membrane and cellular biochemistry, rather than screening large numbers of samples of soil, or toxins. These drugs are "designed", rather than "discovered." Following identification of a likely compound in-vitro, careful studies are performed in animal models to determine the pharmacokinetics and pharmacodynamics of the drug. Once identified as safe and efficacious in animal trials, the pharmacokinetic profile of the drug is determined man. The clinical effects and side effects are then assessed in well controlled, multi-center studies involving large numbers of patients. The challenge to the clinician and pharmacologist participating in these studies has been to translate the exquisite measurements that can be made in animal models to the effective treatment of human disease, despite the lesser data base that can be gathered in man. Issues of dosing schedules, determination of tissue concentration, assessment of the pharmacodynamic effects of therapy, and the effects of therapy on normal and abnormal cell function are less easily assessed in man, because of the need for noninvasive methods.

In the past decade, advances in single photon imaging, PET and magnetic resonance imaging, and the ability to radiolabel a wide variety of compounds for *in vivo* use in humans has created a new technology for making precise physiologic and pharmacologic measurements. Due to the noninvasive nature of this approach, repetitive and/or continuous measurements have become possible. Thus far, these techniques have been primarily used for the one time assessment of a given individual. However, experience suggests that a major use of this methodology will be in the evaluation of new drug therapies. A truism of clinical pharmacology is that the cornerstone of evaluating a new drug is the quality of the measurements made to assess the patient prior to study entrance and the quality of the measurements that can be made to assess the drug effects.

The thesis we propose is that the addition of physiologic measurements with new imaging technologies when coupled with classical pharmacology will provide a powerful new tool for evaluating innovative therapies. Such an approach should be particularly useful in the early phases of drug evaluation both in animals and in humans (Phase I and II) where issues such as dosing schedules, pharmacodynamic effects, tissue delivery, comparative efficacy, and non-idiosyncratic adverse effects are of particular concern. In this review we define those areas in which imaging techniques may be particularly useful in drug evaluation and development.

PHARMACOKINETICS

Radioactive tracers have several roles in drug development and evaluation, ranging from defining the static and time dependent distribution of drug in tissue to, identification of metabolites and routes of excretion. In many cases tracer methods are preferred to chemical determinations because of their innate sensitivity — ability to detect picogram and even femtogram quantities of drug. For most drugs these measures are initially performed using the beta emitting nuclides: tritium (^{3}H) and carbon (^{14}C), because substitution of these radioactive atoms for those native to the molecule will not alter the biological behavior of the drug. Since both these radionuclides have long half-lives compared to the interval of interest for tracing drug behavior, processes requiring hours to days for completion can be readily evaluated. A disadvantage of these tracers is the inability to determine their distribution by external imaging — since the path length of the more energetic beta particles in tissue is < 3 mm. Detecting these tracers requires direct sampling of the tissues of interest. The size of various distribution "compartments", for example, can be determined using simple dilution techniques, while more detailed definition of the location of a drug can be obtained by high resolution autoradiographic imaging of tissues. By repetitive sampling of bodily fluids such as blood, urine, bile, gastric juice or expired air, the relative rate of entry and loss of the drug or its metabolites from various organs can be determined. When these measurements are combined with traditional assay techniques or performed with a drug that is labeled with both ^{3}H and ^{14}C, metabolism can be evaluated. Using conventional assays, the label should be associated with a specific chemical species, while the dual label approach assumes a constant relative concentration of labels in specimens where the drug remains unmetabolized; a decrease in the relative abundance of one tracer signifies catabolism of the drug.

Serial sampling to determine indicator dilution offers substantial information about the kinetics of tracer distribution, but little insight about where in an organ the drug is accumulated or metabolized. To obtain this information, high resolution autoradiography of either tissue specimens or sections of the whole body of the animal can be very enlightening. Figure 1 shows an autoradiograph of a longitudinal section of the body of a mouse injected with the antifungal drug fluconazole that was labeled with ^{14}C [Humphrey, Jevons and Tarbit, 1985]. Under the appropriate experimental conditions, where exposure places the film density in the linear portion of the Heulich and Dryle scale and aliquots containing known amounts of radioactivity are exposed in parallel with the sample, the grain density of the autoradiographs can be quantified to yield absolute concentrations of drug in small volumes of tissue [Yonehura, Brill, Som, et al., 1983, Som, Yonehura, Oster, et al., 1983, Morell, Tompkins, Fischman, et al., 1989, Schnitzer, Morell, Colton, et al., 1987]. To define the kinetics of the drug a series of animals would have to be examined at each point in time.

While the autoradiography is very useful, a major difficulty with the technique is the destructive preparation of the specimen for the measurement, permitting only a single time point to be assessed in each animal. Determining the temporal sequence of drug behavior requires large numbers of animals and is frequently subject to large

150

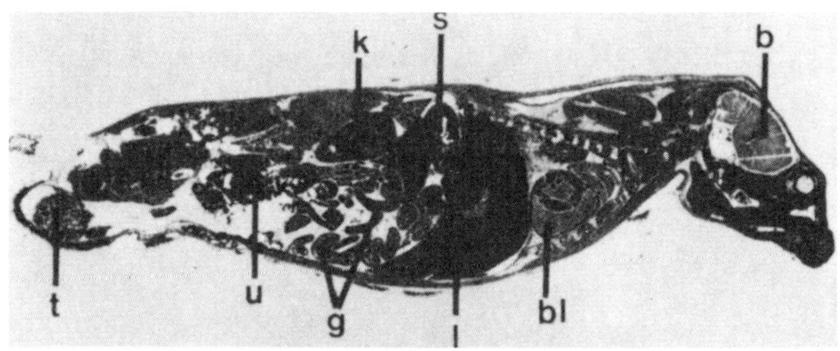

Figure 1. Autoradiograph of a longitudinal section of a mouse injected with ^{14}C-labeled fluconazole. b: brain, bl: blood, l: liver, s: spleen, k: kidney, g: GI tract, u: urinary bladder, t: testis. Reproduced from Humphrey Jevons and Tarbit, 1986.

errors, particularly if the experimental conditions under investigation are difficult to reproduce.

The temporal sequence of drug distribution can be determined in one animal with external imaging, if the drug is labeled with a gamma emitting nuclide. Serial imaging, following a single injection permits simultaneous measurements in multiple organs over an interval of time. From these observations, measures of blood clearance, organ uptake, binding, and excretion can be calculated. The imaging approach offers the advantage of following not only the natural course of drug distribution, but also the changes that are brought about by an intervention — such as a change in blood pressure, renal function or cardiac output. In some cases, it may be possible to label a drug at more than one site with gamma emitting radionuclides of different energies. Under these circumstances, it is possible to perform detailed studies of drug metabolism by dual photon imaging. Most importantly, the imaging approach is the only method of obtaining detailed information on drug distribution, kinetics and metabolism in human subjects under a variety of physiological and pathological conditions.

If the drug under investigation happens to contain iodine, fluorine, gold, cobalt or any of a large number of other atoms with radionuclides that have convenient half-lives and gamma emissions, labeling is a simple matter of substitution. The behavior of the drug can be traced in similar fashion to that with carbon-14 or tritium, except now serial images of the distribution of drug in the animal can be obtained (Table I). Serial measurements allow the animal to behave as its own control, with a reduction in variation of the measurement. If the molecule does not contain a readily exchangeable atom with a convenient radionuclide for imaging, other approaches to radiolabeling are required. When the drug to be studied has a high molecular weight ie. proteins such as growth hormone, antibodies, interleukin-2 etc.; radiolabeling can be accomplished by derivatization with a small functional group that can be labeled. Many functional groups have been described for this purpose i.e, Bolton-Hunter reagent for iodination, diethylenetriaminepentaacetic acid for indium labeling and the hydrazino nicatinamide group for technetium labeling [Bolton and Hunter, 1973, Krejcarek and Tucker, 1977, Abrams, Juweid, tenKate, et al., 1990]. In general, the choice of radionuclide for labeling is dictated by the time frame over which the drug is to be studied. If the molecular weight of the labeling group is significantly lower than the molecule to be labeled, derivatization represents a minor structural perturbation and the biological behavior of the derivative can be assumed to closely approximate that of the native drug. While this assumption is generally true, bioequivalence must always be validated. For example, if derivatization modifies a part of the drug that is critical for receptor binding, biological activity can be totally lost and biodistribution radically altered.

Table I
Some Useful Nuclides for Intrinsic Labeling of Drugs or Their Analogs for Animal Studies

Radionuclide	Major detectable Photon/Particle	Half-life (physical)	Use
Carbon-14	$\beta-$	5730 years	β counting autoradiography
Carbon-11	$\beta+$; τ (511 keV)	20 minutes	PET imaging
Nitrogen-13	$\beta+$; τ (511 keV)	10 minutes	PET imaging
Hydrogen-3	$\beta-$	12.26 years	β counting autoradiography
Iodine-125	Te X-Rays; $\approx$30 keV	60 days	Gamma imaging
Iodine-131	$\beta-$; τ 364 keV	8 days	Gamma imaging
Iodine-123	τ 159 keV	13 hours	Gamma imaging
Gold-198	τ 412 keV	2.7 days	Gamma imaging
Mercury-197	Au X-Rays $\approx$80 keV	2.7 days	Gamma imaging
Manganese-52	τ 1434 keV	5.7 days	τ counting
Iron-52	$\beta+$; τ (511 keV)	8.2 hours	PET imaging
Iron-59	τ 1094 keV	45 days	Gamma imaging
Cobalt-57	τ 122 keV	270 days	τ counting
Fluorine-18	$\beta+$; τ (511 keV)	110 minutes	PET imaging
Potassium-43	τ 374 keV	22 hours	Gamma imaging
Sodium-24	τ 1369 keV	15 hours	τ counting

When the drug to be studied has a low molecular weight, the reagents used for protein labeling cannot be employed. Under these circumstances the molecule may be labeled with the positron emitting nuclides: [11]C (20.4 minute half-life) or [13]N (10.0 minute half-life). These nuclides are useful under circumstances where a rapid synthesis of the molecule is possible, and the process(es) of interest occur rapidly. With these radionuclides, imaging can be performed using positron emission tomography (PET). Compared to conventional single photon gamma camera imaging, PET has the advantages of: higher resolution (5 mm vs 15 mm), sensitivity (~1000 fold) and the possibility for absolute quantitation. A disadvantage of ^{11}C and ^{13}N labeled drugs is the limited time frame over which biological processes can be studied. The very latest time points at which useful data can be obtained with ^{11}C and ^{13}N labeled drugs are 90 and 45 minutes respectively (about 4.5 physical half-lives; typically measurements are limited to about 3 half-lives). Under circumstances in which it is not possible design a synthesis for the rapid incorporation of ^{11}C or ^{13}N in the native structure of a drug it may be necessary to synthesize an analog of the drug for evaluation. With low molecular weight drugs, analogs that are minor perturbations of the native structure can be prepared by fluorination, methylation and amidation. While analogs are acceptable from an imaging perspective, the behavior of the analog relative to the native molecule must be understood for the measurements to provide the desired information.

In addition to radionuclide imaging, drug metabolism can also be studied *in vivo* by nuclear magnetic resonance imaging or spectroscopy. With this technique, it is possible to simultaneously evaluate the time dependence of concentrations of drug and metabolites. The most promising nuclides for these studies are ^{19}F and to a lesser extent ^{13}C. Unfortunately, the low intrinsic sensitivity of this technique ($\sim 10^6$-10^9 lower than radionuclide methods) limits applications to drugs that are administered in very high doses.

Imaging methods have been used to study the *in vivo* pharmacokinetics of a large variety of drugs: including antibiotics, antineoplastic agents, neuroleptics, opiates, angiotensin converting enzyme inhibitors, hormones, etc. The following is a summary of some representative examples.

ANTIMICROBIAL AGENTS

Successful antimicrobial therapy is predicated on the delivery of concentrations of active drug to the site of infection that exceed the level necessary to inhibit the growth or kill the invading microorganism. Although levels of drug may be measured serially in blood and urine without difficulty, other bodily fluids such as cerebro-spinal fluid and bile and most tissues are not easily accessible in man. Therefore, prior to trials in patients with active infection, dosing schedule determinations are usually based on animal tissue data and human blood and urine data. Depending on the nature of the drug in question, there is an incomplete relationship between drug levels in blood and urine and concentrations achieved in such tissues as brain, liver, prostate, and bone. Therefore, approaches in which traditional blood and urine measurements can be correlated with tissue concentrations determined by imaging techniques hold great promise for defining appropriate dosage schedules and in speeding up drug development.

Erythromycin

The macrolide antibiotic erythromycin was the first antibiotic to be labeled with ^{11}C and studied by PET. The molecule was labeled at the dimethylamino group of the D-desosamine portion of the molecule by reductive methylation of N-demethylerythromycin with ^{11}C-formaldehyde and was shown to be identical to authentic erythromycin [Pike, Palmer, Horlock, et al., 1982]. After intravenous injection in human subjects, local concentrations of the antibiotic were measured by PET. In patients with pneumonia, the time course of uptake of drug was measured in normal and pneumonic lung [Wollmer, Rhodes, Pike, et al., 1982]. During the first hour after injection, the extravascular concentration of drug was similar in normal and infected tissue (6.6 ± 2.2 µg/g vs 5.5 ± 2.2 µg/g). These concentrations were achieved within 10 min after injection and were maintained throughout the 60 minutes of observation.

This study was successful, in part, because the rapid equilibration of erythromycin in tissues made the 20 minute physical half-life of ^{11}C appropriate for labeling — i.e. the label and the process were well matched. Unfortunately, many antibiotics require hours to days to equilibrate in tissue, thereby requiring radionuclides with longer physical half-lives to permit sufficient time for observation.

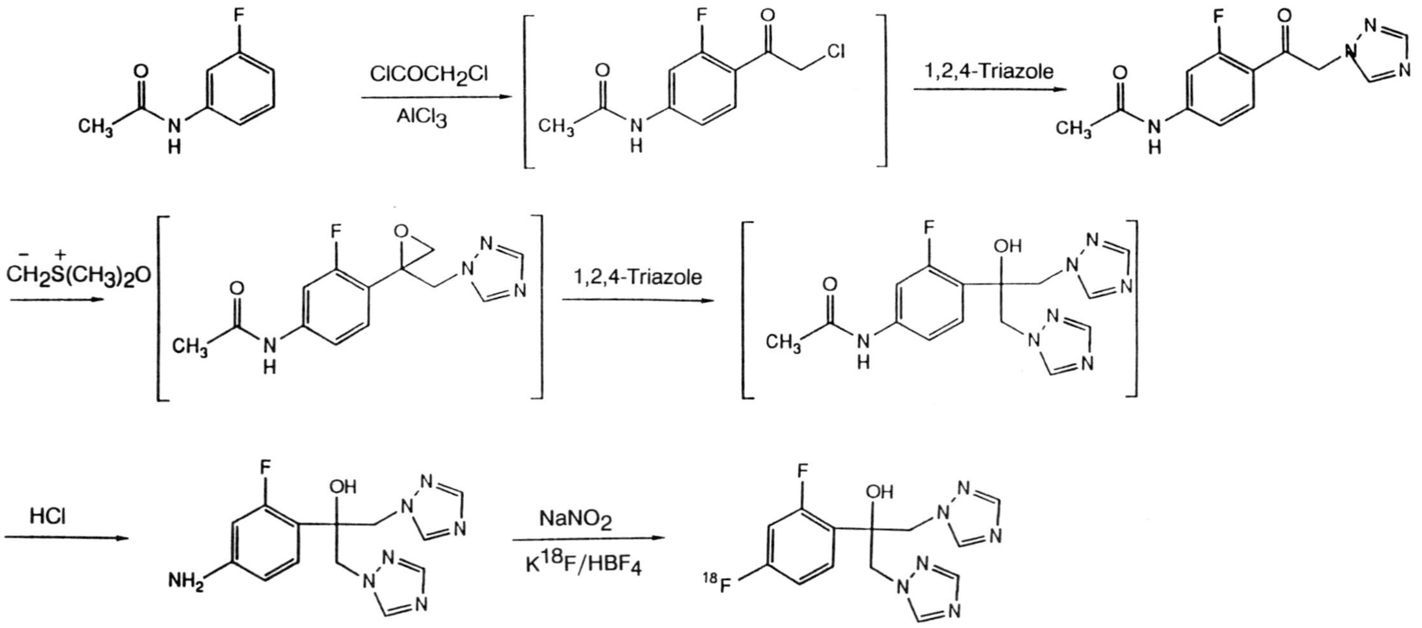

Figure 2. Synthetic scheme for the preparation of [4-18F] fluconazole.

Figure 3. PET image (coronal reconstruction) of the lower extremities of a rabbit, 2 hours after injection of 1.0 mCi of [4-^{18}F] fluconazole. The right thigh was infected with *candida albicans* 24 hours earlier.

Fluconazole

Several antimicrobial agents contain fluorine atoms in their native structure and can be studied over a longer time interval by substitution of the native ^{19}F label with radioactive ^{18}F. The antifungal agent fluconazole is an example of this class of drugs. Recently, we developed method for preparing [4-^{18}F] fluconazole (Figure 2).

The product was identical chemically and microbiologically to the authentic drug. This reagent was used to measure the pharmacokinetics of fluconazole in rabbits by PET. When ^{18}F fluconazole was co-injected with a pharmacological dose of unlabeled drug there was rapid equilibration to a relatively uniform distribution of radioactivity in most organs of the rabbit. When ^{18}F fluconazole was injected alone ("carrier-free"), blood clearance was accelerated, spleen, muscle and heart accumulation was decreased and liver accumulation was increased. Since the "pharmacokinetics" are under investigation, rather than the behavior of the carrier free compound, it is important to administer the drug in amounts known to be pharmacologically active, to avoid misleading results. Quantitative analysis of the images demonstrated that concentrations of ≥ 15 µg/g tissue were present at 2 hours after administering a dose of 5 mg/kg. Based on *in vitro* and animal model experiences, one would predict that this level of drug would be necessary for a therapeutic effect, and that lower doses would be more problematic. It is of interest that clinical experience with different dosage schedules have come to similar conclusions. Even more reassuring is the observation that higher concentrations of drug are achieved in sites of infection compared to normal tissue (Figure 3).

These studies illustrate the power of a noninvasive technique (that is applicable to man) that will permit the determination of the effect of infection, inflammation, congestive heart failure, renal failure, metabolic factors, etc., on the delivery of effective concentrations of an antimicrobial agent to clinically important sites of infection.

ANTINEOPLASTIC AGENTS

Antineoplastic drug therapy is a highly complex undertaking in which potentially toxic compounds are administered to severely ill patients in hope of achieving objective regression of the cancer and increased survival. The newer radionuclide techniques hold great promise for expediting drug evaluation *in vivo* in at least two ways: 1) If the candidate drug can be radiolabeled with a radionuclide suitable for PET imaging, uptake of the drug by the tumor (a necessary step for therapeutic benefit) can be demonstrated. Also, concentrations of drug in tissues such as liver, kidney, and bone marrow can be determined, with important implications for potential toxicity. 2) When the candidate drug cannot be radiolabeled, another approach should be possible. PET scanning, using reagents such as ^{18}FDG (a glucose analog), ^{11}C-methionine (the radiolabeled form of the amino acid), ^{15}O$_2$ (the radiolabeled form of diatomic oxygen), etc. can provide a sensitive measure of the metabolic activity of both normal and abnormal tissue. It has been postulated that long before changes in the size of a tumor mass could be appreciated, metabolic changes could be demonstrated, connoting drug effect. Similarly, metabolic changes in normal tissues could presage significant toxicity.

Platinum compounds

The platinum complexes, cis-dichlorodiammine platinum (II), cis-trans-dichlorodihydroxy-bis-isopropylamine platinum (IV), cis-dichloro-bis-cyclopropylamine platinum (II) and cis-diamino 1,1-cyclobutanedicarboxylate platinum (II) have been radiolabeled with platinum-191 and serial in vivo images acquired in patients with malignant disease [Owens, Thatcher, Sharma, et al., 1985]. The pharmacokinetics of uptake and clearance in liver and kidney were shown to be similar for all 4 drugs. Unfortunately unequivocal evidence of tumor accumulation was not found for any of the agents.

5-Fluoro Uracil (5FU)

In a recent report, patients with colorectal tumors were injected with ^{18}F-labeled 5FU and radioactivity in normal liver and metastasis was monitored for two hours by PET [Port, Strauss and Clorius, 1989]. A kinetic model was developed to calculate the concentration of 5FU and its metabolites. It is possible that in the future, this model can be used to help predict therapeutic response and permit evaluation of the kinetic effects of different methods of drug delivery.

NEUROLEPTICS

Cognitive function is the area in which humans are most strikingly different from lower organisms. There are few reliable animal models of human psychiatric disease. Thus the development of effective neuroleptic agents is in many ways the most complex area of modern drug development. Not only must a drug cross the blood-brain barrier in sufficient quantity to be pharmacologically effective but its regional cerebral distribution must coincide with the distribution of receptors to which it is directed. The observation that binding to neurotransmitter receptors can be used as a predictor of clinical potency of neuroleptic drugs of diverse chemical structure add a rational basis to drug development [Creese, Burt and Snyder, 1976]. External imaging with radiolabeled drugs is the only potentially reliable method of studying these processes in humans and could become an important tool in early phase I drug development.

Studies with the drug raclopride can be considered to be prototypes for methods that can be applied to studying neuroleptic drugs in humans [Farde, Ericksson, Blomquist, et al., 1989, Farde, Wiesel, Stone-Elander, et al., 1990, Farde, Wiesel, Jannson, et al., 1988]. The rate of uptake and equilibration of this drug in the CNS is sufficiently rapid to use ^{11}C as a tracer. Also, as with most drugs of this class of compounds the presence of substitutable alkyl groups makes radiolabeling with ^{11}C relatively straight forward. Not only has it been possible to measure the regional CNS distribution of drug by PET, but by *in vivo* saturation analysis D_2-dopamine receptor density (B_{max}) and affinity (K_d) could be determined. These methods could be extremely useful in screening new neuroleptic drugs. By replacing unlabeled raclopride by a candidate drug, interaction with D_2 receptors could be evaluated. Also, to define possible side effects, the receptor specificity of new drugs could be determined by using other radiolabeled ligands that interact with multiple receptor systems.

ANGIOTENSIN CONVERTING ENZYME INHIBITORS (ACEI)

ACEI's have become important drugs for the treatment of hypertension and congestive heart failure. With the synthesis and *in vitro* testing of numerous new analogs, it has become apparent that different tissues have different ACE's that are more or less affected by different ACEI's. A question that remains to be answered is whether these *in vitro* differences are clinically important; for example, currently available ACEI's cannot be safely used in patients with renovascular hypertension because of their effects on afferent arteriolar pressures. Newer ACEI's are more renal sparing (in terms of their effects on renal ACE) and may be safer in this circumstance. The ability to study these effects in man *in vivo*, particularly combining measurements of drug-enzyme interaction with measurements of the hemodynamic effects of the drug (see below) should greatly facilitate the evaluation of new compounds.

In a recent report, a ^{18}F-labeled analog of captopril was synthesized and its biodistribution was measured in rats [Hwang, Mathias, Welch, et al., 1990]. At 30 minutes after injection there was high uptake in organs known to have high concentrations of ACE: lung, kidney and aorta. Faster clearance was observed for lung and kidney compared to aorta. When the radiolabeled drug was co-administered with unlabeled captopril, uptake in lung and kidney was blocked. These observations establish the feasibility of studying the phamacokinetics of ACEI's *in vivo* using PET.

NUCLEAR MAGNETIC RESONANCE

Applications of NMR to the study of *in vivo* pharmacokinetics are slowly beginning to emerge. Due to the low intrinsic sensitivity of the method it is unlikely to be generally applicable to drug studies. However, in those particular situations in which it can be applied, unique information about the details of drug metabolism can be obtained.

^{19}F-NMRS has been used to study the pharmacokinetics of 5-FU in tumors in both animal models and human subjects [Stevens, Morris, Iles, et al., 1984, Wolf, Albright, Silver, et al., 1987, Wolf, Presant, Servis, et al., 1990]. The results of these studies indicate that free 5-FU is trapped in some tumors after IV administration (600 mg/m^2) and there is a correlation between uptake and response to therapy. While only a limited number of patients have been studied, the technique has great promise. As stated in one of the articles "The key observation is that it is now possible to measure the time course of a therapeutic agent (as an unequivocal chemical species) in its intended target site in humans" [Wolf, Presant, Servis, et al., 1990].

In another study, one dimensional ^{19}F chemical shift imaging of the neuroleptic, fluphenazine, was used to determine the distribution of drug and metabolites in rat brain *in vivo* [Nakada and Kwee, 1989]. The results showed the

Table II
Common Tracer Procedures and Their Application to Pharmacokinetics

Procedure	Application	Radiopharmaceutical	Nature	
			Single	Continuous
Gated scan	Cardiac function	^{99m}Tc red cell		√
Perfusion	Tissue Flow	^{201}Tl, ^{133}Xe, ^{82}Rb	√	
DTPA Scan	Renal Function	^{99m}Tc DTPA		√
IgG Scan	Inflammation	^{111}In IgG	√	
C^{15}O$_2$ Inhalation	Perfusion Ventilation	C^{15}O$_2$	√	
MAA Perfusion	R/L Shunt Lung Perfusion	^{99m}Tc MAA	√	
Ventilation	Gas Exchange	^{133}Xe	√	
FDG Scan	Glucose Metab.	^{18}FDG	√	
Bone Scan	New Bone Formation	^{99m}Tc MDP	√	
Bone Density	Mineral Mass	None	√	
Res. Scan	Liver/Spleen Function	^{99m}Tc Colloid	√	
Vest	Cardiac Function	^{99m}Tc Red Cells		√
Renal Monitor	Renal Function	^{99m}Tc DTPA		√
Gastric Emptying	GI Transit Time	^{99m}Tc Food		√
Biliary Scan	Bile Production	^{99m}Tc DISIDA	√	

expected distribution and verified that the primary fluorinated species in brain is intact drug. Also, a unidentified fluorinated metabolite was detected in subcutaneous tissue.

One therapeutic area in which NMR methods could be particularly useful is in the study of the fluoroquinolones, perhaps the most rapidly growing class of antimicrobial agents. Not only are such issues as therapeutic tissue concentrations important with these agents, but also the possible relationship between tissue concentrations and toxicity is of critical importance. In particular, neurological side effects such as seizures have been of major concern, especially in the elderly, patients receiving theophylline and those with pre-existing epileptogenic foci. The presence of fluorine in the native structure of these drugs makes NMR an ideal technique for studying the uptake and metabolism of fluoroquinolone analogs in the brain. *In vivo* NMR data could greatly facilitate the development of new compounds that are less toxic than currently available members of this class of drugs.

Clearly, the combined use of NMRS and PET could have profound implications for studying drug pharmacokinetics and metabolism. Due to its high sensitivity and quantitation, PET can provide detailed kinetic data on total radiolabeled drug plus metabolites while, the molecular resolution of NMRS can yield information on the distribution of label in specific metabolites.

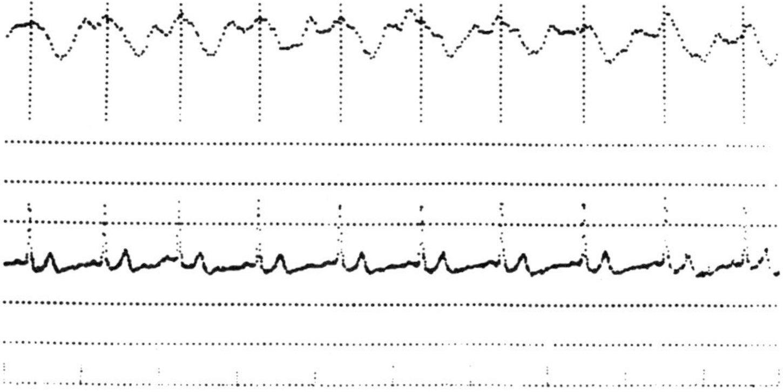

Figure 4. Vest recording of the Beat to beat left ventricular time/activity curve from a patient with sinus rhythm (upper curve). The lower trace is the patients EKG. Each marker at the bottom represents one second.

PHARMACODYNAMICS

The influence of a drug on organ function can be readily determined by radionuclide imaging techniques. Table II lists some measurements of organ function that can be performed with existing equipment and procedures following either intravenous administration or inhalation of the tracer material.

Reviewing specific procedures and their application in the evaluation of a hypothetical drug may help clarify specific applications.

CARDIOVASCULAR

Ventricular Function

Changes in cardiac performance could be assessed by measuring the size, shape and contractile function of the heart before and after administration of the drug using the gated blood pool scan. To record this information the patient's red cells or plasma proteins are labeled. This is usually done with ^{99m}Tc, but other nuclides such as ionic ^{113}In or ^{68}Ga (which bind to the beta globulin transferrin), or ^{11}C labeled carbon monoxide (which binds to hemoglobin) can be used. Following blood pool labeling, data are recorded by "gating" the scintillation camera/computer to record data in synchrony with the cardiac cycle, using the R wave of the electrocardiogram as the physiological trigger. Data are usually recorded over an interval of about 8 to 10 minutes/view (about 6-800 heart beats) to provide a series of images with sufficient resolution to permit analysis of global and regional cardiac function. The strength of this technique is derived from the linear relationship between counts recorded over the cardiac chambers and the volume of blood in the chambers. While absolute calibration can be achieved within 25 ml of the chamber volume, relative changes can be determined with far greater precision. Analyzing serial acquisitions recorded after a single injection permits monitoring of relative left and right ventricular volumes, cardiac output and ejection fraction. These measurements can be made with the subject at rest or during graded exercise to determine the reserve function of the heart. In addition to global function, the images can be assessed for regional wall motion. This information can be of great help in determining the impact

Figure 5. Apparatus for simultaneously monitoring of central and peripheral
blood volume, left ventricular function and pulmonary function.

of a drug on zones of the myocardium prone to ischemia or subject to contractile
dysfunction secondary to myopathy.

Blood pool imaging averages cardiac function over a number of cardiac cycles.
At times, marked changes in function may occur over a few beats, as occurs in
arrhythmia, or with the sudden onset of severe ischemia or dynamic outflow
obstruction. To define these changes in function, data must be recorded on a beat by
beat basis. This can be accomplished with a single radionuclide detector placed over
the left ventricular blood pool. One such instrument, the VEST, consists of a single
radionuclide detector attached to a modified Holter recorder, to continuously measure
the time activity curve of the left ventricle [Wilson, Sullivan, Moore, et al., 1983].
As a result, a beat by beat measure of cardiac performance is recorded in concert with
the patients EKG (Figure 4). The combined measurement is particularly useful to
characterize episodes of ischemia. Prior work has demonstrated a striking incidence
of painless episodes of decreased ejection fraction in patients following acute
myocardial infarction [Kayden, Wackers and Zaret, 1990]. Subjects demonstrating
this behavior had a high incidence of severe recurrent cardiac events in the year
following recovery from their initial infarct. Defining the influence of a specific
therapy, such as a calcium channel blocker or beta blocking drug on these episodes
would be helpful in establishing the drug's efficacy.

A major thrust of cardiovascular drug development in recent years has been the
development of long acting preparations of drugs that have long been in clinical use,
but have required multiple dose regimens. Single long acting preparations of calcium
channel blockers, anti-arrhythmic agents, etc. are coming into clinical use. The
VEST technology provides an ideal way of proving the comparability of long acting
drugs to traditional agents. While the patient is maintained under the conventional
treatment regimen, cardiovascular function could be monitored over a 24 hour period
both at rest and during standardized exercise testing. Comparability of the new
formulation could then be proven by monitoring the patient under the exact
circumstances a few days later. Thus, functional and pharmacodynamic
comparability of the two regimens could be demonstrated. Since each patient serves
as his own control, useful data could be obtained from studying a limited number of
subjects and evaluation of new candidate compounds could be greatly facilitated.

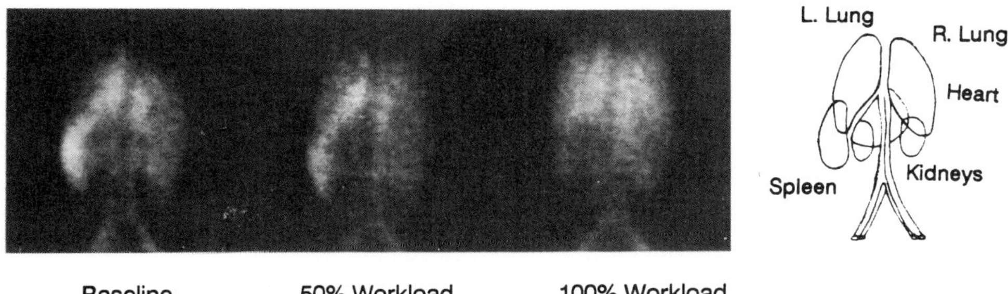

Figure 6. Representative images showing the effect of graded exercise on the blood volume of the abdominal organs.

Blood Volume

The influence of drugs on arterial and/or venous tone can be monitored by measuring the regional changes in blood volume in the lower extremities, abdomen and lungs during rest and physical exertion in a control state and after administration of a drug. In an overall fashion, redistribution of blood into or out of the central circulation can have marked effects on exercise performance. The sequence of changes in blood volume with graduated physical exertion can define the mechanisms employed to increase central volume to maintain cardiac performance with increased demand. These measurements can be made using a large field of view camera to monitor the thoracic and abdominal blood volume and a camera collimated with a pinhole set about 2 meters away, to image the whole body. When this information is combined with data from the VEST and or pulmonary function measures, the influence of the drug on the physiological redistribution of blood volume is well characterized. Changes in pulmonary blood volume are useful indicators of preload, for example, while reduction in splenic blood volume correlates with changes in peripheral hematocrit, white blood cell count, and platelet count.

Recently the feasibility of making these measurements in normal subjects was validated using the experimental apparatus shown in Figure 5 [Flamm, Taki, Moore, et al., 1990]. After injection of ^{99m}Tc-labeled autologous blood cells, measurements were made at rest, zero-load cycling and at 50, 75 and 100 percent of maximum oxygen uptake. In going from rest to zero-load cycling, leg blood volume decreased (32±2%), end-diastolic volume increased (9.6±1.2%) and lung blood volume increased (18±2%). With increasing workload, leg blood volume stabilized, blood volume in the abdominal organs decreased (spleen: 46±2%, kidney: 24±4%, liver: 18±4%) and lung blood volume increased (50±4). Representative images demonstrating the effect of graded exercise on the blood volume of the abdominal organs are shown in Figure 6.

Clearly, these types of measurements (with or without exercise) would be useful in evaluating the effects of a variety of drugs, including: calcium channel blockers, angiotensin converting enzyme inhibitors, ionotropic agents, rheological agents, etc.

Cardiac Output

The distribution of cardiac output may be altered by drugs. Since a number of radiopharamceuticals distribute in proportion to cardiac output, whole body images of these tracers provide a picture of the relative partitioning of perfusion to the

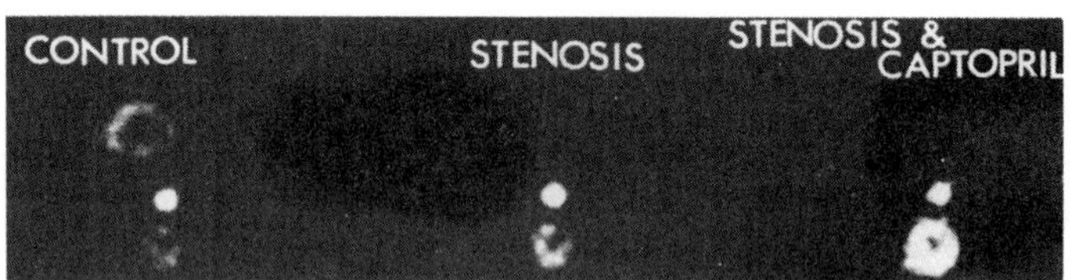

Figure 7. Representative PET images of both kidneys of a dog during
control period, stenosis (upper kidney in figure) and stenosis plus captopril.
Reproduced from Tamaki, Alpert, Rabito, et al., 1988.

muscles of the extremities, bowel, liver, kidneys, and myocardium. Previous work
with thallium-201 demonstrated the changes in cardiac output distribution in patients
with mitral and aortic valve disease [Svensson, Lomsky, Olsson, et al., 1982].
Similar studies can be performed before and after drug administration in both healthy
volunteers and patients with significant heart failure to demonstrate the ability to
effectively mobilize cardiac output after drug administration.

RENAL PERFUSION

A decrease in arterial pressure can reduce renal perfusion, especially in patients
with pre-existing renal artery stenosis. Renal perfusion can be measured on a relative
basis using a number of different radiopharmaceuticals including potassium-43 and
thallium-201 and imaging with a conventional gamma camera, or with more precise
quantitation using rubidium-82 and positron tomography. Recently a report from our
laboratory, demonstrated the use of PET with [82]Rb for measuring the effect of
captopril on renal blood flow in dogs with experimental renal artery stenosis
[Tamaki, Alpert, Rabito, et al., 1988]. In this study, the sequence and magnitude of
acute changes in renal blood flow following injection of captopril were determined
by PET. Data were recorded in nine dogs under baseline conditions, during renal
artery stenosis and during stenosis when captopril was administered intravenously.
In the affected kidney, stenosis resulted in a ~75% decrease in blood flow which was
not significantly affected by administration of captopril. In five dogs, stenosis was
associated with reduced flow to the contralateral kidney, however, following
administration of captopril contralateral perfusion increased and exceeded baseline
flow (Figure 7). In three dogs, stenosis with or without captopril did not have a
significant effect on blood flow. In one dog, contralateral flow was reduced during
stenosis and further reduced after captopril infusion.

Clearly, similar studies can be performed in humans and could be extremely
important in evaluating new angiotensin converting enzyme inhibitors. In fact, due to
the larger size of the human kidney it might be possible to quantify regional renal
blood flow.

In many clinical situations, the evaluation of renal perfusion may not be as
important as the measurement of function. Global and regional renal function can be
determined by imaging the kidneys following an intravenous dose of a chelate
radiopharamceutical, ^{99m}Tc DTPA, an agent that is excreted by glomerular filtration.
When more precise regional quantitation is required, PET with ^{68}Ga DTPA can be
used. To determine overall renal function, a device similar to the VEST, but modified
to measure clearance from the extracellular fluid space of the arm, can be used to
monitor glomerular filtration rate over several hours. These measurements are
particularly helpful in defining the role of specific interventions on renal function in
extremely ill, hypotensive, patients who are being treated in intensive care units.

HEPATOBILIARY AGENTS

Hepatic parenchymal cell function can be assessed by the rate of uptake and clearance of ^{99m}Tc labeled iminoacetic acid derivatives. These agents are concentrated in the hepatic parenchyma and excreted via the biliary tree. Prolongation of hepatic transit time can serve as an indicator of liver injury as a result of drug administration.

ANTISPASMOTICS

Drugs useful for treating bowel dysmotility can be readily evaluated with a variety of tracer labeled meals. Transit through the esophagus, stomach, small bowel and large bowel can be assessed separately for the liquid, solid, and fatty portions of a test meal.

An additional application of imaging technology is evaluating the site of dissolution of orally administered capsules or tablets in the GI tract. Labeling the drug and recording sequential images following its administration (with or without an accompanying test meal) allows the site of absorption to be identified by observing activity outside the GI tract while noting the location of the residual ingested activity in the gut.

ANTIMICROBIAL AGENTS

Antibiotic preparations should reduce the degree of inflammation at a site of infection as the lesion is treated effectively. The use of ^{111}In IgG for imaging focal sites of infection has been extensively validated in both animals and man [Rubin, Fischman, Nedelman, et al., 1989, Fischman, Rubin, Khaw, et al., 1988, Rubin, Fischman, Callahan, et al., 1989]. Recently, we demonstrated that the rate of regression of inflammation can be monitored using serial injections of radiolabeled IgG [Wilkinson, Fischman, Rubin, et al., 1989]. This technique can be employed to assess the relative efficacy of a therapy for the treatment of focal infection. Following infection of a group of animals, for example, one untreated group would define the natural history of the lesion, while another group treated with the clinically favored therapy would define the expected rate of improvement, while a third group treated with the new drug would complete the comparison. Objective evaluation of the relative tracer intensity at the site of the lesion by a region of interest analysis permits assessment of the change in the degree of inflammation as a function of the therapy.

Clinically, inflammation scanning can be very useful in defining the response to therapy. For example, serial ^{111}In-IgG scans have shown that when therapy is incomplete, inflammation will persist, and the patient will promptly relapse if antibiotics are discontinued at this point, even if such traditional clinical markers as fever and peripheral white blood cell count have returned to normal. Such "proof of cure" studies are particularly useful in evaluating the efficacy of therapy in such difficult to treat infection as osteomyelitis, prosthetic joint infections, post-operative infections, and intravascular infections.

NUCLEAR MAGNETIC RESONANCE

In recent years, NMR techniques have begun to be applied to the study of pharmacodynamics. These applications can be divided into two categories: proton imaging and spectroscopy and tracer studies (^{19}F or ^{13}C).

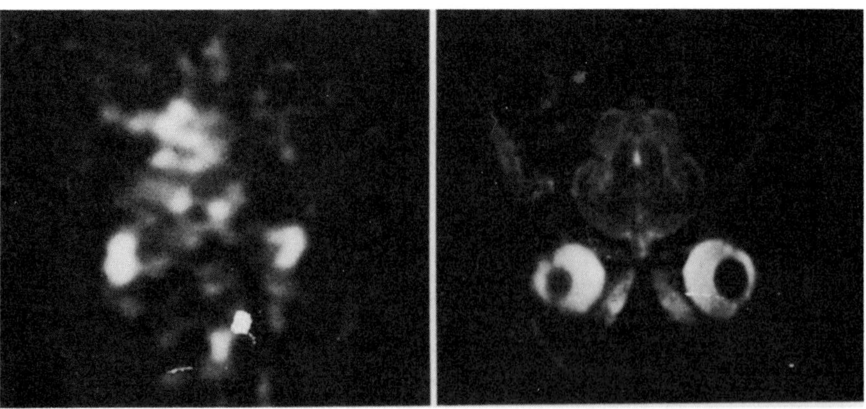

Figure 8. Image of 3-FD-sorbitol distribution in the head of a rabbit (left).
The corresponding proton image is shown on the right. Reproduced from
Kwee, Nakada and Card, 1987.

When the effect of a drug can be monitored by proton imaging or spectroscopic techniques, the problem of low intrinsic sensitivity of NMR is eliminated. Under these circumstances, the exquisite spatial resolution of conventional MRI and the ability to image individual molecular species by chemical shift imaging can provide *in vivo* data that is not available by any other method. An interesting application of MRI to drug evaluation is studying the effect of aldose reductase inhibitors on the water content of peripheral nerves in diabetics. In a recent study, the water content of the sural nerve was compared in a group of patients (n=39) with the following characteristics: symptomatic diabetics with sensory neuropathy, symptomatic diabetics with sensory neuropathy treated with an aldose reductase inhibitor, neurologically asymptomatic diabetics, and normal controls [Griffey, Eaton, Sibbitt, et al., 1988]. Significant increase in water content was detected in more than half the symptomatic diabetics. In 2 of 11 neurologically asymptomatic diabetics increased hydration was detected, suggesting presymptomatic alteration. All symptomatic diabetics treated with aldose reductase inhibitor had normal nerve water content. These results suggest that MRI may be a useful technique for detecting early complications of diabetes and for monitoring response to drug therapy. Clearly, the technique can be extended to monitor any drug effect on water content. With chemical shift imaging other substances can be monitored.

The use of NMR tracer techniques to monitor response to drug treatment is more limited, but can provide unique and useful information. An interesting application of NMR is the use of ^{19}F NMRS to monitor the effect of the aldose reductase inhibitor, sorbinil, on *in vivo* metabolism of 3-fluoro-3-deoxy-D-glucose (3-FDG) in rat brain [Kwee, Nakada and Card, 1987]. Following intravenous infusion of 3-FDG, four resonances were detected and assigned to: the α and β anomers of 3-FDG, 3-fluoro-3-deoxy-D-sorbitol and 3-fluoro-3-deoxy-D-fructose. After oral administration of sorbinil, there was reduced flux of 3-FDG into the aldose reductase sorbitol pathway. The use of 3-FDG to evaluate the aldose reductase sorbitol pathway has been extended to imaging studies. In a recent report, selective excitation ^{19}F MRI was used to obtain individual images of 3-FDG and its metabolites in rabbit head (Figure 8) [Nakada, Kwee, Griffey, et al., 1988]. Images of 3-fluoro-3-deoxy-D-sorbitol showed the distribution of aldose reductase activity. Since 3-FDG has extremely low toxicity, it is possible that in the future similar studies will be performed in humans [Halama, Gatley, DeGrado, et al., 1984].

SUMMARY

The noninvasive physiologic, biochemical, and anatomic measurements that can be performed in humans with modern imaging techniques offer great promise for defining the exact state of a patients disease and its response to therapy. Although the primary application of this technology thus far has been for diagnostic purposes, it is already clear that these methods will greatly facilitate the evaluation of new therapies. In general, there are two critical points in drug development where such innovative techniques are likely to be most useful: 1. In pre-clinical studies, where a new candidate drug can be precisely compared to standard therapies or where one compound from a series of analogs is being chosen for further development on the basis of performance in an appropriate animal model; 2. In Phase I-II human studies, where classical pharmacokinetic measurements can be coupled with imaging measurements to define the proper dosing schedule, the potential utility of interventions in particular clinical situations, and to formulate the design of Phase III studies that are crucial for drug licensure. In general, the types of measurements that should be possible can be grouped into the following categories:

1. In those situations in which the drug can be radiolabeled, the time course of tissue delivery can be determined noninvasively *in vivo* in health and disease. Such information should be most useful for determining dosing schedules, establishing efficacy and predicting possible toxicity.
2. Ligand-receptor binding can be assessed *in vivo* in two ways. The ability of the drug to displace standard radiolabeled ligands from their receptors can be determined; alternatively, labeled drug can be used to more directly assess the distribution and time course of binding. Such measurements will be particularly useful in the study of compounds active in the central nervous system and the cardiovascular system.
3. Measurements of tissue metabolism will be useful in determining the effects of therapies aimed at particular metabolic abnormalities (e.g. aldose reductase inhibition at particular anatomic sites). In addition, such measurements may be useful in defining the viability and function of tissues in such widely disparate clinical situations as cancer chemotherapy and cardiology.
4. Finally, the effect of cardiovascularly active agents can be exquisitely measured with the new techniques for assessing cardiac function, tissue perfusion, and blood volume.

We would suggest that the joining of classical clinical pharmacology to exquisite imaging measurements will help form the basis for 21st century clinical drug development.

REFERENCES

Abrams MJ, Juweid M, tenKate CI, Schwartz DA, Hauser MM, Gaul FE, Fuccello AJ, Rubin RH, Strauss HW, Fischman AJ., 1990. ^{99m}Tc human polyclonal IgG radiolabeled via the hydrazino nicatinamide derivative for imaging focal sites of infection in rats. J Nucl Med **31**:2022.
Bolton AE, Hunter WM., 1973. The labeling to high specific radioactivities by conjugation to a ^{125}I-containing alkylating agent. Biochem. J. **133**:529.
Creese I, Burt DR, Snyder SH., 1976. Dopamine receptor binding predicts clinical and pharmacological potencies of anti-schizophrenic drugs. Science **192**:481.

Farde L, Ericksson L, Blomquist G, Halldin C., 1989. Kinetic analysis of central [^{11}C] raclopride binding to D$_2$-Dopamine receptors studied by PET — a comparison to the equilibrium analysis. J Cereb Blood Flow Metab **9**:696.

Farde L, Wiesel FA, Jannson P, Uppfeldt G, Wahlen A, Sedvall G., 1988. An open label trial of raclopride in acute schizophrenia. Confirmation of D$_2$-dopamine receptor occupancy by PET. Psychopharmacology **94**:1.

Farde L, Wiesel FA, Stone-Elander S, Halldin C, Nordstrom AL, Hall H, Sedvalle G., 1990. D$_2$-dopamine receptors in neuroleptic-naive schizophrenic patients. Arch Gen Psych **47**:213.

Fischman AJ, Rubin RH, Khaw BA, Callahan RJ, Wilkinson R Keech F, Dragotakes S, Kramer P, LaMuraglia GM, Lind S, Strauss HW., 1988. Detection of acute inflammation with ^{111}In-labeled non-specific polyclonal IgG. Semin Nucl Med **18**:335.

Flamm SD, Taki J, Moore R, Lewis SF. Keech F, Maltais F, Ahmad M, Callahan R, Dragotakes S, Alpert N, Strauss HW., 1990. Redistribution of regional organ blood volume and effect on cardiac fubction in relation to upright exercise intensity in healthy human subjects. Circulation **81**:1550.

Griffey RH, Eaton RP, Sibbitt RR, Sibbitt WL, Bicknell JM., 1988. Diabetic neuropathy. Structural analysis of nerve hydration by magnetic resonance spectroscopy. JAMA **260**:2872.

Halama JR, Gatley SJ, DeGrado TR, Bernstein DR, Ng CK, Holden JE., 1984. Validation of 3-deoxy-3-fluoro-D-glucose as a glucose transport analog in rat heart. Amer J Physiol **247**:H754.

Humphrey MJ, Jevons S, Tarbit MH., 1985. Pharmacokinetic evaluation of UK-49,858, a metabolically stable triazole antifungal drug, in animals and humans. öAntimicrob Agents Chemother **28**:648.

Hwang DR, Mathias CJ, Welch MJ, Lloyd J, Petrillo EW, Eckelman WC., 1990. Synthesis and biodistribution of [18-F]-labeled angiotensin converting enzyme inhibitor. [18-F] Fluoro captopril. J Nucl Med **31**:P738.

Kayden DS, Wackers FJ, Zaret BJ., 1990. Silent left ventricular dysfunction during routine activity after thrombolytic therapy for acute myocardial infarction. J Am Coll Cardiol **15**:1500.

Krejcarek GE, Tucker KL., 1977. Covalent attachment of chelating groups to ömacromolecules. Biochem Biophys Res Comm **77**:581.

Kwee IL, Nakada T, Card PJ., 1987. Noninvasive demonstration of in vivo 3-fluoro-3-deoxy-D-glucose metabolism in rat brain by ^{19}F nuclear magnetic resonance spectroscopy: suitable probe for monitoring cerebral aldose reductase activities. J Neurochem **49**:428.

Morrell EM, Tompkins RG, Fischman AJ, Strauss HW, Rubin RH, Wilkinson RA, Yarmush MY., 1989. An autoradiographic method for quantitation of radiolabeled proteins in tissue using Indium-111. J Nucl Med **30**:1538.

Nakada T, Kwee IL., 1989. One-dimensional chemical shift imaging of fluorinated neuroleptics in rat brain in vivo by ^{19}F NMR rotating frame zeumatography. Magn Reson Imaging **7**:543.

Nakada T, Kwee IL, Griffey BV, Griffey RH., 1988. F-19 MR imaging of glucose metabolism in the rabbit. Radiology **168**:823.

Owens SE, Thatcher N, Sharma H, Adam N, Harrison R, Smith A, Zaki A, Baer JC, McAuliffe CA, Crowther D., 1985. In vivo distribution studies of radioactively labelled platinum complexes; cis-dichlorodiammine platinum(II), cis-trans-dichlorodihydroxy-bis-(isopropylamine platinum(IV), cis-dichloro-bis-cyclopropylamine platinum(II), and cis-diamino 1,1-cyclobutanedicarboxylate platinum (II). Cancer Chemother Pharmacol **14**:253.

Pike VW, Palmer AJ, Horlock, Perun TJ, Freiberg LA, Dunnigan DA, Liss RH., 1982. Preparation of carbon-11 labeled antibiotic-erythromycin lactobionate. J Chem Soc Chem Comm , page 173.

Port RE, Strauss LG, Clorius JH., 1989. Positron emission tomography after brief infusion of 5-[^{18}F] uracil: linear model for the kinetics of ^{18}F radioactivity in tumors. Onnkologia **12**:51.

Rubin RH, Fischman AJ, Callahan RJ, Khaw BA, Keech F, Ahmad M, Wilkinson RA, Strauss HW., 1989. The utility of ^{111}In-labeled nonspecific immunoglobulin scanning in the detection of focal inflammation. N Engl J Med **321**:935.

Rubin RH, Fischman AJ, Nedelman M, Wilkinson R, Callahan RJ, Khaw BA, Hansen WP, Kramer P, Strauss HW., 1989. The use of radiolabeled, non-specific polyclonal human immunoglobulin in the detection of focal inflammation by scintigraphy: comparison with gallium-67 citrate and technetium-99m labeled albumin. J Nucl Med **30**:385.

Schnitzer JJ, Morell EM, Colton CK, Smith KA, Stemerman MB., 1987. Absolute quantitative autoradiography of low concentrations of ^{125}I-labeled proteins in arterial tissue. J Histochem Cytochem **35**:1439.

Som P, Yonekura Y, Oster ZH, Meyer MA, Pelletteri ML, Fowler JS, MacGregor RR, Russell JA, Wolf AP, Fand I, McNally, Brill AB., 1983. Quantitative autoradiography with radiopharmaceuticals, Part 2: Application in radiopharmaceutical research: concise communication. J Nucl Med **24**:238.

Stevens AN, Morris PG, Iles RA, Sheldon PW, Martino R., 1984. 5-Fluorouracil metabolism monitored by *in vivo* ^{19}F NMR. Br J Cancer **50**:113.

Svensson SE, Lomsky M, Olsson L, Persson S, Strauss HW, Westling H., 1982. Non-invasive determination of the distribution of cardiac output in man at rest and during exercise. Clin Physiol **2**:467.

Tamaki N, Alpert NA, Rabito CA, Barlai-Kovach M, Correia JA, Strauss HW., 1988. The effect of captopril on renal blood flow in renal artery stenosis assessed by positron Tomography with Rubidium-82. Hypertension **11**:217.

Wilkinson RA, Fischman AJ, Rubin RH, Strauss HW., 1989. Monitoring response to antimicrobial therapy with ^{111}In-labeled polyclonal IgG. J Nucl Med **30**:P890.

Wilson RA, Sullivan PJ, Moore RH, Zielonka JS, Alpert NM, Boucher CA, McKusick KA, Strauss HW., 1983. An ambulatory ventricular function monitor: validation and preliminary clinical results. Am J Cardiol **52**:601.

Wolf W, Albright MJ, Silver MS, Weber H, Reichardt U, Sauer R., 1987. Fluorine-19 NMR spectroscopic studies of the metabolism of 5-fluorouracil in the liver of patients undergoing chemotherapy. Magn Res Imaging **5**:165.

Wolf W, Presant CA, Servis KL, El-Tahtawy A, Albright MJ, Barker PB, Ring R, Atkinson D, Ong R, King M, Singh M, Ray M, Wiseman C, Balayney D, Shani J., 1990. Tumor trapping of 5-fluorouracil: In vivo ^{19}F NMR spectroscopic pharmacokinetics in tumor-bearing humans and rabbits. Proc Natl Acad Sci USA **87**:492.

Wollmer P, Rhodes CG, Pike VW, Silvester DJ, Pride NB, Sanders A, Palmer AJ, Liss RH., 1982. Measurement of pulmonary erythromycin concentration in patients with lobar pneumonia by means of positron emission tomography. Lancet **2**:1361.

Yonekura Y, Brill AB, Som P, Bennett GW, Fand I., 1983. Quantitative autoradiography with radiopharmaceuticals, Part 1: Digital film-analysis system by videodensitometry: concise communication. J Nucl Med **24**:231.

CONTRIBUTION OF POSITRON EMISSION TOMOGRAPHY TO PHARMACOKINETIC STUDIES

B. Mazière, M. Mazière, J. Delforge,
A. Syrota

Service Hospitalier Frédéric Joliot
CEA, Orsay, France

INTRODUCTION

Positron Emission Tomography (PET) is a safe non-invasive visualization technique that provides serial quantitative images of the spatial distribution of a previously administered molecule labelled with a positron emitting radionuclide, in any desired transverse section of the body. By allowing in vivo non-invasive sequential measurements of regional drug concentrations, with sensitivity and specificity equivalent to those of plasma radioimmuno assays (RIA) (nanomolar to picomolar differences can be detected), PET provides an opportunity to follow the kinetics of drugs in humans, in various tissus.

Drug Labelling

Radioisotopes can be utilized as tracers in pharmacokinetic studies in two different ways:
— Direct radiotracer methods which include the various applications in which the drug is made radioactive and is used to determine its tissue distribution and fate.
— Indirect radioactive indicator methods which involve the study of the influence of a non radioactive ("cold") drug on the movements of a radiolabelled probe.

For a non-invasive measurement of the tissular kinetics of a drug, a very small amount (tracer dose) of the drug labelled with an isotope emitting a signal that can be externally detected is injected intravenously. The necessary condition in a pharmacokinetic investigation is that the radioisotope incorporated into the drug does not modify its chemical properties; practically this means that the radiotracer must be an isotope of one of the chemical elements existing in the drug. Hence the radiotracers used should be isotopes of carbon, oxygen or nitrogen (radioactive isotopes of fluorine or bromine are sometimes substituted for hydrogen).

Positron Emitting Radionuclides

^{14}C and ^3H, which are largely used in animal pharmacokinetic studies, cannot be used for external detection measurements because the negative electrons emitted

New Trends in Pharmacokinetics, Edited by A. Rescigno and A.K. Thakur
Plenum Press, New York, 1991

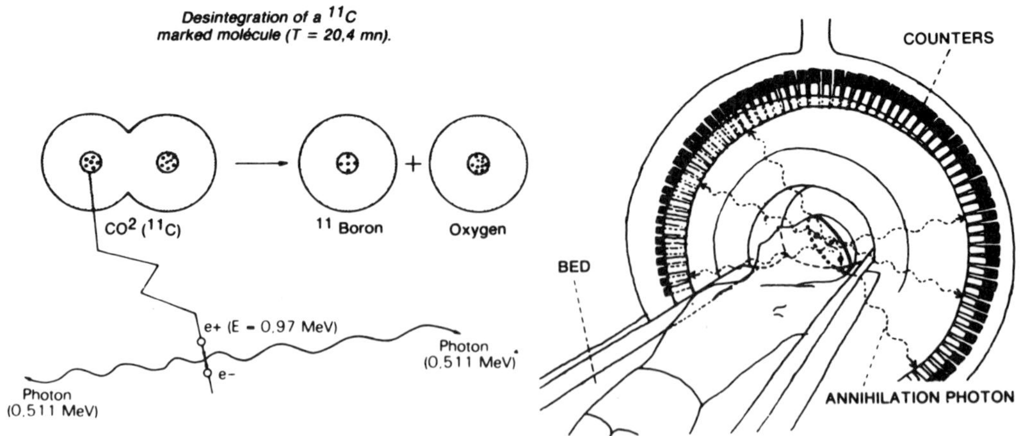

Figure 1. Principle of PET Imaging

during their decay are rapidly absorbed by the surrounding tissues. On the contrary, radioisotopes emitting positively charged electrons, or positrons, can be detected from the outside of the body. As a matter of fact, a positron also travels a very short path in the tissue (2 to 5mm) but finally it combines with an ambient electron. This interaction results in the annihilation of both original particles, with electromagnetic radiation given off. The annihilation radiation is released in the form of two high-energy (511 keV) photons travelling in opposite directions (Fig. 1 left). These photons easily penetrate body tissues and therefore can be recorded by suitable detectors placed on opposite sides of the body. Some common positron emitting radioisotopes used in PET studies are listed in table 1.

Table I

Positron emitting radioisotopes used in clinical studies.

Isotope	Half-life (min)	Target reaction	Available forms
^{15}O	2	$^{14}N(d,n)^{15}O$	$^{15}O_2$, $C^{15}O$, $C^{15}O_2$, $H_2^{15}O$
^{13}N	10	$^{16}O(p,\alpha)^{13}N$	^{13}N, $^{13}NH_3$
^{11}C	20.2	$^{14}N(p,\alpha)^{11}C$	^{11}CO, $^{11}CO_2$, $^{11}CH_4$, $^{11}C_2H_2$
^{18}F	110	$^{20}Ne(d,\alpha)^{18}F$	$H^{18}F$, $^{18}F_2$
^{76}Br	972	$^{75}As(He_3,2n)^{76}Br$	$^{76}Br_2$

These radioisotopes have short half-lives and consequently production and incorporation into molecules must be carried out "on site". However, a short half-life also means that sequential studies can be performed in humans without excessive absorbed radiation.

Positron-emitting isotopes are usually produced relatively free of stable or radioactive contamination (very high specific activity) by means of transmutation nuclear reactions using beams of cyclotron accelerated charged particles.

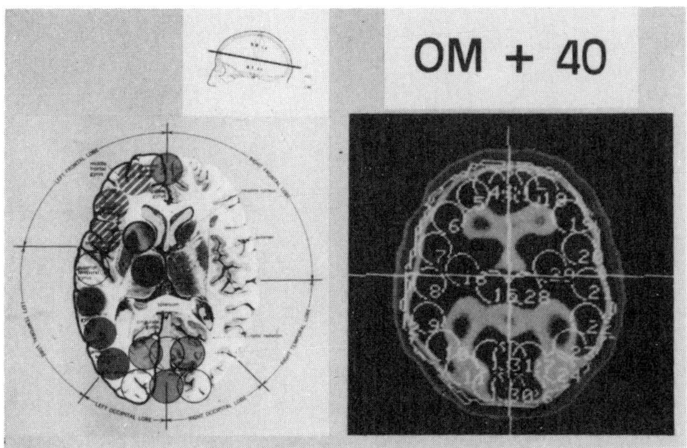

Figure 2. Localization of ROI's on a PET image

Radiochemistry

The isotopes produced in a cyclotron are obtained in very simple molecular forms such as $^{11}CO_2$, $^{13}NH_3$, or $H^{18}F$ and have to be chemically incorporated into the drug. The radiochemical synthesis must be fast and selective and the desired product must be isolated, purified and formulated as a sterile, pyrogen-free injection solution within two or three half-lives of the isotope used [Comar et al., 1982]. This constraint has promoted the development of new methods of remote controlled rapid chemistry [Mazière et al., 1977]. As an illustration, ^{11}C labelling is usually performed with 1 to 2 curies of ^{11}C to obtain 100 to 200 mCi of labelled compound after a 40 min synthesis.

Owing to the short half-lives of the radioisotopes used, radiolabelled drugs are obtained with a very high specific activity (500-4000 mCi/μmol i.e. 20-200 times the specific activity of tritiated molecules). High activities, corresponding to very small masses of drugs (few nanomols) can then be administered safely to subjects with low radiation doses and lack of chemical toxicity.

PET IMAGING

Because two photons are emitted simultaneously after each annihilation and because they travel nearly 180° apart, both photons can be counted by detectors placed on opposite sides of the body. These detectors are electronically connected to a coincidence circuit so that detected radiation will only be recorded if both detectors register the events within a brief timing window. Using this coincidence detection principle (the annihilation process is assumed to have occured somewhere within the tissue volume sustended by the two detectors) and an accurate correction of the attenuation of the γ-rays in the body, PET scanners with several rings of detectors have been designed. Using computer applied filtered retroprojection algorithms (identical to those developed for conventional CT X-ray scanners) PET tomograph produces images that represent, in slices through the body (from 5 to 31 slices), the quantitative, regional and temporal distribution of the administered radiolabelled compound (Fig. 1 right). Multiple, simultaneously obtained slices allow a three-dimensional interpretation of the collected data. The resolution is 4 to 8 mm and successive images can be produced at intervals of a few seconds (4-10 sec).

A PET image can be considered as an accurate mapping of radioactive concentrations in a tomographic slice. These radioactive concentrations can be easily

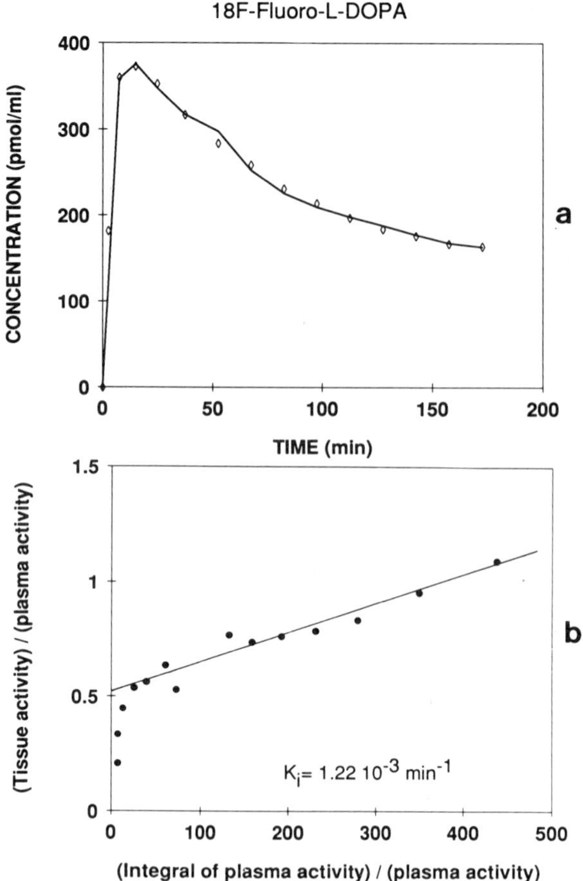

Figure 3. Example of Multiple Time Analysis plot.
a) Time-concentration relationship in striatum region
following the administration of (^{18}F)-6-fluoro-L-dopa
to a baboon.
b) Relationship between specific striatal radioactivity
and integrated plasma (^{18}F)-6-fluoro-L-dopa activity
in the same baboon.

determined in regions of interest (ROI). Usually the ROI's are delineated directly on the PET image using the anatomical information inherent in the functional images (Fig. 2). However, this information is limited by resolution, misrepresentation of the size of small structures, dependance of the type of tracer used, and variation in functional brain activity [Meltzer et al., 1990]. When the planning of the PET studies requires measurements on accurately determined structures, ROI's can be localized on an anatomical image obtained using MRI (retrospectively matched with the PET data set using surface landmarks) before being copied on the PET image.

By taking into account the specific radioactivity of the radiolabelled drug, regional radioactive concentrations (nCi/cm^3) can easily be transformed into regional drug concentrations (nmol/cm^3) if the tissular degradation of the drug during the time course of the PET scans is known. The results obtained in vivo in humans by PET are then quite analogous to those obtained in animals by quantitative autoradiography; the disadvantage is a much lower resolution, but for pharmacokinetic studies PET has the advantage of allowing the acquisition of time course sequential images in a single subject. The duration of such pharmacokinetic PET studies is however limited by the short physical half-lives of the radioisotopes used i.e. 90 to 120 min for ^{11}C- ($T_{1/2}$ = 20.2 min) and a few hours for ^{18}F- ($T_{1/2}$ = 1.83 h) or ^{76}Br- ($T_{1/2}$ = 16.2 h) labelled compounds.

The PET studies are safe, non invasive and, due to the short half-life of the radioisotopes used, weakly irradiating. Radiation doses to organs of the human body after administration of radioligands labelled with ^{11}C, ^{18}F [Harvey et al., 1985; Kilbourn et al., 1989; Herzog et al., 1990; Blin et al., 1990], ^{76}Br [Crawley et al., 1985; Mazière B. et al., 1985] have been estimated. The values obtained show that, with the activity usually injected (10-20 mCi for ^{11}C-, 5-10 mCi for ^{18}F-, 1-1.5mCi for ^{76}Br-ligands), the total body doses are around 200 mrad/study while the resultant doses of radiation absorbed by the critical target organs (i.e. liver, lower larger intestine, bladder) remain within the accepted 5 rad per organ limit per single study.

PET STUDIES

The contributions of PET to drug pharmacokinetic studies result from the possibility of following in the human body either a labelled drug (radioisotope regional tracer kinetics) or the interaction of a cold drug on a labelled probe.

Direct Radiolabelled Drug Studies

Pharmacokinetic data (absorption, distribution, metabolism, elimination) describe the fate of drugs in the body. Many pertinent data are to be found in the scientific literature. More often than not these results are related to animal studies; when they are related to human studies, they deal with absorption and excretion processes, concentration of drugs in various organs being measured only in the course of autopsy studies.

With PET, it is possible to know the fate of a drug in the human body, to follow its concentration in the various structures of various organs [Herzog et al., 1990] and particularly of the target organ, according to time. Organ impairments may alter the pharmacokinetics of drugs in the body; PET also allows to compare the drug kinetics in physiological and pathological conditions. However, as with all the in vivo techniques, the PET approach suffers certain limitations as different serial barriers are present between the site of administration (a brachial vein in humans) and the target itself. The first barrier is the pulmonary filter: In a single passage through the lung circulation, lipophilic molecules can be totally extracted by the pulmonary endothelial or epithelial cells; in this situation the amount of drug which will reach the investigated organ will depend on its clearance from the lungs. The second barrier is the capillary barrier and this problem is very acute in the brain where the intercellular junctions are very tight. Permeability of the blood-brain barrier is higher for molecules which enter the lipid matrix of the endothelial cell membrane more readily and the octanol/water partition coefficient is often used to predict this permeability. Another obstacle to the brain uptake of a tracer comes from the binding of the labelled molecule to plasma proteins, the degree of which appears to be independent of the partition coefficient in octanol and water.

Another limitation of PET studies comes from the fact that, as with all the investigations done by external detection, PET measurements are related to the total radioactivity and make no distinction between the various forms of the drug: unchanged or metabolized, free or bound.

Drug Distribution

After administration of the radiolabelled drug itself, in vivo tissue pharmacokinetics can be measured in organs such as liver, kidney, heart or brain, supplementing the data available from pharmacokinetic studies of body fluids, such as plasma and urine. To illustrate this PET unique ability to measure the achieved drug tissular concentrations, some applications will be reviewed.

One of the first studies of PET tissue pharmacokinetics is related to the measurements of the uptake and concentration of an antibiotic labelled with ^{11}C (erythromycin) in infected lungs in human beings [Wollmer et al., 1982]. The results indicated that erythromycin penetrates rapidly into areas of acute pneumonia, the effective concentration being reached within 10 min.

5,5-Diphenylhydantoïn (DPH) is an antiepileptic drug, which in animal studies has been well characterized in terms of plasma kinetics and therapeutic steady-state concentrations, brain penetration and metabolism. Using DPH labelled with ^{11}C, similar results were obtained in humans [Baron et al., 1983] demonstrating that DPH distribution in the brain, which is non-specific, is essential for its effectiveness.

Spiperone is a neuroleptic drug which acts by blocking the central dopamine D_2 receptors. Using an analog (ethylspiperone) labelled with ^{18}F, its whole-body distribution was studied in man. Only 1% of the total administered drug was found in the brain 2 h post-injection; urinary bladder, gall bladder, and liver were the organs with the highest drug concentration ranging from 6% to 25% [Herzog et al., 1990].

When antineoplasic agents used for chemotherapy of brain tumors are labelled (e.g. ^{11}C-BCNU), their selective uptake in the tumor can be visualized by PET. This provides the opportunity for improving methods of administering chemotherapy: For ^{11}C-BCNU, it has been shown that its specific penetration in the tumor is 50 times higher after intracarotid [Tyler et al., 1986] than after intravenous administration.

When the transfer of a drug from blood to brain is unidirectional, the transfer constant can be easily evaluated using a normalized graphical method, also called the ratio method. This method was first proposed to analyse deoxyglucose brain uptake, a phenomenon which is supposed to be irreversible [Patlak et al., 1983, 1985]. The model consists of a blood-plasma compartment, a reversible tissue region and one irreversible tissue region. The reversible region consists of an arbitrary number of compartments, which freely communicate with the plasma; the drug may move from the plasma to any of these compartments and vice versa. The irreversible regions communicate with the plasma and/or the reversible region; the drug can enter but cannot leave the irreversible region. The solution of the equations for the previous model shows that a graph of the ratio of the total tissue drug concentration at the times of sampling to the plasma concentration at the respective times versus the "normalized time" defined as the integral of plasma activity from time 0 to t divided by the plasma activity at time t will yield a curve that eventually becomes linear during steady state with a slope corresponding to the influx or transfer constant.

To illustrate the potentialities of this general model, the transfer constant of the drug L-dopa (the physiological precursor of dopamine) has been evaluated in baboons. The drug, labelled with ^{18}F was administered i.v. and the time course curve of the drug concentration in the striatum was registered by PET (Fig. 3a). The rate constant K_i for uptake of L-dopa from blood to striatum during steady state (i.e. when all the fluxes and velocities are constant with respect to time) is given by the asymptotic slope of the normalized graphical plot (Fig. 3b).

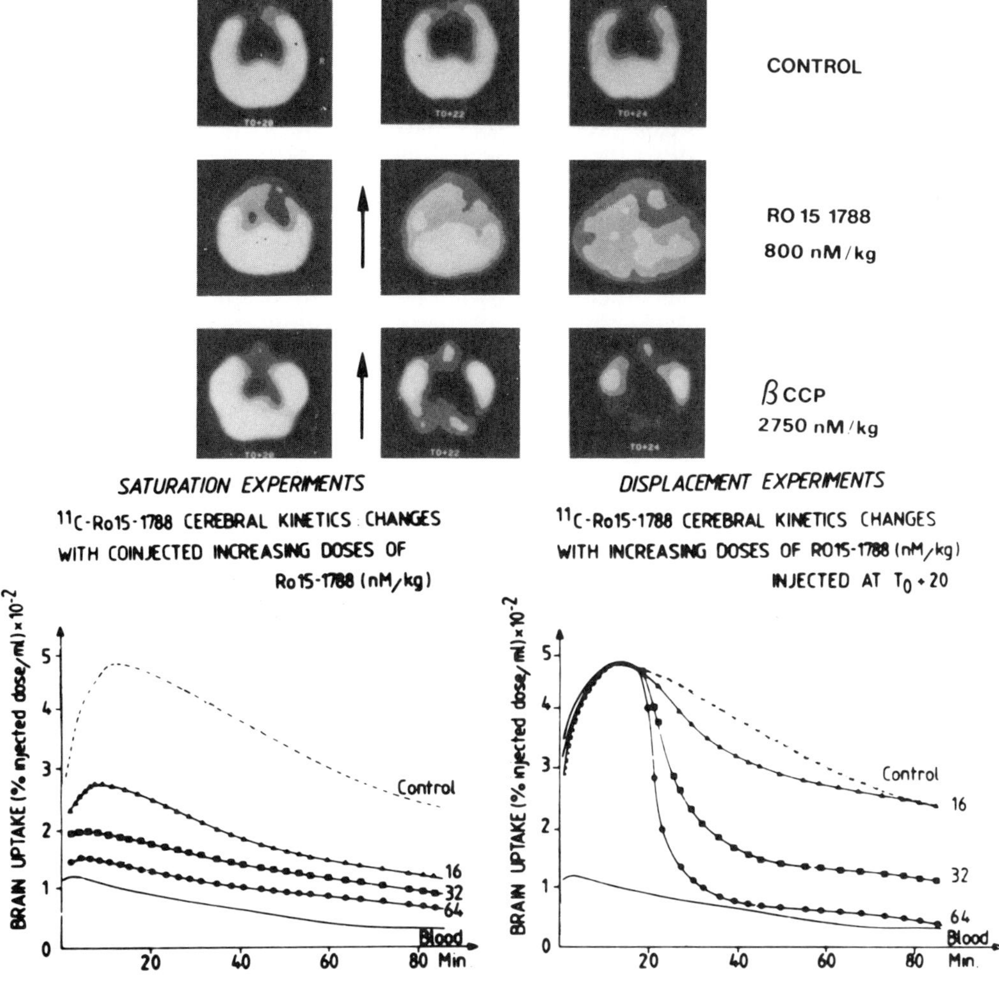

Figure 4. Saturability of ^{11}C-flumazenil in a baboon brain
(presaturation and displacement experiments)

Drug-Receptor Interactions

PET is well suited for quantification of the availability and affinity of recognition sites (receptors, enzymes), making it possible to examine in the various organs of living humans the effects of drugs that act by blocking (antagonists) or stimulating (agonists) recognition sites [Mazière et al., 1990].

Complexities of the PET receptor study using a radiolabelled drug is related to the non-specific binding of the drug (the lower the lipophilicity, the higher the specific to non-specific ratio) and to its removal process within the tissue: Uptake by different cells, binding to different receptors or different receptor subtypes, enzymatic or chemical degradation, intracellular trapping. The binding of a ligand by a receptor exhibits two unique properties, molecular specificity and saturability,

which will be used for ligand characterization. Molecular specificity describes the behavior of ligand binding in terms of affinity and ability to recognize a particular molecule while saturability is related to receptor density. Brain tissue contains a minute concentration of receptors (approximately 10^{-12} mole per gram) and if a ligand is delivered in excess, the receptors will become saturated. Very low concentrations of ligands must then be injected which implies the preparation of radioligands labelled with very high specific activities (curies per nanomole). Receptor-mediated localization of a drug in an organ or a structure can be validated in vivo by demonstrating the following criteria of the ligand-receptor complex:
— Specific regional distribution of the specific binding: The radiolabelled drug must be found in higher amounts in structures known for their high receptor densities.
— Saturability: Two kinds of competition experiments allow this demonstration (Fig. 4):

> a) In the displacement experiments, excess of "cold" (non-radioactive) agonists or antagonists, belonging preferably to different chemical classes, is intravenously administered at a time when the tissue radioligand concentration reaches its maximum. The radioactive concentration then rapidly decreases with time because of the competitive inhibition between the tracer and the excess of unlabelled drug.
>
> b) In the pre-saturation experiments, the receptors are blocked by an excess of unlabelled ligand administered prior to the administration of the radioligand. In this case, the regional tracer radioactive concentration remains lower than that measured in the absence of the "cold" drug. All these saturability experiments are dose-dependent. However, the possibility to apply this characterization technique in humans can be limited by the pharmacological or the toxicological effects of the large amounts of drug which have to be administered.

— Stereospecificity: Stereoselectivity appears to be a powerful proof for the characterization of receptor binding. When the two enantiomers of a drug have been radiolabelled, the one with pharmacological activity (the eutomer) displays a higher specific accumulation than the one without pharmacological activity (the distomer) [Fowler et al., 1987; Farde et al., 1988].

When the two isomers of a drug are available but have not been radiolabelled, the displacement of a radiolabelled probe from the receptor is preferentially obtained with the eutomer [Hantraye et al., 1984].

— High affinity of the radiolabelled drug for the receptor: Receptors have generally lower affinity for endogenous transmitters or exogenous agonists (10^{-6} M) than for antagonists (10^{-9} M). Therefore, in PET studies, better results are often obtained with labelled antagonists than with labelled agonists [Mazière et al., 1983].
— Biological effect: Correlation between binding affinity or receptor occupancy and potency in biological effect is essential to distinguish between the binding of the drug to an acceptor site, with no signal transmission, or to a receptor site, associated with a pharmacological response.

When it has been in vivo demonstrated that the drug actually binds receptors the ultimate goal of a PET pharmacokinetic investigation is the in vivo determination of the parameters governing the fate of the drug in terms of equilibrium constant (K_d), association and dissociation rate constants and maximum concentration of receptor (B_{max}).
The probability of binding of a ligand to a receptor depends on the concentrations of both free ligand and empty receptors. The kinetic description of the

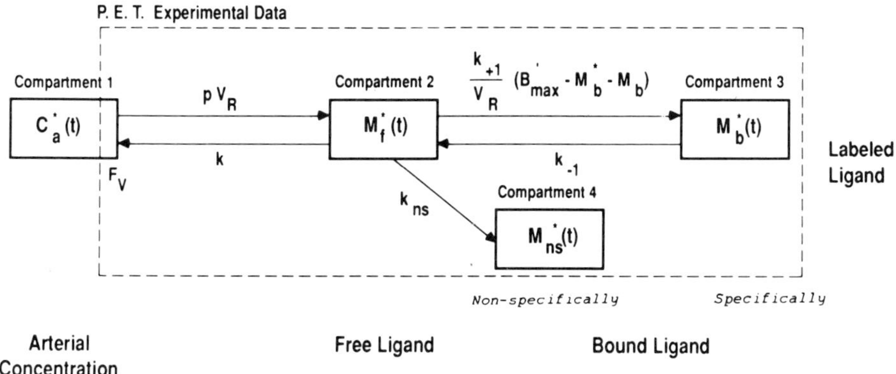

Figure 5. Compartmental drug-receptor model used in analysing the myocardial binding of a muscarinic antagonist MQNB labelled with [11]C. In this possible model describing the kinetics of the radiolabelled drug, compartment 1 represents the intravascular drug, compartment 2 the free tissular drug, compartment 3 the specifically bound drug and compartment 4 the irreversibly non-specifically bound drug. All transfer probabilities of drug between compartments are linear except for the binding probability which depends on the bimolecular association rate constant and on the local concentration of free receptors.

tissular uptake follows the bimolecular rate law. Modeling receptor-ligand interactions helps to describe the forward and reverse fluxes of the binding process: A biomathematical model, most of the time a compartmental model, is used to describe the fate of the tracer injected in the body.

Taking into account the individual characteristics of the radiolabelled drugs, suitable quantitative approaches have been applied to interpret the regional uptake kinetic curves in terms of parameters such as receptor density (B_{max}), affinity (K_d) and exchange rate constants:

— For drugs which dissociate rapidly from the receptors, allowing equilibrium of binding to be established in vivo within the time span of a PET experiment [Farde et al., 1986, 1989; Huang et al., 1986; Mazière et al, 1990], an equilibrium analysis of specific binding can be performed. A regional time-activity curve is used to define a point in time that represents equilibrium of specific binding to receptors. Series of PET scans are then performed in the same individual using the drug labelled with various specific activities. On the basis of regional radioactivity at equilibrium time, B_{max} and K_d values can be calculated from saturation curves or Scatchard plots.

— For drugs with a slow dissociation rate from the receptor, a kinetic analysis according to a compartmental model provides valuable information. Under those circumstances, the mathematical model will attempt to simulate the measured tissue response. The relationship between the data acquired from PET (and from other simultaneous measurements such as arterial sampling) and the resulting physiological information depends directly upon the kinetic model used in the data analysis.

Four different main steps can be characterized in such a kinetic analysis:

— Measurement of the variation of the tissue activity versus time in various ROI's: A classic dynamic PET experiment designed to investigate the binding of a radiolabelled drug to receptors in any human organ can be described as follow: A single injection of a β+-labelled drug is performed and the time-activity curves in some regions of interest in the tomographic slices are recorded. However, the binding parameters are not directly observable from these curves and it is necessary to use a mathematical models that transforms values of radioactive concentrations

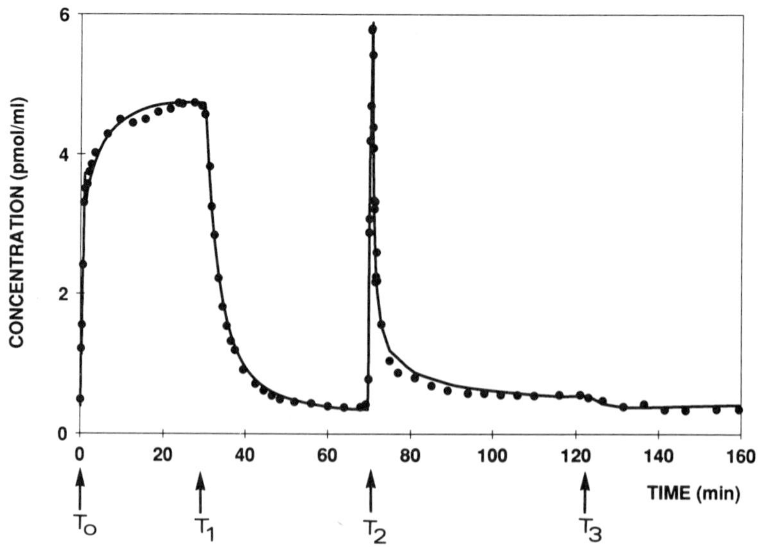

Figure 6. Experimental protocol (and corresponding PET data) used to evaluate the binding parameters of ^{11}C-MQNB to myocardial muscarinic receptors in dogs. The protocol included four injections of radiolabelled and/or cold MQNB. At time T_0 a tracer dose of ^{11}C-MQNB (high specific activity) was i.v. administered. At time T_1, an additionnal intraveinous injection of an excess of cold (unlabelled) MQNB was given (first displacement). At time T_2, a third injection of a mixture of labelled and cold MQNB was performed (coinjection experiment). At time T_3, a fourth injection consisting of a much larger amount of cold MQNB was injected (second displacement). Each injection was designed to follow the drug receptor interaction in a particular situation: The first injection mainly reflected the input kinetics; the first displacement revealed the dissociation and reassociation kinetics; the coinjection allows the investigation of the input-output kinetics when all receptors are occupied; the second displacement allows measurement of the non specific rate constant. The dots correspond to the experimental PET data and the solid line shows the values obtained with the model simulation.

measured in the selected regions of interest on PET images into values of pharmacokinetic interest.

— Definition or choice of a mathematical model describing the fate of the radioligand in the various tissues: Under those circumstances, the mathematical model (most of the time a nonequilibrium, nonlinear compartmental model) will attempt to elucidate the measured tissue response. The unknown parameters in the set of differential equations defining the model are the receptor density (B_{max}) and the various kinetic rate constants. The data that are usually included in the model are blood volume, regional rates of blood flow, rate of transport across the blood-brain barrier, specific activity of the radiolabelled drug, amount of free drug in tissue, extent of nonspecific binding, extent of metabolic breakdown of the radiolabelled drug in blood and tissue, and time measurement. Several models have recently been proposed as a framework for analysing kinetic data of drug-receptor interactions with PET [Perlmutter et al., 1986; Wong et al., 1986; Mintun et al., 1984; Syrota et al., 1984; Huang et al., 1986].

In these models, the reactions are usually considered to be a two-step process; the free drug is first transported from blood to tissue where binding sites are located

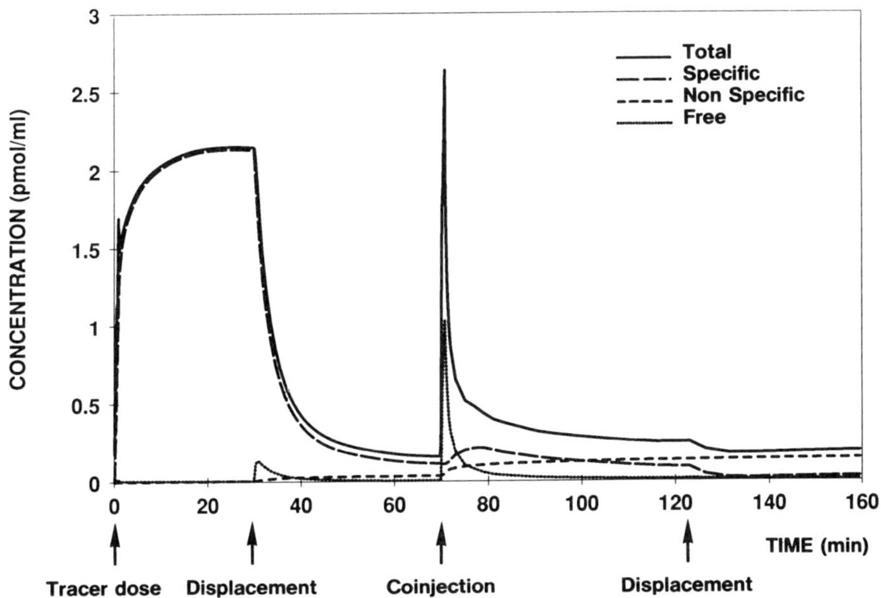

Figure 7. Computer simulations of MQNB myocardial muscarinic binding. Each curve represents the variation of the myocardial activity vs time for the free, specifically and non specifically bound, and total MQNB, computed using the previously estimated parameters.

and second, the drug-receptor complex is then formed. However, the model structure may differ depending on the human organ, the drug or the experimental procedure used. For example, in the nonlinear (the receptors are significantly occupied during the experiment) model describing the binding of MQNB labelled with ^{11}C to the myocardial muscarinic receptors, a nonspecifically bound drug compartment had to be introduced (Fig. 5). Thus the number of model parameters used in the various models can vary from 4 to 8 and must be identified from a single curve.

— Identification of the model parameters: This identification is performed by fitting (using nonlinear least squares fitting routines) the data predicted by the equations describing the model to the experimental data measured by PET. However, it is well known that when the number of parameters is too large in comparison with the available experimental data, two difficulties may appear. First, the uncertainties of parameters may be so large that the identified values are meaningless. Second, the identification problem may have several distinct solutions, i.e. several very distinct sets of parameter values will lead to similar theoretical curves. In both cases, the identification procedure may not lead to a valid solution. The necessary balance between the respective complexities of a model structure and experimental data corresponds to a general identifiability problem [Delforge et al., 1990 a]. To improve parameter estimation, more sophisticated PET protocols have to be designed: Along with the first injection of the labelled drug a second injection of the "cold" drug is administered (displacement experiment) [Huang et al., 1989 Delforge et al., 1990 a]; in other cases, the biologically valid solution can only be determined with a protocol including both a displacement and a co-injection experiment [Delforge et al., 1990 b] (Fig. 6).

— Computer Simulations: The theoretical tissue activity versus time curves for the various compartments corresponding to the free, bound and total ligands can then be constructed using the set of differential equations defining the model and the parameter values previously identified (Fig. 7).

The pharmacokinetic studies so explored by PET are often related to drugs which act by binding to neuroreceptors,or to peptide hormone receptors or to enzymes involved in substrate transport processes.

A highly intriguing potential of the PET technology is the possibility of directly imaging the binding of a psychoactive drug on brain receptors in an individual patient. Millions of dollars have presumably been invested in studies that attempted to explore the relationships between antipsychotic effects in patients and objective variables such as drug pharmacokinetic parameters in body fluids. Such studies produced a wide scatter of data occasionnally demonstrating significant relationships that were difficult to reproduce. As receptors are currently believed to be the targets of these drugs, PET allows, in the brain of living patients, a direct observation and measurement of drug receptor binding pharmacokinetic parameters. Thus, maps of B_{max}, k_{on}, k_{off} and K_d in the living human brain have been displayed for a neuroleptic (raclopride) and a benzodiazepine antagonist (flumazenil) labelled with [11]C [Blomqvist et al., 1990].

To illustrate the ability of PET of directly imaging the binding of drugs to neuroreceptors, some applications will be described.

Tranquillizers

— The first in-vivo imaging of the binding of a drug to brain receptors by PET was demonstrated more than a decade ago when it was shown that a trace amount of ^{11}C-labelled flunitrazepam could be displaced specifically from a baboon's brain by a therapeutic load of an unlabelled competitor benzodiazepine [Comar et al., 1979].
— The pharmacokinetic properties of the benzodiazepine antagonist flumazenil (RO 15 1788) labelled with ^{11}C have been studied in baboons [Mazière et al., 1983] and in humans [Samson et al., 1985; Person et al., 1985; Pappata et al., 1988] with PET. These studies have shown that this powerful antagonist, used as an andidote against benzodiazepine intoxication:
. crosses easily and intensively the blood-brain barrier,
. is located mainly in cerebral structures known for their rich concentrations in benzodiazepine receptors (cerebral and cerebellar cortices),
. is for the major part (90%) rapidly, specifically and selectively bound to receptors from which it can be stereoselectively displaced by stereoisomeric benzodiazepine receptor ligands.
. is rapidly eliminated from the brain (biological half-life in humans: about 50 min).

Moreover, by combining the PET experiments with EEG recordings, it has been shown that the convulsant or anti-convulsant actions of drugs (β-carbolines or diazepam) acting on the GABA receptor complex was mediated by the ^{11}C-flumazenil labelled binding sites [Hantraye et al., 1987].
— Pharmacokinetic properties of suriclone, a cyclopyrrolone without structural similarity to classical benzodiazepines, labelled with ^{11}C have been studied by PET. Although equilibrium and kinetic binding studies have suggested that ^{3}H-suriclone binds to a site distinct from that of ^{3}H-diazepam, the cerebral distribution of ^{11}C-suriclone observed by PET was similar to that of ^{11}C-RO 15 1788 [Frost et al., 1986].

Antidepressant

The brain pharmacokinetic of imipramine labelled with ^{11}C was studied as early as 1977 [Mazière et al., 1977].

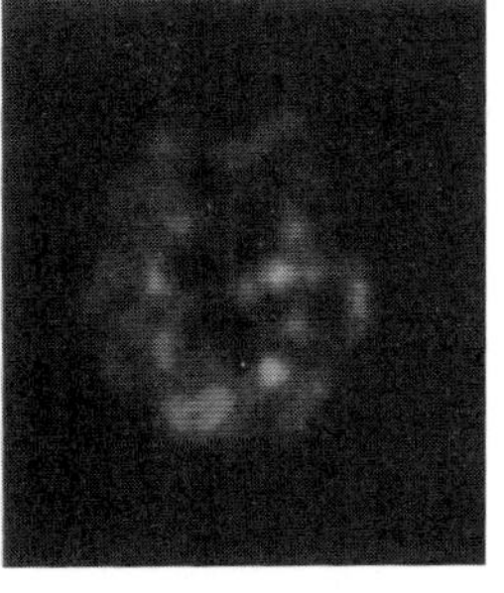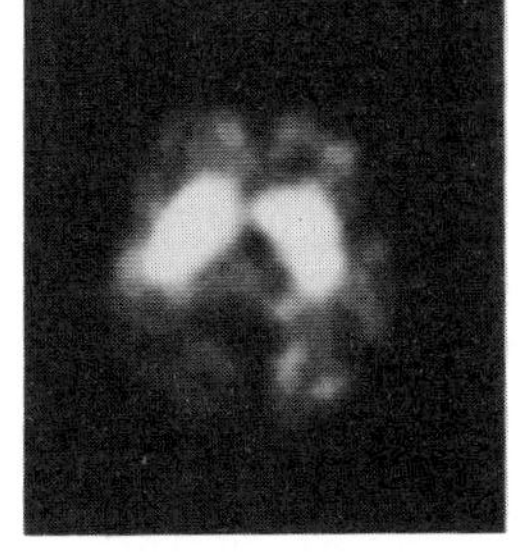

Figure 8. Occupancy of D_2 receptors during
and after an oral neuroleptic treatment.

Neuroleptic

PET investigations of Chlorpromazine [Comar et al., 1979] and Pimozide [Crouzel et al., 1980] labelled with ^{11}C, showed that the drug distributions were essentially in the gray matter and in structures such as caudate nucleus. However the in vivo nonspecific bindings of these drugs were too high to enable an identification of their supposed sites of action, the D_2 dopamine receptors.

When the drug is known to act through a specific binding to receptors or recognition sites, PET can be of particular value in assessing stereoselective binding of enantiomers in various organs [Farde et al., 1988]:

Deprenyl, an inhibitor of the enzyme monoamine oxidase B (MAO B), is used to treat Parkinson's disease (Dopamine, as other endogenous neurotransmitter amines, is metabolized by MAO B). A comparison, using PET, of the magnitude of brain uptake and retention of both ^{11}C-labelled inactive (D-) and active (L-) enantiomers of deprenyl exemplified the stereospecificity effect:

. L-deprenyl is bound by sites containing MAO B to a degree 25 times that of D-deprenyl

. rapid clearance of the inactive enantiomer and retention of the active enantiomer are observed within MAO B-rich brain structures such as corpus striatum and thalamus [Fowler et al. 1987].

INDIRECT RADIOACTIVE PHARMACOKINETIC STUDIES

For some technical reasons, it can be tedious, expensive, or impossible to label a drug with a positron emitter. In these cases, it is still possible to study in vivo the kinetics of the possible interaction of the drug with certain classes of receptors. Then, the strategy consists in performing an "in vivo radioreceptor assay" in using a specific positron emitting radioligand for labelling the receptors (a radiolabelled probe) and in testing the cold drug as a competing agent on this binding. We do not have presently at our disposal specific radiolabelled ligands for each type of receptors (Table 2), but we can expect that in the near future, the vast majority of new drugs could be tested using this approach.

Table II

Labelled probes for "in vivo radioreceptor assays".

RECEPTORS	PROBES	SUBTYPES	REFERENCES
DOPAMINERGIC	^{11}C-SCH 23390	D_1	Farde et al., 1986
	^{11}C-methylspiperone	D_2, $5HT_2$	Wagner et al., 1983
	^{76}Br-bromospiperone	D_2, $5HT_2$	Mazière et al., 1984
	^{11}C-raclopride	D_2	Farde et al., 1986
	^{76}Br-bromolisuride	D_2	Mazière et al., 1986
SEROTONINERGIC	^{18}F-setoperone	$5HT_2$	Blin et al,. 1988
	^{11}C-methylbromo LSD	$5HT_2$	Wong et al., 1987
OPIOID	^{11}C-carfentanyl	μ	Frost et al., 1987
	^{11}C-diprenorphine	μ, K	Jones et al., 1988
	^{18}F-acetylcyclofoxy	μ	Pert et al., 1984
BENZODIAZEPINE	^{11}C-flumazenil	BZ_1, BZ_2	Mazière et al., 1985
	^{11}C-PK 11195	periph.	Charbonneau et al., 1986
MUSCARINIC	^{11}C-methyl QNB	M_1, M_2	Syrota et al., 1984
	^{11}C-levetimide	M_1, M_2	Dannals et al., 1988
	^{11}C-scopolamine	M_1, M_2	Frey et al., 1988
	^{11}C-cogentin	M_1, M_2	Dewey et al., 1990
ADRENERGIC	^{11}C-CGP 12 177	β_1, β_2	Syrota et al., 1989
ESTROGEN	^{18}F-fruoroestradiol		Mathias et al., 1987

Blockade of the central dopamine receptors is held to be the essential mechanism in the pharmacological action of neuroleptics. Hence, to measure in vivo, in patients treated with neuroleptics, the actual dopamine D_2 receptor blockade (receptor occupancy) by the therapeutical treatment should provide an estimate of the effective neuroleptic tissue level. Treatment of schizophrenic patients with conventional doses of chemically distinct antipsychotic drugs (haloperidol, flupenthixol, sulpiride, clozapine, levopromazine, thioproperazine, propericiazine, thiethyperazine) almost completely abolished the imaging of dopamine D_2 receptor binding by specific radioligands such as spiperone labelled with ^{11}C [Wong et al., 1986] or ^{76}Br [Cambon et al., 1987], or ^{11}C-raclopride [Farde et al., 1986].

PET "in vivo radioreceptor assays" indicate that during conventional clinical treatments with such neuroleptics the receptor occupancy is total or subtotal, rapidly induced and dose-related; following drug withdrawal, recovery to normal or supranormal receptor availability appears to only be a matter of days (Fig. 8).

As another example of this indirect strategy used for the measurement of the duration of action of a drug, we can mention the study done with naltrexone. This drug is a long acting antagonist of opiate receptors used in the treatment of opiate dependance. For this investigation, the opioid receptors were labelled with a potent mu-type agonist, ^{11}C-carfentanil. The results of the PET studies showed that the half-time of the duration of blockade by naltrexone in the brain, ranges from 72 to 108 h, which is greater than the fast component of clearance of the drug and its principal metabolite from plasma [Lee et al., 1988].

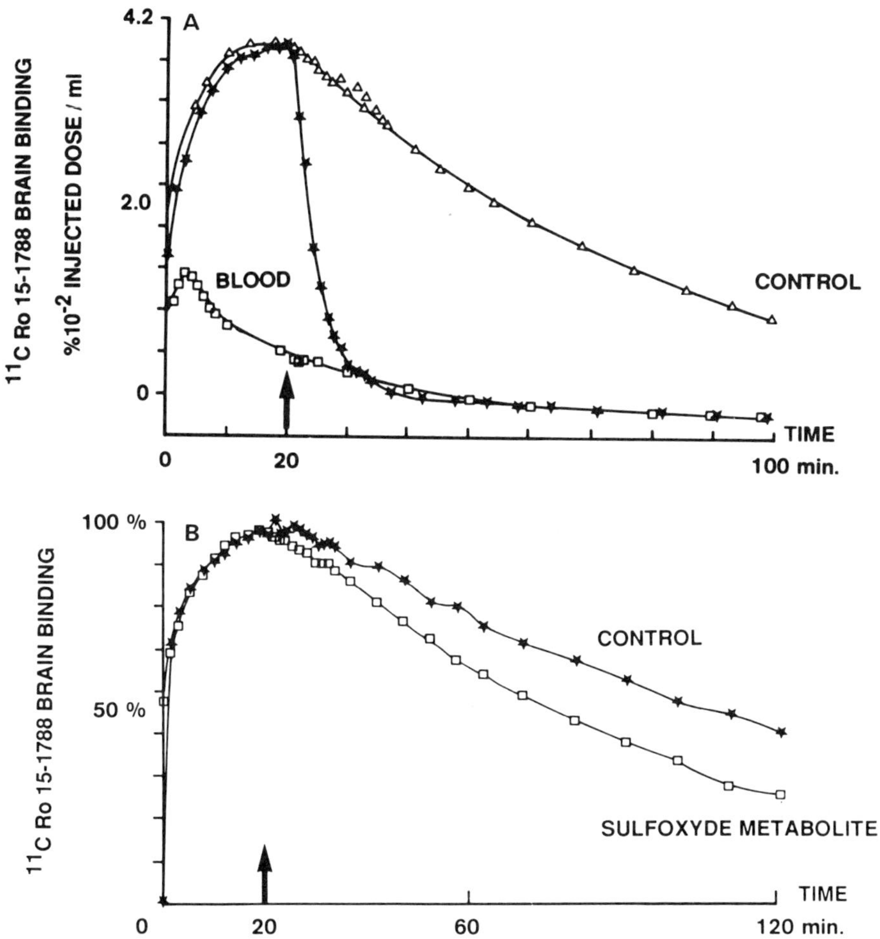

Figure 9. In vivo interaction of suriclone and its metabolite with central benzodiazepine receptors (labelled with the probe ^{11}C-RO 15 1788). The radiolabelled probe was injected at time T_0. Typical kinetics under control conditions showed that the maximal binding obtained at time $T_0 + 20$ min was followed by a slow decrease of the brain bound radioactivity.

Upper: When suriclone (0.5 mg/kg) was administered at $T_0 + 20$ min (arrow), the wash-out of the radioligand was dramatically accelerated and the specifically bound probe was totally displaced.

Lower: When a high dose (2 mg/kg) of the suriclone sulfoxyde metabolite was administered at $T_0 + 20$ min (arrow), the wash-out of the radioligand was only slightly accelerated and less than 15 % of the bound probe was displaced.

The kinetic of the binding of various unlabelled metabolites to receptors can also be studied using this indirect strategy. For example, the tranquillizer suriclone is known to be mainly metabolized in demethylated and sulfoxide derivatives. Although the interactions of these metabolites were reported in vitro, the in vivo interactions remained to be investigated in order to evaluate their possible involvement in the pharmacological activity of suriclone. A PET kinetic study, using ^{11}C-RO 15 1788 for the labelling of benzodiazepine receptors, demonstrated that these metabolites (which only have a low potency for inhibiting the probe binding) should not affect seriously the in vivo GABAergic transmission [Brouillet et al., 1990] (Fig. 9).

CONCLUSION

PET methodology has already been proved to be a useful tool for following the kinetics of drug actions in the living human brain. PET allows close analysis of the relationship between receptor occupancy and the onset of desired pharmacological effects and unwanted side effects. The pursuit of such studies is likely to lead to a greater understanding of the relationship between drug pharmacokinetic properties (receptor occupancy) and therapeutic effects. The possibility of labelling drugs, or precursors, or metabolites, by a radioisotope ethically administrable to human being opens up entirely new prospects.

PET has opened a new era in human biochemistry, and promises to have several important applications in pharmacokinetic studies, clinical pharmacology and drug development.

REFERENCES

Baron J-C., Roeda D., Crouzel C. et al., 1983. Brain regional pharmacodynamics of ^{11}C-labeled diphenylhydantoin: Positron emission tomography in humans. Neurology **33**:580.

Blin, J., Pappata, S., Kiyosawa et al., 1988. (^{18}F)Setoperone: a new high affinity ligand for positron emission tomography study of the serotonin-2 receptors in baboon brain in vivo. Eur. J. Pharmacol. **147**:73.

Blin J., Sette G., Fiorelli M. et al., 1990. A method for the in vivo investigation of the serotoninergic S_2 receptors in the human cerebral cortex using Positron Emission Tomography and ^{18}F-labeled Setoperone. J. Neurochem. **54**:1744.

Blomqvist G., Pauli S., Farde L. et al., 1990. Maps of receptor binding parameters in the humain brain — A kinetic analysis of PET measurements. Eur. J. Nucl. Med. **16**:257.

Brouillet E., Chavoix C., Hantraye P. et al., 1990. Interaction of suriclone with central type benzodiazepine receptors in living baboons. Eur. J. Pharmacol. **175**:49.

Charbonneau P., Syrota A., Crouzel C. et al., 1986. Peripheral-type benzodiazepine receptors in the living heart characterized by positron emission tomography. Circulation **73**:476.

Cambon H., Baron J-C., Boulenger J-P. et al., 1987. In vivo assays for neuroleptic receptor binding in the striatum. Positron Emission Tomography in humans. Brit. J. Psychiatry **151**:824.

Comar D., Zarifian E., Verhas M. et al., 1979. Brain distribution and kinetics of ^{11}C-chlorpromazine in schizophrenics: Positron emission tomography studies. Psychiatry Research **1**:23.

Comar D., Berridge M., Mazière B. et al., 1982. Radiopharmaceuticals labelled with positron-emitting radioisotopes. In: "Computed Emission Tomography" (Ell P.J., Holman B.L., eds), page 42. Oxford University Press, Oxford, U.K.

Crawley J.C., Smith T., Veall N. et al., 1985. Distribution, retention and radiation dosimetry of 77Br-p-Bromospiperone. Radiat. Protect. Dosimetry 8:147.

Crouzel C., Mestelan G., Kraus E. et al., 1980. Synthesis of a ^{11}C-labelled neuroleptic drug: pimozide. Int. J. Appl. Radiat. Isot. 31:545.

Dannals R.F., Långström B., Frost J.J. et al., 1988. Synthesis of radiotracers for studying muscarinic cholinergic receptors in the living human brain using positron emission tomography: (^{11}C)dexetimide and (^{11}C)levetimide. Appl. Radiat Isot. 39:291.

Delforge J., Syrota A., Mazoyer B.M., 1990a. Identifiability analysis and parameter identification of an in-vivo receptor model for PET data. IEEE Biomed. Engineer. 37:653.

Delforge J., Janier M., Syrota A. et al., 1990b. Non-invasive quantification of muscarinic receptors in vivo with Positron Emission Tomography in the dog heart. Circulation 82:1494.

Dewey S.L., Macgregor R.R., Brodie J.D. et al., 1990. Mapping muscarinic receptors in human and baboon brain using [N-^{11}C-methyl]-benztropine. Synapse 5:213.

Farde L., Hall H., Ehrin E. et al., 1986. Quantitative analysis of D_2 dopamine receptor binding in the living human brain by PET. Science 231:258.

Farde L., Pauli S., Hall H. et al., 1988. Stereoselective binding of ^{11}C-raclopride in the living human brain — A search for extrastriatal central D_2 dopamine receptors by PET. Psychopharmacology 94:471.

Farde L., Eriksson L., Blomquist G. et al., 1989. Kinetic analysis of ^{11}C-raclopride binding to D_2 dopamine receptors studied by PET — A comparison to equilibrium analysis. J. Cereb. Blood Flow Metab. 9:696.

Frey K.A., Koeppe R.A., Mulholland G.K. et al., 1988. Muscarinic receptor imaging in human brain using (^{11}C)scopolamine and Positron Emission Tomography. J. Nucl. Med. 29:808.

Fowler J.S., MacGregor R.R., Wolf A.P. et al., 1987. Mapping human brain monoamine oxydase A and B with ^{11}C-labelled suicide inactivators and PET. Science 235:481.

Frost J.J., Wagner H.N., Dannals R.F. et al., 1986. Imaging benzodiazepine receptors in man with (^{11}C)suriclone by Positron Emission Tomography. Eur. J. Pharmacol. 122:381.

Frost J.J., Mayberg H.S., Douglas K.H. et al., 1987. Alteration of cerebral mu-opiate receptors in the temporal lobe epilepsy and following electroconvulsive therapy. J. Cereb. Blood Flow Metab. 7 (suppl.1):421.

Hantraye Ph., Kaijima M., Prenant C. et al., 1984. Central type benzodiazepine binding sites: A Positron Emission Tomography study in the baboon's brain. Neurosci. Lett. 48:115.

Hantraye P., Chavoix C., Guibert B. et al., 1987. Benzodiazepine receptors studied in living primates by positron emission tomography: antagonist interactions. Eur. J. Pharmacol. 138:239.

Harvey J., Firnau G., Garnett E.S., 1985. Estimation of the Radiation Dose in Man due to 6-[^{18}F]Fluoro-L-dopa. J. Nucl. Med. 26:931.

Herzog H., Coenen H.H., Kuwert T. et al., 1990. Quantification of the whole-body distribution of PET radiopharmaceuticals, applied to 3-N-([^{18}F]fluoroethyl)-spiperone. Eur. J. Nucl. Med. 16:77.

Huang S-C., Barrio J.R., Phelps M.E., 1986. Neuroreceptor assay with Positron emission tomography: equilibrium versus dynamic approaches. J. Cereb. Blood Flow Metab. **6**:515.

Huang S-C, Bahn M.M., Barrio J.R., et al., 1989. A double-injection technique for in vivo measurement of dopamine D_2-receptor density in monkeys with 3-(2'(^{18}F)fluoroethyl)spiperone and dynamic positron emission tomography. J. Cereb. Blood Flow Metab. **9**: 850.

Jones A.K.P., Luthra S.K., Mazière B. et al., 1988. Regional cerebral opioid receptor studies with (^{11}C)diprenorphine in normal volunteers. J. Neurosci. Meth. **23**:121.

Kilbourn M.R., Carey J.E., Koeppe R.A. et al., 1989. Biodistribution, dosimetry, metabolism and monkey PET studies of ^{18}F-GBR 13119. Imaging the dopamine uptake system in vivo. Nucl. Med. Biol. **6**:569.

Lee M.C., Wagner H.N., Tanada S. et al., 1988. Duration of occupancy of opiate receptors by Naltrexone. J. Nucl. Med. **29**:1207.

Mathias C.J., Welch M.J., Katznellenbogen J.A. et al., 1987. Characterization of the uptake of 16a-([^{18}F]fluoro)-17-estradiol in DEMBA-induced mammary tumors. Nucl. Med. Biol. **14**:15.

Mazière B., Loc'h C., Hantraye P. et al., 1984. ^{76}Br-Bromospiroperidol: A new tool for quantitative in-vivo imaging of neuroleptic receptors. Life Sci. **35**: 1349.

Mazière B., Loc'h C., Baron J-C. et al., 1985. In vivo imaging of dopamine receptors in human brain using Positron Emission Tomography and ^{76}Br-Bromospiperone. Eur. J. Pharmacol. **114**:267.

Mazière B., Loc'h C., Stulzaft O. et al., 1986. ^{76}Br-Bromolisuride: A new tool for quantitative in vivo imaging of D_2 dopamine receptors. Eur. J. Pharmacol. **127**:239.

Mazière B., Mazière M., 1991. Positron Emission Tomography studies of brain receptors. Fundam. Clin. Pharmacol. **5**:61.

Mazière M., Todd-Pokropek A.E., Berger G. et al., 1977. Carbon-11 labelled compounds in dynamic imaging studies of the brain. In: "Medical Radionuclide Imaging", vol.II, page 203. SM 210/155, A.I.E.A., Vienna.

Mazière M., Prenant C., Sastre J. et al., 1983. Etude "in vivo" des récepteurs aux benzodiazépines par tomographie d'émission de positons. L'encéphale **IX**:151b.

Mazière M., Hantraye P., Kaijima M. et al., 1985. Visualization by positron emission tomography of the apparent regional heterogeneity of central type benzodiazepine receptors in the brain of living baboons. Life Sci. **36**:1609.

Meltzer C., Bryan R., Holcomb H. et al., 1990. Anatomical Localization for PET using MR imaging. J. Comput. Assist. Tomog. **14**:418.

Mintun M.A., Raichle M.E., Kilbourn M.R. et al., 1984. A quantitative model for the in vivo assessment of drug binding sites with positron emission tomography. Ann. Neurol. **15**:217.

Pappata S., Samson Y., Chavoix C. et al., 1988. Regional specific binding of [^{11}C]RO 15 1788 to central type benzodiazepine receptors in human brain: Quantitative evaluation by PET. J. Cereb. Blood Flow Metab. **8**:304.

Patlak C.S., Blasberg R.G., Fensternmacher J.D., 1983. Graphical evaluation of blood-to-brain transfer constants from multiple-time uptake data. J. Cereb. Blood Flow Metab. **3**:1.

Patlak C.S., Blasberg R.G., 1985. Graphical evaluation of blood-to-brain transfer constants from multiple-time uptake data. Generalizations. J. Cereb. Blood Flow **5**:584.

Perlmutter J., Larson K.B., Raichle M.E. et al., 1986. Strategies for in vivo measurement of receptor binding using positron emission tomomgraphy. J. Cereb. Blood Flow Metab. **6**:154.

Persson A., Ehrin E., Eriksson L. et al., 1985. Imaging of ^{11}C-labelled Ro 15 188 binding to benzodiazepine receptors in the human brain by positron emission tomography. J. Psychiatr. Res. **19**:609.

Pert C.B., Danks J.A., Channing M.A. et al., 1984. 3-[^{18}F]Acetylcyclofoxy: A useful probe for the visualization of opiate receptors in living animals. FEBS Letters **177**:281.

Samson Y., Hantraye P., Baron J-C. et al., 1985. Kinetics and displacement of (^{11}C)RO 15 1788, a benzodiazepine antagonist, studied in vivo in human brain by positron tomography. Eur. J. Pharmacol. **110**:247.

Syrota A., Paillotin G., Davy J.M. et al., 1984. Kinetics of in vivo binding of antagonist to muscarinic cholinergic receptor in the human heart studied by positron emission tomography. Life Sci. **35**:937.

Syrota A., 1989. In vivo investigation of myocardial perfusion, metabolism and receptors by positron emission tomography. Int. J. Microcirc: Clin. Exp. **8**:411.

Tyler J.L., Yamamoto Y.L., Diksic M. et al., 1986. Pharmacokinetics of superselective intra-arterial and intravenous [^{11}C]BCNU evaluated by PET. J. Nucl. Med. **27**:775.

Wagner H.N., Burns H.D., Dannals R.F.et al., 1983. Imaging dopamine receptors in the humain brain by positron emission tomography. Science **221**:1264.

Wollmer P., Pride N.B., Rhodes C. et al., 1982. Measurement of pulmonary erythromycin, concentration in patients with lobar pneumonia by means of positron tomography. Lancet 1361.

Wong D.F., Gjedde A., Wagner H.N. Jr., 1986a. Quantification of neuroreceptors in living human brain. I. Irreversible binding of ligands. J. Cereb. Blood Flow Metab. **6**:137.

Wong D.F., Gjedde A., Wagner H.N. Jr. et al., 1986b. Quantification of neuroreceptors in living human brain. II. Inhibition studies of receptor density and affinity. J. Cereb. Blood Flow Metab. **6**:147.

Wong D.F., Lever J.R., Hartig P.R. et al., 1987. Localization of serotonin 5-HT$_2$ receptors in living human brain by positron emission tomography using N1-(^{11}C-methyl)-2-Br-LSD. Synapse **1**:393.

PITFALLS IN PHARMACOKINETIC MODELING OF MONOCLONAL ANTIBODY BIODISTRIBUTION IN MAN

Giuliano Mariani,[1] Luigi Ferrante,[2] Aldo Rescigno[3]

[1]CNR Institute of Clinical Physiology and
Second Medical Clinic of the University of Pisa
Pisa, Italy
[2]Medical School of the University of Ancona
Ancona, Italy
[3]School of Pharmacy
University of Parma
Parma, Italy

INTRODUCTION

Immunoscintigraphy employing radiolabeled monoclonal antibodies (MoAbs) specific for certain selected tumor-associated antigens (TAAs) is becoming increasingly accepted as a routine diagnostic procedure, particularly for staging patients with cancer. In fact, even though the diagnostic potentials of immunoscintigraphy have been explored for diverse applications including cardiovascular diseases (for the detection of thrombi or of damaged myocardial cells), oncology is certainly the field where clinical validation of this new imaging method has been most extensive [for a brief overview on the clinical usefulness of immunoscintigraphy and on some pharmacokinetic aspects concerning this new diagnostic procedure, see Mariani and Strober, 1990]. It is now widely acknowledged that tumor immunoscintigraphy is of great value in terms of both specificity (close to 100%) and sensitivity (about 75-80% on the average). In addition, one of the major advantages of this procedure has been shown to be the ability of detecting distant tumor lesions in about one-third of patients who are apparently "tumor-free" as classified by other, noninvasive diagnostic techniques; therefore, for some specific forms of cancer, immunoscintigraphy may be considered as a mandatory diagnostic procedure for correctly staging and monitoring patients, complementary to other techniques already well established in the clinical routine [Mariani et al., 1989].

Development of MoAbs for Tumor Immunoscintigraphy

The starting point for producing MoAbs potentially useful for tumor immunoscintigraphy is usually a human-derived TAA which is employed as the heterologous antigen to challenge the immune system of the experimental animal [i.e.

the mouse in the classical procedure introduced by Kohler and Milstein in 1975]. The antigen can be an already well characterized marker such as the carcino-embryonic antigen (CEA), alpha-fetoprotein (AFP), or other similar markers utilized in a pure chemical form. Alternatively, crude homogenates or extracts from human tumors are employed as a complex antigenic mosaic, and the resulting immune response is utilized for the process itself of identifying and characterizing previously unknown TAAs. Novel TAA-MoAb systems are therefore identified, defined and screened as to their potential use for tumor immunoscintigraphy in patients.

Screening of TAA-MoAb Systems for Immunoscintigraphy

The requirements for the *in vivo* use of MoAbs for tumor immunoscintigraphy in patients are much more stringent than those for their use *in vitro* (such as for the immunohistochemical identification of tumor lesions in tissue sections). Accordingly, the screening process goes through several steps, first *in vitro*, then *in vivo*. First of all, there is a screening based on the immunoglobulin class of the MoAbs produced against certain TAAs so that, without exceptions, only IgG or IgM MoAbs are to be employed for administration to humans; this serves to avoid possible adverse reactions due to specific immunoglobulin functions (as those, for instance, of IgE).

As indicated in Table I, some of the requirements for the preliminary *in vitro* screening of prospective TAA-MoAb systems for tumor immunoscintigraphy have to do with the MoAb, others with the TAA by itself [see also Larson et al., 1984; Buraggi, 1985].

Table I

Preliminary In Vitro Characterization of Prospective TAA-MoAb Systems for Tumor Immunoscintigraphy

MoAb (IgG, IgM)	TAA
- Cross-reactivity with normal adult tissues	- Membrane-bound *vs.* cyto-plasmic antigen
- Binding to circulating blood cells	- Density of antigenic sites >100,000/cell
- Affinity for TAA >10^{-9} Liters/Mole	- Expressed by over 90% of the tumor cells
- (Cross-reactivity with other TAAs)	- Modulated by the MoAb
	- (Shed *vs.* non-shed TAA)

Assuming a certain TAA-MoAb system has successfully gone through all the steps of *in vitro* characterization listed in Table I, the preliminary *in vivo* screening phase is performed utilizing experimental laboratory models (typically, human tumor xenografts in nude mice). These *in vivo* studies are designed to determine the extent of radiolabeled MoAb uptake at the tumor implant, the nonspecific radioactivity accumulation at some non-tumoral sites (for instance in the liver, spleen and bone marrow) and, in general, what is often called "pharmacokinetics". However, the vast majority of these studies are merely "biodistribution" experiments, whereby the organ distribution of radioactivity at preselected times (*e.g.*, 24 and/or 48 hours) is estimated without actually calculating the kinetic parameters of clearance, exchange, and so on. On the other hand, these kinetic parameters do become essential when proceding to the really final screening phase which is invariably carried out in patients, in the clinical setting.

In fact, it should be reemphasized once and for all that, however promising the results can be from the *in vivo* screening phase in animals, they usually bear little relation with the actual results in patients as concerns especially cross-reactivity with normal tissues, nonspecific uptake in the reticuloendothelial system, percentage of injected dose bound to the tumor, and tumor/non-tumor ratios [Mariani et al., 1985; Working Group Meeting, 1986; Mariani, 1987]. These differences derive primarily from biological factors such as utilizing a heterologous protein (the murine MoAb) in humans (as opposed to murine MoAb in the mouse), physiological differences in the vascular architecture (affecting the access through circulation of the radiolabeled MoAb to the cell-bound TAA) between spontaneously growing tumors in man and the experimental implants in animals, and so on. Additional factors affecting the parameters of kinetic distribution *in vivo* of radiolabeled MoAbs have to do with the mode of radiolabeling itself and with the type of radioisotope employed, with the possible presence of TAA shed in circulating fluids, and finally with possible reinjection (for monitoring and follow-up purposes) of the monoclonal tracer in the same patients (presence of human anti-mouse antibodies at repeat injection).

The Role of Pharmacokinetics in the Development of Tumor Immunoscintigraphy Agents

Pharmacokinetic studies in humans have a major role in the development of new systems for tumor immunoscintigraphy and they may well be performed at the same time of completing the screening of new immunoscintigraphy tracers through the assessment of tissue biodistribution in a relatively limited number of patients, prior to the actual clinical trials in patients affected by the particular form of cancer for which the MoAb tracer is supposed to be specific. The radioisotope of choice for such studies is Iodine-131 [Mariani, 1987; Mariani and Strober, 1990]; however, the final choice of the isotope to be used in the actual clinical trials (either it be Iodine-131 itself, or Indium-111, Iodine-123, or finally Technetium-99m) depends on a variety of factors that are elucidated mostly through pharmacokinetic studies. These studies allow one to correct the accumulation of isotope at a specific site for the contribution of circulating blood radioactivity and thus make it possible to predict the concentration of radiolabeled MoAb tracers at some selected targets at any given time point after administration. The information thus obtained is useful both for choosing the optimal time window for imaging and for estimating the unitary radiation dose to the tumor and to normal tissues.

It should be noted, however, that most of the human biodistribution studies published so far for the development of new immunoscintigraphy systems are represented by a mere description and/or plotting of the whole-body retention and plasma disappearance curves of the radiolabeled MoAb tracers, possibly relating these data to the curves of radioactivity accumulation at some particular organ sites such as tumor lesions or the liver and spleen. No actual analysis whatsoever is usually attempted to derive kinetic parameters from such data (besides the common half-life determination for the plasma curve), and the conclusions of interest for setting up the new immunoscintigraphy system are based on simple determinations such as the tumor-to-blood or tumor-to-background ratios. Instead, a true pharmacokinetic modeling would allow one to draw some general laws about the physiologic structure of the system under study, and therefore some systematic properties and predictability about the kinetic parameters characterizing the system.

Indeed, although the need for testing new MoAb immunoscintigraphy tracers through pharmacokinetic analysis has been repeatedly emphasized [Gobuty et al., 1985; Goodwin, 1987; Mariani, 1987], surprisingly few studies of this kind have been performed [Eger et al., 1987; Griffin et al., 1989; Halpern et al., 1988; Hayes et al., 1986; Hnatowich et al., 1987; Koizumi et al., 1986; Mariani et al., 1986; Murray et al., 1988; Rosenblum et al., 1985; 1987; Rescigno et al., 1991]. Moreover, a protocol that could be considered reliable from the pharmacokinetic

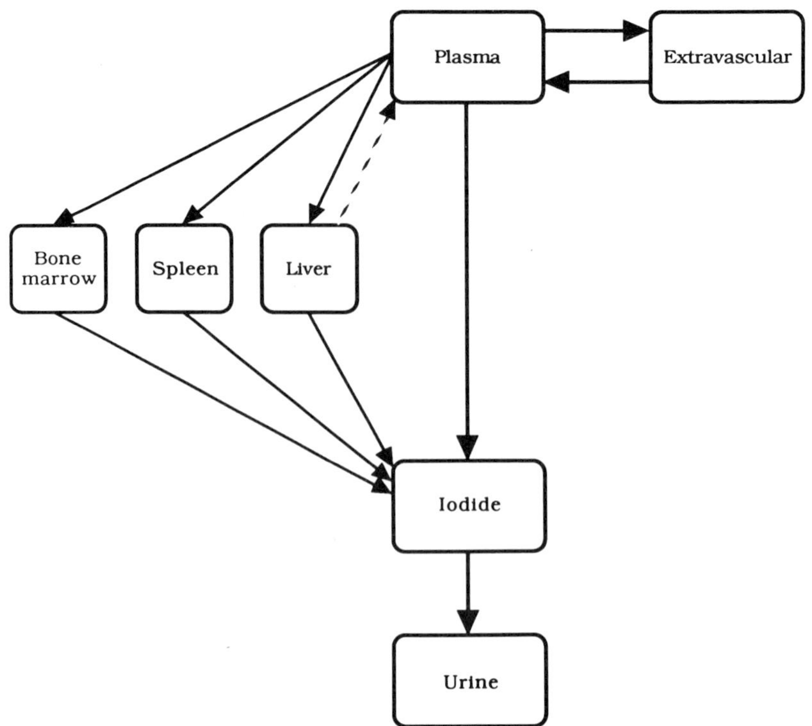

Figure 1. Multicompartmental model utilized as a first approach for the pharmacokinetic analysis of tissues biodistribution data in man of the anti-adenocarcinoma MoAb KC4 radiolabeled with Iodine-131 [modified from Mariani et al., 1986]. The liver, spleen and bone marrow compartments represent nonspecific uptake (assumed to occur as a first-order process) in the reticuloendothelial system, followed by degradation at a constant rate at those sequestration sites (the dashed arrow indicates reversible exchange of the radiolabeled MoAb between plasma and liver, a hypothesis that was not actually tested in the analysis). The quota of "normal" MoAb catabolism is assumed to occur at sites rapidly exchanging with the plasma space.

point of view concerning both the experimental design and the analysis of data has been followed in an even smaller number of studies [Eger et al., 1987; Koizumi et al., 1986; Hnatowich et al., 1987; Mariani et al., 1986; Rescigno et al., 1990], whereas a few other papers have just approached the surface of pharmacokinetic modeling [Griffin et al., 1989; Rosenblum et al., 1985; 1987].

The above considerations help explain how the matter of pharmacokinetic modeling for MoAb biodistribution in humans is far from being settled. With this rationale in mind, we will now illustrate, with two specific examples, the problems most commonly encountered in this type of pharmacokinetic analysis, utilizing data for which we have a deeper knowledge than is usually possible for data published in the literature by other groups.

We would like to stress the fact that, given the complexity of pharmacokinetic modeling for the biodistribution of radiolabeled MoAbs in man and the necessary ongoing interaction between the modeling phase itself and the experimental design of the biodistribution studies, the two examples that follow must not be considered as

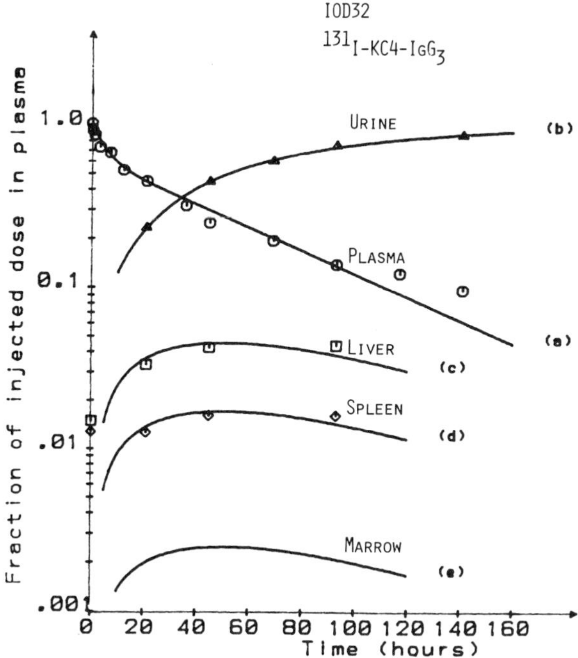

Figure 2. Theoretical curves generated by the model depicted in Fig. 1 to fit the experimental measurements obtained in the first patient studied with MoAb KC4 (case IOD32) (see text for further comments on this case and on the kinetic parameters determined in the entire group of nine patients submitted to the study).

definitive conclusions about the structure of the metabolic system under study, but rather as "work-in-progress" reports that open the way towards improving our understanding of the explored biological phenomena.

PHARMACOKINETIC MODELING FOR THE BIODISTRIBUTION OF MOAB KC4

MoAb KC4 is an IgG3 originally developed against a TAA present in most human adenocarcinomas. The biodistribution studies analyzed by pharmacokinetic analysis have been performed utilizing the intact MoAb radiolabeled with Iodine-131 as the tracer, in a total of nine patients affected by cancer other than adenocarcinoma [Mariani et al., 1986]. For several reasons, this may be considered as one of the simplest immunoscintigraphy systems to approach with pharmacokinetic modeling. In fact, radiolabeling with Iodine-131 is the least likely to alter, by itself, the metabolic behavior *in vivo* of plasma proteins and, on the other hand, the metabolic fate of radioiodine *per se* has already been quite well characterized in the past. Furthermore, the TAA recognized by MoAb KC4 is not shed in the circulating fluids, therefore there is no appreciable production of radiolabeled immune complexes circulating in blood (and, on the other hand, there is no actual "tumor compartment" for KC4 in this particular group of patients). Moreover, the overall clearance rate of ^{131}I-labeled KC4 is rather slow, thus reducing recirculation through the plasma space of free radioiodine released in the course of degradation of the MoAb KC4; this translates into a definite experimental simplification, since the

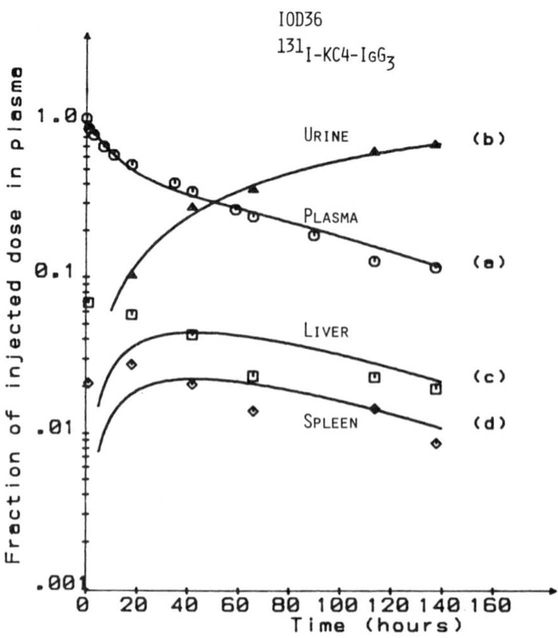

Figure 3. Results of kinetic modeling obtained in case IOD36. This case is a good example of the general behavior observed in the vast majority of the patients studied, particularly concerning the poor fitting of the liver and spleen curves.

low proportion of free ^{131}I relative to protein-bound [131]I (about 3.2% at the peak time point for free radioiodine) makes it unnecessary to process all the plasma samples through chromatographic analysis of the two radioactive species (total plasma radioactivity is therefore assumed to represent, with a good degree of approximation, undegraded radiolabeled KC4.

Based on *a priori* knowledge of the system under study, the multicompartmental model depicted in Figure 1 was employed as a first approach for analyzing, for each patient, a set of experimental data constituted by: i) the plasma radioactivity curve, ii) the cumulative urinary radioactivity excretion curve, iii) the radioactivity accumulation curve in the liver, iv) the radioactivity accumulation curve in the spleen (although possible based on the scintigraphic acquisitions recorded during this study, no actual determinations were available for pharmacokinetic analysis concerning the radioactivity accumulation curve in the bone marrow).

According to classical notations, the sequence of arrangements among the different compartments was arbitrarily chosen so as to indicate the most likely sequence of events that any tracer molecule introduced into the plasma space may undergo, up to complete degradation of the labeled MoAb and generation of a new radioactive species (free radioiodine) representing the only exit of activity from the body, through urinary excretion. Thus, in this model the "one-way" arrows from the plasma space to liver, spleen, and bone marrow indicate irreversible binding of the MoAb due to nonspecific accumulation in the reticuloendothelial structures (the dashed arrow returning from the liver to the plasma space indicates a possible reversible exchange between the two compartments that was, however, not tested in this analysis). Furthermore, such nonspecific binding was initially assumed to follow a first-order, constant rate kinetics, as was the rate of metabolic degradation both at these accumulation sites and at other sites in rapid exchange with the plasma space. It should also be emphasized that the compartments depicted in Figure 1 are meant to represent actual physical spaces within the body.

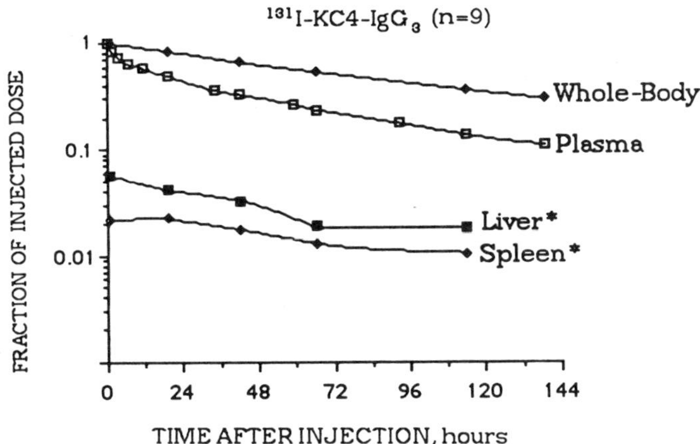

Figure 4. Mean curves of whole-body radioactivity retention, plasma disappearance, liver and spleen activity determined in all nine patients studied with radiolabeled MoAb KC4; the liver and spleen curves have been corrected for contribution due to radioactivity circulating in blood [see Mariani et al., 1986].

As a first step towards fitting the experimental values with the activity curves predicted by the model, a preliminary non-compartmental analysis of the plasma curve was performed [Rescigno and Gurpide, 1973; Sheiner, 1981; Rescigno et al., 1983], in order to derive some initial estimates from which to initiate the iteration procedure for the entire model. An *ad hoc* program based on the algorithm proposed by Marquardt [1963] for least-square fitting was then utilized for multicompartmental analysis of the experimental data according to the model described.

The results of fitting the experimental values observed in the first patient studied (case IOD32) are depicted in Figure 2, as is the curve simulating radioactivity accumulation in the bone marrow (a parameter for which no experimental data are available). The fit of theoretical values to the experimentally measured values was judged to be quite satisfactory, with the possible exception of the first experimental time point for the spleen accumulation curve and the later portion of the plasma curve. This favorable impression was further supported by a remarkable consistency of the kinetic parameters determined (although without actually plotting the individual curves as done for case IOD32) in all nine patients included in the pharmacokinetic study. In fact, the percent coefficient of variation (% CV) calculated for some critical kinetic parameters (such as the mean transit time in the plasma space and the rates of degradation occurring in liver, spleen and bone marrow) was markedly low (about 15-20% of the mean values for the entire group of patients), while it was relatively higher, but still quite acceptable, for the rates of irreversible binding in the liver and spleen (about 50-60% of the mean values) [Mariani et al., 1986].

However, when reconsidering with a critical mind the results of the above pharmacokinetic modeling and, above all, when plotting all the fits obtained in the individual patients, it soon became obvious that case IOD32 was to be considered the exception rather than the typical fit of theoretical curves (as generated by this model) to the observed values. In fact, the results of kinetic analysis obtained in all the other patients were similar to the plot depicted in Figure 3 for case IOD36, clearly indicating that some of the assumptions underlying the model were unrealistic, at least concerning the behavior of radioactivity accumulation in the irreversible nonspecific binding sites (liver and spleen).

Furthermore, looking back at the average curves calculated from the experimental measurements obtained in the nine patients studied, one can realize that radioactivity accumulation in the liver and spleen does do not appear to occur at a

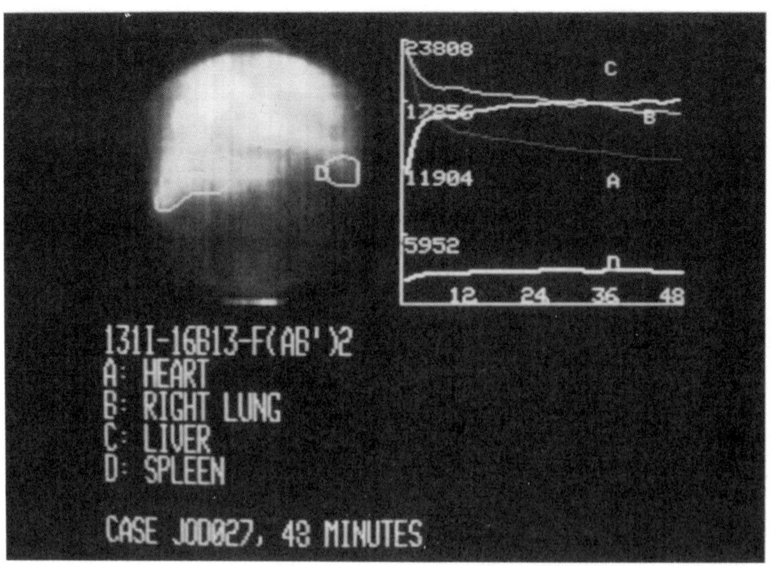

Figure 5. Radioactivity curves experimentally determined in specific regions of interest pertaining to the heart, right lung, liver and spleen upon bolus injection of a radiolabeled MoAb (continuous recording over 48 minutes). The liver and spleen curves reach a plateau at about 30 minutes, thus showing a behavior that can be assimilated to a fast kinetics, saturable mechanism. This kinetic pattern heavily affects also the plasma curve which (after semilogarithmic transformation) does not exhibit a monoexponential decline as radiolabeled proteins do over such early time after bolus injection.

constant rate throughout the experimental interval, but rather as a very fast process in the early part of the experiment followed by a steady decline up to about 72 hours, then (but only for the liver curve) by a later, slight increase in activity (Fig. 4). On the other hand, an explanation to the behavior of radioactivity accumulation in the liver and spleen following the i.v. bolus injection of a radiolabeled MoAb may be found in the results obtained in an experiment where the regional radioactivity changes occurring in critical organs (such as the heart chambers, liver and spleen) were dynamically recorded during the very early period (over 48 minutes) upon injection of a radiolabeled MoAb (Fig. 5).

The radioactivity curves determined for regions of interest selected over such organs demonstrate in fact that both the liver and the spleen curves exhibit a very fast rise, reaching a plateau as early as about 30 minutes. It is therefore clear that, given the overall duration of pharmacokinetic studies such as these (about six days, or 144 hours), the 30-minute time point is not appreciably different from the zero time point. Thus, radioactivity accumulated in the liver and spleen at the time of recording the first whole-body scintigraphic map (just about 30-45 minutes) may well be considered, in terms of kinetic modeling, as the initial condition in those compartments. In addition, available experimental evidence indicates that nonspecific binding of radiolabeled MoAbs in the reticuloendothelial system follows a saturable mechanism kinetics [Murray et al., 1988]; it may thus be inferred that such phenomenon is virtually complete early after administration of the MoAb tracer, representing a behavior that may be assimilated to the process of clearance through a

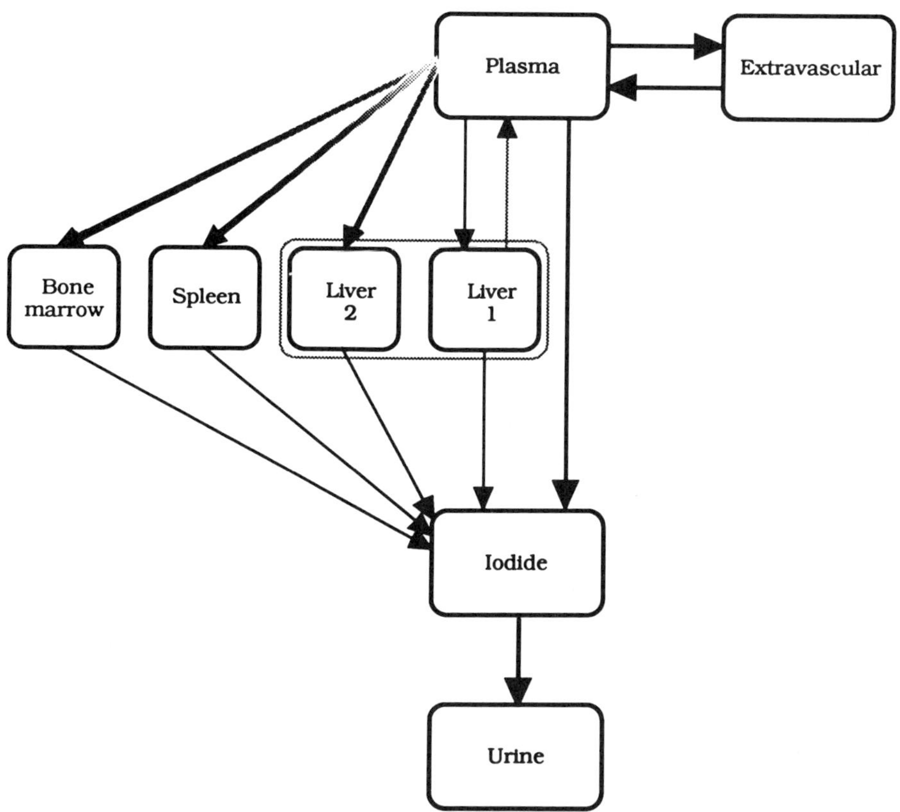

Figure 6. New model proposed for pharmacokinetic analysis of ^{131}I-labeled KC4 tissue biodistribution studies in man (and possibly of other radioiodinated MoAb tracers); in case of radioactivity uptake at tumor sites, a tumor compartment may easily be added in rapid exchange with the plasma space (reversibly or irreversibly). The thick, shadowed arrows from plasma to the reticulo-endothelial components of various organs are meant to indicate a very fast, saturable binding mechanism (see text for complete description of the model).

highly efficient "filter" system of a fraction of denatured tracer present in the injected bolus.

Based on the above considerations, a further model can then be developed to take into account all available knowledge on the radiolabeled MoAb KC4, as shown in Figure 6. In this model, the thick, shadowed arrows joining the plasma compartments with the compartments representing Liver-2, spleen and bone marrow stand to indicate the fast, saturable binding of a certain fraction of the radioiodinated MoAb in the nonspecific sequestration sites of the reticuloendothelial system. It should also be noted that the values of radioactivity accumulation experimentally measured for the liver region actually include two compartments that can be considered as separate from the biological point of view. In fact, while Liver-2 represents binding of radioactivity in this organ due to the reticuloendothelial component of nonspecific binding (the thick, shadowed arrow from plasma indicating the fast, saturable kinetics), Liver-1 represents the 15-30% quota of IgG immunoglobulin catabolism normally occurring in the liver (regular, constant rate arrow from the plasma space) [for a review on the sites of immunoglobumin catabolism, see Mariani and Strober, 1990]. From the kinetic point of view, this

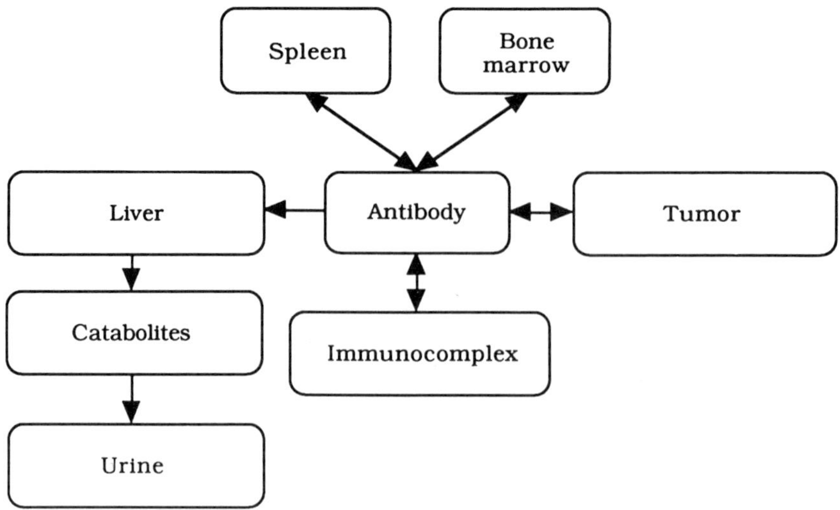

Figure 7. Simplified version of the model proposed by Rescigno et al. [1990] for pharmacokinetic analysis of the anti-CEA MoAb C110 radiolabeled with Indium-111.

fraction of normal degradation is reflected by the later, slow increase (or slowered decrease) in the total liver radioactivity accumulation curve. From the pharmacokinetic analysis standpoint, the behavior of extremely fast nonspecific uptake in liver, spleen and bone marrow can be solved either by some non-linear, non-time-invariant kinetics (adequately taking into account saturability of this mechanism), or by considering the experimental activity at the first time point as the initial condition in the corresponding compartments (of course, adequately adjusting the injected dose in the plasma space to take this assumption into account). The validity of the model proposed in Figure 6 for the pharmacokinetic analysis of tissue biodistribution in man of ^{131}I-labeled MoAb KC4 and of other radioiodinated MoAbs for immunoscintigraphy is currently under evaluation.

PHARMACOKINETIC MODELING FOR THE BIODISTRIBUTION OF MoAb C110

In a recent paper, Rescigno and coworkers [1990] propose a complex model for the pharmacokinetic analysis of tissue biodistribution in man of an anti-CEA monoclonal tracer for tumor immunoscintigraphy, *i.e.* MoAb C110 radiolabeled with Indium-111. Due to several factors, the system under study is indeed characterized by a high degree of complexity, quite higher than for the MoAb KC4 previously discussed. These factors can be briefly summarized as follows: i) MoAb C110 has a faster clearance rate than KC4, therefore there is a significant recirculation in the blood of the radioactive label released in the course of catabolism (thus making mandatory the use of chromatographic techniques to identify the different radioactive species in each plasma sample); ii) the metabolic fate of Indium-111 *per se* in the body follows a quite complex kinetics involving different tissues and organs; iii) the TAA recognized by MoAb C110 (that is, CEA) is indeed shed in the body fluids, thus leading to the formation of a circulating immune complex; iv) the patients included in this study are affected by CEA-secreting tumor lesions that exhibit a

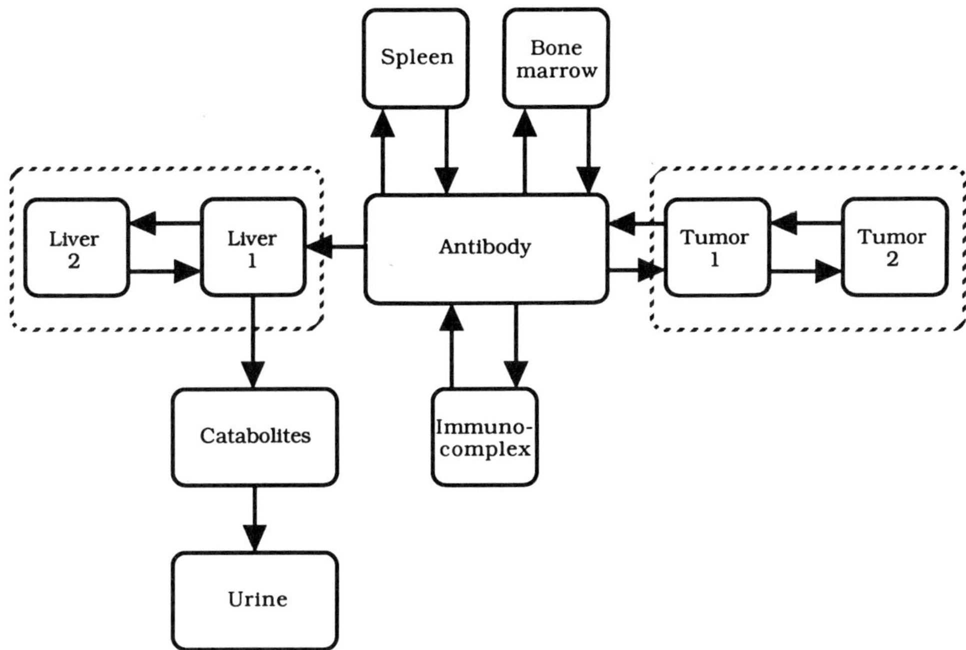

Figure 8. Detailed version of the pharmacokinetic model utilized for the analysis of experimental data obtained in one of the patients receiving ^{111}In-C110 (see text for additional details).

certain kinetic of radioactivity binding and release and must therefore be taken into account as specific compartments.

From the experimental point of view, the authors resort to HPLC analysis of the plasma samples to identify the different radioactive species mentioned above. However, poor resolution between intact MoAb and immune complex on one side and radiolabeled catabolites on the other side precludes utilization of the latter parameter as one of the experimentally measured values for pharmacokinetic analysis.

Kinetic analysis is therefore based on the following experimental values: i) plasma disappearance curve of the radiolabeled MoAb; ii) plasma curve of the radiolabeled immune complex; iii) whole-body radioactivity retention curve (*i.e.*, the reciprocal of the cumulative urinary activity excretion curve); iv) radioactivity accumulation curve in tumor lesions; v) radioactivity accumulation curve in the liver; vi) radioactivity accumulation curve in the spleen; vii) radioactivity accumulation curve in the bone marrow (indirectly estimated on the basis of limited regional measurements and extrapolations from theoretical normal values).

Based on *a priori* knowledge and on a preliminary analysis of data, the authors follow a sophisticated computational approach that can be represented in the form of the multicompartmental model depicted in Figure 7. It should be noted that not all the compartments of this model are meant to represent physically separate spaces in the body. For instance, the undegraded MoAb, the immuno complex and, to some extent, catabolites are all to be found within the physical distribution space of plasma. However, they are indeed to be considered as separate entities from the kinetic point of view (different chemical forms of radioactivity in the same physiological space). It should also be added that particular care is taken by Rescigno and coworkers in describing in details some of the critical parameters included in this model, such as the rates of formation and dissociation of the immune complex.

Moreover, it should also be pointed out that, while the experimental part of the study includes a total of eight patients, the pharmacokinetic analysis of one patient only is reported and discussed by the authors, as an example of methodological approach to the problem.

Finally, the model published in the cited paper (see Fig. 7) is a simplified version of the model developed and actually utilized for pharmacokinetic analysis of the experimental data, a more complex model that is depicted in Figure 8. When considering the complete model, it should be noted that the subdivision of the liver and tumor compartments into Liver-1 and Liver-2 and, respectively, Tumor-1 and Tumor-2 was made necessary in order to obtain a satisfactory fit of the experimental data. Therefore, as the authors rightfully point out, when a single organ is represented by more than one compartment, the transfer rates between them do not have an immediate physiological meaning if not supported by other types of experimental evidence.

The pharmacokinetic study on MoAb C110 is now being continued by extending the analysis to other patients, especially those given different amounts of antibody in order to determine the influence of dose on biodistribution and retention times of the radiolabeled MoAb in normal and tumor-containing organs.

CONCLUSIONS

The examples discussed in this Chapter have been chosen to illustrate the complexity and some of the problems most frequently encountered in pharmacokinetic modeling as applied to human studies on the tissue biodistribution of radiolabeled MoAb tracers for tumor immunoscintigraphy. This complexity of analysis arises from the complexity of the biological events involved in such studies, including not only the variable physiology of different normal tissues and of tumoral tissues, but also various radiochemical and radiopharmaceutical factors *per se*.

From the experimental point of view, some practical indications can be derived from the examples discussed. A very important indication concerns the initial phase of distribution in the plasma space of a radiolabeled MoAb tracer immediately following its i.v. injection as a bolus. First of all, such injection should actually be a bolus (that is, taking place as a flash of a few second duration) and not, as it is most commonly done, an infusion lasting several minutes. Furthermore, given the very fast events of nonspecific binding in the reticuloendothelial structures taking place upon injection (however important such binding might be for the individual monoclonal tracers), it is not advisable to rely solely on the radioactivity concentration of the 5-minute sample for estimating plasma volume (a critical value to which several other volumes are subsequently referred during pharmacokinetic analysis). In fact, an initial volume of distribution so determined may easily be spuriously overestimated because of the early exit from plasma of a sizable fraction of the injected dose (even worse is to utilize standard formulas to estimate plasma volume in patients). Therefore, it would be useful to determine plasma volume by an independent, reliable technique, such as the dilution of a bolus of radiolabeled albumin coinjected with the MoAb tracer. It should also be noted that the accuracy achieved in determining the injected dose (another critical parameter in tracer kinetic studies) is usually much poorer in radiolabeled MoAb biodistribution studies than that commonly achieved in classical tracer turnover experiments (where the error in such determination is usually in the order of 0.5%, or even lower).

It is also obvious that a major advantage in kinetic analysis is made possible by the availability of laboratory procedures adequate enough for an accurate quantitative identification of the different radioactive species circulating in blood (undegraded MoAb tracer, radioactive label released in the course of catabolism and possible labeled catabolites, immune complex with the shed TAA and/or with possible human antimurine antibodies upon repeat administration, transchelation compounds for some of the labels).

From the kinetic point of view, a *caveat* should be issued regarding the identifiability of complex multicompartmental models on the basis of experimental measurements in a relatively small number of compartments and/or with a poor temporal definition for each compartment.

Finally, it should be emphasized that the goodness of fit achieved through the introduction of whatever number of additional compartments in the model used for analysis should not distract the investigator from the main goal of assigning specific physiological and biological meanings to the kinetic parameters determined with the analysis.

REFERENCES

Buraggi G.L., 1985. Radioimmunodetection of cancer. J. Nucl. Med. Allied Sci. **29**:261.

Eger R.R., Covell D.G., Carrasquillo J.A., Abrams P.G., Foon K.A., Reynolds J.C., Schroff R.W., Morgan A.C., Larson S.M., Weinstein J.N.,.1987. Kinetic model for the biodistribution of an ^{111}In-labeled monoclonal antibody in humans. Cancer Res. **47**:3328.

Gobuty A.H., Kim E.E., Weiner R.E., 1985. Radiolabeled monoclonal antibodies: radiochemical pharmacokinetic and clinical challenges. J. Nucl. Med. **26**:546.

Goodwin D.A., 1987. Pharmacokinetics and antibodies. J. Nucl. Med. **28**:1358.

Griffin T.W., Bokhari F., Collins J., Stochl M., Bernier M., Gionet M., Siebecker D., Wetheimer M., Giroves E.S., Grenfield L., Houston L.L., Doherty P.W., Wilson J., 1989. A preliminary pharmacokinetic study of ^{111}In-labeled 260F9 anti-(breast cancer) antibody in patients. Cancer Immunol. Immunother. **29**:43.

Halpern S.E., Haindl W., Beauregard J., Hagan P., Clutter M., Amox D., Merchant B., Unger M., Mongovi C., Bartholomew R., Jue R., Carlo D., Dillman R., 1988. Scintigraphy with In-111-labeled monoclonal antitumor antibodies: kinetics, biodistribution, and tumor detection. Radiology **168**:529.

Hayes D.F., Zalutsky M.R., Kaplan W., Noska M., Thor A., Colcher D., Kufe D.W., 1986. Pharmacokinetics of radiolabeled monoclonal antibody B6.2 in patients with metastatic breast cancer. Cancer Res. **46**:3157.

Hnatowich D.J., Gionet M., Rusckowski M., Siebecker D.A., Roche J., Shealy D., Mattis J.A., Wilson J., McGann J., Hunter R.E., Griffin T., Doherty P.W., 1987. Pharmacokinetics of ^{111}In-labeled OC-125 antibody in cancer patients compared with the 19-9 antibody. Cancer Res. **47**:6111.

Kohler G., Milstein C., 1975. Continuous cultures of fused cells secreting antibody of predefined specificity. Nature **256**:495.

Koizumi K., DeNardo G.L., DeNardo S.J., Hays M.T., Hines H.H., Scheibe P.O., Peng J.-S., Macey D.J., Tonami N., Hisada K., 1986. Multicompartmental analysis of the kinetics of radioiodinated monoclonal antibody in patients with cancer. J. Nucl. Med. **27**:1243.

Larson S.M., Carrasquillo J.A., Reynolds J.C., 1984. Radioimmunodetection and radioimmunotherapy. Cancer Invest. **2**:363.

Mariani G.,.1987. A protocol for screening tumour radioimmunoscintigraphy agents. In: "Lectures and Symposia, 14th International Cancer Congress, vol. 6: Epidemiology, Prevention, Diagnosis." Lapis K. and Eckardt S., eds.), page 329. Akademiai Kiadò, Budapest .

Mariani G., Ferrante L., Rescigno A., 1986. Kinetic modeling for the analysis of digital images from tissue distribution studies of tumor radioimmunoscintigraphy agents. In: "Radioaktive Isotope in Klinik und Forschung, 17th (II)." (Höfer R. and Bergmann H., eds.), page 741. Verlag H. Egerman, Wien.

Mariani G., Mazzuca N., Molea N., Bianchi R., Donato L., 1985. Kinetic distribution studies in vivo of radioiodinated monoclonal preparations as a screening procedure for tumour immunoscintigraphy agents. In: "Immunoscintigraphy." (Donato L. and Britton K., eds.), page 83. Gordon & Breach, New York and London.

Mariani G., Rosa C., Donato L.,.1989. Comparison of the diagnostic sensitivity of tumor radioimmunoscintigraphy by means of an anti-CEA monoclonal antibody with other noninvasive diagnostic techniques. In: "Nuclear Medicine. Trends and Possibilities in Nuclear Medicine." (Schmidt H.A.E. and Buraggi G.L., eds.), page 527. Schattauer Verlag, Stuttgart.

Mariani G., Strober W., 1990. Immunoglobulin metabolism. In: "F c Receptors and the Action of Antibodies." (Metzeger H., ed.), page 94. American Society for Microbiology, Washington, D.C.

Marquardt D.W., 1963. An algorithm for least squares estimation of nonlinear parameters. SIAM J. **11**:431.

Murray J.L., Lamki L.M., Shanken L.J., Blake M.E., Plager C.E., Benjamin R.S., Schweighardt S., Unger M.W., Rosenblum M.G., 1988. Immunospecific saturable clearance mechanisms for Indium-111-labeled anti-melanoma monoclonal antibody 96.5 in humans. Cancer Res. **48**:4417.

Rescigno A., Bushe H., Brill A.B., Rusckowski M., Griffin T.W., Hnatowich D.J., 1990. Pharmacokinetic modeling of radiolabeled antibody distribution in man. Am. J. Physiol. Imaging **5**:141.

Rescigno A., Gurpide E., 1973. Estimation of average times of residence, recycle, and interconversion of blood-borne compounds using tracer methods. J. Clin. Endocrinol. Metab. **36**:263.

Rescigno A., Lambrecht R.M., Duncan C.C., 1983. Mathematical methods in the formulation of pharmacokinetic models. In: "Tracer Kinetics and Physiologic Imaging." (Lambrecht R.M. and Rescigno A., eds.), page 59. Springer Verlag, Berlin.

Rosenblum M.G., Murray J.L., Haynie T.P., Glenn H.J., Jahns M.F., Benjamin R.S., Frincke J.M., Carlo D.J., Hersh E.M., 1985. Pharmacokinetics of ^{111}In-labeled anti-p97 monoclonal antibody in patients with metastatic malignant melanoma. Cancer Res. **45**:2382.

Rosenblum M.G., Murray J.L., Lamki L., David G., Carlo D.J., 1987. Comparative clinical pharmacology of [111In]-labeled murine monoclonal antibodies. Cancer Chemother. Pharmacol. **20**:41.

Sheiner L.B., 1981. Extended least square fit (ELSFIT). Technical Report, Division of Clinical Pharmacology, University of California.

Working Group Meeting, 1986. Radioimmunoimaging - Report. Hybridoma **5**:166.

INTERSPECIES SCALING IN PHARMACOKINETICS

Kannan Krishnan and Melvin E. Andersen

Chemical Industry Institute of Toxicology
Research Triangle Park, NC 27709

INTRODUCTION

Scaling is the process of utilizing structural and functional features of one system as a basis to predict those of another. "Interspecies scaling in pharmacokinetics" signifies the prediction of *in vivo* chemical disposition behavior in untested species from the experimental observations made in one or more species. Interspecies scaling of the pharmacokinetic processes (i.e., uptake, distribution and clearance) of chemicals can be performed by (1) allometry and (2) physiological modeling. Whereas the allometric approach involves estimation of the pharmacokinetic parameters — clearance, half-life, volume of distribution etc. — in untested species based on their relationship to body mass in several test animal species, physiological modeling involves computer simulation of pharmacokinetics first, in one species, and then extrapolation to other species by scaling the appropriate critical biological determinants of disposition (e.g., blood flow rates, tissue volumes, rates of metabolism).

ALLOMETRY

The allometric approach is empirical, primarily utilizing relationships between body mass and physiological or pharmacokinetic parameters. This is based on the assumption that physiological parameters vary in a regular manner across species and, therefore, these and the pharmacokinetic processes can be described as a power function of the body mass.

According to the allometric approach, a parameter of interest variable among the various species can be quantitatively related to their body mass as follows:

$$Y = aW^b$$

where   Y is the parameter of interest,
            W is the body weight of the organism,
            a is an empirically-derived constant,
            b is an empirically-derived exponent.

New Trends in Pharmacokinetics, Edited by A. Rescigno and A.K Thakur
Plenum Press, New York, 1991

In allometric equations of this form, the sign and magnitude of the exponent (b) indicate how the parameter of interest (Y) varies across biological species with body weight W. The closer b is to 1, the more closely proportional Y is to W. Allometric analyses have generated a number of empirical equations that describe the relationship between W and various anatomical/physiological parameters (Table 1). Similarly, some simple mathematical relationships have been derived to describe the pharmacokinetics of chemicals; the exponent (b) for clearance is approximately 0.75, for volume of distribution (V_d) is 1 and for biological half-life of chemicals ($t_{1/2}$) is about 0.25.

Table 1
Relationship between Anatomical/Physiological Parameters and W
[Mordenti and Cappell, 1989].

Value of exponent	Pattern of change in the parameters of interest (Y)	Example
$b < 0$	A decrease with increasing W	Number of heartbeats (b=−0.25)
$b = 0$	No change regardless of W	Body core temperature (b=0)
$0 < b < 1$	An increase with, but not as fast as W	Blood volume circulation time (b=0.21); blood flow to organs, oxygen consumption, GFR, clearance (b=0.65 to 0.8)
$b = 1$	A proportional increase with W	Blood volume, tissue weights (b≅1)
$b > 1$	An increase with, but at a faster rate than, W	Bone weight (b=1.083)

The single term power function based on W is not always sufficient to scale parameters of interest across all biological species accurately. For example, based solely on W, allometric analysis predicts a maximum lifespan of only 27 years for a 70 kg man [Yates and Kugler, 1986]! The known longevity and several other distinct features (e.g., longer gestational period, delayed sexual maturation, higher degree of socialization, extensive parental investment, smaller brain size, reduced mixed function oxidase levels) of man have been proposed to be related to neoteny, the retention of the formerly juvenile characteristics by adult descendents [Gould, 1977; Yates and Kugler, 1986; Boxenbaum and D'Souza, 1987]. Accounting for the consequences of neoteny results in a more appropriate scaling of several parameters across species. The scaling of the maximum lifespan potential (MLP) for mammals, for example, is adequately described by a two term power function that contains both brain weight (B) and body weight W as follows [Sacher, 1959]:

$$MLP = 10.839 \cdot W^{-0.225} \cdot B^{0.636}.$$

The allometric approach is empirical and generates mathematical relationships that are consistent with the data on pharmacokinetic behavior or other parameters of interest in all test species. The allometric equations are derived by linear regression of the data on the parameter of interest, obtained after extensive experimentation in several animal species. The number of power functions would be determined by the consistency of the relationship in predicting experimental observations in all test

204

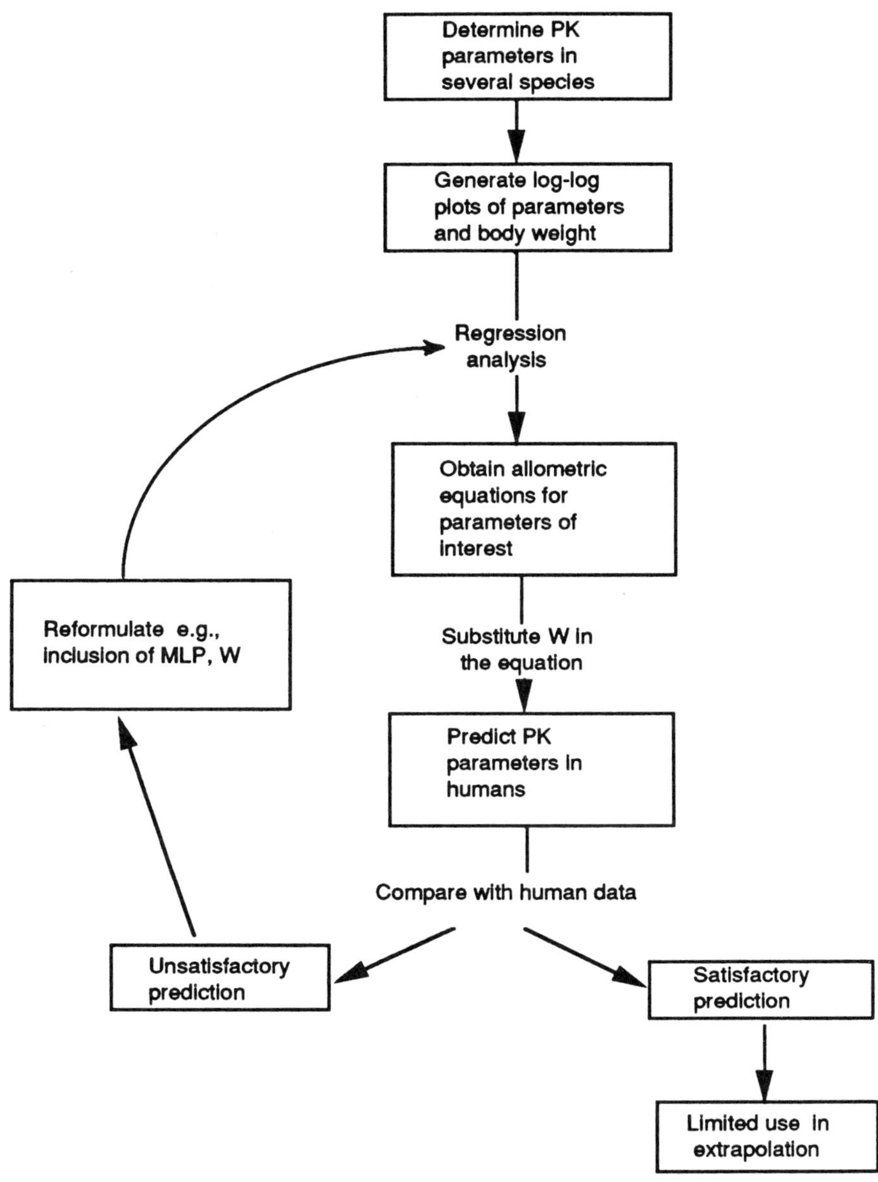

Fig. 1. Schematic representation of the use of allometric approach in interspecies scaling of pharmacokinetic parameters (PK = pharmacokinetic, W = body weight, MLP = maximum lifespan potential).

species (Figure 1). After validation, such mathematical expressions can be, to a limited extent, used for interspecies scaling of pharmacokinetic parameters of chemicals.

Scaling Uptake

One mode of uptake of non-volatile chemicals is through the consumption of water or food. The empirical relationships for these processes have been published. For example, interspecies scaling of daily water consumption in liters (Y) across species can be performed with the following equation [Adolph, 1949]:

$$Y = 0.01 \cdot W^{0.88}$$

The pulmonary uptake of volatile organic chemicals, on the other hand, is primarily determined by the blood solubility of the chemical, cardiac output, and alveolar ventilation rate of the host species. In a simple case, considering the body as a single compartment with a volume of distribution V_d, the rate of change in the amount of a chemical in the body (dA_{body}/dt) can be described as follows:

$$\frac{dA_{body}}{dt} = \frac{V_d dC}{dt} = Q_c C_a - Q_c C_v,$$

where
$\dfrac{dC}{dt}$ = Rate of change in the concentration of the chemical in the body,

Q_c = Blood flow rate (L/hr),

C_a = Concentration in arterial blood (mg/L),

C_v = Concentration in venous blood (mg/L).

Therefore,

$$\frac{dC}{dt} = \frac{Q_c C_a}{V_d} - \frac{Q_c C_v}{V_d}$$

Since [Ramsey and Andersen, 1984]

$$C_a = \frac{Q_p C_{inh} + Q_c C_v}{(Q_c + Q_p/P_b)}$$

where
P_b = Blood:air partition coefficient,

Q_p = Alveolar ventilation rate (L/hr),

C_{inh} = Concentration of the chemical in the inhaled air (mg/L),

the rate of change in the concentration of the chemical in the body can be rewritten as:

$$\frac{dC}{dt} = \frac{Q_c P_b Q_p C_{inh}}{(P_b Q_c + Q_p) V_d} - \frac{Q_p Q_c C_v}{(P_b Q_c + Q_p) V_d}$$

or,

$$\frac{dC}{dt} = K (P_b C_{inh} - C_v)$$

where

$$K = \frac{Q_p Q_c}{V_d (P_b Q_c + Q_p)}$$

At steady state,

$$C_v = P_b \cdot C_{inh}.$$

Considering the rate constant K, it is evident that K will be determined by $\frac{Q_p}{V_d P_b}$ when P_b is large ($\gg 1$), and by $\frac{Q_c}{V_d}$ if P_b is very small ($\ll 1$). The former is a case of ventilation-limited uptake while the latter is representative of the perfusion-limited uptake process.

The equation for the rate constant of ventilation-limited uptake can be rewritten, using the allometric equations for Q_p and V_d as follows:

$$Q_p = a_1 W^{0.75}$$

$$V_d = a_2 W^{1.0}$$

Therefore,

$$K = \frac{a_1 W^{0.75}}{a_2 W^{1.0} P_b} = \frac{a_1}{a_2 W^{0.25} P_b}$$

Similarly, the perfusion-limited uptake process can be described by substituting the allometric equations for blood flow and volume of distribution. Thus,

$$K = \frac{a_1}{a_2 W^{0.25}}$$

where
$$Q_c = a_1 \cdot W^{0.75}$$
$$V_d = a_2 \cdot W^{1.0}$$

The uptake processes result in some concentration of the parent chemical in the blood. Figure 2a is a stylized plot of the time course of venous blood concentration ($C_{ss} = P_b \cdot C_{inh}$) resulting from exposure to some concentration of a hypothetical chemical in a number of mammalian species. The time unit in this graph represents normal or chronological time. Since the relative rate constants are different among the various species (mouse = 7.51, rat = 4.15, dog = 1.66, humans = 1.0), the time course blood concentrations are variable too. However, the areas under the curve ($t = 0 \rightarrow \infty$) for various species are equal and the shapes of the curves are expected to be similar when C_v is plotted against physiological time (Figure 2b). The physiological time, a species-dependent time unit of chronological time required to complete a species-independent physiological event [Boxenbaum, 1982], is derived empirically using the relationship: chronological time/$W^{0.25}$. The rationale for this relationship can be examined using a simple case of a plasma profile described by a monoexponential equation of the following form:

$$C = C_1 \exp(-\lambda t)$$

where
C is the plasma concentration at time t,
C_1 is the exponential coefficient,
λ is the exponential constant.

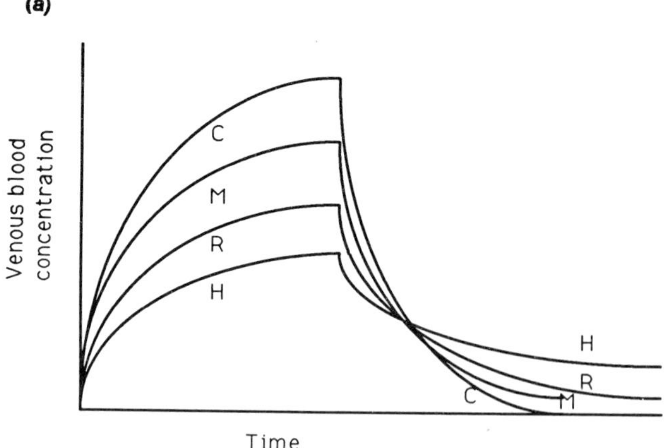

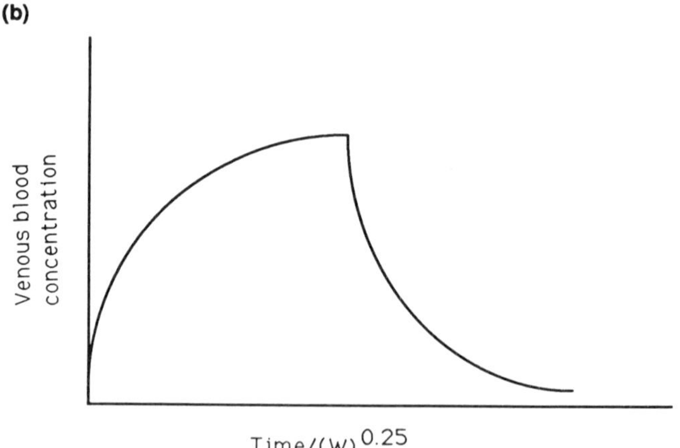

Fig. 2. Interspecies comparison of the venous blood levels of a hypothetical chemical resulting from inhalation exposure (H = human, C = canary, R = rat, M = mouse) as a function of chronological (a) and physiological time (b).

Note that,

$$\lambda = \frac{CL}{V_d} \text{ and } C_1 = \frac{D}{V_d}$$

where   CL = Clearance (L/hr)
           V_d = Volume of distribution (L)
           D = Dose (mg)
By substitution,

$$C = \frac{D}{V_d} \exp\left(-\frac{CL}{V_d} t\right)$$

208

However, allometrically

$$V_d = a_1 \cdot W^{1.0}$$

$$CL = a_2 \cdot W^{0.75}$$

where a_1 and a_2 are allometric coefficients.

Therefore,

$$C = \frac{D}{a_1 W^{1.0}} \exp\left[-\left(\frac{a_2}{a_1}\right) W^{(0.75-1.0)} t\right]$$

or rearranging,

$$\frac{C}{D/W^{1.0}} = \frac{1}{a_1} \exp\left[-\left(\frac{a_2}{a_1}\right) \frac{t}{W^{0.25}}\right]$$

A Naperian semilogarithmic plot of $[C/D/W^{1.0}]$ versus $[t/W^{0.25}]$ produces a straight line with a slope $-\frac{a_2}{a_1}$ and an intercept of $\frac{1}{a_1}$. This is an example of an "elementary Dedrick plot" [Dedrick et al., 1970], in which one unit of physiological time is equivalent to $W^{0.25}$ of chronological time. The area-under-the-curve (AUC) will then be equal in all species. Since,

$$AUC = \int_0^\infty \frac{C}{D/W^{1.0}} d\left(\frac{t}{W^{0.25}}\right)$$

$$= \frac{1}{D/W} \cdot \frac{1}{W^{0.25}} \cdot \int_0^\infty C \, dt$$

and by definition,

$$\int_0^\infty C \, dt = \frac{D}{CL} = \frac{D}{a_2 W^{0.75}},$$

the preceding equation can be simplified such that,

$$AUC = \left(\frac{1}{D/W^{1.0}}\right) \left(\frac{1}{W^{0.25}}\right) \left(\frac{D}{a_2 W^{0.75}}\right) = \frac{1}{a_2}.$$

Since the physiological time is equivalent to $t/W^{0.25}$, which itself is a function of the clearance allometric exponent, it is logical that a plot based on $t/W^{0.25}$ produces equivalent AUCs in different species [Boxenbaum and Ronfield, 1983]. However, the interspecies superimposability will not occur if the allometric exponent for volume of distribution is not equal to 1. In the above example, the consequences of neoteny have not been considered; this might become more important for low-clearance drugs and chemicals especially when extrapolating to humans (*vide infra*).

Scaling Distribution

The volume of distribution of a chemical can be expressed allometrically with reference to W. The steady-state volume of distribution (V_{dss}) can be described as follows [Boxenbaum, 1984]:

$$V_{dss} = V_p + V_T \left(\frac{fu}{fu_T} \right)$$

where V_p = Volume of plasma,
$\quad\quad V_T$ = Physiological volume outside the plasma into which drug distributes,
$\quad\quad fu$ = Unbound fraction of drug in plasma,
$\quad\quad fu_T$ = Mean, weighted, unbound fraction of drug outside the plasma (i.e., in the tissues).

Frequently, $V_{dss} \gg V_p$, and fu does not often vary greatly between species, implying that,

$$V_{dss} \propto \left(\frac{V_T}{fu_T} \right)^{\alpha}$$

Since α tends to cluster about unity, it appears that for a number of drugs, the ratio $\frac{V_T}{fu_T}$ is proportional to W.

Scaling Clearance

The extent of hepatic clearance of a chemical is dependent on its tissue solubility characteristics, the metabolic capacity of the organism, and blood flow to the metabolizing organ(s). Assuming liver to be the sole metabolizing organ, clearance (CL_H) can be expressed as follows [Wilkinson and Shand, 1975]:

$$CL_H = \frac{Q_H \, CL_{int}}{(Q_H + CL_{int})}$$

where Q_H = Blood flow to liver (L/hr)
$\quad\quad CL_{int}$ = Intrinsic clearance (L/hr) = V_{max}/K_m

Hepatic clearance of a chemical may either be blood flow-limited (if $CL_{int} \gg Q_H$, then $CL_H = Q_H$), or be metabolism-limited (if $CL_{int} \ll Q_H$ then $CL_H = CL_{int} = V_{max}/K_m$). For flow-limited clearance (compounds with high extraction ratios), coherent relationships will be expected between clearance and W. This allometric relationship is simply a consequence of the primary relationship between hepatic blood flow and W. A reasonable correlation has been found to exist between the clearance of a number of compounds and W for various animal species, including humans. However, these compounds tend to be cleared either renally or by metabolic reactions not involving the hepatic mixed function oxidase system.

For compounds cleared primarily by low affinity phase I metabolic reactions in the liver (low clearance compounds) the elimination process is independent of blood flow and is determined by the magnitude of biochemical constants such as V_{max} and K_m, which cannot be expected a priori to vary coherently across species. Even though allometric scaling based on W shows good correlation, the CL_{int}, predicted for a number of chemicals from other mammalian species based solely on W, is

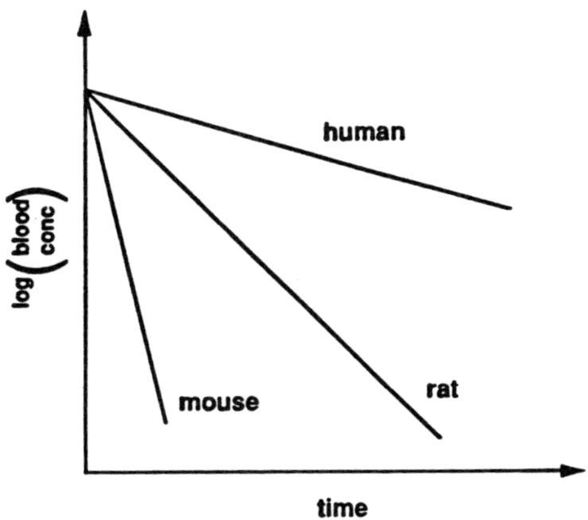

Fig. 3. A semi-logarithmic plot of the blood-time course of a
hypothetical chemical in mouse, rat, and humans.

consistently higher than what has been observed in humans. The 'outlier' feature of
man has been explained by the lower intrinsic clearance, a consequence of neoteny
[Boxenbaum and D'Souza, 1987]. Thus, when longevity, also a consequence of
neoteny, in addition to W, is taken into consideration, a better prediction of the
clearance of compounds metabolized primarily by the hepatic mixed function
oxidases could be attained [Boxenbaum, 1980, 1983; Campbell and Ings, 1988]:

$$CL = \frac{aW^b}{MLP}$$

where   MLP = Maximum documented life-span for the species of interest (years).
   It is also the reason why the allometric equations based on both W and brain
weight adequately predict the hepatic clearance of certain compounds [Ings, 1990]:

$$CL = a\ W^b\ B^c$$

where   B = Brain weight (g),
      c = An empirically-derived exponent of B.
   Renal clearance, a measure of excretory functions, has also been calculated for
chemicals in a number of species using allometric equations. The clearance of several
drugs has been found to be related to W [Adolph, 1949]. The allometric equations
based on W all had an exponent value of less than 1, usually between 0.69-0.89. All
species are apparently similar in excreting xenobiotics when the rate of elimination is
expressed as a function of internal factors contributing to creatinine excretion. Thus,
the interspecies ratio of $CL_{methotrexate}/CL_{creatinine}$ has been found to be constant
regardless of the species and body weight [Dedrick et al., 1970].
   The clearance of compounds with high extraction ratios (e.g., methotrexate)
varies as a fractional power of W; but the volume of distribution is constant across
the species. Therefore, the AUC would be expected to increase with increasing W
for a constant dose of a chemical (Fig 3).

   This can be examined with the following expression:

$$t_{1/2} = 0.693 \, \frac{V_d}{CL}$$

Since both V_d and CL are allometrically related to W, the terminal half-life ($t_{1/2}$) becomes:

$$t_{1/2} = 0.693 \, \frac{a_1 W^{1.0}}{a_2 W^{0.75}}$$

Therefore,

$$t_{1/2} \propto W^{0.25}$$

Apparently, the systemic availability of many organic chemicals is more prolonged in larger animals (such as man) than in smaller animals (such as rodents). This trend of variation can also be explained with the following observations:

1. Smaller mammals have proportionally larger organs responsible for clearance (e.g., liver, kidney) than do larger mammals [Adolph, 1949; Boxenbaum, 1980].

2. Smaller animals receive greater blood flows per unit volume of tissue than do larger animals [Adolph, 1949; Holt and Rhode, 1976; Boxenbaum, 1980]; and

3. The mean blood volume circulation time scales to $W^{0.21}$. Thus, smaller animals generally eliminate drugs more rapidly than larger animals by virtue of rapidity of blood circulation to organ(s) of elimination.

This is especially true for chemicals, such as methotrexate that are primarily cleared by the kidneys and for which clearance is flow dependent. The clearance rates for all species can be fit to a single regression line, if the chronological time is adjusted to physiological time, to decelerate the rates in smaller species in comparison to larger species. For low clearance compounds, which are eliminated by flow-independent processes, there still exists good correlation, but such relationships consistently predict a higher clearance rate for humans (*vide supra*).

Since the allometric approach characterizes the regularities without necessarily understanding why, its use in interspecies extrapolation of pharmacokinetic behavior of chemicals can be misleading. Whereas in the allometric approach equations are derived to describe a phenomenon such that they are consistent with the data set, in physiological modeling the physico-chemical characteristics of the chemical and physiology of the test species define the system structure and account for the pharmacokinetic behavior. Whereas the allometric approach is empirical, the physiological approach is mechanistic and provides more confidence for extrapolating pharmacokinetic behavior of chemicals to a range of exposure doses, administration routes, and target species.

PHYSIOLOGICAL MODELING

Physiological modeling involves computer simulation of the pharmacokinetics of chemicals based on their tissue solubility characterisitics, metabolism/binding rates, and physiology of the test species. A physiologically-based pharmacokinetic model consists of a series of tissue compartments (Figure 4) each of which receives the chemical via arterial blood and loses the free chemical via venous blood leaving the tissue. These biologically relevant compartments, arranged in an anatomically

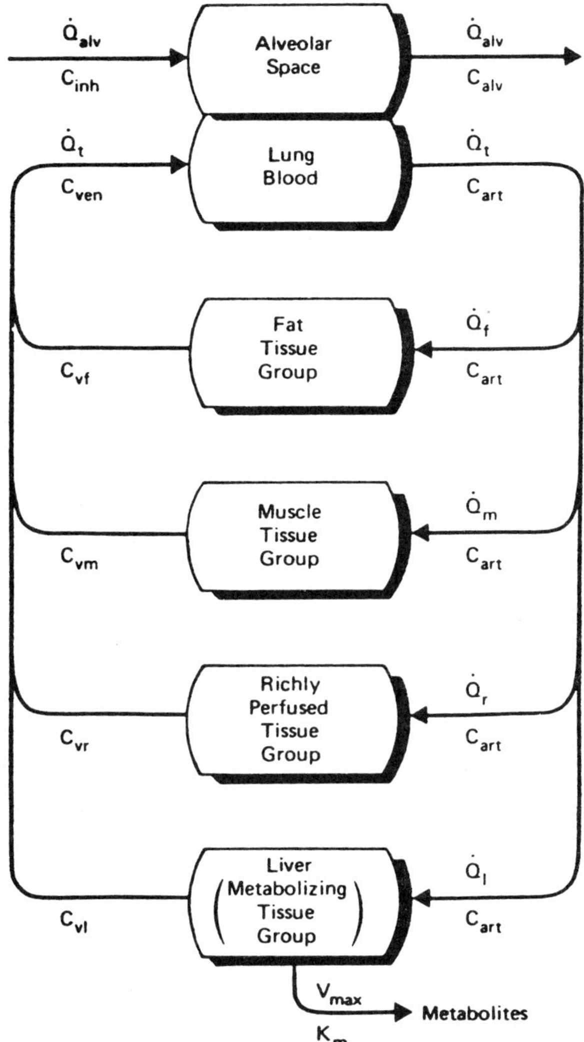

Fig. 4. The structure of a particular physiological model used to describe the pharmacokinetics of volatile chemicals. Q terms are air and blood flow rates; C terms are concentrations. These are indexed to individual tissue compartments - fat (f), muscle (m), richly perfused tissues (r) and liver (l). Effluent venous concentrations have a double lettered subscript. Q_{alv} and Q_t are alveolar ventilation and cardiac output. The subscripts inh, alv, art, and ven signify inhaled air, exhaled air, arterial blood and venous blood. Kinetic constants for liver metabolism are V_{max} (maximum rate of metabolism) and K_m (binding affinity of the substrate with metabolizing enzyme). From Ramsey and Andersen [1984].

accurate manner, are defined with appropriate physiological characteristics. The compartments may represent a single tissue or a tissue group that consists of several tissues with similar blood flow and solubility characteristics (e.g., adrenals, kidney, thyroid, brain, heart, and hepatoportal system can be pooled into one compartment, referred to as "richly perfused tissues").

Literature review
In vitro experiments
In vitro/In vivo experiments
physiological parameters
partition coefficients
biochemical rate constants
Formulation of a rodent PBPK model
Simulation
Eg., Time-course data on blood levels, GSH depletion, tissue concentrations, urinary metabolite levels
obtain from litr. (or) in. animal expts
Adequate simulation
Model scale-up
Inadequate simulation
Refine estimates

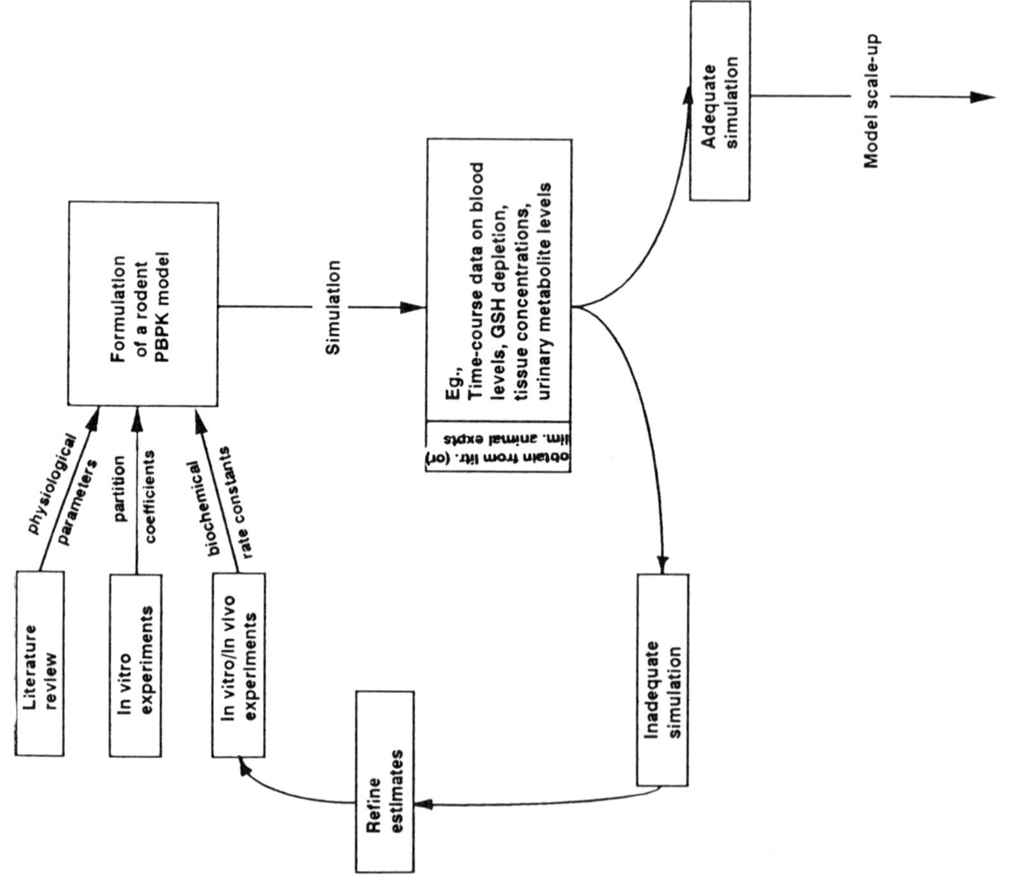

214

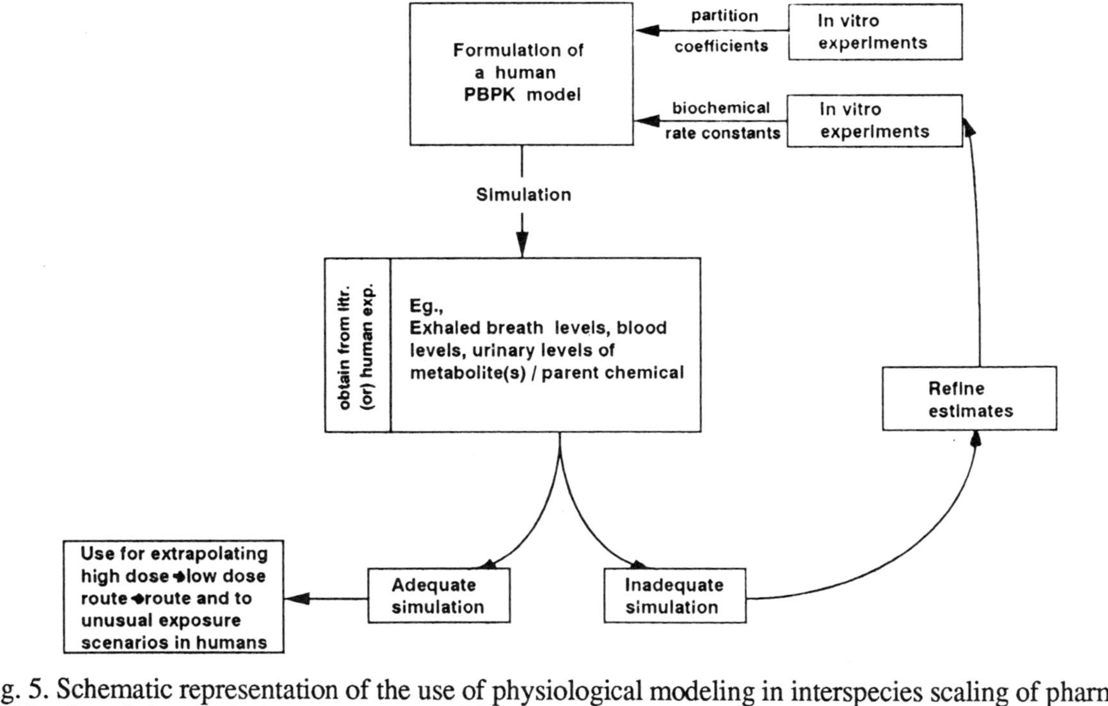

Fig. 5. Schematic representation of the use of physiological modeling in interspecies scaling of pharmacokinetics of chemicals (exp = exposure, GSH = glutathione, litr = literature, PBPK = physiologically based pharmacokinetic).

The use of physiological modeling approach for interspecies scaling of pharmacokinetic behavior of chemicals involves several steps (Figure 5). First, a rodent model is developed to describe the uptake and disposition of the chemical. The physiological parameters in the rodent model are then scaled, and the chemical-specific parameters are determined for the species of interest (e.g., humans). The description of the various pharmacokinetic processes (i.e., uptake, distribution and clearance) in both models essentially remains the same, the rodent model serving to identify the critical biological determinants of the kinetic processes. The following is an account of some of the descriptions frequently employed in physiological pharmacokinetic modeling. This is not intended as an exhaustive account of the field of physiological modeling.

Uptake

The equilibration of volatile chemicals in the lung is described as follows [Ramsey and Andersen, 1984]:

$$C_a = \frac{Q_p C_i + Q_c C_v}{Q_c + (Q_p/P_b)}$$

Dermal absorption of vapors can be accounted for, by introducing a diffusion-limited compartment to represent skin as a portal of entry [McDougal et al., 1986]:

$$V_{sk} \cdot \frac{dC_{sk}}{dt} = K_p \, A \, (C_{air} - \frac{C_{sk}}{P_{sk/air}}) + Q_{sk} \, (C_a - \frac{C_{sk}}{P_{sk/b}})$$

where V_{sk} = Volume of skin (L),
K_p = Permeability constant (cm/hr),
C_{sk} = Concentration of the chemical in skin compartment (mg/L),
A = Exposed surface area (cm^2),
C_{air} = Concentration of the chemical in air contacting skin (mg/L),
Q_{sk} = Blood flow to skin (L/hr),
$P_{sk/b}$ = Skin:blood partition coefficients,
$P_{sk/air}$ = Skin:air partition coefficients.

Oral uptake can be modeled by introducing a first-order uptake rate constant [Ramsey and Andersen, 1984; Fisher et al., 1990]:

$$\frac{dA_a}{dt} = K_a D_o \, e^{-K_a t}$$

where $\dfrac{dA_a}{dt}$ = Rate of appearance of the chemical in the liver (mg/hr),

K_a = First order uptake rate constant (hr^{-1}),
D_o = Oral dose of the chemical (mg).

In this case, the compound is assumed to appear in the liver after absorption. Therefore, the amount in the liver becomes:

$$\frac{dA_l}{dt} = Q_l(C_a - C_{vl}) + \frac{dA_a}{dt}$$

where $\dfrac{dA_l}{dt}$ = Rate of change in the amount of a chemical in the liver compartment (mg/hr),

Q_l = Blood flow rate to the liver (L/hr)

C_{vl} = Concentration of the chemical in the venous blood leaving liver (mg/L).

If the intragastric absorption occurs at a constant rate, then a zero order uptake rate constant is added to the liver mass balance equation.

Entry of a chemical via the intravenous route can be described by including a term representing the rate of infusion into the mixed venous blood as follows:

$$C_v = \frac{[K^o + (Q_i C_{vi})]}{Q_c}$$

where K^o = Zero order infusion rate constant (mg/hr),

Q_c = Cardiac output (L/hr).

The infusion rate can be adjusted for very long or very short time periods by adjusting the infusion time parameter T (hr), where

$$K^o = [\, K^o;\ 0 < t < T,\ o;\ t > T\,]$$

and t represents elapsed time (hr). The total infused dose in mg is then given by $(T)(K^o)$.

Distribution

Tissue distribution of chemicals is considered to be a function of both the solubility of the chemical and the rate of blood flow through the organ. Thus, the rate of change in the amount of a chemical in a tissue compartment $\dfrac{dA_i}{dt}$ can be described as the product of the organ's perfusion rate (Q_i) and the arteriovenous concentration difference ($C_a - C_{vi}$):

$$\frac{dA_i}{dt} = Q_i\,(C_a - C_{vi}) = Q_i\left(C_a - \frac{C_i}{P_i}\right)$$

where C_i = Concentration of the chemical in the tissue "i" (mg/L)

P_i = Tissue:blood partition coefficient.

The amount of the parent chemical remaining in a tissue compartment is equal to the total amount delivered minus that metabolized.

Clearance

Metabolic clearance in individual tissues can be described by introducing one or more terms, to account for the amount lost by metabolism which may be first order, second order, or a saturable process.

The rate of the amount of chemical metabolized dA_{met}/dt can be calculated as follows:

$$\frac{dA_{met}}{dt} = K_f C_{vi} V_i \qquad \text{(First order process)}$$

$$= K_s C_{vi} V_i\, C_{cf} \qquad \text{(Second order process)}$$

$$= V_{max} C_{vi}/(K_m + C_{vi}) \qquad \text{(Saturable process)}$$

where
K_f = First order rate constant (hr^{-1}),
K_s = Second order rate constant (L/mg·hr),
K_m = Michaelis affinity constant for enzymatic reaction (mg/L),
V_i = Volume of the tissue "i" (L),
V_{max} = Maximal velocity of enzymatic reaction rate (mg/hr),
C_{cf} = Concentration of cofactor in the tissue (mg/L).

Thus the total amount of the chemical in a metabolizing organ (e.g., liver) becomes:

$$\frac{dA_l}{dt} = Q_l\,(C_a - C_{vl}) + \frac{dA_a}{dt} - \frac{dA_{met}}{dt}.$$

Clearance by exhalation is an important process for volatile organic chemicals. Although, metabolic clearance (e.g., oxidation) in liver is often capacity-limited and therefore follows zero order kinetics at sufficiently high concentrations, clearance by pulmonary exhalation is a first order process at all concentrations. The sum of hepatic metabolic clearance (Cl_{met}) and pulmonary clearance (Cl_{exh}), which represents the total body clearance (CL_{tot}) of a volatile metabolized chemical, can be expressed as follows:

$$CL_{tot} = CL_{met} + CL_{exh} \quad \text{or,}$$

$$CL_{tot} = \frac{Q_l V_{max}/(K_m + C_{vl})}{(Q_l + V_{max})/K_m C_{vl}} + \frac{Q_p Q_p}{(Q_p + Q_c/P_b)}$$

As blood concentration of a chemical decreases, the concentration in the term CL_{met} becomes smaller than K_m, and the percentage of chemical exhaled reaches a constant, minimum value. As the liver concentration increases, the metabolism becomes saturated, hepatic clearance diminishes and the percentage exhaled increases. Metabolic clearance occurring in organs other than liver can also be included [Andersen et al., 1987; Krishnan et al., 1991].

Renal clearance of the unbound chemical can be modeled by accounting for the excretory processes of interest. For example, glomerular filtration of unbound chemical that is neither secreted nor reabsorbed can be described as follows [Rowland, 1986]:

$$\frac{d\,Arc}{dt} = fu\cdot GFR$$

where
$\dfrac{d\,Arc}{dt}$ = Rate of the amount of chemical cleared renally (mg/min),
fu = Fraction of the chemical in plasma unbound (mg/ml),
GFR = Glomerular filtration rate (ml/min).

Reabsorption can be modeled by assuming that a fixed portion of the chemical, which is either filtered and/or secreted, is then reabsorbed [Levy, 1980], or by accommodating the known dependence of reabsorption on urine flow [Wesson 1954; Tangliu et al., 1983; Hall and Rowland, 1983].

Scaling

The physiological model once formulated, by integrating information on animal physiology (e.g., cardiac output, ventilation rates, tissue volumes, blood flow rates), rate constants for metabolism (K_f, K_s, K_m, V_{max}) and partition coefficients of a chemical (blood:air, tissue:blood), can be used to simulate its kinetic behavior in the test species. After validation in one test species, it can be used for interspecies scaling of the pharmacokinetic behavior of chemicals, by scaling/determining the model parameters for the species of interest, as follows.

Physiological Parameters

The physiological parameters in the model can be scaled allometrically. Thus, the organ volumes are scaled across the species to the first power of the body weight. The cardiac output, blood flow to organs, and alveolar ventilation rates are scaled to the $3/4^{th}$ power of the body weight.

Metabolic Constants

Whereas the physiological parameters vary coherently from species to species, the kinetic constants for metabolizing enzymes do not necessarily follow any type of readily predictable patterns, making interspecies extrapolation of xenobiotic metabolism difficult. The large majority of drugs exhibit linear (dose/concentration-dependent) elimination kinetics over the concentration range of interest, because the K_m for overall metabolism is much greater than C_{vl}. In these instances, intrinsic clearance is approximated by the ratio of V_{max}/K_m and the clearance is a constant. For drugs that have nonlinear kinetics, the K_m for one or more routes of metabolism may not be sufficiently greater than C_{vl}, and therefore the clearance will vary depending on the concentration.

Thus, with inhalants with high affinity (low K_m) for the metabolizing enzymes, interspecies extrapolation is possible because physiological parameters (particularly liver blood flow rates) are rate-limiting for metabolism at low inhaled concentrations. While maximal metabolic rate (V_{max}) has in such cases been scaled to the $3/4^{th}$ power of W, as the basal metabolic rate, the half-saturation constants (K_m) have been considered to be species-invariant [Ramsey and Andersen, 1984]. Assignment of this fractional power of scaling enables the maximum rates of inherently capacity-limited metabolic processes to achieve approximate equivalence among the various species. This may be useful as a crude approximation but should be used only when other direct measurements of metabolic parameters are not available.

It is preferable, therefore, that the metabolic rate constants for xenobiotics be determined for the species of interest. The *in vivo* metabolic rate constants for the human model, for example, can be estimated from the *in vitro* rate constants. This is based on a parallelogram approach in which the relationship between the *in vivo* and *in vitro* rate constants in rodent species is examined to estimate the *in vivo* rate constants from *in vitro* data obtained with human liver tissues [Reitz et al., 1988].

Partition Coefficients

Tissue:air partition coefficients of chemicals tend to be relatively constant across species, while blood:air partition coefficients show some species-dependent variability. Therefore, the tissue:blood partition coefficients for the species (e.g., humans) to which the pharmacokinetic data are to be scaled can be calculated by dividing the rodent tissue:air partitions by the appropriate (e.g., human) blood:air partition values.

UTILITY AND PROBLEMS OF INTERSPECIES SCALING METHODOLOGIES

Allometric scaling (particularly from rodents to humans) of pharmacokinetics presents some serious problems. For example, some aspects such as the following, cannot be adequately described/accomodated:

1. Effect of interspecies variation in fat:W ratios on chemical disposition,
2. Individual differences in pharmacokinetics among humans, and
3. Interspecies differences in metabolism.

The effect of varying per cent fat between species can have a profound effect on the distribution of lipophilic chemicals. Further, post-exposure metabolism of stored xenobiotics can play an important role in causing interspecies differences in their disposition in such cases. This can be addressed with the physiological modeling approach [Clewell and Andersen, 1985; Medinsky, 1990].

The general rules of thumb generated by the allometric approach are not adequate to predict a specific behavior in individuals of human population. In such cases, physiological modeling offers a major advantage in that the quantitative changes in the pharmacokinetic behavior of chemicals due to physiological and pathological alterations can be predicted by perturbation of the appropriate model parameter(s) (Table 2).

Species difference in metabolism and binding is another important factor that needs to be considered in interspecies scaling of pharmacokinetics. When metabolism produces active metabolites, the factors considered in allometry are often not relevant [Mordenti and Chappell, 1989]. However, physiological modeling offers the flexibility of accomodating specific information on the species-specific cofactor levels and metabolic rate constants such that quantitative and qualitative differences in metabolism can be accounted for.

The ability to conduct interspecies extrapolations with physiological models arises from the fact that these models are developed with a mechanistic understanding of the factors which determine the disposition of a chemical. These models, once developed in rodents, lead to the identification of the critical biological determinants of uptake and disposition of a particular chemical. The integration of these determinants within the physiological structure permits scaling to any other species of interest, including humans.

INTERSPECIES SCALING OF EXPOSURE CONCENTRATIONS

The use of allometric approach to estimate safe exposure levels of drugs/chemicals has been criticized for it yields a range of values depending on which parameter is considered or given weight. For instance, if we consider the allometric coefficient for body volume $W^{1.0}$ to be the appropriate measure, then the dose predicted for humans would be 3,500 times that for a mouse; if we use metabolic rate $W^{0.75}$ it would be 400 times greater; if we use surface area $W^{0.67}$, it would be 200 times greater; and if we use the heart beat interval $W^{0.21}$, the predicted equivalent safe dose for humans would be 8 times greater than that for a mouse [Yates and Kugler, 1986]. Thus, there exists an enormous range of possible doses depending upon the parameter chosen. This ambiguity is unconvincing. However, the allometric approach can be useful when the pharmacokinetic processes of a drug/chemical remain first order at all doses in all species, and the plasma concentration of the chemical is the lone parameter of interest [Ings, 1990].

Mordenti and Chappell [1989] have proposed a methodology (using a power

Table 2
Summary of Factors Causing Interindividual Variation of Critical
Biological Determinants of Pharmacokinetics (I - increase; D-
decrease; GFR - glomerular filtration rate). From: Krishnan and
Andersen, 1991.

Model parameter	Type of change	Influencing factors
I. Physiological		
1. Alveolar ventilation rate	I	Physical activity, pregnancy
	D	Obesity, ventilatory disorders, certain medications (e.g., Hypnotics), living at high altitudes
2. Cardiac output	I	Physical activity, hot ambient temperatures
	D	Gender (women)
3. Blood flow rates		
Muscle	I	Physical activity, thin persons
Kidney	D	Physical activity
Several organs	I	Pregnancy, high altitudes
	D	Disease state, cardiac failure
4. Tissue volumes		
% Fat	I	Obesity, aging, pregnancy, women
% Muscle	D	Aging
5. GFR	I	Pregnancy
	D	Aging
II. Biochemical		
1. Hepatic metabolic capacity	I	High protein diet, alcohol intake, beef, cruciferous vegetables, chronic pulmonary diseases, coexposure to certain xenobiotics
	D	Oversupply of dietary carbohydrate thiamin and iron; medication, genetic polymorphism, cold ambient temperature, photoperiod, coexposure to xenobiotics
2. Plasma protein binding	D	Fasting, high fat food, disease states

function of 0.7 as a compromise between body surface area and blood flow rates) to calculate equivalent doses between animals and humans. This approach is based on allometric scaling of clearance to achieve the same average steady-state levels of drugs in the species of interest:

$$D_h = D_r \, \frac{F_r}{F_h} \, \frac{T_h}{T_r} \, \frac{CL_h}{CL_r}$$

Allometrically,

$$CL = a \, W^{0.7}$$

Therefore,

$$D_h = D_r \frac{F_r}{F_h} \frac{T_h}{T_r} \left(\frac{W_h}{W_r}\right)^{0.7}$$

where D = Dose (mg),
T = Dosing interval (min),
F = Bioavailable fraction of the drug (0 to 100%),
h = human,
r = rodent.

Thus,

$$N_h = N_r \frac{F_r}{F_h} \frac{D_r}{D_h} \left(\frac{W_h}{W_r}\right)^{0.7},$$

where N_h = Number of daily doses in humans,
N_r = Number of daily doses in rodents.

Again, if metabolism by the hepatic mixed function oxidases is a major process of clearance, a term to account for the brain weight or maximum lifespan potential should be included.

Recently, Ings [1990] has suggested the use of a modified Hill equation to calculate the maximum allowable exposure level in animals, and in turn, used it to estimate the safe dose for humans. Thus,

$$CL = \frac{F \cdot D}{AUC}$$

or rearranging,

$$F\,D = CL \cdot AUC$$

If clearance for people is not known, it can be estimated allometrically; since $CL = a \cdot W^b \cdot B^c$,

$$D = \frac{a \cdot W^b \cdot B^c \cdot AUC}{F}$$

Maximum allowable safe dose for humans can be calculated with this equation and compared with the therapeutic dose after correcting for any bioavailability differences. This approach has not yet been tested for its validity and usefulness.

If it is sufficient to know the plasma concentration of the chemical alone, the allometric approach might be adequate for interspecies scaling of kinetically-equivalent doses. However, several conditions such as the following should also be met: (1) first order pharmacokinetics in all species, (2) a similar percentage of linear protein binding over the concentration range and species of interest, (3) the elimination processes should be physical (i.e., renal or biliary), and (4) enough data must be available for satisfactory linear regressions [Mordenti, 1986].

For several chemicals, the presumed proportional rise in tissue or blood levels of a metabolite with increasing exposure concentrations does not occur due to the limited capacity for processes such as metabolism, excretion and binding [Levy, 1968; Gehring et al., 1976; van Ginneken and Russel, 1989]. Especially considering the very high doses used in qualitative toxicology studies (e.g., carcinogenicity bioassays), the responses seen are frequently not simple functions of the exposure concentration but are more complex due to the dose-dependencies in pharmacoki-

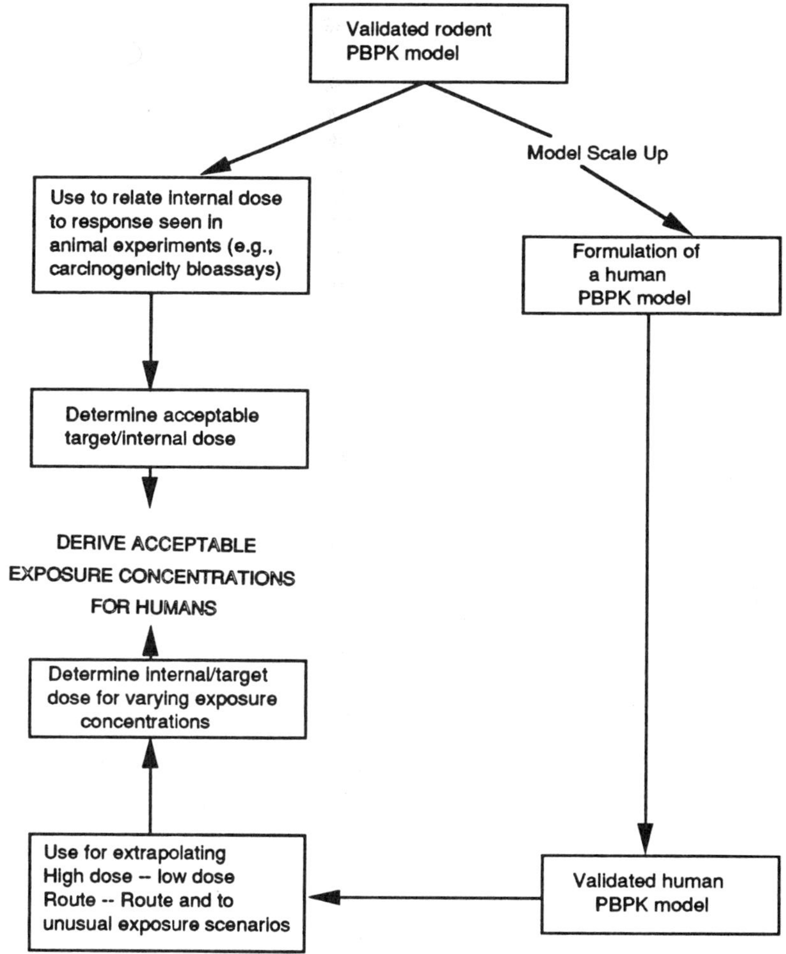

Fig. 6. Schematic representation of the use of physiologically based pharmacokinetic (PBPK) models in risk assessment.

netics. The description and prediction of the pharmacokinetic behavior in such cases, where the allometic approach can not be relied upon, are adequately done with physiological modeling.

The principal application of the physiological pharmacokinetic models is to predict the tissue dosimetry of the toxic parent chemical or its reactive metabolite. Integrating the internal dose of the toxic form of a chemical at its target site can undoubtfully provide a better basis of relating to the observed toxic effects than the external or exposure concentration of the parent chemical. Because they allow the prediction of target tissue dosimetry in people, physiological models can help reduce the uncertainty of the extrapolation procedures adopted in conventional risk assessment approach. The process of utilizing physiological pharmacokinetic models to predict tissue dosimetry in humans for use in risk assessment is schematically depicted in Figure 6.

INTERSPECIES SCALING IN RISK ASSESSMENT

Conventionally in health risk assessment, equivalent safe exposure concentration or dose of a chemical for humans is predicted from that of the rodents using a conversion factor, usually W to the power of 2/3 or 1. Whereas the former is considered scaling on the basis of body surface area, the latter is direct body weight scaling. Based on the fact that the metabolic rates for different animal species are not directly related to their respective W but more closely related to body surface area [Kleiber, 1947], the 2/3 scaling has been used by U. S. Environmental Protection Agency for interspecies scaling of equivalent exposure doses. Thus,

$$D_h = D_r \left(\frac{W_h}{W_r}\right)^{2/3},$$

where D_h = Dose for humans (mg),
D_r = Dose for rodents (mg),
W_h = Body weight of humans (kg),
W_r = Body weight for rodents (kg).

The problem, however, is that this approach is used universally for all chemicals regardless of the toxic moiety (parent or metabolite) and mechanism of toxicity. This is a sensible approach only if (1) the parent chemical is responsible for toxicity and (2) total exposure is the appropriate correlate of toxicity [National Research Council, 1986]. Whereas the body surface scaling is appropriate for direct-acting toxicants, the body weight scaling is appropriate for chemicals which produce stable metabolites. In these cases, the AUC is the appropriate correlate of toxicity. Since the AUC of direct-acting toxicants is dependent on clearance which at low inhaled concentrations is primarily influenced by blood flow to the metabolizing organs, a 2/3 scaling would be appropriate. For chemicals that produce stable metabolites, both processes - metabolite production and elimination - are likely to be related to the fractional power of body weight; thus in such cases, a direct body weight scaling is appropriate for deriving kinetically-equivalent doses in humans. However, for compounds producing reactive, short-lived metabolites, the appropriate surrogate of tissue exposure would be the integral of the amount of metabolite produced per unit volume of the target tissue. Since the metabolite production is related to a fractional power of W, and the volume of the target tissue varies directly with W, a 1/3 scaling would appear to be appropriate [National Research Council, 1986] with inhalation or dietary intake.

In summary, most physiological parameters vary predictably with W across animal species. Pharmacokinetic behaviors of chemicals which depend on the physiological parameters, also are expected to show a consistent relationship to W. However, biochemical constants are not necessarily expected to show any particular dependence on W. These should, therefore, be treated cautiously in interspecies scaling. It is preferable to determine the metabolic rate constants for each chemical and biological species of interest, and to describe the pharmacokinetic behavior with a physiological model. In the absence of such models, interspecies scaling of kinetically- equivalent doses using a single scaling factor for all chemicals is not appropriate, and due consideration should be given to the toxic moiety and the mechanisms of toxicity.

REFERENCES

Adolph, E. F., 1949. Quantitative relations in the physiological constitutions of mammals. Nature **109**:579.

Andersen, M.E., Clewell, H.J.III., Gargas, M.L., Smith, F.A. and Reitz, R.H., 1987. Physiologically based pharmacokinetics and the risk assessment process for methylene chloride. Toxicol. Appl. Pharmacol. **87**:185.

Boxenbaum, H., 1980. Interspecies variation in liver weight, blood flow and antipyrine intrinsic clearance: Extrapolation of data to benzodiazepines and phenytoin. J Pharmacokin Biopharm **8**:165.

Boxenbaum, H., 1982. Interspecies scaling, allometry, physiological time, and the ground plan for pharmacokinetics. J. Pharmacokinet. Biopharm. **10**:201.

Boxenbaum, H., 1984. Interspecies pharmacokinetic scaling and the evolutionary-comparitive paradigm. Drug Metab. Rev. **15**:1071.

Boxenbaum, H. and D'Souza, R., 1987. Physiological models, allometry, neoteny, space-time and pharmacokinetics. In: "Pharmacokinetics: Mathematical and statistical approaches to metabolism and distribution of chemical and drugs" (A. Pecile and A. Rescigno, editors), page 191. Plenum Press, New York.

Boxenbaum, H. and Ronfield, R., 1983. Interspecies pharmacokinetic scaling and the Dedrick plots. Am. J. Physiol. **245**:R768.

Campbell, D.B. and Ings R.M.J., 1988. New approaches to the use of pharmacokinetics in toxicology and drug development. Human Toxicol. **7**:469.

Dedrick, R.L., Bischoff, K.B. and Zaharko, D.Z., 1970. Interspecies correlation of plasma concentration history of methotrexate (NSC-740). Cancer Chemother. Rep. (Part 1) **54**:95.

Fisher, J.W., Whittaker, T.A., Taylor, D.H., Clewell, JH.J.III. and Andersen, M.E., 1990. Physiologically based pharmacokinetic modeling of the pregnant rat: a multiroute exposure model for trichloroethylene and trichloroacetic acid. Toxicol. Appl. Pharmacol. **99**:395.

Gehring, P.J., Watanabe, P.G. and Blau, G.E., 1976. Pharmacokinetic studies in evaluation of the toxicological and environmental hazard of chemicals. In: "New concepts in safety evaluation" (M.A. Mehlman, R.E. Shapiro and H. Blumenthal, eds.), page 193. Hemisphere, New York.

Gould, S.J., 1977. "Neoteny and phylogeny." The Belknap press of Harvard University Press, Cambridge, MA.

Hall, S. and Rowland, M., 1983. Relationship between renal clearance, protein binding, and urine flow for digitoxin, a compound of low clearance in isolated perfused rat kidney. J. Pharmacol. Exp. Ther. **227**:174.

Holt, J.P. and Rhode E.A., 1976. Similarity of renal glomerular hemodynamics in mammals. Am. Heart J. **92**:465.

Ings, R.M.J., 1990. Interspecies scaling and comparisons in drug development and toxicokinetics. Xenobiotica **20**:1201.

Kleiber, M., 1947. Metabolic turnover rate: a physiological meaning of the metabolic rate per unit body weight. J. Theor. Biol. **53**:199.

Krishnan, K. and Andersen, M.E., 1991. Pharmacokinetics, individual differences. In: "Handbook of Hazardous Materials" (M.Corn, ed.). Academic Press, New York (in press).

Krishnan, K., Gargas, M.L., Fennell, T.R. and Andersen, M.E., 1991. A physiologically based description of ethylene oxide dosimetry in the rat. Toxicologist **11**:33.

Leung, H.W., Paustenbach, D.J., Murray, F.J. and Andersen, M.E., 1990. A physiologically based pharmacokinetic description and enzyme inducing properties of 2,3,7,8-tetrachlorodibenzo-p-dioxin in the rat. Toxicol. Appl. Pharmacol. **103**:399.

Levy, G., 1968. Dose dependent effects in pharmacokinetics. In: "Importance of fundamental principles in drug evaluation" (D.H. Tedeschi and R.E. Tedeschi, eds.), page 141. Raven, New York.

Levy, G., 1980. Effect of plasma protein binding on renal clearance of drugs. J. Pharmaceut. Sci. **69**:482.

McDougal, J.N., Jepson, G.W., Clewell, H.J.III., McNaughton, M.G. and Andersen, M.E., 1986. A physiological pharmacokinetic model for dermal absorption of vapors in the rat. Toxicol. Appl. Pharmacol. **85**:286.

Medinsky, M., 1990. Critical determinants in the systemic availability and dosimetry of volatile organic chemicals. In: "Principles of route-to-route extrapolation for risk assessment" (T. R. Gerrity and C.J. Henry, eds.). Elsevier, New York (in press).

Mordenti, J., 1986. Man versus beast: Pharmacokinetic scaling in mammals. J. Pharmaceut. Sci. **75**:1028.

Mordenti, J. and Chappell, W., 1989. The use of interspecies scaling in toxicokinetics. In: "Toxicokinetics and new drug development" (A. Yacobi, JP Skelly and VK Batra, eds.), page 42. Pergamon Press, New York.

National Research Council, 1986. "Drinking water and health", volume 6. NAS, Washington, D.C.

Ramsey, J.C. and Andersen, M.E., 1984. A physiologically based description of the inhalation pharmacokinetics of styrene in rats and humans. Toxicol. Appl. Pharmacol. **73**:159.

Reitz, R.H., Mendrela, A.L., Park, C.N., Andersen, M.E. and Guengerich, F.P., 1988. Incorporation of in vitro enzyme data into physiologically based pharmacokinetic model for methylene chloride: implications for risk assessment. Toxicol. Lett. **43**:97.

Rowland, M., 1986. Physiological pharmacokinetic models and interanimal species scaling. Pharmacol. Ther. **29**:49.

Sacher, G.A., 1959. Relationship of lifespan to brain weight and body weight in mammals. Ciba Foundation Colloquim on Aging **5**:115.

Stahl, W.R., 1963. The analysis of biological similarity. Adv. Biol. Med. Phys. **9**:355.

Tangliu, D.D., Tozer, T.N. and Riegelman, S., 1983. Dependence of renal clearance on urine flow: a mathematical model and its application. J. Pharmaceut. Sci. **72**:154.

van Ginneken, C.A.M. and Russel, F.G.M., 1989. Saturable pharmacokinetics in the renal excretion of drugs. Clin. Pharmacokinet. **16**:38.

Wesson, L.G., 1954. A theoretical analysis of urea excretion by the mammalian kidney. Am. J. Physiol. **179**:364.

Wilkinson, G.R. and Shand, D.G., 1975. A physiological approach to hepatic drug clearance. Clin. Pharmacol. Ther. **18**:377.

Yates, E. and Kugler, P.N., 1986. Similarity principles and intrinsic geometries: contrasting approaches to interspecies scaling. J. Pharmaceut. Sci. **75**:1019.

226

STEREOSELECTIVE PHARMACOKINETICS

Alberto Tajana

Department of Pharmacokinetics
Recordati S.p.A.
Milano, Italy

INTRODUCTION

Many drugs are marketed as racemates, i.e. equimolar mixtures of molecules with the same molecular formula (isomers) that differ only in the arrangement of their atoms in space (stereoisomers) and are related as two non-superimposable mirror images (enantiomers).

Under "ordinary conditions" defined as an achiral (symmetrical) environment, in which only achiral reagents are used, no difference whatsoever can be discerned between a pair of non- superimposable mirror images, but the biological environment is chiral, the macromolecules can distinguish between a pair of enantiomeric drugs, and consequently pharmacokinetics and pharmacodynamics are enantioselective. This means that disposition and activity of enantiomers may be, and in general are, different. The aim of this work is to present information on the stereoselective pharmacokinetics issue, its nature and implications for development of new therapeutic agents.

TERMINOLOGY

Compounds that have identical molecular formulas but differ in the nature of sequence of binding of their atoms or in arrangement of their atoms in space are termed *isomers*.

The constitution of a compound of given molecular formula defines the nature and sequence of bonding of the atoms.

Isomers differing in constitution, that is the position and arrangement of the same set of atoms in the molecule, are termed *constitutional isomers*. Constitutional isomers are different compounds and will have different

$$H_3C - O - CH_3 \quad \text{and} \quad H_3C - CH_2 - OH$$

methyl ether   **ethanol**

$$CH_3CONH-\text{⟨benzene⟩}-OCH_2CHCH_2NHCH(CH_3)_2 \quad \text{practolol}$$
$$\overset{|}{OH}$$

$$H_2NCOCH_2-\text{⟨benzene⟩}-OCH_2CHCH_2NHCH(CH_3)_2 \quad \text{atenolol}$$
$$\overset{|}{OH}$$

New Trends in Pharmacokinetics, Edited by A. Rescigno and A.K. Thakur
Plenum Press, New York, 1991

H₃C   CH₃ and H₃C   H

2-butene

CHO ... and ... CHO

glyceraldehyde

chemical, physical, and biological properties.

Isomers that differ only in the arrangement of their atoms in space are termed *stereoisomers*.

Stereoisomers that differ only in the position of atoms relative to a specified plane in cases where these atoms are, or are considered to be, part of a rigid structure, are termed *cis/trans* isomers or *Z* (zusammen)/*E* (entgegen) isomers. In the Z isomers the highest priority substituents on each of the atoms in the double bond are on the same side of the double bond; in the E isomers they are on the opposite sides. The broken line denotes a bond projecting behind the plane of the paper, the thickened line denotes a bond projecting in the front of the paper; a line of normal thickness denotes a bond lying in the plane of paper.

As with constitutional isomers, geometrical isomers are different compounds and have different properties.

When stereoisomerism is due to the orientation of groups around a tetrahedral carbon atom, the two stereoisomers are called *enantiomers* (optical

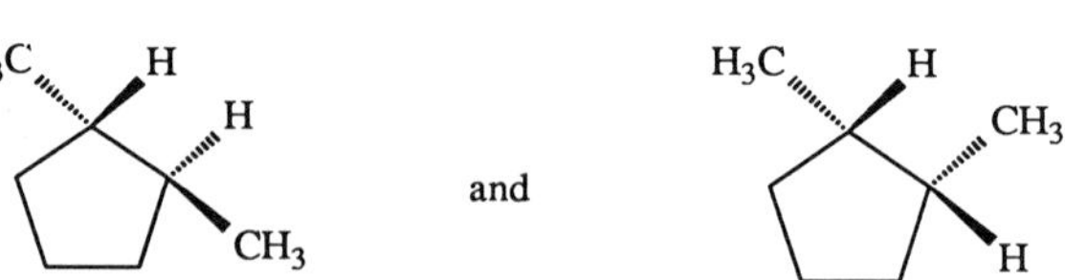

maleic acid   **fumaric acid**

1,2-dimethylcyclopentane

isomers, optical antipodes) and are mirror images of each other. The criterion of non-superimposability of a structure and its mirror image is a necessary and sufficient one for the existence of enantiomers.

The property of non-identity of an object with its mirror image is termed *chirality*. Chirality is equivalent to handed-

configurational isomers:

CHO ... and ... CHO

glyceraldehyde

conformational isomers:

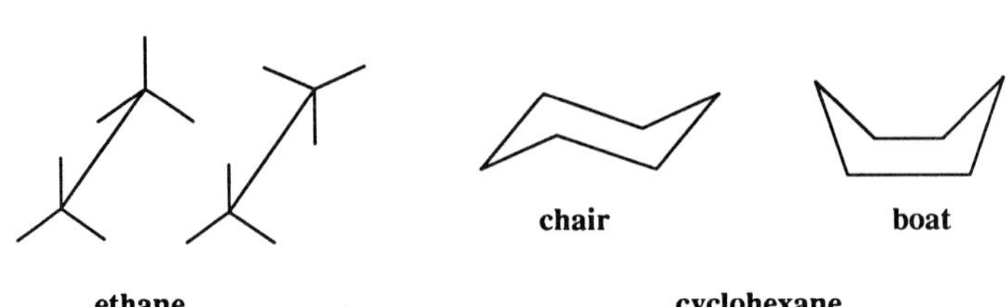

ethane   **chair**   **boat**

cyclohexane

228

ness, the term being derived from the Greek χειρ = hand.

A molecule in a given *configuration* (the configuration of a molecule of definite constitution is the arrangement of its atoms in space without regard to arrangements that differ only as after rotation about one or more single bonds; configuration cannot be altered without breaking bonds) or *conformation* (the conformations of a molecule of defined configuration are the various arrangements of its atoms in space that differ only as after rotation about single bonds; they can be altered without breaking bonds) is termed *chiral* when it is not identical with its mirror image.

Such a molecule is said to be *asymmetric* and this asymmetry is generally conferred on the molecule by a carbon which has four different atoms or groups attached to it (asymmetric carbon).

The term asymmetric was chosen by van't Hoff because there is no plane of symmetry through a tetrahedron whose corners are occupied by four atoms or groups that differ in scalar properties.

Enantiomers will have identical properties in a symmetric environment, since they are mirror images, except in chiral systems: they will have identical melting or boiling points, refractive indices, absorption spectra, chemical reactivity and identical solubilities in achiral solvents but will rotate the plane of plane-polarized light in opposite directions, with the same magnitude.

Where the rotation of plane-polarized light by a solution of one particular stereoisomer under particular experimental conditions is right-handed (clockwise), that isomer is said to be *dextro* (*d* or (+))-isomer, the other is the *levo* (*l* or (−))-isomer.

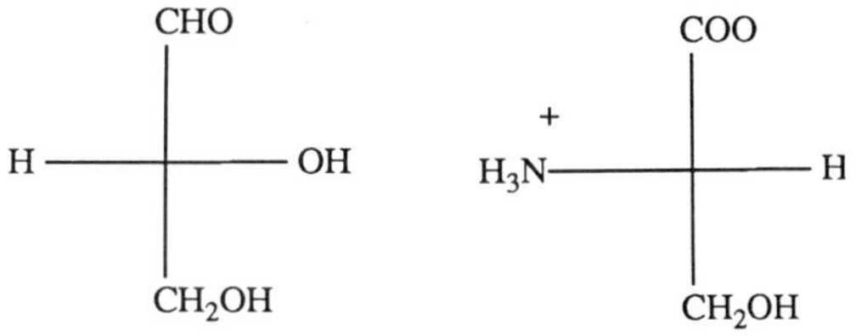

Optical rotation is not an independent variable; direction and magnitude of the rotation are affected by sample concentration, solvent, pH, temperature, light wavelength. Changes in optical activity as a result of changing one or more of these variables are not the result of a change in the spatial orientation of the atoms in the molecule i.e. its configuration, which can be altered only by breaking and making of the atomic bonds. Consequently, optical rotation does not reveal anything about the configuration of a molecule.

$$
\begin{array}{ccc}
\text{CHO} & & \text{COO} \\
| & & | \\
\text{H}-\!\!\!\!-\!\!\!\!-\text{OH} & \overset{+}{\quad} & \text{H}_3\text{N}-\!\!\!\!-\!\!\!\!-\text{H} \\
| & & | \\
\text{CH}_2\text{OH} & & \text{CH}_2\text{OH} \\
\text{D-glyceraldehyde} & & \text{L-serine}
\end{array}
$$

If the absolute configuration, i.e. the actual orientation of the atoms in space has been determined, then the molecule may be related to the structure of natural glyceraldehyde (D-configuration) or serine (L-configuration) and the prefix D or L is applied to describe the chirality.

This method for designation of configuration, still widely used for amino acids and carbohydrates, proposes that the Fisher projection of a compound (the Fisher projection representation is a two-dimensional rendering of the three-dimensional molecule oriented in space so that two of the substituents attached to chiral center are vertical and the other two are horizontal, the vertical groups projecting away from the viewer and the horizontal groups projecting toward the viewer) be oriented so that the most oxidized carbon is at the top and the group of highest priority is either on the right or on the left. If the group of highest priority lies on the right, the molecule is called D, if on the left, it is called L. D and L have no relation to the rotation of plane-polarized light.

Names of chiral compounds whose absolute configuration is known are better differentiated by prefixes R (for *rectus*, Latin for right) and S (for *sinister*, Latin for left), assigned by the Cahn-Ingold-Prelog sequence-rule procedure [Cahn, Ingold and Prelog, 1966]. According to this procedure, a priority number is assigned to each of the atoms or groups of atoms bonded to the asymmetric carbon atom. Priority (or sequence) rules are that atoms of highest atomic weight have highest pri-

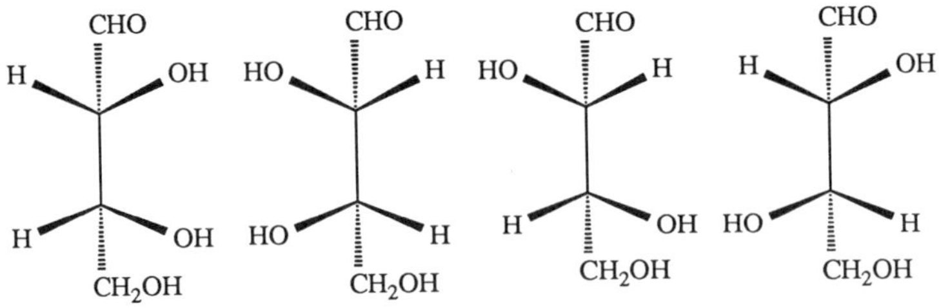

CHO — CHO — CHO — CHO structures

levo-Erythrose — dextro-Erythrose — levo-Threose — dextro-Threose

(a) — (b) — (c) — (d)

ority and if this does not distinguish between two atoms attached to the asymmetric carbon atom, then the priority of the atoms next but one from the asymmetric carbon atom are determined and so on.

The chiral center (asymmetric carbon) is then viewed from the side opposite to the group of lowest priority. If the other groups are arranged in a clockwise order of highest to lowest priority groups, then the center is designated R, if they are arranged anticlockwise, the center is S.

A mixture of equimolar amounts of enantiomeric molecules is termed *racemate*.

Thermodynamically, pure enantiomers are unstable, and equilibrium is attained when there is an equal mixture of the two enantiomers.

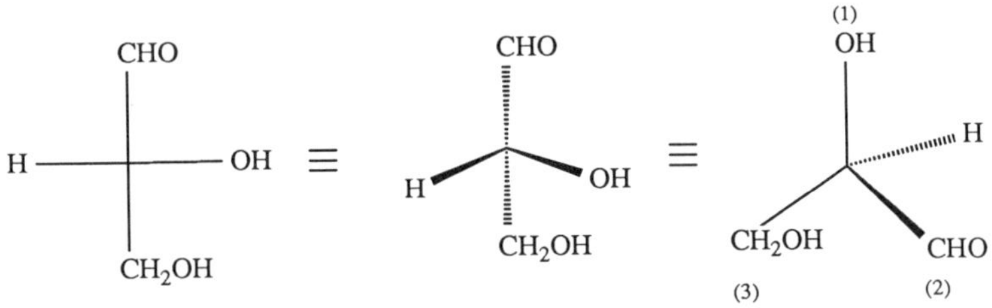

D-glyceraldehyde

R-glyceraldehyde

Priority : -OH > -CHO > -CH₂OH > H

Hydrogen, the group of lowest priority, oriented toward the back

It is essential to recognize that a racemate is a mixture of compounds, which, in a chiral system like the human body, may present enantioselective pharmacokinetic and pharmacodynamic properties.

If there is one chiral center there are two stereoisomers, R and S, at the chiral center. If there are two chiral centers, then there are $2^2=4$ stereoisomers, namely RR, RS, SR, SS. There are two pairs of enantiomers RR/SS and RS/SR. Other pairs of isomers, i.e. RR/RS, SR/SS are not mirror images since one center is the same. Such pairs of isomers are termed *diastereoisomers* or *diastereomers*.

Forms (a) and (b) above bear a non-superimposable mirror image relationship to each other, as do for (c) and (d): thus (a) and (b) constitute an enantiomeric pair, and (c) and (d) constitute a second enantiomeric pair. The properties of (a) and (b) are identical in an achiral environment. The relationship of (a) to (c), however, is not a mirror image one; similarly (d) is not the mirror image of (a), but is the mirror image only of (c): a real object can have only one mirror image. The relationship between (a) and (c), (a) and (d), (b) and (c), (b) and (d) is a diastereomeric one, and these compounds are said to be diastereoisomers.

Diastereisomers can be chiral or achiral.

Compounds containing an equal number of structurally identical chiral groups of opposite chirality, and no other chiral groups, lead to an achiral molecule.

230

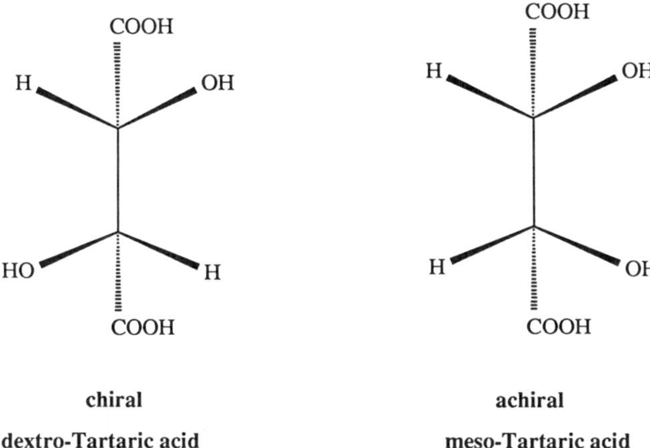

chiral

dextro-Tartaric acid

achiral

meso-Tartaric acid

Unlike enantiomers, diastereoisomers differ from each other in physical and chemical properties under ordinary achiral conditions; this is because whereas all intramolecular distances between corresponding groups are the same in enantiomers, they are different in diastereoisomers.

It is the creation of diastereomeric relationship that provided the basis for most of the methods for resolving optically inactive mixtures of enantiomers, i. e. racemates, into their optically active forms.

The antihypertensive labetalol is an approximately equimolar mixture of two racemates that are in diastereomeric relation to each other, hence labetalol is a mixture of four stereoisomers.

Labetalol

Where there are three chiral centers there will be $2^3=8$ stereoisomers, comprised of four enantiomeric pairs of isomers RRR/SSS; RSR/SRS; RRS/SSR; SRR/RSS. Where two stereoisomers differ only in the configuration at one chiral center, RRS/RRR, the compounds are termed *epimers*.

When considering a particular biological activity, the most active isomer is termed the *eutomer*; the least active, the *distomer*. The ratio of activities of eutomer and distomer, the *eudismic ratio*, is a measure of the degree of stereoselectivity. The Pfeiffer rule [Pfeiffer, 1956] states that "The greater the difference in the pharmacological effect of the optical isomers of a drug, the lower its effective dose".

Natural products are synthesized in a chiral environment, consequently they are generally found as a single stereoisomer. Pharmaceuticals of natural origin are optically active, e.g. *levo*-morphine from Papaver Somniferum and *dextro*-digitoxin from Digitalis Purpurea.

Contrary to the natural products, synthetics are usually obtained as isomeric mixtures, such as racemates; about 25% of all marketed pharmaceutical agents are such kind of mixtures.

Generic names do not reveal unequivocally the pharmacological agents present in pharmaceutical products; dealing with stereoisomers, a single name may designate single compounds or mixtures.

Simonyi, Gal and Testa [1989] suggested to use the SIGNS nomenclature, the acronym SIGNS standing for "Stereochemically Informative Generic Name System".

To characterize unequivocally the composition of drugs that can exist in stereoisomeric forms, the prefixes listed in the following Table 1 are linked to the generic names of the drugs.

The prefixes will be printed in italicized lower-case letters; *dextro, levo, cis* and *trans* refer to single stereoisomer drugs, *rac, diam* and *mep* designate mixtures.

The prefixes *dextro* and *levo* refer to the direction of the rotation of plane-polarized light by the enantiomers and apply to any single enantiomer, regardless of the number of asymmetric centers; e.g. the R:R-stereoisomer of labetalol, levorotatory, is designated as *levo*-dilevalol.

The prefixes *cis* and *trans* refer to stereoisomerism arising from the presence of a double bond or cyclic structures. If a drug structure displays *cis/trans* isomerism and is also chiral, the latter form of isomerism takes the precedence e.g. diltiazem is the dextrorotatory *cis*-isomer, and will be designated *dextro*-diltiazem.

The prefix *rac* designates a single racemate, composed of a 1:1 mixture of stable enantiomers (half-life of racemization longer than 24 hours in aqueous solution, 37°C, pH 7.4).

The prefix *diam* designates a single racemate, composed of stable enantiomers in unspecified proportions.

The prefix *mep* designates a mixture of two epimers (stereoisomers containing several centers of asymmetry and differing in the configuration of one chiral center only) not implying a well defined proportion of the stereoisomers.

Table 1

Summary of Proposed prefixes for SIGNS [Simonyi, Gal and Testa, 1989]

	Prefixes	
Type of drug	**in SIGNS**	**not to be used in SIGNS**
Single drugs		
Enantiomers	*dextro/levo*[a]	(+)/(−) (R)/(S) *d/l*
Diastereomers	*cis/trans*	(Z)/(E) *syn/anti*
Agents with no stereoisomers	none	
Mixtures		
Racemates[b]	*rac*	(±) (R,S) *dl*
Diastereomeric mixtures[c]	*diam*	(ZE) *cis/trans* *rac*[2]
Mixtures of epimers[d]	*mep*	*dia*

[a]In ambiguous cases, defined by the optical rotation in the more polar solvent.
[b]Involving stable enantiomers.
[c]Involving achiral diastereomers or mixtures of racemates.
[d]Involving stable epimers.

STEREOSELECTIVE ADME

Absorption

Most drugs diffuse passively through biological membranes, although absorption involves diffusion through chiral barriers. Enantiomers do not differ in their lipid/water partition coefficient, consequently stereoselectivity is not expected. When an active or receptor-mediated process is involved, the two enantiomers may differ in their absorption characteristics. Carrier transport systems involved in the absorption of natural amino acids, sugars and other endogenous substances, are stereoselective. Consequently drugs with structures analogous to naturally occurring compounds may be absorbed stereoselectively. L-dopa is absorbed from rat small intestine rapidly by an active process; D-dopa is also absorbed, but more slowly by passive diffusion [Wade, Mearrick and Morris, 1973]. L-methotrexate is actively and completely absorbed by G.I. tract up to 30 mg/m2 BSA while only 3% of D-methotrexate is absorbed [Hendel and Brodthagen, 1984]. A 15% difference in the bioavailability of the enantiomers of atenolol has been reported [Boyd et al., 1989].

Chiral β-lactam antibiotics are actively absorbed via the intestinal dipeptide transport system which is present in the intestinal brush border membrane. In the rat, after oral administration, only D-cephalexin could be detected in plasma; the L-isomer has a higher affinity for the carrier site than the D-form but is more susceptible to hydrolysis by peptidases in the intestinal wall [Tamai et al., 1988]. Amino-β-lactam antibiotics are commercially available as D-enantiomers.

Enantioselectivity was reported for folinic acid (Schilsky et al., 1989). A greater oral availability of the more active *levo*-isomer of terbutaline (0.15 vs 0.075 for the *dextro*-form) was accounted for partly by selectivity in first-pass metabolism but also for the suggestion that this enantiomer selectively increases membrane permeability [Borgstrom et al., 1989].

Stereoselective absorption arises when enantiomers differ in their ability to constrict or dilate blood vessels at sites of administration. Many local anaesthetics are optically active, e.g. mepivacain and bupivacaine; the enantiomers have different effects on local blood flow, which in turn account for differences in rates of systemic absorption [Akerman, Persson and Tegner, 1967; Luduena, 1969; Aberg, 1972; Aps and Reynolds, 1978].

Finally, the aqueous solubilities and crystal forms of racemates can differ from those of the individual enantiomers, and this may give rise to corresponding differences in dissolution rates at sites of administration. *rac*-Talidomide, for example, is 5-9 times less soluble than the separate enantiomers in many solvents. Because the racemate exists as a complex in solution, it may also have different biological properties from the isomers [Hague and Smith, 1988].

Distribution

The extent of distribution of drugs is determined by plasma and tissue protein binding and partition coefficient. The latter is a physical property, identical for the two enantiomers, therefore not enantioselective. The extent of binding of enantiomers may be different, and enantioselectivity expressed at the recognition interaction by macromolecules present in tissues and plasma can be an important determinant of the pharmacokinetics of enantiomers.

According to Testa [1989], three levels of interaction can be considered:

i) penetration: the distribution of drug between body compartments, cells and fluids will be controlled primarily by physicochemical characteristics of the molecule such as its relative solubility in different body compartments. This is essentially a non chiral process.

ii) recognition: involves specific binding of the enantiomer to a macromolecule (proteins, nucleic acids); the affinity of this binding is measured by its equilibrium dissociation constant Kd for receptors or its Michaelis-Menten constant Km for enzymes; stereoselective differences in binding may be quantitative and qualitative.

iii) activation: process that may or may not follow upon recognition of the enantiomer. If activation occurs, if the observed pharmacological effect is enantioselective, it could be due to differences in binding affinities; if activation does not occur for one of the enantiomers, this bound enantiomer may function as antagonist.

The discrimination of enantiomers by proteins is not surprising. Proteins consisting of chiral amino acid units are macromolecules possessing chiral secondary structure by forming asymmetric cavities or helical sequences, thus providing chiral environments for small bioactive molecules. Consequently, the fitting is unequal.

This is often manifested in differential binding strength for the enantiomers, giving rise to different distribution between bound and free states for the optical antipodes if the drug is applied in racemic form [Simonyi, Fitos and Visy, 1986].

The degree of drug protein binding affects the volume of distribution and total body clearance, especially for drugs with low extraction ratios, since clearance is then proportional to the free fraction in plasma $CL_{total} = f_u \cdot CL_{free}$; measurement of the free fraction in plasma is consequently necessary before trying conclusions on stereoselectivity in these parameters.

$$CL_H = Q_H \frac{f_u CL_i}{Q_u + f_u CL_i} \quad \text{"well stirred model"}$$

$$V = V_p + \frac{f_u}{f_{uT}} \cdot V_T$$

where CL_H = hepatic clearance; Q_H = liver blood flow; fu = fraction free in blood; CL_i = intrinsic clearance; V = volume of distribution; V_p = plasma volume; V_T = volume of tissue the drug distributes into; f_{uT} = fraction free in tissue.

For instance, mean clearance values for *levo*- and *dextro*-pentobarbital in humans were found to be 0.031 and 0.039 l/h/kg respectively [Cook et al., 1987]. The free plasma protein fraction of *levo*- and *dextro*-pentobarbital were 0.265 and 0.366 and CL_{free} 0.117 and 0.107 l/h/kg, respectively; the last values are similar and suggest that the apparent enantioselectivity in the clearance of total enantiomers is due to stereoselective protein binding. For disopyramide in humans [Lima, Boudoulas and Shields, 1985], the two enantiomers had similar pharmacokinetics when assessed using the total concentration of each enantiomer; the clearance of the free *dextro*-enantiomer was significantly higher than that of *levo*-enantiomer, 26 vs 19 l/min, but the more extensive plasma protein binding of the *dextro*-enantiomer, 73% vs 61%, compensated the higher clearance of the unbound fraction of this enantiomer, resulting in similar clearances for both enantiomers with respect to their total concentrations.

The extent of stereoselective plasma binding of drugs ranges up to a factor of about 1.5 [Tucker and Lennard, 1990], reflecting a diastereoisomeric association with both albumin, the major binding protein for most acids, and with α_1-acid glycoprotein, which binds predominantly basic compounds. Propranolol illustrates a case where stereoselectivity in binding occurs in opposite directions for the different proteins.

Whereas the *dextro*-enantiomer bind less to human α_1-acid glycoprotein, f_u = 0.162, than the *levo*-enantiomer, f_u = 0.127, its binding to human albumin is greater, f_u = 0.607 vs 0.649. In whole plasma the binding to α_1-acid glycoprotein predominates such that the *dextro*-isomer binds less than the *levo*-isomer, f_u = 0.203 vs 0.176 [Walle et al., 1988]. The enantio selectivity of plasma binding increases with the total extent of binding of propranolol.

For drugs which are highly protein bound, such as the Non-Steroidal Anti-Inflammatory Agents (NSAIDs) and the coumarin anticoagulants, even small enantioselective differences in binding can result in significant differences in their distribution profiles. Thus, the partitioning of ibuprofen enantiomers between plasma and synovial fluid was found to be enantioselective, dependent upon enantioselective binding to plasma albumin which affects the concentration gradient of diffusible free drug [Day et al., 1989]. Tissue distribution of chiral drugs may be determined by enantioselectivity in uptake and storage mechanisms.

The comparison of the volumes of distribution of propranolol isomers based on unbound drug in plasma suggests stereoselective tissue binding of this drug. In dogs the values were greater for the *levo*-enantiomer [Bai et al., 1983] whereas in man the opposite was indicated [Walle et al., 1988].

The more active *levo*-enantiomers of propranolol and atenolol undergo selective uptake, storage, and secretion by adrenergic nerve endings in cardiac and other tissues [Walle et al., 1988; Webb et al., 1988]; this specific binding is consistent with higher tissue: plasma concentration ratios for the *levo*-enantiomer compared to the *dextro*-isomer of propanolol and timolol in rats [Kawashima, Levy and Spector, 1976; Tocco et al., 1976; Takahashi et al., 1988]

The *levo*-forms of some NSAIDs are selectively taken up into adipose tissue where they may replace one of the fatty acid moieties in natural triglycerides.

A much greater uptake of both isomers is seen when the racemate or the *levo*-isomer rather than the *dextro*-isomer is administered [Williams et al., 1986; Sallustio, Meffin and Knights, 1988].

In this process, metabolism is also involved: the uptake in adipose tissue appears to occur during the *levo*-isomer to *dextro*-isomer inversion when the coenzyme A thioester of the *levo*-enantiomer is formed; inversion to the *dextro*-isomer is followed by incorporation of both esters as unnatural triglycerides; these hybrid triglycerides could disrupt normal lipid metabolism and membrane function and therefore have a potential for toxicity. Some of the CNS side effects of clinical NSAIDs may be attributed to abnormal lipid uptake [Caldwell and Marsh, 1983].

Stereoselective pharmacokinetics of some chiral NSAIDs depends on differences in the volume of distribution of the enantiomers [Jamali, 1988]).

The higher plasma concentration of *dextro*-indoprofen in humans [Bjorkman, 1985] and of *dextro*-ketoprofen in rabbits (Abas et al., 1987) have been attributed to smaller volumes of distribution as opposed to their *levo*-isomers. Etodolac is an exception among NSAIDs in that the plasma concentration of the active *dextro*-isomer in humans is lower than that of the *levo*-isomer; this is attributed to an almost 10 fold greater volume of distribution of the active enantiomer [Jamali et al., 1988a].

Metabolism

Much of the stereoselectivity observed in pharmacokinetics is due to metabolism. Both phase I and phase II metabolic reactions may be stereoselective. We can observe selectivity in the biotransformation of chiral drugs (substrate stereoselectivity), in the production of chiral metabolites from prochiral drugs (product stereoselectivity) or in both instances (substrate/product stereoselectivity) [Jenner and Testa, 1973].

Drugs are metabolized predominantly in the liver and their hepatic clearances are determined to different extents by hepatic blood flow, plasma protein binding, and intrinsic enzyme activity. Intrinsic clearance, CL_i, is the maximal ability of the liver to irreversibly remove drugs by all pathways in the absence of any flow limitation. For drug enantiomers with high hepatic extraction ratio, given parenterally, hepatic blood flow is of primary importance.

Protein binding and intrinsic enzyme activity are important for drugs with low extraction ratios given parenterally and for all drugs given orally [Wilkinson and Shand, 1975]. The extent of any difference in enantiomers hepatic clearance may be route dependent.

Classical examples of drugs for which stereoselective metabolism is observed are verapamil and propranolol. *Levo*-Verapamil is 8-10 times more potent than *dextro*-verapamil with regard to its negative dromotropic activity on atrioventricular conduction in man [Echizen et al., 1985].

Total verapamil plasma concentrations three times higher are required after oral administration to produce the same pharmacodynamic effect as after i.v. administration. The result depends on the differences in the *dextro/levo* isomer ratio of plasma verapamil in relation to the route of administration. After i.v. administration, the *dextro* to *levo* isomer ratio is approximately 2 (the systemic clearances are 10.2 and 18.1 ml/min/kg for *dextro* and *levo* isomers respectively), after oral administration the ratio is about 5 (oral bioavailability for *dextro*-verapamil 50%, for *levo*-verapamil 20%). [Eichelbaum, Mikus and Vogelgesang, 1984; Vogelgesang et al., 1984].

The metabolism of verapamil involves several N-dealkylation and O-demethylation reactions. In studies carried out "in vitro" with human hepatic microsomal fraction and "in vivo" after oral administration of the enantiomers, no differences in K_m and V_{max} could be observed for the N-dealkylated metabolites. A three-fold difference in clearance of these metabolites was observed "in vivo": 118 l/h for the *levo*-isomer, 40 l/h for the *dextro*.

The formation of the O-dealkylated metabolites exhibited a high degree of enantioselectivity: V_{max} (pmol/mg/min) and K_m (mcM) were 333 and 60 respectively for the *levo*-isomer, and 153 and 48 for the *dextro*-isomer. The metabolic clearance of *levo*- verapamil was 30-fold higher compared to *dextro*-verapamil (43.6 l/h vs. 1.32 l/h), [Eichelbaum, 1988].

Propranolol is a β receptor blocking agent, on the market as a racemate.

A clear stereoselectivity in the oral clearance of this drug is observed in man, the *levo* to *dextro* plasma concentrations ratio being 1.4-1.9, with large (about 4-fold) intersubject variability [Walle et al., 1988].

Because of the principal effects of β blockers, the antianginal effects mediated via cardiac β receptor blockade and management of hypertension involve the *levo*-enantiomer, at equal total propranolol plasma concentrations, racemic propranolol is 2-3 fold more potent after oral than intravenous administration [Coltart and Shand, 1970]. The difference in oral clearance is due to stereoselective hepatic metabolism. The following tables summarize some pharmacokinetic characteristics of propranolol enantiomers.

Table 2

Pharmacokinetic parameters (± S.D.) of propranolol enantiomers [Walle et al., 1988].

	dextro	*levo*	*dextro/levo*
$CL_{i.v.}$ l/min	1.21 (0.15)	1.03 (0.12)	1.17 (0.02)
$CL_{u\ i.v.}$ l/min	6.0	5.9	1.02
V_d l/kg	4.82 (0.34)	4.08 (0.33)	1.18 (0.02)
$t_{1/2}$ h	3.57 (0.25)	3.53 (0.21)	1.01 (0.02)
f_u	0.203 (0.008)	0.176 (0.007)	1.15 (0.01)
V_{du} l/kg	20.3 (1.7)	18.9 (1.7)	1.07 (0.02)
a_1-AGP_u	0.162 (0.017)	0.127 (0.013)	1.28 (0.01)
HSA_u	0.607 (0.013)	0.649 (0.011)	0.94 (0.01)
$RBC/plasma_u$	3.16 (0.24)	3.05 (0.28)	1.04 (0.01)
$CL_{p.o.}$ l/min	2.78 (1.36)	1.96 (0.87)	1.42
$CL_{u\ p.o.}$ l/min	13.7	11.1	1.23

Table 3
Propranolol enantiomer clearances and partial metabolic clearances (l/min) after
single oral racemic doses in 17 subjects [Ward et al., 1986]

	dextro	*levo*	
Propranolol	2.78 (1.36)	1.96 (0.87)	P < 0.01
Propranolol glucuronide	0.24 (0.10)	0.27 (0.12)	NS
Naphthoxylactic acid	0.38 (0.28)	0.31 (0.10)	NS
4-Hydroxypropranolol glucuronide+sulfate	0.88 (0.70)	0.35 (0.30)	P < 0.001

After i.v. administration, a lowering of hepatic blood flow by the active *levo*-isomer contributes to its lower systemic clearance, compared to that of the inactive *dextro*-isomer [Branch, Nies and Shand, 1973; Nies, Evans and Shand, 1973]. After oral administration, clearance is practically independent of liver blood flow and the lower value for the *levo*-isomer reflects the net balance of enantioselectivity in plasma binding and intrinsic clearance.

The relative contributions of enantioselectivity in plasma binding and enzymatic transformation are evident when plasma clearance values are divided by free fraction to give clearance based upon unbound drug [Tucker and Lennard, 1990]. The selectivity in plasma binding could account for virtually all the enantiomeric difference in the volume of distribution of propranolol. The enantioselective plasma binding is due to stereoselective binding of *levo*-propranolol to α_1-acid glycoprotein. It is interesting to note the very slight but opposite stereoselectivity in the binding of propranolol to HSA. The enantioselectivity of this plasma binding increases with greater total binding.

Stereoselectivity in intrinsic clearance may be a consequence of different situations:

i) Drugs may be metabolized by different pathways and stereoselectivity is the net result of selectivity in different enzymes. For propranolol, ring oxidation (i.e. 4-hydroxylation) and side chain oxidation favour the *dextro*-enantiomer, direct glucuronation tends to favour the *levo*-isomer [Ward et al., 1989].

ii) Several isozymes may be involved in the formation of the same product and each may exhibit different steric preference. For propranolol, 4-hydroxylation is mediated by two different enzymes, only one of which is stereoselective [Otton et al., 1989].

iii) Specific preference for completely different pathways, e.g. *levo*-warfarin is mainly 7-hydroxylated, *dextro*-warfarin is 6-hydroxylated and reduced at the keto group [Lewis et al., 1974].

Of the three primary pathways involved in propranolol metabolism, glucuronidation, side chain oxidation, and ring oxidation, the higher oral clearance of *dextro*-enantiomer is practically solely due to selective ring oxidation, that is 2.5-fold greater for this enantiomer as compared to *levo*-enantiomer. Both conjugation pathways appear to be stereoselective in man, with glucuronic acid conjugation favouring the levo-enantiomer [Ward et al.,1986; Walle et al., 1984] and sulfoconjugation favouring the dextro-isomer [op.cit.; Christ and Walle, 1985].

An interesting example of stereoselective metabolism is the metabolic chiral inversion of 2-arylpropionic acids. These drugs constitute an important group of non-steroidal anti-inflammatory agents (NSAIDs). They contain a chiral center, are generally administered as racemates, but the anti-inflammatory activity resides in the

237

dextro-enantiomers. In vivo some of these drugs undergo a metabolic inversion of the chiral center with a unidirectional reaction that transforms the *levo*-enantiomer to its *dextro*-isomer. The reaction proceeds via the formation of the acyl-CoA thioesters of the 2-arylpropionates, which can undergo either chiral inversion about the α-carbon atom or hydrolysis [Hutt and Caldwell, 1983]:

$$\textit{dextro}\text{-Ar-CH(CH}_3\text{)-COOH}$$

$$\textit{levo}\text{-Ar-CH(CH}_3\text{)-COOH} \rightarrow \text{Ar-CH(CH}_3\text{)-COSCoA} \nearrow \searrow$$

$$\textit{levo}\text{-Ar-CH(CH}_3\text{)-COOH}$$

Ibuprofen, benoxaprofen, fenoprofen, cicloprofen and thioxaprofen give rise to inversion; indoprofen, flurbiprofen, ketoprofen, carprofen and tiaprofenic acid do not invert significantly. There are species differences in the extent of inversion: ketoprofen undergoes 80% inversion in the rat [Foster and Jamali, 1988a], while in humans its inversion is negligible [Foster et al., 1988b,c]. Assessment of extent of inversion is not easy because the kinetics could be dose and absorption rate-dependent and could also be due to interaction between enantiomers. The site of inversion is not clear. Studies with benoxaprofen [Simmonds et al., 1980] and ketoprofen [Foster and Jamali, 1988a] suggest that G.I.tract is one of the major sites of inversion in rats; a similar situation has been suggested for ibuprofen in humans [Jamali et al., 1988b]. Liver and kidney may be involved in systemic inversion of 2-phenylpropionic acid in the rat [Yamaguchi and Nakamura, 1987].

Stereoselective oxidation of drugs in the liver by cyt.P-450 is influenced by genetically determined factors. Since enzyme activity is deficient in a certain percentage of the population, some metabolic pathways may exhibit polymorphism. Thus stereoselectivity in metabolism may be affected by the genetic phenotype of the subject.

After an oral dose of *rac*-metoprolol to phenotyped extensive metabolizers of debrisoquine, the plasma concentrations of *levo*-enantiomer were higher than for the *dextro*-form (*levo*AUC/*dextro*AUC = 1.4); systemic availability of enantiomers in poor metabolizers was similar. Practically, these data indicate that at equal total metoprolol plasma concentrations, a greater degree of β-blockade is expected in extensive metabolizers [Lennard et al., 1983].

Excretion

Renal Excretion

Renal drug clearance is the net result of passive glomerular filtration, active secretion, passive and active reabsorption, and renal drug metabolism.

Stereoselectivity in glomerular filtration and passive reabsorption is apparent, secondary to any difference in the plasma protein binding of the isomers. The other processes may be expected to be stereoselective.

The stereoselective renal clearance of metoprolol, 75 ml/min for *dextro*-isomer, 70 ml/min for *levo*-isomer [Lennard et al., 1983] is probably due to the difference in the unbound fraction of the enantiomers [Lee et al., 1990]. Active renal tubular secretion, not excluding active reabsorption and renal drug metabolism, is believed to be responsible for the stereoselective renal clearance of pindolol [Hsyu and Giacomini, 1985], chloroquine [Ofori-Adjei et al., 1986] and disopyramide [Lima, Boudoulas and Shields, 1985; Giacomini et al., 1986; Le Corre et al., 1988].

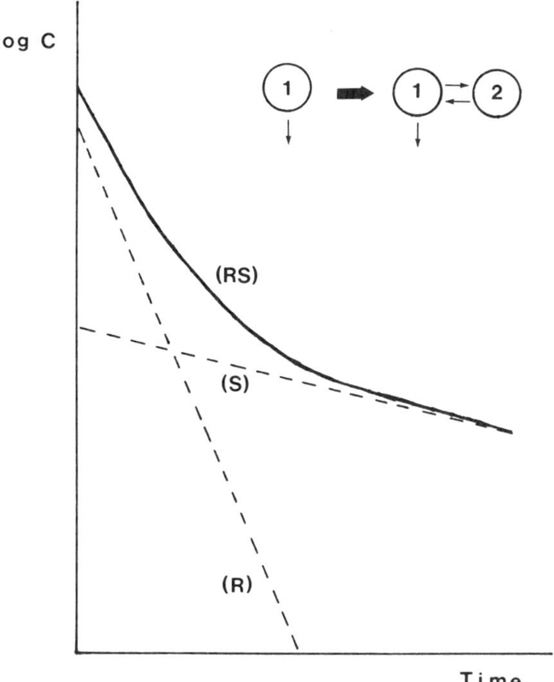

Fig. 1. Plasma concentration-time profiles (log scale) of a hypothetical chiral compound showing how monoexponential disposition kinetics for both isomers result in a biexponential time-course for unresolved drug. A one compartment pharmacokinetic model for each isomer gives rise to a spurious two compartment model for the mixture [Tucker and Lennard, 1990]

Selectivity has been found also for the diastereoisomers quinidine and quinine, the unbound renal clearance of the former being 4 times greater [Notterman et al., 1986].

Biliary Excretion

Both active and passive transport mechanisms are involved in the excretion of drugs and their metabolites into the bile. Stereoselective excretion into the bile has been reported for ketoprofen enantiomers in rats [Foster and Jamali, 1988a] and in cholecystomized patients [Foster et al., 1988b].

For acenocoumarol (nicoumalone) in rats the existence of stereoselective excretion has been suggested [Thijssen and Baars, 1987], the amount of *dextro*-enantiomer in bile being four times higher than that of the *levo*-isomer; because in another study [Thijssen, Baars and Drittij-Reijnders, 1985] plasma AUC ratio for *dextro* and *levo* enantiomers was four, the higher amount of *dextro* acenocoumarol in the bile may simply reflect greater plasma concentration of this enantiomer [Jamali, Mehvar and Pasutto, 1989].

THE STUDY OF TIME-COURSE OF INDIVIDUAL ENANTIOMERS AS A TOOL FOR THE CORRECT INTERPRETATION OF DISPOSITION DATA

There are several benefits for carrying out enantiospecific pharmacokinetic studies [Tucker et al., 1990].

239

Calculation of "True" Pharmacokinetic Parameters

Pharmacokinetic models could be misspecified: if the enantiomers decay monoexponentially at different rates, the kinetics of the mixture appears biexponential and the derived model parameters are meaningless.

Drug clearance and bioavailability, calculated from parameters derived for the unresolved drug, are complex functions of the clearance and bioavailability of the individual enantiomers:

$$CL_{RS} = \frac{Dose}{AUC_{RS}} = \frac{2 \cdot CL_S \cdot CL_R}{CL_S + CL_R}$$

$$F_{RS} = \frac{CL_{RS} \cdot AUC_{RS}}{Dose_{RS}} = \frac{F_R \cdot CL_S + F_S \cdot CL_R}{CL_S + CL_R}$$

If the bioavailabilities of the isomers differ and one of them is much more active than the other, the calculation of bioavailability from the measurement of unresolved drug, especially for highly cleared drugs, does not give a true assessment of therapeutic effect.

Renal clearance and the free fraction in plasma of unresolved drugs may show time-dependence if the plasma concentrations of enantiomers, which have constant renal clearances CL_R, CL_S and constant f_{uR}, f_{uS}, change at different rates.

$$CL_{rRS} = \frac{C_R}{C_S + C_R} CL_{rR} + \frac{C_S}{C_S + C_R} CL_{rS}$$

$$f_{uRS} = \frac{C_R}{C_S + C_R} f_{uR} + \frac{C_S}{C_S + C_R} f_{uS}$$

If the total clearance depends upon plasma binding, the total clearance of the unresolved drug will also be time dependent.

Evaluation of Dose Potency Ratio of Enantiomers (Eudismic Ratio)

The eudismic ratio, the ratio of the activity of enantiomers (ED_{50}'s) determined *in vivo* reflects contribution to stereoselectivity in both pharmacokinetics and pharmacodynamics. It is generally calculated relating a pharmacological response to the unbound plasma concentration at steady-state after separate administration of each isomer.

In vitro experiments allow the calculation of the intrinsic potency ratio, the ratio of unbound concentrations of the isomers necessary to produce 50% of the effect (EC_{u50}), that reflects only the pharmacodynamic phase.

Comparing the eudismic ratio and the intrinsic potency ratio it should be possible to estimate true contribution of stereoselective pharmacokinetics to the eudismic ratio. A particular example is the inhibition of prostaglandin synthesis by NSAID's. For those compounds whose *levo*-enantiomer is inverted to the active form *dextro* "in vivo", large differences are observed between "in vitro" and "in vivo" eudismic ratio values.

Table 4
Differential Stereoselectivity of Action of Profens NSAIDs "in vitro" and "in vivo"
[Caldwell, Hutt and Fournel-Gigleux, 1988].

| Drug | Eudismic ratio (dextro/levo) | | Inhibitory actions tested *in vitro* |
	in vitro	*in vivo*	
Ibuprofen	160	1.3	PG synthesis inhibition (bovine)
Flurbiprofen	878	8	SRS-A antagonism (guinea pig
Indoprofen	100	25	PG synthesis inhibition
Naproxen	130		PG synthesis inhibition (sheep)
	70	21	PG synthesis inhibition (bovine)
Carprofen	>16		PG synthesis inhibition (sheep)
	>23	15	Platelet aggregation
Fenoprofen	35	1	PG synthesis inhibition Human

Explanation of Shifts in Concentration-Effect Relationship

Shifts as function of time

If elimination half-lives of enantiomers differ, the composition of the racemic mixture will vary with time; and if one isomer is more active than the other, the relationship between plasma concentration of unresolved drug and effect is difficult to interpret. *Levo*-tocainide is three times more active as an antidysrhythmic agent than the *dextro*-isomer [Block, Merrill and Smith, 1988], and has a shorter half-life, 9.6 vs 15.1 hours [Edgard et al., 1984]. The mean *dextro/levo* plasma concentration ratio changes progressively during i.v. infusion of *rac*-tocainide in man [Thomson et al., 1986] so that despite increasing concentrations of racemate with time, little change in pharmacological effect may be observed, a fact that could be misinterpreted as being due to tolerance or tachyphylaxis [Tucker and Lennard, 1990].

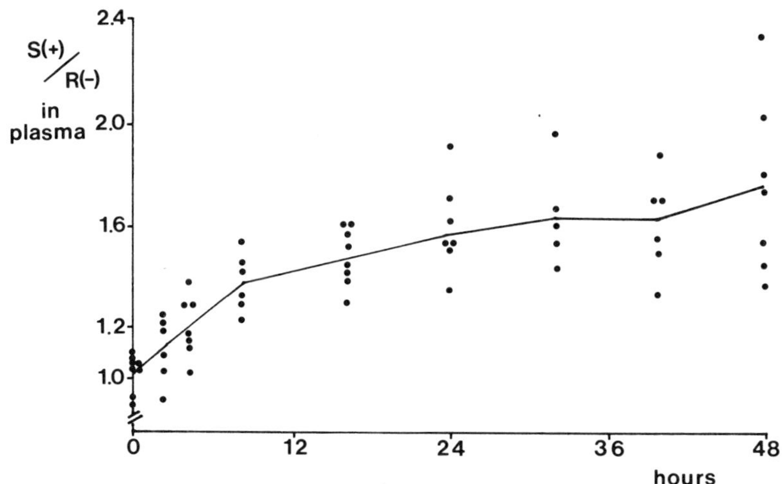

Fig.2 Ratio of *dextro* to *levo*-tocainide in plasma from patients during i.v. infusion of the racemate. The points are data from individual subjects [Thomson et al., 1986]

241

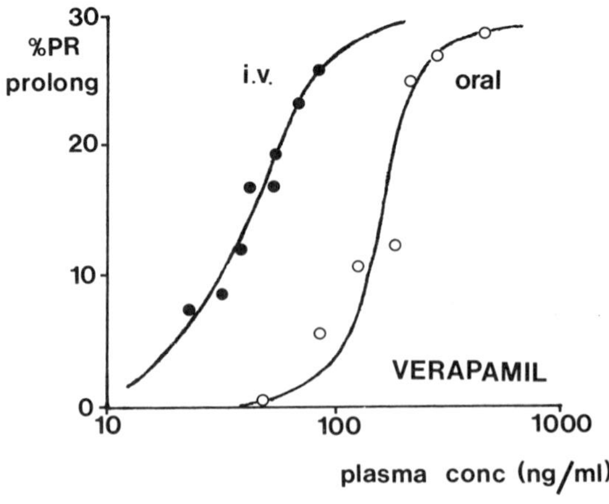

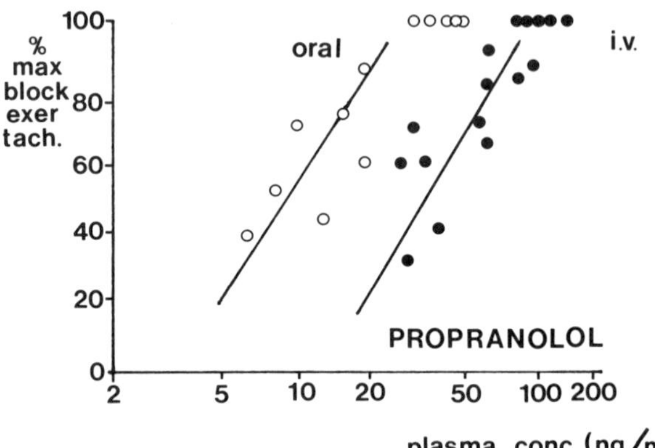

Fig. 3. Effect of route of administration on plasma-concentration-response relationship based upon measurement of unresolved verapamil and propranolol [Eichelbaum, Mikus and Vogelgesang, 1984; Vogelgesang et al., 1984; Coltart and Shand, 1970]

Shifts as function of route of administration

Based upon measurements of plasma concentrations of the unresolved drugs, verapamil appears to be more effective when given i.v. than orally and propranolol appears to be more effective when given orally than i.v. Stereoselective first-pass metabolism is the reason.

Shifts as function of diseases

Oral verapamil in patients with liver cirrhosis is effective at lower total plasma concentrations than in normal subjects for a reduction of stereoselective first-pass effect as a consequence of extra and intra-hepatic shunting of portal blood and of impaired enzyme activity [Eichelbaum, 1988].

Shifts as function of genetics

Rac-metoprolol plasma concentration-β-blockade relationship in extensive metabolizers is displaced to the left of that in poor metabolizers, owing to the stereoselective metabolism of the less active dextro-enantiomer by the polymorphic enzyme [Lennard et al., 1983].

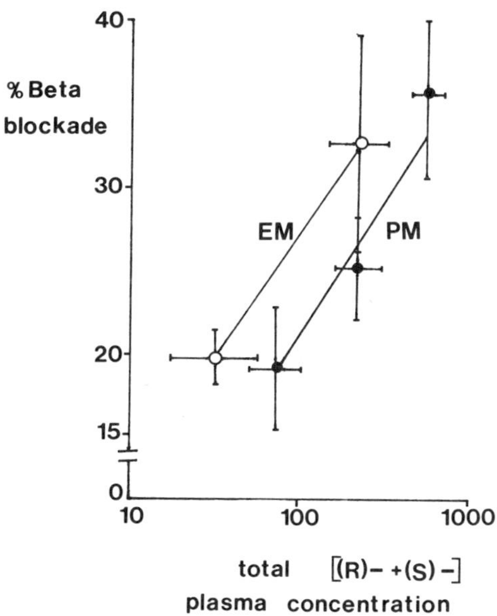

Fig. 4. Relationship between percentage reduction in exercise hearth rate(% β blockade) and the log plasma concentration of total (*dextro* + *levo*) metoprolol in four EM and six PM of debrisoquine. Each point represents mean (± S.D.) data obtained at 2, 12 or 24 h after a 200 mg oral dose [Lennard et al., 1983]

Selection of Species for Toxicity Testing

Species differences in enantioselectivity have been observed for many drugs. Drug toxicity tests performed in animal species which show different stereoselectivity as compared to man are not adequate to predict the potential risks for human use.

Clearance of *levo*-propranolol is higher than that of *dextro*-propranolol in the dog, but lower in man [Walle and Walle, 1979; Bai et al., 1983; Walle et al., 1988]. Clearance of *dextro*-warfarin is higher than that of the *levo*-isomer in the rat, but lower in man [Breckenridge and Orme, 1972; Yacobi and Levy, 1974; Toon et al., 1987].

The rate of chiral inversion shows a considerable interspecies variability: inversion half-life of benoxaprofen in man is 40 times larger than that in rat [Simmonds et al., 1980]; for ketoprofen, the extent of inversion is 80% in the rat [Foster and Jamali, 1988a], 10% in rabbits [Abas and Meffin, 1987] and small but significant in humans [Jamali, Mehvar and Pasutto, 1989]. *Rac*-oxazepam undergoes stereoselective glucuronidation in rhesus monkey, rabbit, dog and humans. The *dextro*-oxazepam predominates in the human, dog and rabbit, while the *levo*-conjugate prevails in rhesus monkey [Sisenwine et al., 1982; Ruelius et al., 1979]. The conjugates are present in approximately equal amounts in miniature swine [Ruelius et al., 1979].

N-methylation of nicotine enantiomers is species dependent: in rats N-methylation is not apparent, in guinea pig it is specific for R-enantiomer, in humans it occurs for both enantiomers, with preference for the R-enantiomer [Crooks and Godin, 1988].

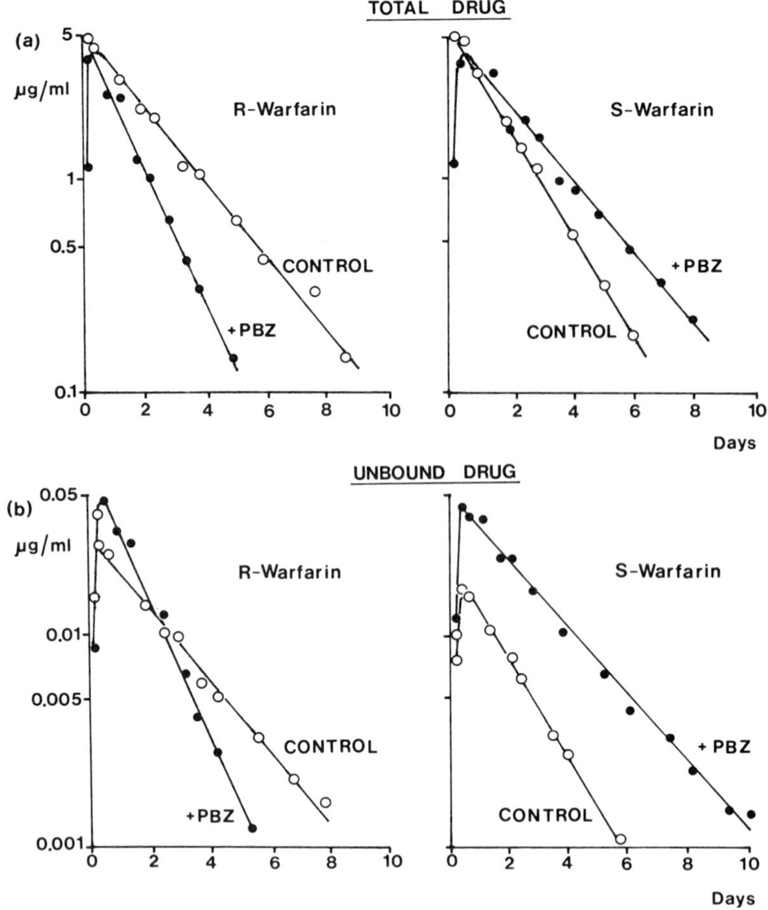

Fig. 5. Total and unbound concentration of R-(*dextro*) and S-(*levo*)-warfarin following oral administration of 1.5 mg/kg of *rac*-warfarin, alone (○) and 4 days into a regimen of 100 mg phenylbutazone 3 times daily (●) to a subject [Banfield et al., 1983]

Rational Therapeutic Drug Monitoring

To better define the "therapeutic window", when there are differences in activity or disposition of enantiomers of commonly monitored drugs, it is better to measure plasma concentrations of the individual enantiomers.

To Better Understand Interactions

Drug-drug interactions

The anticoagulant effect of *rac*-warfarin is potentiated by coadministration of phenylbutazone. No changes of clearance or half-life of the racemic warfarin are observed [O'Reilly et al., 1980], although total warfarin plasma concentrations decreased. Only when the kinetics of the individual enantiomers was studied [Banfield et al., 1983] an explanation was found. After warfarin+phenylbutazone, the plasma clearance of the less active *dextro*-enantiomer increases and that of *levo*-isomer decreases as compared to clearance after warfarin alone (see Fig. 5).

The total warfarin plasma concentration is reduced but inhibition of the metabolism of the more active isomer results in augmented hypothrombinemic activity.

When the plasma kinetics of the unbound isomers is examined, it becomes apparent that there is inhibition of unbound clearance of both the enantiomers, more so in the case of *levo*-isomer.

Displacement of warfarin from binding sites in the plasma and tissues by phenylbutazone, leaving more of the unbound *dextro*-warfarin available for hepatic elimination, obscures the inhibitory effect on the metabolism of *dextro*-enantiomer by increasing its elimination (increased clearance, decreased half-life). In contrast, the marked reduction of the clearance of the more active *levo*-isomer due to a stereoselective inhibition of metabolism of this enantiomer overrides the decreased plasma and tissue binding. The plasma concentrations of the unbound *levo*-enantiomer are consequently markedly increased, explaining the potentiation of the anticoagulant activity. An increased anticoagulant effect of *rac*-warfarin is observed also with coadministration of sulphinpyrazone [Toon et al., 1986].

When cimetidine is coadministered with oral verapamil, increased bioavailabilities of both enantiomers are observed, but the bioavailability increase of the more active *levo*-verapamil is twice that of *dextro*-verapamil; consequently at equal total plasma concentrations of verapamil, the effect is more pronounced in the presence of cimetidine [Mikus et al., 1988].

Enantiomer-enantiomer interactions

Enantiomers may interact pharmacokinetically by numerous mechanisms: the kinetics of one enantiomer may differ when given as such and when administered as part of a racemic mixture. They may displace each other from protein binding sites, as in the case of ibuprofen [Evans et al., 1989]: the areas under the total plasma drug concentration-time curves of both isomers were found to be less when administered as racemate than when given alone [Lee et al., 1985]. Competition for concentration-dependent plasma binding also accounts for interactions between disopyramide isomers [Giacomini et al., 1986]. Total clearance and volume of distribution were higher for *levo*-isomer when the *dextro*-isomer was present, whereas the clearance and volume of distribution of the *dextro*-isomer were higher when it was administered alone.

The f_u value for *levo*-enantiomer is greater in the presence of *dextro*-isomer, but although the *levo*-enantiomer also displaces the *dextro*-isomer, the f_u value of the *dextro*-isomer is actually greater at a given total disopyramide concentration when present by itself.

Enantiomers may compete for active transport processes: *dextro*-terbutaline enhances the renal clearance of *levo*-terbutaline competing for tubular reabsorption [Borgstrom et al., 1989].

For parachloroamphetamine enantiomers, an interaction at the recognition step by competing for the active site of the enzyme has been suggested [Ames and Frank, 1982].

SYNTHESIS OF OPTICALLY ACTIVE COMPOUNDS

The ever increasing demand for more selective drugs, which are more targeted in their action and show less side effects, is providing an important stimulus to pharmaceutical companies to market these products as pure enantiomers. Consequently there is an increasing demand for methods for industrial-scale synthesis of optically active compounds.

Synthetic routes to optically active compounds can be divided into three groups [Sheldon, 1990]:

i) chiral pool: it refers to readily available natural products such as carbohydrates and amino acids, which can be transformed into synthetic products with retention or inversion of configuration or chirality transfer;

ii) racemate resolution: still constitutes the main method for industrial-scale synthesis; is obtained by chemical or enzymatic kinetic resolution or diastereomer crystallization;

iii) catalytic asymmetric synthesis: starting from prochiral substrates, involves introduction of the asymmetric center into the molecule.

Fermentation is a biocatalytic technique that can be applied to the kinetic resolution or enantioselective biotransformation of prochiral substrates. Kinetic resolution involves the reaction of a mixture of enantiomers the molar ratio of which varies with conversion whilst asymmetric synthesis pertains to the reaction of a single (prochiral) compound and enantioselectivity is independent of conversion.

Kinetic resolution requires at least one extra step, the racemization of the unwanted isomer. Sometimes conditions can be found where the unwanted enantiomer undergo spontaneous racemization, leading to a theoretical yield of 100%.

Enzymatic kinetic resolution has been applied to the commercial production of D-p-hydroxyphenylglycine (Recordati, S.p.A., Milano, Italy) using a microorganism able to produce both a D-hydantoinase and a second enzyme, the n-carbamoyl-D-amino acid amidohydrolase which catalyzes the subsequent hydrolysis of the N-carbamoylderivative [Olivieri R et al., 1981]. Spontaneous racemization occurs under alkaline reaction conditions, due to the acidic character of the C-N bond in the hydantoin ring, thus allowing a one-step conversion in a theoretical yield of 100%.

Diasteromers crystallization generally involves a reaction of the racemate with an optically active pure acid or base, the resolving agent, to give a mixture of two diastereomeric salts whose physical properties are different, e.g. D-phenylglycine using camphorsulphonic acid as the resolving agent. When the diastereomer remaining in solution undergoes spontaneous epimerization, it is possible to obtain yields >50%.

ENANTIOMERS RESOLUTION BY HPLC

The possibility to realize the individual analysis of enantiomers is important both in industrial pharmaceutical analysis and in biopharmaceutical one. High performance liquid chromatography (HPLC) has become the method of choice for the separation of enantiomers [Karnes and Sarkar, 1987]. Separation can be achieved by different mechanisms: precolumn derivatisation with chiral reagents, addition of chiral agents to the mobile phase, selective interactions with chiral stationary phases.

Precolumn Derivatisation

The racemic mixture is reacted with a single, pure enantiomer of the chiral derivatising agent to generate diastereomers which can be separated by conventional liquid chromatographic techniques. NSAIDs of the 2-arylpropionic acid type have been esterified with *dextro*-2-octanol [Johnson et al., 1979] and *levo*-2-butanol [Kamerling et al., 1981].

Racemic mixtures of propranolol have been resolved making an isothiocyanate derivative [Sedman and Gal, 1983].

The advantages of this procedure are that a wide variety of derivatisation reagents are available and that conventional chromatographic conditions are requested. The limitations are: the necessity of enantiomeric pure derivatising reagents, which are stereochemically stable, and that different proportions of diastereomers may be generated for the different reaction rates.

Chiral Mobile Phase Additives

By adding a chiral agent to the mobile phase, by interaction with the analyte, diastereomeric complexes are formed. Resolution is produced by differences in stabilities of the diastereomeric complexes and solubilities in the mobile phase, or the degree of binding of the complexes to an achiral stationary phase [Karnes and Sarkar, 1987].

Diastereomeric tertiary complexes are formed between a transition metal Cu(II), Ni(II), Zn(II), and a single enantiomer of a chiral molecule, e.g. an amino acid ligand and the racemic analyte.

$$(levo\text{-Pro-Cu-}levo\text{-Pro})^- + \text{R-A}^- \rightarrow (levo\text{-Pro-Cu-R-A})^- + \text{levo-Pro}^-$$

$$(levo\text{-Pro-Cu-}levo\text{-Pro})^- + \text{S-A}^- \rightarrow (levo\text{-Pro-Cu-S-A})^- + \text{levo-Pro}^-$$

Resolution is caused by one of the ligands forming a stronger mixed complex. This approach has been utilized for separation of racemic mixtures of α-amino acids [Nakazawa and Yoneda, 1978; Lepage et al., 1979]. Resolution based on formation of inclusion complexes have been obtained with cyclodextrins.

Advantages of the chiral mobile phase additives are the possibility to employ aqueous mobile phases; limitations are the costs of additives and for cyclodextrins, the necessity of an appropriate spatial configuration of the analyte for inclusion-complex formation.

Chiral Stationary Phases

Pirkle-type are chiral stationary phases, e.g. R-N-(3,5-dinitrobenzoyl) phenylglycine which is bound to α-aminopropyl packing through ionic or covalent bonds. These phases are designed on the basis of the three-point chiral recognition model proposed by Dalgliesh [1952] who postulated that chiral recognition requires a minimum of three simultaneous interactions between the chiral stationary phase and solute. At least one of these interactions must be sterically controlled and may be either attractive or repulsive. The relative strengths of the resulting distereomeric complexes determine the resolution and elution order of the two enantiomers.

Enantiomer I interacts with the chiral stationary phase at sites A—A', B—B' and C—C' whereas its mirror image ,enantiomer II, lacks the C—C' interaction. If the C—C' interaction is an attractive interaction, enantiomer I will be retained on the column longer than II. If the C—C' interaction is repulsive, the diastereomeric complex involving enantiomer II is more stable and I will be eluted first.

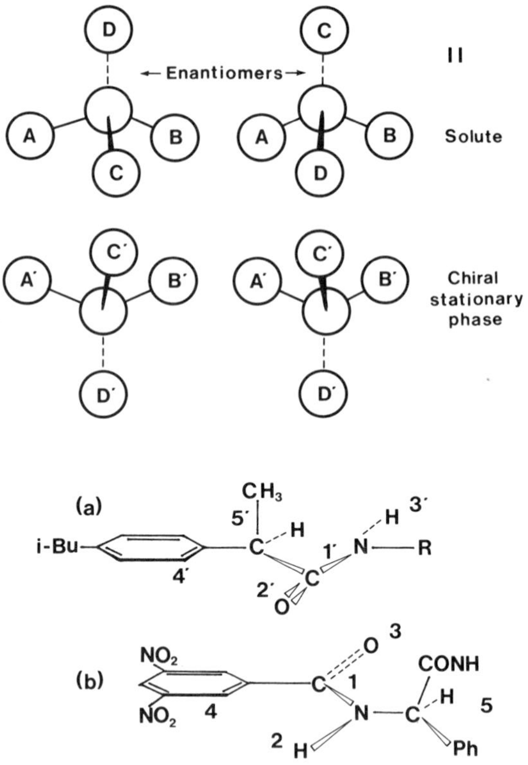

Examples of interactions are hydrogen bonding between carbonyl oxygen of a keto group of the stationary phase and an hydroxyl group of the analyte; Van der Waals interactions between alkyl groups, π-bond donor/acceptor interaction between an aromatic group on the analyte and an electron withdrawing group on the chiral stationary phase.

Pirkle-type chiral columns are generally used in the normal phase mode, using hexane-isopropanol mixtures as mobile phases. Traditional Pirkle stationary phases are not compatible with aqueous mobile phases; covalently bonded columns can be used with polar mobile phases but selectivity and column stability are sacrificed [Armstrong, 1987]. Pre-column derivatisation is generally required for acids and bases.

Inclusion Complexation

Stationary phases are available which have a seven to nine atom spacer between the support and a cyclodextrin molecule. Cyclodextrins are cyclic oligosaccarides linked by α-glycosidic bonds. There are three forms of cyclodextrins, α, β and γ, which contain six, seven and eight glycopyranose units respectively. The β-cyclodextrins have been more widely applied because of their optimum size for inclusion-complex formation. Separation are obtained in the reverse-phase mode but the organic phase may compete with enantiomers for location in the hydrophobic cavity. Satisfactory resolutions have been obtained for enantiomers of metadone, verapamil, metoprolol [Armstrong et al., 1986].

248

Protein-type

There are columns with albumin, for binding acidic drugs, or α_1-acid-glycoprotein, for binding basic drugs, as irreversibly bound phases. The separation takes place through differential interaction of enantiomers by hydrogen bonding and Van der Waals mechanism.

Careful adjustement of pH, ionic strength and organic solvent in the mobile phase must be done. The columns are used in the reverse-phase mode, with aqueous buffered phases containing isopropanol as organic modifier, at pH 3-7.5. The enantiomers of several commonly used drugs have been successfully resolved with α_1-acid-glycoprotein columns [Wainer, Barkan and Schill, 1986].

REGULATORY CONSIDERATIONS

As the authorities realized the differences in activity, binding, or toxicity of enantiomers in racemic mixtures and their potential for adverse reactions in particular patients or groups of patients (e.g. the elderly), they began to emphasize the importance of data on the individual enantiomers [Cartwright, 1990].

For new active substances with one or more chiral centers information must be given as to the form of the active ingredient to be used in the product to be marketed. Details should be provided on the chemical separation of different chiral forms used in the various tests reported in the application for marketing authorization.

The Notice to Applicants summarizes the possible problems that should be discussed in the appropriate Expert Report:

– Batch-to-batch consistency of stereoisomer ratio;
– The toxicological issues;
– Pharmacological aspects (including evidence on which isomer(s) have the desired pharmacological properties);
– Pharmacokinetics (including information of relative metabolism of stereoisomers);
– Extrapolation of preclinical data (with particular reference to possible problems in species handling of stereoisomers);
– The significant clinical issues.

In the Pharmaceutical Expert Report reference is made to the situation where an application is being made for one isomer where previously a mixture of stereoisomers has been marketed. It indicates that in this case full quality data on this isomer need to be provided. The Notice to Applicants does not indicate what data if any are needed for safety and efficacy in these circumstances.

Thus, the Notice to Applicants does not presently require companies to file applications for single isomers, or to provide systematically data on all aspects of the isomers present in a mixture. At present, it merely requires the appropriate expert to comment and justify on the basis of the data given that the product is of suitable quality, safety, and efficacy.

The US Food, Drug and Cosmetic Act, under which the U.S. Food and Drug Administration operates, and its regulations have as their objective that human drugs consumed by the public will be safe and effective for the conditions for which the drugs are labeled. The Act does not specifically address the situation concerning drug molecules containing stereoisomers.

The decision whether to market a specific isomer or a racemate is one that is primarily under the control of the pharmaceutical companies. Many companies now attempt to resolve a chiral drug substance or to synthesize its pure enantiomers in very early stage of drug development, with the objective of determining the primary pharmacological profile of each of the enantiomers. If the individual enantiomer cannot be obtained, or if neither of the enantiomers has the major advantages with respect to primary pharmacological profile, the racemate may then be developed. The current FDA position is that if a racemate is submitted, if there is adequate information to support safety and efficacy, and if the product can be reproduced uniformly lot by lot, then the product will normally be accepted and approved [Kunkumian, 1990].

In Japan, for registration of a racemic mixture, information will be required on the toxicity, general pharmacology, efficacy and disposition of each isomer. Also, when the active compound is either of the two isomers, information will be required on the extent of interconversion of the two isomeric forms. This will apply probably to new antiinflammatory 2-arylpropionic acids. Lastly, information will be required on isomeric purity [Smith and Caldwell, 1988].

CONCLUSIONS

Biological macromolecules are highly asymmetric. Most of the synthetic chiral pharmaceuticals are administered as racemates, equimolar mixtures of enantiomers. Enantiomers are handled differently by the body: their pharmacokinetic and pharmacodynamic results are different. Consequently, when new chemical entities are developed, the study of individual isomers as well as the racemates becomes essential.

Analytical methods capable to measure separately the two or more isomers are needed. When possible it seems more useful to develop new drugs as single enantiomers.

ACKNOWLEDGEMENTS

The assistance of Drs. Alberto Catto and Enzo Dolfini is gratefully acknowledged.

REFERENCES

Abas A, Meffin PJ., 1987. Enantioselective disposition of 2-arylpropionic acid nonsteroidal anti-inflammatory drugs. IV. Ketoprofen disposition. J. Pharmacol. Exp. Ther. **240**: 637.

Aberg G., 1972. Toxicological and local anaesthetic effects of optically active isomers of two local anaesthetic compounds. Acta Pharmacol. Toxicol. **31**: 273.

Akerman B, Persson H, Tegner C., 1967. Local anesthetic properties of the optically active isomers of prilocaine (Citanest). Acta Pharmacol. Toxicol. **25**: 233.

Ames MM, Frank SK., 1982. Stereochemical aspects of para-chloramphetamine metabolism. Biochem. Pharmacol. **31**: 5.

Aps C, Reynolds F., 1978. An intradermal study of the local anaesthetic and vascular effects of the isomers of bupivacaine. Br. J. Clin. Pharmacol. **6**: 63.

Armstrong DW., 1987. Optical isomer separation by liquid chromatography. Anal. Chem. **59**: 84A.

Armstrong DW, Ward TJ, Armstrong RD, Beesley TE., 1986. Separation of drug stereoisomers by the deformation of β-cyclodextrin inclusion complexes. Science. **232**: 1132.

Bai SA, Walle UK, Wilson MJ, Walle T., 1983. Stereoselective binding of the (−)-enantiomer of propranolol to plasma and extravascular binding sites in the dog. Drug Metab. Dispos. **11**: 394.

Banfield C, O'Reilly R, Chan E, Rowland M., 1983. Phenylbutazone-warfarin interaction in man: Further stereochemical and metabolic considerations. Br. J. Clin. Pharmacol. **16**: 669.

Bjorkman S., 1985. Stereoselective disposition of indoprofen in surgical patients. Br. J. Clin. Pharmacol. **20**: 463.

Block AJ, Merrill D, Smith ER., 1988. Stereoselectivity of tocainide pharmacodynamics in vivo and in vitro. J. Cardiovasc. Pharmacol. **11**: 216.

Borgstrom L, Nyberg L, Jonsson S et al., 1989. Pharmacokinetic evaluation in man of terbutaline given as separate enantiomers and as the racemate. Br. J. Clin. Pharmacol. **27**: 49.

Boyd RA, Chin SK, Don-Pedro O et al., 1989. The pharmacokinetics of the enantiomers of atenolol. Clin. Pharmacol. Ther. **454**: 403.

Branch RA, Nies AS, Shand DG., 1973. The disposition of propranolol. VII. Drug Metab. Dispos. **1**: 687.

Breckenridge AM, Orme MLE., 1972. The plasma half-lives and the pharmacological effect of the enantiomers of warfarin in rats. Life Sci. **11**: 337.

Cahn RS, Ingold C, Prelog V., 1966. Specification of molecular chirality. Angev. Chem. Intern. Ed. Engl. **5**: 385.

Caldwell J, Marsh MV., 1983. Interrelationships between xenobiotic metabolism and lipid biosynthesis. Biochem. Pharmacol. **32**: 1667.

Caldwell J, Hutt AJ, Fournel-Gigleux S., 1988. The metabolic chiral inversion and dispositional enantioselectivity of the 2-arylpropionic acids and their biological consequences. Biochem. Pharmacol. **37**: 105.

Cartwright AC., 1990. Stereochemistry and safety, efficacy and quality issues: genesis of new regulations. Drug Informat. J. **24**: 115.

Christ DD, Walle T., 1985. Stereoselective sulfate conjugation of 4-hydroxypropranolol in vitro by different species. Drug Metab. Dispos. **13**: 380.

Coltart DJ, Shand DG., 1970. Plasma propranolol levels in the quantitative measurement of β-adrenergic blockade in man. Brit. Med. J. **3**: 731.

Cook CE, Seltzman TB, Tallent CR et al., 1987. Pharmacokinetics of pentobarbital enantiomers as determined by enantioselective radioimmunoassay after administration of racemate to humans and rabbits. J. Pharmacol. Exp. Ther. **241**: 779.

Crooks PA, Godin CS., 1988. N-Methylation of nicotine enantiomers by human liver cytosol. J. Pharm. Pharmacol. **40**: 153.

Dalgliesh CE., 1952. The optical resolution of aromatic amino-acids on the paper chromatograms. J. Chem. Soc. **137**: 3940.

Day RO, Williams KM, Graham GG et al., 1989. Stereoselective disposition of ibuprofen enantiomers in synovial fluid. Clin. Pharmacol. Ther. **435**: 480.

Echizen H, Brecht T, Niedergesaess S et al., 1985. The effect of dextro, levo and racemic verapamil on atrioventricular conduction in man. Am. Heart J. **109**: 210.

Edgar B, Heggelund A, Johansson L et al., 1984. The pharmacokinetics of R- and S-tocainide in healthy subjects. Br. J. Clin. Pharmacol. **16**: 216P.

Eichelbaum M, Mikus G, Vogelgesang B., 1984. Pharmacokinetics of (+)-, (−)- and (±)-verapamil after intravenous administration. Br. J. Clin. Pharmacol. **17**: 453.

Eichelbaum M., 1988. Pharmacokinetic and pharmacodynamic consequences of stereoselective drug metabolism in man. Biochem. Pharmacol . **37**: 93

Evans AM, Nation RL, Sansom LN et al., 1989. Stereoselective plasma protein binding of ibuprofen enantiomers. Eur. J. Clin. Pharmacol. **36**: 283.

Foster RT, Jamali F., 1988 a. Stereoselective pharmacokinetics of ketoprofen in the rat. Influence of route of administration. Drug Metab. Dispos. **16**: 623

Foster RT, Jamali F, Russell AS, Alballa SR., 1988 b. Pharmacokinetics of ketoprofen enantiomers in healthy subjects following single and multiple doses. J. Pharm. Sci. **77**: 70.

Foster RT, Jamali F, Russell AS, Alballa SR., 1988 c. Pharmacokinetics of ketoprofen enantiomers in young and elderly arthritic patients following single and multiple doses. J. Pharm. Sci. **77**: 191.

Giacomini KM, Nelson WL, Pershe RA et al., 1986. In vivo interaction of the enantiomers of disopyramide in human subjects. J. Pharmacokinet. Biopharm. **14**: 335.

Hague D, Smith RL., 1988. Enigmatic properties of (+)-thalidomide; an example of a stable racemic compound. Br. J. Clin. Pharmacol. **26**: 632P.

Hendel J, Brodthagen H., 1984. Entero-hepatic cycling of methotrexate estimated by the use of the D-isomer as a reference marker. Eur. J. Clin. Pharmacol. **26**: 103.

Hsyu P-H, Giacomini KM., 1985. Stereoselective renal clearance of pindolol in humans. J. Clin. Invest. **76**: 1720.

Hutt AJ, Caldwell J., 1983. The metabolic chiral inversion of 2-arylpropionic acids. A novel route with pharmacological consequences. J. Pharm. Pharmacol. **35**: 693.

Jamali F., 1988. Pharmacokinetics of enantiomers of chiral non-steroidal anti-inflammatory drugs. Eur. J. Drug Metab. Pharmacokinet. **13**: 1.

Jamali F, Mehvar R, Lemko C, Eradiri O., 1988 a. Application of a stereospecific high-performance liquid chromatography assay to a pharmacokinetic study of etodolac enantiomers in humans. J. Pharm. Sci. **77**: 963.

Jamali F, Singh NN, Pasutto FM et al., 1988 b. Pharmacokinetics of ibuprofen enantiomers in human following oral administration of tablets with different absorption rates. Pharmacol. Res. **5**: 40.

Jamali F, Mehvar R, Pasutto FM., 1989. Enantioselective Aspects of Drug Action and Disposition: Therapeutic Pitfalls. J. Pharm. Sci. **78**: 695.

Jenner P, Testa B., 1973. The Influence of Stereochemical Factors in Drug Disposition. Drug Metab. Rev. **2**: 117.

Johnson DM, Reuter A, Collins JM, Thompson GF., 1979. Enantiomeric purity of naproxen by liquid chromatographic analysis of its diastereomeric octyl esters. J. Pharm. Sci. **68**: 112.

Kamerling JP, Duran M, Gerwig GJ et al., 1981. Determination of the absolute configuration of some biologically important urinary 2-hydroxydicarboxylic acids by capillary gas-liquid chromatography. J. Chromatogr. **222**: 276.

Karnes HT, Sarkar MA., 1987. Enantiomeric resolution of drug compounds by liquid chromathography. Pharmaceut. Res. **4**: 285.

Kawashima R, Levy A, Spector S., 1976. Stereospecific radioimmunoassay for propranolol isomers. J. Pharmacol. Exp. Ther. **196**: 517.

Kunkumian CS., 1990. Regulatory considerations concerning stereioisomers in drug products. Drug Informat. J. **24**: 125.

Le Corre P, Gibassier D, Sado P, Le Verge R., 1988. Stereoselective metabolism and pharmacokinetics of disopyramide enantiomers in humans. Drug Metab. Disp. **16**: 858.

Lee EJD, Williams K, Day R. et al., 1985. Stereoselective disposition of ibuprofen enantiomers in man. Brit. J. Clin. Pharmacol. **19**: 669.

Lee EJD, Williams KM., 1990. Chirality. Clinical Pharmacokinetic and Pharmacodynamic Considerations. Clin. Pharmacokinet. **18**: 339.

Lennard MS, Tucker GT, Silas JH et al., 1983. Differential stereoselective metabolism of metoprolol in extensive and poor debrisoquin metabolisers. Clin. Pharmacol. Ther. **34**: 732.

Lepage J, Lindner W, Davies G. Karger B., 1979. Resolution of the optical isomers of dancyl amino acids by reversed phase liquid chromatography with optically active metal chelate additives. Anal. Chem. **51**: 433.

Lewis RJ, Trager WF, Chan KK et al., 1974. Warfarin: Stereochemical aspects of its metabolism and the interaction with phenylbutazone. J. Clin. Invest. **53**: 1607.

Lima JJ, Boudoulas H, Shields B., 1985. Stereoselective pharmacokinetics of disopyramide enantiomers in man. Drug Metab. Dispos. **13**: 572.

Luduena FP., 1969. Duration of local anesthesia. Ann. Rev. Pharmacol. **9**: 503.

Mikus G, Kroemer HK, Klotz U Eichelbaum M., 1988. Stereochemical considerations of the cimetidine-verapamil interaction. Clin. Pharmacol. Ther. **43**: 134

Nakazawa H, Yoneda H., 1978. Chromatographic study of optical resolution. II. Separation of optically active cobalt (III) complexes using potassium antimony d-tartrate as eluent. J. Chromatogr. **160**: 89.

Nies AS, Evans GR, Shand DG., 1973. Regional hemodynamic effects of β-adrenergic blockade with propranolol in the unanesthetized primate. Amer. Heart J. **85**: 97.

Notterman DA, Drayer DE, Metakis L, Reidenberg MM., 1986. Stereoselective renal tubular secretion of quinidine and quinine. Clin. Pharmacol. Ther. **40**: 511.

O'Reilly RA, Trager WF, Motley CH, Howald W., 1980. Stereoselective interaction of phenylbutazone with (12C/13C) warfarin pseudoracemates in man. J. Clin. Invest. **65**: 746.

Ofori-Adjei D, Ericsson O, Lindstrom B et al., 1986. Enantioselective analysis of chloroquine and desethylchloroquine after oral administration of racemic chloroquine. Ther. Drug Monit. **8**: 457.

Olivieri R, Fascetti E, Angelini L , Degen L., 1981. Enzymic conversion of N-carbamoyl-D-amino acids to D-amino acids. Biotechnol. Bioeng. **23**: 2173.

Otton SV, Lennard MS, Tucker GT, Woods HF., 1989. Cumene hydroperoxide-supported oxidation of propranolol enantiomers by human liver microsomes. Abstractc No. PP 02.66. IV World Conference on Clinical Pharmacology, Mannheim-Heidelberg. Eur. J. Clin. Pharmacol. 36 (Suppl.).

Pfeiffer CC., 1956. Optical isomerism and pharmacological action, a generalization. Science **124**, 29.

Ruelius HW, Tio CO, Knowles JA et al., 1979. Diastereoisomeric glucoronides of oxazepam. Isolation and stereoselective enzymic hydrolysis. Drug Metab. Dispos. **7**: 40.

Sallustio BC, Meffin PJ, Knights KM., 1988. The stereospecific incorporation of fenoprofen into rat hepatocyte and adipocyte triacylglycerols. Biochem. Pharmacol. **37**: 1919.

Schilsky RL, Choi KE, Vokes EE et al., 1989. Clinical pharmacology of the stereoisomers of leucovorin during repeated oral dosing. Cancer. **63** (Suppl. 6): 1018.

Sedman AJ,Gal J., 1983. Resolution of the enantiomers of propranolol and other β-adrenergic antagonists by high-performance liquid chromatography. J. Chromatogr. **278**: 199.

Sheldon RA., 1990. The industrial synthesis of pure enantiomers. Drug Inform. J. **24**: 129.

Simmonds RG, Woodage TJ, Duff SM, Green JN., 1980. Stereospecific inversion of (R)-(−)-benoxaprofen in rat and man. Eur. J. Drug Metab. Pharmacokinet. **5**: 169.

Simonyi M, Fitos I, Visy J., 1986. Chirality of bioactive agents in protein binding storage and transport processes. TIPS **7**: 112.

Simonyi M, Gal J and Testa B., 1989. Signs of the times: the need for a stereochemically informative generic name system. TIPS **10**: 349.

Sisenwine SF, Tio CO, Knowles JA et al., 1982. Species-related differences in the stereoselective glucuronidation of oxazepam. Drug Metab. Dispos. **10**: 605.

Smith RL, Caldwell J., 1988. Racemates: towards a New Year solution? TIPS **9**: 75.

Takahashi H, Kanno S, Ogata H et al., 1988. Determination of propranolol enantiomers in human plasma and urine and rat tissues using chiral stationary-phase liquid chromatography. J. Pharm. Sci. **77**: 993.

Tamai I, Ling H-Y, Timbul S-M et al., 1988. Stereospecific absorption and degradation of cephalexin. J. Pharm. Pharmacol. **40**: 320.

Testa B., 1989. Mechanism of chiral recognition in xenobiotic metabolism and drug-receptor interactions. Chirality. **1**: 7.

Thijssen HHW, Baars LGM, Drittij-Reijnders MJ., 1985. Stereoselective aspects in the pharmacokinetics and pharmacodynamics of acenocoumarol and its amino and acetamido derivatives in the rat. Drug Metab. Dispos. **13**: 593.

Thijssen HHW, Baars LGM., 1987. The biliary excretion of acenocoumarol in the rat: Stereochemical aspects. J. Pharm. Pharmacol. **39**: 655.

Tocco DJ, Hooke KF, Deluna FA, Duncan AEW., 1976. Stereospecific binding of timolol, a β-adrenergic blocking agent. Drug Metab. Dispos. **4**: 323.

Thomson AH, Murdoch G, Pottage A et al., 1986. The pharmacokinetics of R- and S-tocainide in patients with acute ventricular arrhythmias. Brit. J. Clin. Pharmacol. **21**: 149.

Toon S, Low LK, Gibaldi M et al., 1986. The warfarin-sulfinpyrazone interaction: stereochemical considerations. Clin. Pharmacol. Ther. **39**: 15.

Toon S, Hopkins KJ, Garstang FM, Rowland M., 1987. Comparative effects of ranitidine and cimetidine on the pharmacokinetics and pharmacodynamics of warfarin in man. Eur. J. Clin. Pharmacol. **32**: 165.

Tucker GT, Lennard MS., 1990. Enantiomer specific pharmacokinetics. Pharm. Ther. **45**: 309.

Vogelgesang B, Echizen H, Schmidt E, Eichelbaum M., 1984. Stereoselective first-pass metabolism of highly cleared drugs: studies of the bioavailability of L- and D-verapamil examined with a stable isotope technique. Br. J. Clin. Pharmacol. **18**: 733.

Wade DN, Mearrick PT, Morris JL., 1973. Active transport of L-dopa in the intestine. Nature. **242**: 463.

Wainer IW, Barkan SA, Schill G., 1986. α_1-Acid glycoprotein chiral stationary phase. HPLC application to the resolution of enantiomeric drugs. LC/GC Mag. **5**: 422.

Walle T, Walle UK., 1979. Stereoselective bioavailability of ($\pm$) propranolol in the dog. A GC-MS study using a stable isotope technique. Res. Commun. Chem. Pat. Pharmacol. **23**: 453.

Walle T, Walle UK, Wilson MJ et al., 1984. Stereoselective ring oxidation of propranolol in man. Br. J. Clin. Pharmacol. **18**: 741.

Walle T, Webb JG, Bagwell EE et al., 1988. Stereoselective delivery and actions of beta receptor antagonists. Biochem. Pharmacol. **37**: 115.

Ward S, Branch RA, Walle T, Walle UK., 1986. Cosegregation of propranolol metabolism with debrisoquine and mephenytoyn polymorphism. Pharmacologist **28**: 137.

Ward SA, Walle T, Walle UK et al., 1989. Propranolol's metabolism is determined by both mephenytoin and debrisoquine hydroxylase activities. Clin. Pharmacol. Ther. **45**: 72.

Webb JG, Street JA, Bagwell EE et al., 1988. Stereoselective secretion of atenolol from PC12 cells. J. Pharmacol. Exp. Ther. **247**: 958.

Wilkinson GR, Shand DG., 1975. A physiological approach to hepatic drug clearance. Clin. Pharmacol. Ther. **18**: 377.

Williams K, Day R, Knihinicki R, Duffield A., 1986. The stereoselective uptake of ibuprofen enantiomers into adipose tissue. Biochem. Pharmacol. **35**: 3403.

Yacobi A, Levy G., 1974. Pharmacokinetics of the warfarin enantiomers in rats. J. Pharmacokinet. Biopharm. **23**: 239.

Yamaguchi T, Nakamura Y., 1987. Stereoselective metabolism of 2-phenylproprionic acid in rat. II. Studies on the organ responsible for the optical isomerization of 2- phenylpropionic acid in rat in vivo. Drug Metab. Dispos. **15**: 535.

STEREOSELECTIVITY IN DRUG DISPOSITION AND METABOLISM: CONCEPTS AND MECHANISMS

Bernard Testa

Institut de Chimie Thérapeutique
Ecole de Pharmacie
Université de Lausanne
Lausanne, Switzerland

INTRODUCTION

A proper grasp of the stereoselectivity of pharmacological processes must begin at the molecular level, and particularly with the various classes of stereoisomers encountered among bioactive compounds. These notions then lead to a direct understanding of chiral recognition and its manifold expressions.

Classification of stereoisomers

Stereoisomers (Figure 1) are characterized by having identical atoms that are identically connected but have different arrangements in space. Subdivision of stereoisomers is possible according to two completely independent and unrelated criteria, namely the geometry criterion and the energy criterion [Testa, 1979, 1990a].

The *geometry criterion* separates with complete rigour and mutual exclusiveness enantiomers from diastereomers. *Enantiomers* are stereoisomers that are related to each other like non-superimposable mirror images, while by definition *diastereomers* are stereoisomers that do not mirror each other. Because two enantiomers have identical intramolecular relationships (interatomic distances, etc), their energy content is identical, and so are all their chemical and physicochemical properties. In contrast, diastereomers do not have identical intramolecular relationships and all their properties will differ, however small the difference.

The second criterion is one of energy, more accurately the *energy barrier* that separates two isomers and must be overcome during their interconversion. When the barrier is relatively high and the two stereoisomers interconvert very slowly (e.g. when the $t_{1/2}$ of the unimolecular reaction is in the order of years), they are called *configurational isomers*. When in contrast the interconversion is fast (e.g. $t_{1/2}$ in the order of minutes), they are called *conformers* (conformational isomers). Since a continuum of intermediate cases exists between rapidly and slowly interconverting stereoisomers, the discrimination between conformers and configurational isomers cannot be a very sharp and accurately defined one.

New Trends in Pharmacokinetics, Edited by A. Rescigno and A.K. Thakur
Plenum Press, New York, 1991

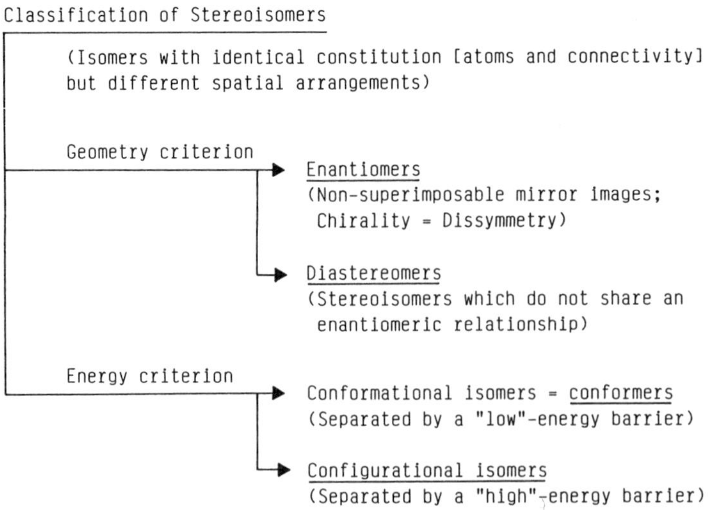

Classification of Stereoisomers

(Isomers with identical constitution [atoms and connectivity]
but different spatial arrangements)

Geometry criterion

Enantiomers
(Non-superimposable mirror images;
Chirality = Dissymmetry)

Diastereomers
(Stereoisomers which do not share an
enantiomeric relationship)

Energy criterion

Conformational isomers = conformers
(Separated by a "low"-energy barrier)

Configurational isomers
(Separated by a "high"-energy barrier)

Figure 1. Scheme showing the classification of stereoisomers according
to the two independent criteria of geometry and energy [Testa, 1979].

Because as stated above the two criteria of geometry and energy are unrelated,
stereoisomers can in fact be separated into four classes:

A) Enantiomeric configurational isomers;
B) Diastereomeric configurational isomers;
C) Enantiomeric conformers;
D) Diastereomeric conformers.

Classes A, B and D are well known and do not need to be exemplified. As for class C, the fact that it is of pharmacological relevance is illustrated by oxazepam (**I**) which exists as two very rapidly interconverting enantiomers (estimated $t_{1/2}$ of racemization under physiological conditions 2-4 min) [Aso et al., 1988].

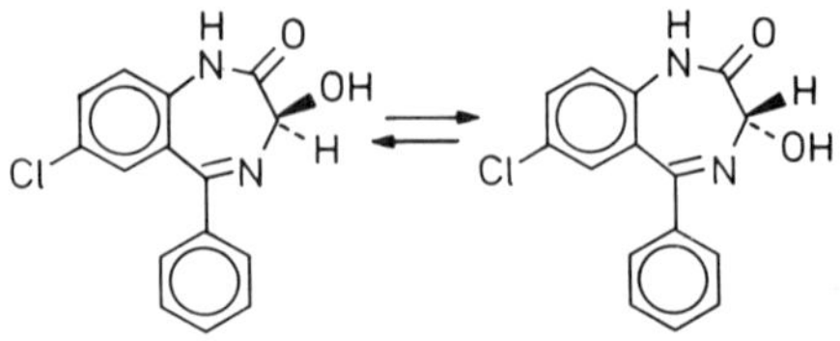

Chiral Discrimination

Since enantiomers have identical properties, they can be discriminated or physically separated only by applying a "chiral handle", as illustrated in Figure 2 with a simple case pertaining to a two-dimensional universe. When a mixture of two enantiomers reacts with an enentiometrically pure reagent, two diastereomers are formed that can be distinguished spectroscopically and separated by chromatography or other techniques. Note that the diastereomers may be produced by the formation of covalent bonds, ionic bonds (diastereomeric salts), or other types of intermolecular forces (diastereomeric complexes involving hydrogen bonds, Van der Waals forces and/or hydrophobic interactions). When such weak forces are involved, the diastereomeric complexes will not be of the highest stability and may dissociate depending on conditions. This does not prevent *chiral recognition* provided that the "exposure time" of the analytical method is short compared to the half-life of the complexes. The recognition resulting from diastereomeric complexes is vividly illustrated by a verse taken from "Through the Looking-Glass" (Lewis Carroll):

> "Or madly squeeze a right-hand foot
> Into a left-hand shoe".

Biological macromolecules (be they for example proteins, receptors, enzymes or nucleic acids) are enentiometrically pure. As a consequence, the binding of two enantiomers to a biomacromolecule will result in diastereomeric complexes and, at least in principle, in chiral recognition [Testa, 1990 a]. This phenomenon is of fundamental significance in biology and accounts for enantioselectivity in pharmacodynamic and pharmacokinetic processes. Furthermore, the reaction of two enantiomers with an enzyme will result in two diastereomeric transition states, affording another mechanism of chiral recognition as discussed below.

GENERAL CONCEPTS IN STEREOSELECTIVE DRUG DISPOSITION AND METABOLISM

Stereoselective Processes in Drug Disposition

A number of examples document differences, usually but not always modest ones, in various processes of drug disposition [Campbell, 1990 a, 1990 b; Jamali et al., 1989; Tucker and Lennard, 1990]. These phenomena of *enantioselectivity* were recently illustrated by Campbell [1990 a] in a table summarizing a large body of information (Table 1). It can be seen that enantiomeric ratios may vary considerably; while low ratios usually correspond to passive processes, high ratios are believed to result from active (carrier-mediated) transport and hence from effective chiral recognition by the carrier enzyme. Exceptions to this schematic view are probably numerous.

Besides biotransformation to be discussed below, the pharmacokinetic process most actively investigated from the viewpoint of chiral recognition is *binding to plasma proteins*. Much information is available in the literature [Campbell, 1990 b;

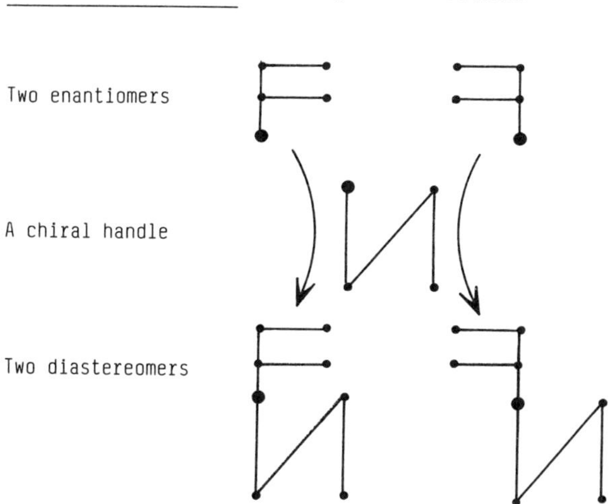

Figure 2. Reaction scheme outlining the principle of chiral recognition in a two-dimensional universe.

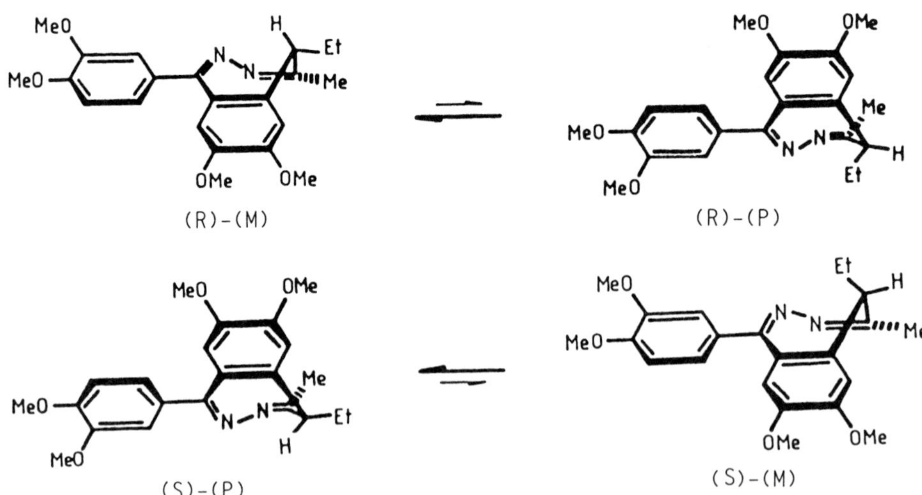

Figure 3. The two enantiomers of tofisopam, each existing as two diastereo-meric conformers of (*M*) and (*P*) helicity (Modified from Visy and Simonyi, 1989, and reproduced with the permission of Wiley-Liss Inc., New York).

Müller, 1988], but no study to date appears to address the molecular mechanisms underlying chiral recognition by such blood macromolecules as serum albumin and alpha₁-acid glycoprotein. One is thus faced with a wealth of descriptive results that document, for example, species variations, differences between proteins and binding sites, and differences in binding mode between enantiomers. For example, the two enantiomers of propranolol were shown to bind with different affinities and by different modes to a common high-affinity, saturable site on human alpha₁-acid glycoprotein; in addition, *dextro*-propranolol did bind to a low-affinity, non-saturable site [Oravcova et al., 1989].

Table 1
Ranges of enantiomeric ratios in various processes of drug disposition
(excluding all ratios of 1.0 where there is no detectable enantioselectivity)
(Taken from Campbell 1990a, 1990b)

Process	Range of enantiomer ratios
Absorption	1.1 to >100
Binding to plasma proteins	
— acidic	1.1 to 3.0
— basic	1.1 to 100
Tissue distribution	1.2 to 13
Hepatic clearance	1.1 to 28
Renal clearance	1.1 to 50

One fundamental aspect of protein binding worthy of mention is *recognition of chiral conformers*, as described for bilirubin. This biological molecule exists in a dynamic equilibrium between two intramolecularly H-bonded enantiomeric conformations, only one of which is strongly (but not covalently) bound to human serum albumin (HSA) [Lightner et al., 1988]. As a result, bilirubin in human plasma will exist mainly as a single, HSA-bound conformer.

Among drugs, a number of benzodiazepines have revealed a comparable behaviour. Tofisopam and analogous 2,3-benzodiazepines contain a centre of chirality at C(5); in addition, the 2,3-diazepine nucleus exists in two boat

Table 2
Enantiomeric ratios of free fractions (f_u) of various drugs in human plasma at
therapeutic concentrations (Modified from Müller, 1988)

Drug	f_u ratio	lesser bound enantiomer
Amphetamine	1.0	
Fenfluramine	1.04	(–)
Propranolol	1.1	(+)
Tocainide	1.2	(R)
Methadone	1.3	(–)
Warfarin	1.3	(R)
Latamoxef	1.6	(R)
Phenprocoumon	1.6	(R)
Verapamil	1.7	(–)
Disopyramide	2.5	(S)
Oxazepam glucuronide	3.5	(R)

conformations ($t_{1/2}$ of interconversion 3.6 h for tofisopam). As a result, each
enantiomer of tofisopam exists as two *diastereomeric* conformers described by their
(*M*) and (*P*) helicity (Figure 3), the conformers with the 5-substituent in a quasi-
equatorial position predominating (79:21 ratio) over the conformers with this
substituent in a quasi-axial position [Visy and Simonyi, 1989]. Affinity
chromatography on immobilized HSA revealed three peaks whose assignement
indicated the following order in affinities: (*S*)-(*P*) > (*R*) > (*S*)-(*M*). Comparable
results with a number of analogues led to the conclusion that the stereochemical
features favoured by the binding site were a (*P*)-helicity and a quasi-equatorial
orientation of the 5-substituent [Visy and Simonyi, 1989].

Substrate Stereoselectivity in Drug Metabolism

Because they can produce more than one response (in the form of more than
one metabolite), biotransformation reactions are more complex than other processes
of drug disposition. One consequence of this complexity is the existence of two basic
types of selectivity, namely substrate and product selectivity. In a stereochemical
context, the two types of selectivity become *substrate stereoselectivity* and *product
stereoselectivity*, respectively [Jenner and Testa, 1973, Testa and Mayer, 1988].
These concepts have proven their broad applicability and their power in extracting
valuable additional information from experimental studies [Jamali et al., 1989; Testa,
1986, 1988, 1990 a; Trager and Testa, 1985; Vermeulen and Breimer, 1983].
Substrate stereoselectivity is seen when stereoisomers are metabolized:
— differently (in quantitative and/or qualitative terms),
— by the "same" biological system, and under identical conditions.
Substrate stereoselectivity, and principally enantioselectivity, is a well known
and abundantly documented phenomenon, be it under *in vivo* [Walle and Walle,
1986] or *in vitro* conditions. Apparent and not absolute substrate stereoselectivity is
usually observed. Biological factors, inhibition, competition between several
metabolic routes, concentration effects, all these factors can influence the observed
selectivity. Two recently published examples will illustrate this phenomenon.
One of the metabolic reactions undergone by nicotine is methylation of its
pyridyl nitrogen to yield the N-methylnicotinium ion. The reaction is mediated by an

N-methyltransferase and in human liver cytosol shows a marked preference for (+)-(*R*)-nicotine (*R/S* ratio of reaction rates about 8/1) [Crooks and Godin, 1988]. Interestingly, the reaction appears totally enantioselective (i.e. enantiospecific) for (*R*)-nicotine in guinea-pig liver, and does not occur in rat liver. This species variation may be due to differences in isozyme populations (see also below).

The involvement of different isozymes appears to account for the opposed enantiospecificity seen in the glucuronidation of (*E*)-10-hydroxynortriptyline in human liver and intestine [Dahl-Puustinen et al., 1989]. Indeed, human hepatic preparations (microsomes and homogenates) were shown to specifically glucuronidate the (+)-enantiomer, while the opposite enantiospecifity was seen in human intestinal homogenates.

Product Stereoselectivity and Substrate-Product Stereoselectivity in Drug Metabolism

Product stereoselectivity occurs when stereoisomeric metabolites are generated:
— differentially (in quantitative or qualitative terms),
— from a single chiral, prochiral or proachiral substrate.

Note that only apparent product stereoselectivity can be observed under a given set of experimental conditions.

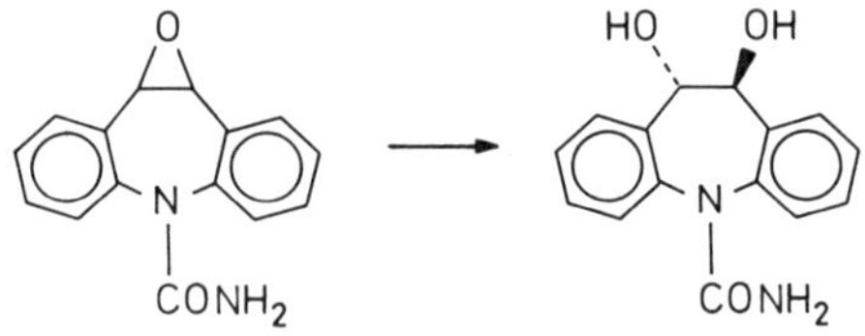

All documented cases of product enantioselectivity, and most cases of product diastereoselectivity, result from the metabolic generation of elements of chirality from a suitable prochiral centre or face. The most representative redox reactions creating elements of chirality are ketone reduction, methylene hydroxylation, oxidation of enantiotopic substituents, N-oxygenation and sulfoxidation (Figure 4). Additional cases are reactions of glutathione conjugation [e.g. Cobb et al., 1983], and the hydrolysis of prochiral diesters and oxiranes. Thus, carbamazepine 10,11-oxide (**III**), an active metabolite of the anticonvulsant drug carbamazepine, is very resistant to hydrolysis mediated by epoxide hydrolase in a number of animal species. In humans however, the reaction is a rapid one, yielding the 10,11-diol metabolite. While the epoxide is symmetric, the diol can exist as three stereoisomers, namely two enantiomers and a *meso*-form; formation of the latter is excluded since epoxide hydrolase only produces *trans*-diols, but the human metabolite has recently been shown to be the (−)-(10*S*, 11*S*)-enantiomer (**IV**) produced with 80% enantiomeric excess [Bellucci et al., 1987].

The question may be asked as to the influence of an existing element of chirality (e.g. a centre of chirality) on the metabolic generation of a second such element in a given molecule. The phenomenon is known in synthetic chemistry as *asymmetric induction* and implies the preferential formation of one diastereomeric product from a chiral reactant. In asymmetric induction there is only one chiral tool, namely the preexisting chiral centre. This contrasts with xenobiotic metabolism where an enzyme is most frequently operative as a second chiral tool, implying that the concept of asymmetric induction must take a special meaning when applied to enzymatic reactions. Yet it is a common and amply documented observation that product stereoselectivity can be different for two enantiomeric substrates, in other words that product stereoselectivity is itself substrate-enantioselective. The term *substrate-product stereoselectivity* has been proposed a number of years ago to designate such cases [Jenner and Testa, 1973].

Table 3
Substrate-product stereoselective 3'-hydroxylation of pentobarbital
(Holtzman and Thompson, 1975; Palmer et al., 1969, 1970)

Substrate	Metabolite	Percent of dose in urine of dogs	Rat liver microsomes	
			K_m (mM)	V_{max} (nmol [mg prot]$^{-1}$ min^{-1})
(+)-(R)	(1'R;3'S)	16.6	0.195	0.581
	(1'R;3'R)	16.2	0.087	0.654
(−)-(S)	(1'S;3'R)	6.2	0.198	0.296
	(1'S;3'S)	31.6	0.097	1.017

A highly illustrative example of substrate-product stereoselectivity is seen in the 3'-hydroxylation of pentobarbital (**II**). Upon administration of (+)-(*R*)-, (−)-(*S*)- and *rac*-pentobarbital to dogs, the two enantiomers yielded comparable amounts (35% ± 3%) of 3'-hydroxylated pentobarbital, and substrate enantioselectivity regarding overall 3'-hydroxylation was found to be non-existent [Palmer et al., 1969, 1970]. However, the diastereomeric ratio varied considerably with the substrate; while the (*R*)-enantiomer gave equal amounts of the (1'*S*;3'*S*)- and (1'*R*;3'*R*)-diastereomers, (*S*)-pentobarbital gave the (1'*S*;3'*R*)- and (1'*S*;3'*S*)-diastereomers in a 1/5 ratio. Thus, no product stereoselectivity exists for (*R*)-pentobarbital, as opposed to (*S*)-pentobarbital (Table 3). These *in vivo* results much resemble *in vitro* results obtained with rat liver microsomes (Table 3) [Holtzman and Thompson, 1975]. The K_m values for the (1'*R*;3'*S*)- and (1'*S*;3'*R*)-enantiomers were the same, as were the K_m values for the (1'*R*;3'*R*)- and (1'*S*;3'*S*)-enantiomers. These two pairs of values, however, clearly differed, perhaps suggesting that each pair of enantiomeric products was generated by a distinct isozyme. It also appears from Table 3 that the V_{max} values paralleled the *in vivo* product stereoselectivity.

While the concepts of substrate, product, and substrate-product stereoselectivity are indeed useful in ordering and clarifying many observations, they do not add to our understanding of underlying molecular mechanisms. Here other approaches are necessary, as discussed below.

MECHANISMS OF ENANTIOSELECTIVE DRUG METABOLISM

Conceptual Steps in Biological Processes

All biological processes can, at least conceptually, be subdivided into three steps. There is first penetration, then binding, and finally activation [Testa, 1984, 1990 b]. In xenobiotic metabolism, activation is the catalytic step. Thus, chiral recognition can be postulated to occur at any of these steps, but mainly at the *binding step* and/or at the *catalytic step*, resulting in different affinities and/or reactivities of the two enantiomers. In other words, both binding and reactivity factors can be expected *a priori* to contribute to substrate enantioselectivity and to product enantioselectivity. Here, the question is whether and to what extent these two contributions can be disentangled to reveal some aspects of the mechanisms of enantioselective drug metabolism [Testa, 1988].

Substrate Enantioselectivity in Binding and Catalysis

The contribution of *binding* to substrate enantioselectivity, in other words the differential binding of enantiomers to an enzyme, is due to the chirality of the substrate binding site, usually a hydrophobic pocket in the enzyme protein. But is this binding enantioselectivity sufficient to account for all cases of substrate enantioselectivity resulting from enzymatic reactions? The answer is a negative one, the *catalytic step* contributing in a number of cases to substrate enantioselectivity. Indeed, the chirality of the binding site should not conceal the chirality of the active site, which can be deduced from the fact that enzyme reactivity rests of the proper positioning and alignment of the catalytic and target groups; only so can the reaction proceed with adequate rapidity and stereoelectronic control.

Despite its limitations, *Michaelis-Menten analysis* offers a potentially interesting but insufficiently explored approach to evaluate the binding and catalytic components of substrate enantioselectivity. In a recent compilation [Testa and Mayer, 1988], some metabolic examples have shown two enantiomers to have comparable affinities but different velocities. Other cases show the opposite behaviour, namely analogous rates but markedly different affinities. And finally, cases also exist of substrate enantioselectivity caused by differences in both affinity and rate. In other words,

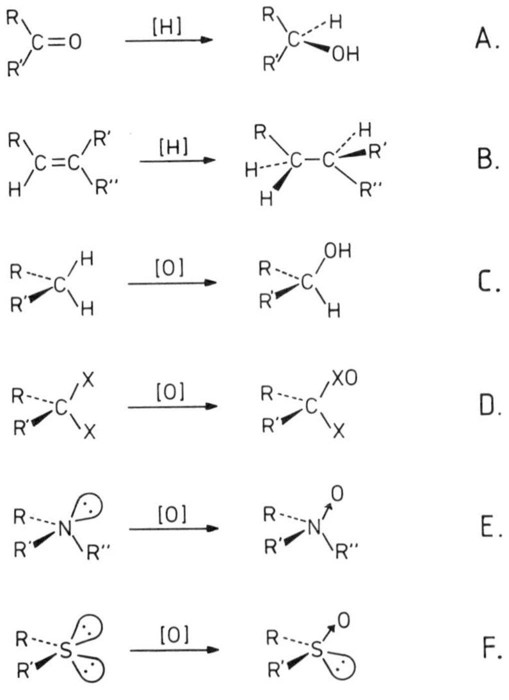

Figure 4. Some reductive and oxidative pathways producing new centres of
chirality in substrate molecules:
A: ketone reduction;
B: reduction of carbon-carbon double bonds;
C: hydroxylation of prochiral methylene groups;
D: oxygenation of enantiotopic moieties;
E: oxygenation of a configurationally labile tertiary amine to an N-oxide;
F: oxygenation of a sulfide to a configurationally stable sulfoxide.
The configurations depicted are arbitrary (modified from Testa, 1986, and
reproducedwith the permission of Elsevier Trends Journals, Cambridge).

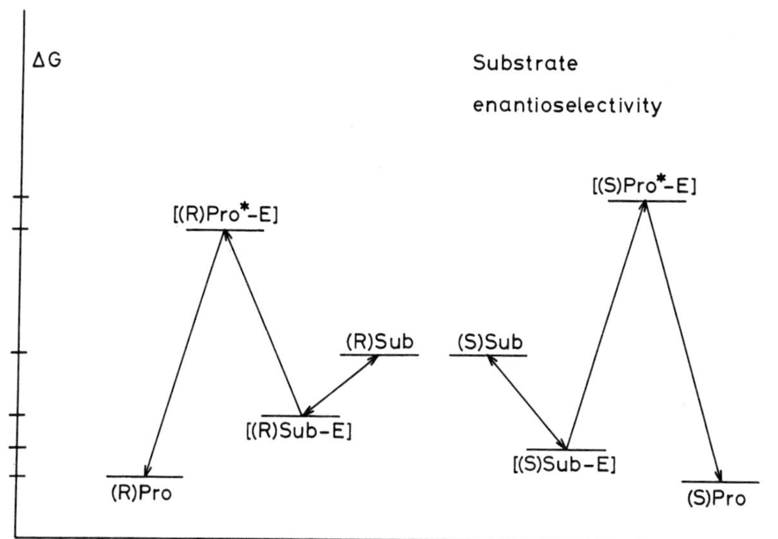

Figure 5. Hypothetical thermodynamic profile of substrate enantioselective reactions in drug metabolism. E: enzyme; (R)Sub and (S)Sub: enantiomeric substrates; (R)Pro and (S)Pro: enantiomeric products; * : transition state (Reproduced from Testa 1989 with the permission of Taylor & Francis, London).

Michaelis-Menten analysis suggests that the mechanism of chiral recognition underlying substrate enantioselectivity can be found in the binding step (different affinities), in the catalytic step (different reactivities), or in both steps.

Metabolic inhibition between enantiomers may also yield information on molecular mechanisms of substrate enantioselectivity [Testa, 1988]. To return to the example of nicotine N-methylation discussed above, guinea-pig lung aromatic azaheterocycle N-methyltransferase only methylates $(+)$-(R)-nicotine (K_m 14.2 μM), while $(-)$-(S)-nicotine is totally unreactive but acts as a strong competitive inhibitor of (R)-nicotine N-methylation (K_i 62.5 μM) [Cundy et al., 1985]. Obviously (S)-nicotine is a good ligand of the active site, but its binding is a non-productive one.

The thermodynamic scheme shown in Figure 5 is offered as a summary of the above discussion and an explanation of the mechanisms of recognition of chiral substrates. Indeed, binding of the two enantiomers leads to diastereomeric complexes. If their difference in free energy is large enough, it will be characterized as a difference in affinity. The catalytic step involves two diastereomeric transition states whose energy difference may be detectable as a difference in velocities [Testa, 1989].

Mechanisms of Product Enantioselectivity

Schematically, product stereoselectivity may be due to the action of distinct isozymes. Alternatively, it may results from *different binding modes* of the substrate molecule to the active site of a single isozyme. In this case, each *productive* binding mode will bring another of the two enantiotopic target groups in the vicinity of the catalytic site, resulting in diastereomeric enzyme-substrate complexes. Thus, the ratio of products may first depend on the relative probability of the two binding modes. In addition, the catalytic step, which involves diastereomeric transition states, may also influence or control product enantioselectivity.

Recent studies have confirmed the critical role of the catalytically productive orientation of the target groups, the relative importance of which is governed by the equilibrium constants between the various binding modes. Thus, the hydroxylation of the enantiotopic methyl groups in 2-phenylpropane (cumene, **V**) occurred with high product enantioselectivity. The latter was found to afford a good measure of the equilibrium constant between two binding modes, each of which brings one of the two methyl groups in contact with the reactive oxygen in cytochrome P450 [Sugiyama and Trager, 1986]. In contrast, methoxychlor (**VI**) contains two bulky enantiotopic anisyl groups which interchanged only slowly in the enzyme-substrate complex, resulting in a large degree of incomplete equilibrium in the reaction of O-demethylation to the two enantiomeric monophenols [Ichinose and Kurihara, 1987].

The second possible mechanism of product stereoselectivity is *binding to distinct isozymes*. Consider as an example the glutathione conjugation of the prochiral substrate phenanthrene 9,10-oxide, which produces 9-S-glutathionyl-10-hydroxy-9,10-dihydrophenanthrene as a chiral metabolite [Cobb et al., 1983]. Interestingly, the $(9S,10S)/(9R,10R)$ enantiomeric ratio was shown to be respectively 43/57, 73/27 and 95/5 for isozymes A, B and C of glutathione S-transferase. What is not known however is whether the product enantioselectivity of each isozyme results from a single binding mode, or if an equilibrium between two binding modes exists and is different for each isozyme. It thus appears that the two mechanisms discussed here are not mutually exclusive and may well occur simultaneously except in studies where a single isozyme is being used.

In analogy with Figure 5, a hypothetical thermodynamic scheme can be offered to summarize the mechanism of product enantioselectivity when a single isozyme is involved (Figure 6) [Testa, 1989]. Indeed, the ratio of enantiomeric products may

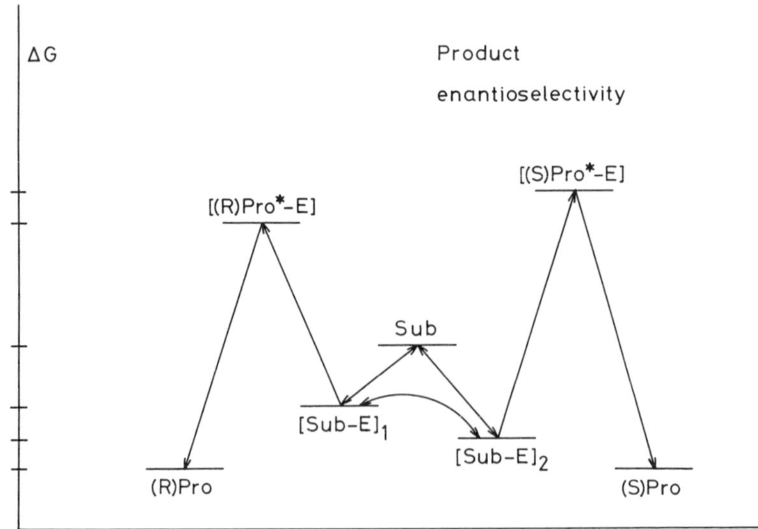

Figure 6. Hypothetical thermodynamic profile of product enantioselective reactions in drug metabolism. E: enzyme; Sub: prochiral substrate; (R)Pro and (S)Pro: enantiomeric products; * : transition state (Reproduced from Testa 1989 with the permission of Taylor & Francis, London).

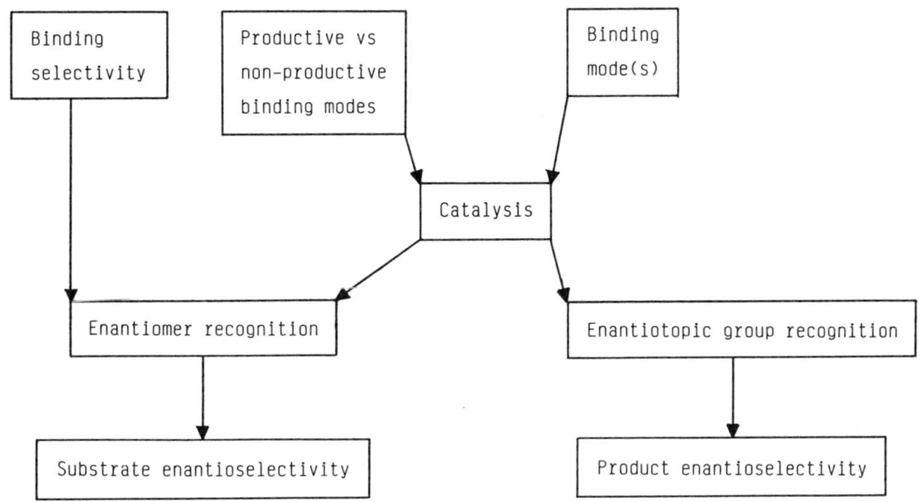

Figure 7. Possible contributions of binding selectivity, binding modes, and catalysis, to the phenomena of substrate enantioselectivity and product enantioselectivity.

first depend on the relative probability of the two binding modes. In addition, the catalytic step and its diastereomeric transition states may also influence or control the stereochemical course of the reaction.

Mechanistic Outlook

In brief, substrate enantioselectivity (or stereoselectivity) is the result of enantiomer (or stereoisomer) recognition at the binding and/or catalytic step, the relative contribution of the two factors varying from case to case. Product enantioselectivity (or stereoselectivity) on the other hand results from enantiotopic (or diastereotopic) group recognition (proximity of target group and catalytic centre) originating in the binding mode(s) of the substrate (Figure 7). The combined existence of stereoisomer recognition and stereotopic group recognition leads to complex cases of substrate-product stereoselectivity. These facts, together with structural differences in binding sites, offer a rationale for the variation in substrate and product stereoselectivity seen among isozymes. They also contribute significantly to a better understanding of structure-metabolism relationships and may lead to improved metabolic predictions.

REFERENCES

Aso Y., Yoshioka S., Shibazaki T. and Uchiyama M., 1988. The kinetics of the racemization of oxazepam in aqueous soultion. *Chem. Pharm. Bull.* **36**:1834.
Bellucci G., Berti G., Chiappe C., Lippi A. and Marioni F., 1987. The metabolism of carbamazepine in humans: steric course of the enzymatic hydrolysis of the 10,11-epoxide. *J. Med. Chem.* **30**:768.
Campbell D.B., 1990a. The development of chiral drugs. *Acta Pharm. Nord.* **2**:217.
Campbell D.B., 1990b. Stereoselectivity in clinical pharmacokinetics and drug development. *Eur. J. Drug Metab. Pharmacokin.* **15**:1109.
Cobb D., Boehlert C., Lewis D. and Armstrong R.N., 1983. Stereoselectivity of isozyme C of glutathione S-transferase toward arene and azaarene oxides. *Biochemistry* **22**:805.

Crooks P.A. and Godin C.S., 1988. N-Methylation of nicotine enantiomers by human liver cytosol. *J. Pharm. Pharmacol.* **40**:153.

Cundy K.C., Crooks P.A. and Godin C.S., 1985. Remarkable substrate-inhibitor properties of nicotine enantiomers towards a guinea pig lung aromatic azaheterocycle N-methyltransferase. *Biochem. Biophys. Res. Commun.* **128**:312.

Dahl-Puustinen M.L., Dumont E. and Bertilsson L., 1989. Glucuronidation of E-10-hydroxynortriptyline in human liver, kidney and intestine. Organ-specific differences in enantioselectivity. *Drug Metab. Disposit.* **17**:433.

Holtzman J.L. and Thompson J.A., 1975. Metabolism of R-(+)- and S-(–)-pentobarbital by hepatic microsomes from males rats. *Drug Metab.Disposit.* **3**:113.

Ichinose R. and Kurihara N., 1987. Intramolecular deuterium isotope effect and enantiotopic differentiation in oxidative demethylation of chiral [monomethyl-D_3]methoxychlor in rat liver microsomes. *Biochem. Pharmacol.* **36**:3751.

Jamali F., Mehvar R. and Pasutto F.M., 1989. Enantioselective aspects of drug action and disposition: Therapeutic pitfalls. *J. Pharm. Sci.* **78**:695.

Jenner P. and Testa B., 1973. The influence of stereochemical factors on drug disposition. *Drug Metab. Rev.* **2**, 117.

Lightner D.A., Wijekoon W.M.D. and Zhang M.H., 1988. Understanding bilirubin conformation and binding. Circular dichroism of human serum albumin complexes with bilirubin and its esters. *J. Biol. Chem.* **263**:16669.

Müller W.E., 1988. Stereoselective plasma protein binding of drugs. In: "Drug Stereochemistry, Analytical Methods and Pharmacology" (I.W. Wainer and D.E. Drayer, eds), page 227. Dekker, New York.

Oravcova J., Bystricky S. and Trnovec T., 1989. Different binding of propranolol enantiomers to human alpha$_1$-acid glycoprotein. *Biochem. Pharmacol.* **38**:2575.

Palmer K.H., Fowler M.S., Wall M.E., Rhodes L.S., Waddell W.J. and Baggett B., 1969. The metabolism of R(+)- and RS-pentobarbital. *J. Pharmacol. Exp. Therap.* **170**:355.

Palmer K.H., Fowler M.S. and Wall M.E., 1970. Metabolism of optically active barbiturates. II. S(–)-pentobarbital. *J. Pharmacol. Exp. Therap.* **175**:38.

Sugiyama K. and Trager W.F., 1986. Prochiral selectivity and intramolecular isotope effect in the cytochrome P-450 catalyzed ω-hydroxylation of cumene. *Biochemistry* **25**:7336.

Testa B., 1979. "Principles of Organic Stereochemistry." Dekker, New York.

Testa B. , 1984. Drugs? Drug research? Advances in drug research? Musings of a medicinal chemist. In "Advances in Drug Research (B. Testa, ed.), Vol. 13, page 1. Academic Press, London.

Testa B., 1986. Chiral aspects of drug metabolism. *Trends Pharmacol. Sci.* **7**, 60

Testa B., 1988. Substrate and product stereoselectivity in monooxygenase-mediated drug activation and inactivation. *Biochem. Pharmacol.* **37**, 85.

Testa B. , 1989. Conceptual and mechanistic overview of stereoselective drug metabolism. In: "Xenobiotic Metabolism and Disposition" (R. Kato, R.W. Estabrook and M.N.Cayen , eds), page 153. Taylor & Francis, London.

Testa B., 1990a. Definitions and concepts in biochirality. In: "Chirality and Biological Activity" (B. Holmstedt, H. Frank and B. Testa, eds), page 15. Wiley-Liss, New York.

Testa B., 1990b. Mechanisms of chiral recognition in pharmacology. The Easson-Stedman model revisited. *Acta Pharm. Nord.* **2**:137.

Testa B. and Mayer J.M., 1988. Stereoselective drug metabolism and its significance in drug research. In: "Progress in Drug Research" (E. Jucker ed.), Vol. 32, page 249. Birkhäuser, Basel.

Trager W.F. and Testa B., 1985. Stereoselective drug disposition. In: "Drug Metabolism and Disposition: Considerations in Clinical Pharmacology" (G.R. Wilkinson and M.D. Rawlins , eds), page 35. MTP Press, Lancaster.

Tucker G.T. and Lennard M.S., 1990. Enantiomer specific pharmacokinetics. *Pharmacol. Ther.* **45**:309.

Vermeulen N.P.E. and Breimer D.D., 1983. Stereoselectivity in drug and xenobiotic metabolism. In: "Stereochemistry and Biological Activity of Drugs" (E.J. Ariëns, W. Soudijn and P.B.M.W.M. Timmermans, eds), page 33. Blackwell, Oxford.

Visy J. and Simonyi M., 1989. The role of configuration and conformation in the binding of 2,3-benzodiazepines to human serum albumin. *Chirality* **1**:271.

Walle T and Walle UK , 1986. Pharmacokinetic parameters obtained with racemates. *Trends Pharmacol. Sci.* **7**, 155.

IMPLICATIONS OF STEREOSELECTIVITY IN CLINICAL PHARMACOKINETICS

M. Eichelbaum and A.S. Gross

Fischer-Bosch-Institut für Klinische Pharmakologie
Stuttgart, Federal Republic of Germany

INTRODUCTION

Approximately 50% of all drugs in therapeutic use have a chiral center and hence exhibit stereoisomerism. Chiral drugs derived from natural sources are stereochemical pure due to the stereospecificity of biological synthesis. By contrast, most of the drugs with a chiral center produced by chemical synthesis are only available as racemates, since chemical synthesis usually leads to a racemate unless stereospecific synthesis is employed [Roth & Kleemann, 1982; Simonyi, 1984].

Enantiomers contained in racemates quite often differ in the pharmacological activity in either qualitative or quantitative terms. In order to characterize enantiomers with regard to their pharmacological potency Lehmann et al. [1976], have introduced the terms eutomer and distomer. The eutomer is the more active enantiomer. The distomer is the enantiomer which demonstrates the lesser pharmacological activity. This description pertains only to the pharmacological effect being discussed. It is not unusual for the enantiomers of a drug to have a high degree of stereoselectivity for one action, but no stereoselectivity for an other action. For instance in the case of β-adrenoreceptor antagonist propranolol, with regard to β-blockade the S-enantiomer is the eutomer which is at least 100 times more potent than the R-enantiomer. However, other effects such as membrane stabilizing activity are not stereoselective [Jaillon et al. 1980; Stensrud et al. 1976].

STEREOSELECTIVITY IN DRUG ACTION

Several examples of the importance of stereoisomerism in drug activity will be discussed focusing on the therapeutically relevant action.

1. The enantiomers are equipotent. This situation applies for the antiarrhythmic activity of class 1 antiarrhythmic drugs where both enantiomers have similar electrophysiologic activity [Hill et al. 1988; Kroemer et al. 1989; Lie-A-Huen et al. 1989; Block et al. 1988; Uprichard et al. 1988].

New Trends in Pharmacokinetics, Edited by A. Rescigno and A.K Thakur
Plenum Press, New York, 1991

2. The enantiomers have the same quality of action but differ in their potency. This situation is encountered with racemic Ca^{2+} antagonists where a difference in potency of at least an order of magnitude exists between the eutomer and distomer [Bayer et al. 1975; Nawrath et al. 1981; Ferry et al. 1985; Satoh et al. 1980; Echizen et al. 1988].
3. The activity is mainly associated with one enantiomer. This is the case for β-blocking drugs where the β-adrenoceptor antagonistic activity is at least 50 to 100 times greater for the eutomer [Jaillon et al. 1980; Stensrud et al. 1976; Heyna et al. 1980].
4. Both enantiomers are active but have qualitatively different actions. Enantiomers of a racemate can be agonists or antagonists for different receptors. For instance, the (–)-enantiomer of dobutamine is a α-adrenergic agonist, whereas the (+)-enantiomer is a β-adrenergic agonist [Ruffolo et al. 1985].

STEREOSELECTIVE DRUG DISPOSITION

Whereas differences in activity between enantiomers have been known for many years, only recently has the importance of stereoselectivity in drug disposition been realised and the possible therapeutic consequences of this phenomenon appreciated.

The processes of drug absorption, distribution, metabolism and excretion may all involve an interaction between the drug molecule and chiral biological macromolecules. Therefore, it is not surprising that in analogy to the differences in pharmacological activity, the interactions of enantiomers with the macromolecules associated with their absorption, distribution, metabolism and excretion will exhibit a certain degree of handedness. As a consequence stereoselective drug disposition occurs [Ariens et al. 1984; Williams et al. 1985; Drayer 1986; Tucker & Lennard 1990].

Since a racemate is not a single compound but a 50:50 mixture of two enantiomeric drugs it is therefore mandatory to apply stereochemical aspects in clinical pharmacokinetics. There can be no doubt that the application of stereospecific analysis to the study of the disposition of racemic drugs has been very helpful in understanding phenomena which at first sight appeared to be rather paradoxical [Eichelbaum 1988; Tucker & Lennard 1990]. It is often overlooked that the stereoselectivity in the activity of enantiomers as determined *in vitro* cannot be extrapolated to the *in vivo* situation since stereoselective drug disposition can lead to an enantiomer ratio *in vivo* which differs substantially from that in the dosage form administered. Stereoselectivity in drug disposition seems to be the rule rather than the exception and, depending on whether the active or less active enantiomer is preferentially affected, there may be amplifaction or attenuation of *in vivo* as compared to the *in vitro* drug potency [Eichelbaum 1988; Tucker & Lennard 1990].

It is biotransformation by the drug metabolizing enzymes of the liver which exhibits the greatest degree of stereoselectivity of all the processes contributing to drug disposition. Numerous pathways of drug metabolism display stereoselectivity. The consequences of stereoselective drug metabolism for the *in vivo* activity of racemic drugs as compared to their *in vitro* potency depends on whether the enantiomers differ in potency. If the two enantiomers of a racemic drugs have the same potency preferential metabolism of one enantiomer is probably of no consequence. However, if the enantiomers differ in potency stereoselective metabolism will lead to greater or lesser potency *in vivo* depending on whether the eutomer or distomer is metabolized more rapidly. Drugs with low oral bioavailability due to extensive heaptic first-pass metabolism represent a special case. For these drugs differences in total drug concentration required to elicit a given

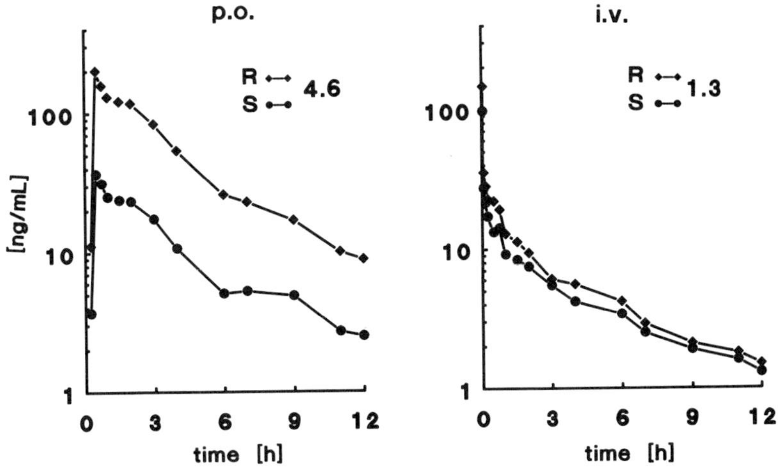

Fig 1. Metabolism of R-, S-verapamil is enantioselective, S-verapamil being cleared more rapidly. The R to S ratio in plasma following i.v. administration is 1.3 to 1.8. The hepatic extraction ratio (E) of the R-enantiomer is 0.55 and of the S-enantiomer 0.85. After oral administration pronounced differences in bioavailability (F) will be observed, since F = 1 − E. As a consequence of stereoselective first-pass metabolism the R to S ratio in plasma following oral administration is 4 to 5.

pharmacological effect have been described in relation to the route of administration. Pronounced amplification or attenuation of the response can follow oral compared with intravenous administration, depending on whether the distomer or eutomer is preferentially metabolized [Eichelbaum 1988; Tucker & Lennard 1990].

A much higher dose of verapamil is required after oral administration than after i.v. administration to elicit a comparable antiarrhythmic effect. This has been attributed to the low bioavialability of verapamil as a result of hepatic first-pass metabolism. However, as the absolute bioavailability of verapamil is 20 to 30% an oral dose of 25 to 50 mg should elicit an antiarrhythmic effect comparable to 5 to 10 mg of intravenous verapamil. In practise a dose exceeding that predicted is required. Measurement of verapamil plasma concentrations revealed that higher concentrations of verapamil are necessary after oral than after intravenous administration in order to elicit a comparable effect. For example in normal subjects after intravenous administration of 10 mg of racemic verapamil a plasma concentration of 40 ng/ml is associated with 50% of the maximal prolongation of the PR intervall, whereas this effective plasma concentration is 140 ng/ml after oral administration of 160 mg racemic verapamil. After oral administration 3 to 4 times higher verapamil concentration are thus required to elicit the same negative dromotropic response, indicating that verapamil is more potent after intravenous administration [Eichelbaum et al. 1980]. This observation at first appears to contradict the principle that the same drug concentration should elicit the same response irrespective of the route of administration. In order to explain this discrepancy it was hypothesised that enantiomerselective first pass metabolism may lead to a lower proportion of the more active S-enantiomer in plasma after oral than after intravenous verapamil administration [Eichelbaum et al. 1980]. In agreement with this hypothesis it could be demonstrated that after intravenous administration plasma concentrations of the less active R-enantiomer were twice as high as those of the more active S-enantiomer indicating that there was stereoselective metabolism [Eichelbaum et al. 1984]. After oral administration, however, the plasma concentration of the less active R-

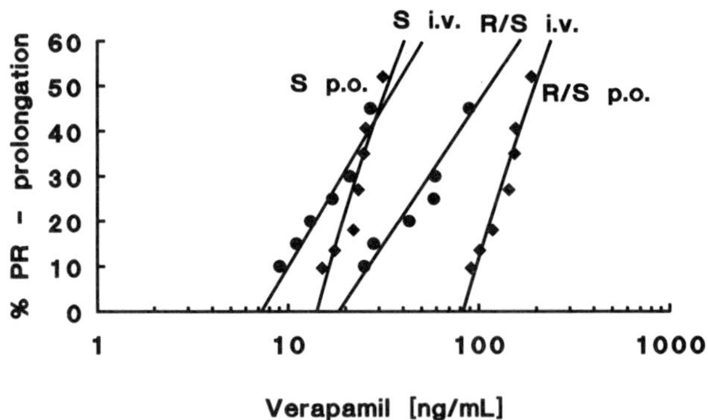

Fig 2. Impact of stereoselective first-pass metabolism which affects mainly the eutomer on the concentration-effect curve in relation to the route of administration. The effect curve is based on total plasma concentration. The S-enantiomer of verapamil is an order of magnitude more potent than R-verapamil. Stereoselective first-pass metabolism affects mainly the S-verapamil. Compared to i.v. administration the oral concentration effect curve is shifted to the right since the proportion of S-verapamil of total verapamil concentration is substantially lower after oral administration.

enantiomer was 5 times higher than the plasma concentration of the more active S-enantiomer [Vogelgesang et al. 1984] (Fig.1). This is a consequence of enantioselective first-pass metabolism of the S-enantiomer leading to lower bioavailability of it. Therefore after oral administration the active S-enantiomer accounts for smaller proportion of the total verapamil plasma concentration than after intravenous administration explaining why higher total plasma concentrations of verapamil are required after oral therapy than after intravenous administration to attain comparable concentrations of the active S-enantiomer and to produce the pharmacological effect desired (Fig.2). If one related PR-interval prolongation to the plasma concentration of S-verapamil, which is the enantiomer mainly responsible for antiarrhythmic efficacy, then there is no longer a difference in verapamil potency between oral and intravenous administration [Echizen et al. 1985] (Fig.2).

The reverse, i.e. amplification of the effect after oral as compared to the intravenous route of administration is seen if stereoselective first-pass metabolism affects mainly the distomer. In this case, following oral administration, a greater proportion of the total plasma concentration will consist of the eutomer.

IMPACT OF LIVER DISEASE

In this context, another aspect of stereoselective metabolism has to be considered i.e. the impact of liver cirrhosis on enantiomerselective first-pass metabolism. Liver cirrhosis has pronounced consequences on first-pass metabolism. In liver cirrhosis two major functional changes affect drug metabolism.

1. Due to a decrease in functional liver cell mass, the drug metabolizing capacity is decreased. As a consequence of the decreased extraction ratio following oral administration, bioavailability will increase in relation to the decrease of the extraction ratio.
2. Portal hypertension leads to extra- and intra-hepatic shunting of portal blood

supply and therefore a substantial proportion of an oral dose will bypass the liver, thereby escaping first pass metabolism. The part of the dose which bypasses the liver becomes completely available. The proportion of the racemate which is delivered to the liver will be subjected to first-pass metabolism. Although the fraction of the dose which is metabolized during first-pass through the liver will be decreased as a consequence of the lower extraction ratio, the stereoselectivity of this process, however, will be retained. In contrast, the proportion of the dose which bypasses the liver becomes completely available and hence the bioavailability of the eutomer and distomer will be 100%. Compared to the situation in subjects with normal liver function in liver cirrhosis an amplification of drug effect will be observed which by far exceeds the increase in effect predicted from the increase of the bioavailibility based on total drug concentration, if stereoselective metabolism preferentially affects the eutomer.

CHIRAL INVERSION

A unique and interesting pathway of stereoselective metabolism has been established for the nonsteroidal anti-inflammatory 2-arylpropionic acid derivatives [Caldwell et al. 1988]. With the exception of naproxen, which is administered as the S-enantiomer, all 2-arylpropionic acids are used as racemates. *In vitro*, the S-enantiomers of these drugs are between 10 and 100 times more potent in inhibiting prostaglandin synthesis than the R-enantiomers. However, this degree of stereoselectivity in drug action is not always observed *in vivo*. For example, S-ibuprofen is 160 times more potent than R-ibuprofen *in vitro*, but only 1.4 times more potent *in vivo*. This decrease in relative potency *in vivo* is due to the unique undirectional metabolic inversion of the chiral center from the inactive (−)-R isomer to the active (+)-S isomer. This chiral inversion has been demonstrated to occur in man for ibuprofen [Kaiser et al. 1976; Lee et al. 1985], benoxaprofen [Bopp et al. 1979; Simmonds et al. 1980], cicloprofen [Hutt et al 1983], fenoprofen [Rubin et al. 1985] and thioxaprofene [Diekmann et al. 1979], and involves the formation of an acyl coenzyme-A-thioester of the R-enantiomer [Knights et al. 1988]. A proportion of acyl CoA-R-ibuprofen is racemised by a CoA racemase and then hydrolyzed by CoA hydrolase liberating the S-enantiomer. Since the S-enantiomer is not a substrate for CoA, the acyl-CoA-thioester of this enantiomer is not formed and the chiral inversion is unidirectional from the R- to the S-enantiomer [Nakamura et al.; Yamaguchi et al. 1987]. Up to 71% of an orally administered dose of R-ibuprofen is converted to the S-enantiomer in humans *in vivo* [Lee et al. 1985].

DRUG INTERACTIONS

Stereoselectivity in action and disposition may have important consequences for meaningful interpretation of pharmacokinetic data in drug interaction studies where differences in potency exist between enantiomers and the pharmacokinetics of the enantiomers are affected in different ways by the causative drug.

For example, a drug may mainly inhibit the metabolism of the eutomer and may induce or not affect the metabolism of the distomer. Therefore, despite an unchanged or even decreased total plasma concentration, there may be an increase in activity of the drug. On the other hand, if mainly the distomer is affected, the higher total drug concentration that results is of no or only minor consequence for the therapeutic effect. The relevance of this phenomenon has been clearly demonstrated for a number of drug interactions with warfarin [Lewis et al. 1970; O'Reilly et al. 1980; Banfield et al. 1983; Toon et al. 1986; O'Reilly 1976; O'Reilly 1980; O'Reilly 1982; Toon et al. 1986; Choonara et al. 1986].

BIOAVAILABLITY STUDIES

Another area of clinical pharmacokinetics for which stereochemical aspects should be considered are bioavailability studies. Bioavailability is defined as the rate and extent to which the active drug ingredient or therapeutic moiety is absorbed from a drug product and becomes available at the side of drug action. If the enantiomers of a racemate differ by at least an order of magnitude in potency than the definition of bioavailability which relates to the active drug ingredient or therapeutic moiety would imply the measurement of the active enantiomer since this is the active therapeutic principle.

THERAPEUTIC DRUG MONITORING

Due to the large inter-individual variations in drug disposition, therapeutic drug monitoring has been used to select the drug dosage required for individual patients in order to achieve a plasma concentration within a defined therapeutic range. In view of the differences in disposition and pharmacological activity both in qualitative and quantitative terms that exist between the enantiomers in racemic drug preparations, the value of measuring total plasma concentrations as a guide to drug therapy must be questioned [Drayer 1988]. The definition of a therapeutic range is in particular a problem for drugs for which the enantiomers have pronounced differences in potency and, in addition, there is substantial interindividual variability in the proportion of the more active enantiomer in the total plasma concentration. This is certainly a factor confounding the establishment of a meaningful therapeutic range for some racemic compounds. For the therapeutic monitoring of racemic drugs to be more clinically useful, it would be prudent in the future to take stereoselective differences in activity and disposition into account when determining the therapeutic range of a compound. Initially, however, information on the activity and disposition of the enantiomers must be available.

CONCLUSIONS

Although only a limited number of the many racemic drugs in therapeutic use have been investigated in regard to stereoselectivity of action and disposition, the results clearly demonstrate that it is not valid to consider these drugs to be single compounds. As proposed by Ariens [1988], a policy should be adopted whereby drug companies and the authors of scientific reports clearly state the composite nature of racemic drugs. There can be no doubt that application of this aspect in drug action and disposition will facilitate our understanding of the concentration-effect relationship and pharmacokinetics of racemic drugs.

REFERENCES

Ariens EJ., 1984. Stereochemistry, a basis for sophisticated nonsense in pharmacokinetics and clinical pharmacology. Eur J Clin Pharmacol **26**: 663.

Banfield C, O'Reilly R, Chan E, Rowland M., 1983. Phenylbutazone-warfarin interaction in man: further stereochemical and metabolic considerations. Br J Clin Pharmacol **16**: 669.

Bayer R, Kaufmann R, Mannhold R., 1975. Inotropic and electrophysiological actions of verapamil and D-600 in mammalian myocardium. II. Pattern of inotropic effects of the optical isomers. Naunyn Schmiedebergs Arch Pharmacol **290**: 69.

Block AJ, Merrill D, Smith ER., 1988. Stereoselectivity of tocainide pharmacodynamics in vivo and in vitro. J Cardiovasc Pharmacol **11**: 216.

Bopp RJ, Nash JF, Ridolfo AS, Shepard ER., 1979. Stereoselective inversion of (R)-(–)-benoxaprofen to the (S)-(+)-enantiomer in humans. Drug Metab Dispos **7**: 356.

Caldwell J, Hutt AJ, Fournel-Gigleux S., 1988. The metabolic chiral inversion and dispositional enantioselectivity of the 2-aryl propionic acids and their biological consequences. Biochem Pharmacol **37**: 105.

Choonara IA, Cholerton S, Haynes BP, Breckenridge AM, Park BK., 1986. Stereoselective interaction between the R enantiomer of warfarin and cimetidine. Br J Clin Pharmacol **21**: 272.

Diekmann H, Garbe A, Steiner K, Bühring KU, Faro HP, Nowak H., 1979. Disposition of thioxaprofene: kinetics of the stereoisomers. Naunyn Schmiedeberg's Arch Pharmacol **307**(Supp): R3.

Drayer DE., 1986. Pharmacodynamic and pharmacokinetic differences between drug enantiomers in humans: an overview. Clin Pharmacol Ther **40**: 125.

Drayer DE., 1988. Problems in therapeutic drug monitoring: the dilemma of enantiomeric drugs in man. Ther Drug Mon **10**: 1.

Echizen H, Manz M, Eichelbaum M., 1988. Electrophysiologic effects of dextro- and levo-verapamil on sinus node and AV node function in humans. J Cardiovasc Pharmacol **12**: 543.

Echizen H, Vogelgesang B, Eichelbaum M., 1985. Effects of d,l-verapamil on atrioventricular conduction in relation to its stereoselective first-pass metabolism. Clin Pharmacol Ther. **41**:71.

Eichelbaum M., 1988. Pharmacokinetic and pharmacodynamic consequences of stereoselective drug metabolism in man. Biochem Pharmacol **37**: 93.

Eichelbaum M, Mikus G, Vogelgesang B., 1984. Pharmacokinetics of (+), (–) and (±) verapamil after intravenous administration. Br J Clin Pharmacol **17**:453.

Eichelbaum M, Birkel P, Grube E, Gütgemann U, Somogyi A., 1980. Effects of verapamil on P-R intervals in relation to verapamil plasma levels following single I.V. and oral administration and drug chronic treatment. Klin Wochenschr **58**: 919.

Ferry DR, Glossmann H, Kaumann AJ., 1985. Relationship between the stereoselective negative inotropic effects of verapamil enantiomers and their binding to putative calcium channels in human hearts. Br J Pharmacol **84**: 811.

Heyma P, Larkins RG, Higgonbotham L, Ng KW., 1980. D-Propranolol and DL-propranolol both decrease conversion of L-thyroxine to L-triiodothyronine. Br Med J. page 24.

Hill RJ, Duff HJ, Sheldon RS., 1988. Determinants of stereospecific binding of type I antiarrhythmic drugs to cardiac sodium channels. Mol Pharmacol **34**: 659.

Hutt AJ, Caldwell J., 1983. The metabolic chiral inversion of 2-arylpropionic acids — a novel route with pharmacological consequences. J Pharm Pharmacol **35**: 693.

Jaillon P, Heckle J, Weissenburger J, Cheymol G., 1980. Cardiac electrophysiologic properties of dl-propranolol, d-propranolol, l-propranolol and dl-pindolol in anesthetized dogs. J Pharmacol Exp Ther **212**: 347.

Kaiser DG, Vangiessen GJ, Reischer RJ, Wechter WJ., 1976. Isomeric inversion of ibuprofen (R)-enantiomer in humans. J Pharm Sci **65**: 269.

Knights KM, Drew R, Meffin PJ., 1988. Enantiospecific formation of fenoprofen coenzyme A thioester in vitro. Biochem Pharmacol **37**: 3539.

Kroemer HK, Funck-Brentano C, Silberstein DJ et al., 1989. Stereoselective disposition and pharmacologic activity of propafenone enantiomers. Circulation **79**: 1068.

Lee EJD, Williams K, Day R, Graham G, Champion D., 1985. Stereoselective disposition of ibuprofen enantiomers in man. Br J Clin Pharmacol **19**: 669.

Lehmann FPA, Rodrigues de Miranda JF, Ariens EJ., 1976. Stereoselectivity and affinity in molecular pharmacology. Progress in Drug Research **20**:101.

Lewis RJ, Trager WF, Chan KK et al., 1974. Warfarin: Stereochemical aspects of its metabolism and the interaction with phenylbutazone. J Clin Invest **53**: 1607.

Lie-A-Huen L, van den Akker J, den Hertog A, Meijer DKF., 1989. The action of flecainide acetate and its enantiomers on mammalian non-myelinated nerve fibres. Pharmaceut Weekbl [Sci] **11**: 92.

Nakamura Y, Yamaguchi T., 1987. Stereoselective metabolism of 2-phenylpropionic acid in rat . Drug Metab Dispos **15**: 529.

Nawrath H, Blei I, Gegner R, Ludwig C, Zong X., 1981. No stereospecific effects of the optical isomers of verapamil and D-600 on the heart. In: "Calcium antagonism in cardiovascular therapy: experience with verapamil" (Zanchetti A, Krikler DM eds.), page 52. Amsterdam, Excerpta Medica.

O'Reilly RA, Trager WF, Motley CH, Howald W.1980. Stereoselective interaction of phenylbutazone with [^{12}C/^{13}C] warfarin pseudoracemates in man. J Clin Invest **65**: 746.

O'Reilly RA., 1976. The stereoselective interaction of warfarin and metronidazole in man. New Engl J Med **295**: 354.

O'Reilly RA., 1980. Stereoselective interaction of trimethoprim-sulfamethoxazole with the separated enantiomorphs of racemic warfarin in man. New Engl J Med **302**: 33.

O'Reilly RA. Ticrynafen-racemic warfarin interaction: hepatotoxic or stereoselective? Clin Pharmacol Ther **32**: 356.

Powell JR, Ambre JJ, Ruo TI., 1988. The efficacy and toxicity of drug stereoisomers. In: "Drug Stereochemistry" (Wainer IW and Drayer DE, eds.), page 245. Marcell Dekker Inc., New York.

Roth HJ, Kleemann A., 1982. "Arzneistoffsynthese", pages 16-20. Georg Thieme Verlag, Stuttgart.

Rubin A, Knadler MP, Ho PPK, Bechtol LD, Wolen RL., 1985. Stereoselective inversion of (R)-fenoprofen to (S)-fenoprofen in humans. J Pharm Sci **74**: 82.

Ruffolo RR, Messick K., 1985. Effects of dopamine, (±)-dobutamine and the (+)- and (−)-enantiomers of dobutamine on cardiac function in pithed rats. J Pharmacol Exp Ther. **235**: 558.

Satoh K, Yanagisawa T, Taira N., 1980. Coronary vasodilator and cardiac effects of optical isomers of verapamil in the dog. J Cardiovasc Pharmacol **2**: 309.

Simmonds RG, Woodage TJ, Duff SM, Green JN., 1980. Stereospecific inversion of R-(−)-benoxaprofen in rat and man. Eur J Drug Metab Pharmacokinet. **5**: 169.

Simonyi M., 1984. On chiral drug action. In "Medicinal Research Reviews", Vol.4, page 359. John Wiley & Sons, Inc, New York.

Stensrud P, Sjaastad O., 1976. Short-term clinical trial of propranolol in racemic form (Inderal), d-propranolol and placebo in migraine. Acta Neurol Scandinav **53**: 229.

Toon S, Hopkins KJ, Garstang FM, Diquet B, Gill TS, Rowland M., 1986. The warfarin-cimetidine interaction: stereochemical considerations. Br J Clin Pharmacol **21**: 245.

Toon S, Hopkins KJ, Garstang FM, Aarons L, Sedman A, Rowland M., 1987. Enoxacin-warfarin interaction: Pharmacokinetic and stereochemical aspects. Clin Pharmacol Ther **42**: 33.

Toon S, Low LK, Gibaldi M et al., 1986. The warfarin-sulfinpyrazone interaction: stereochemical considerations. Clin Pharmacol Ther **39**: 15.

Tucker GT, Lennard MS., 1990. Enantiomer specific pharmacokinetics. Pharmac Ther **45**: 309.

Uprichard ACG, Allen JD, Harron DWG., 1988. Effects of tocainide enantiomers on experimental arrhythmias produced by programmed electrical stimulation. J Cardiovasc Pharmacol. **11**: 235.

Vogelgesang B, Echizen H, Schmidt E, Eichelbaum M., 1984. Stereoselective first-pass metabolism of highly cleared drugs: studies of the bioavailability of l- and d-verapamil examined with a stable isotope technique. Br J Clin Pharmacol **18**: 733.

Williams K, Lee E., 1985. Importance of drug enantiomers in clinical pharmacology. Drugs **30**: 333.

Yamaguchi T, Nakamura Y., 1987. Stereoselective metabolism of 2-phenylpropionic acid in rat. Drug Metab Dispos. **15**:535.

PHARMACOKINETIC STRATEGIES IN THE DEVELOPMENT OF PRODRUGS

L.P. Balant,[1,2] E. Doelker[2] and P. Buri[2]

[1]Clinical Research and Clinimetrics Unit
   Psychiatric University Institutions of Geneva
[2]School of Pharmacy
   University of Geneva, Switzerland

1. INTRODUCTION

The term prodrug was first introduced by Albert in 1958 to describe compounds which undergo biotransformation prior to exhibiting their pharmacological effects. Since then, many papers have been published on this subject [Higuchi & Stella, 1975; Roche, 1977; Higuchi et al, 1983; Stella et al, 1985; Waller & George, 1989; Balant et al, 1990a]. Usually, the use of the term prodrug implies a covalent link between an "active moiety" and a "carrier moiety", but some authors also use this term to characterize some form of salts of the active principle. In this review, only covalently bound moieties will be considered.

1.1. The biopharmaceutical point of view

If prodrugs are considered from a biopharmaceutical point of view, they can be defined as precursors of active principles that can be utilized to modify a variety of both pharmaceutical and biological properties including modification of the pharmacokinetics of the drug in vivo to improve absorption, distribution, metabolism and excretion. This may lead to improvement of bioavailability by increasing aqueous solubility, increase of drug product stability, enhancement of patient acceptance and compliance by minimizing taste and odor problems, elimination of pain on injection, and decrease of gastro-intestinal irritation [Sinkula, 1977]. It is usually accepted that a prodrug should not possess any relevant pharmacological activity. It would thus appear that in this context the term prodrug is used essentially when chemical modifications are deliberately introduced to improve an unsatisfactory situation when the active compounds are administered per se. The classical example is the synthesis in 1899 of aspirin in an attempt to improve therapy with salicylic acid.

The present review deals essentially with enhancement of gastro-intestinal availability of drugs through prodrug administration, but some other aspects of the prodrug concept are also discussed briefly. The prodrug approach can be applied, for example, in connection with the depot neuroleptics for which esterification with

New Trends in Pharmacokinetics, Edited by A. Rescigno and A.K. Thakur
Plenum Press, New York, 1991

decanoic acid and intramuscular injection in Viscoleo allow the drug to be given only once or twice a month. It is thus not so much a problem of poor absorption that is involved, but one of timing. The synthesis of the precursor permitted a new approach to long term treatment of schizophrenia. Compounds available in depot prodrug formulations include fluphenazine, flupenthixol and zuclopentixol [Balant-Gorgia & Balant, 1987].

1.2. The pharmacokinetic point of view

Even if, according to the accepted definition, the couple "prodrug-and-its-active-derivative" should not be considered as the pair "parent-compound-and-metabolite", from a pharmacokinetic point of view the two entities are strictly identical. Accordingly, the equations used for the determination of the kinetics of metabolites (active or not) are to be used [Balant & McAinsh, 1980; Gibaldi & Perrier, 1982; Rowland & Tozer, 1989]. The basic equations will be presented below.

1.3. The regulatory point of view

Presently there is no clear opinion about the requirements for the development of a prodrug. In general, regulatory agencies are reluctant to register this type of products. Of particular concern is the fact that toxicological studies might not be relevant for the human use of the drug because of differences in the rate and/or extent of formation of the active moiety.

This potential problem was raised by Aungst et al. [1987] concerning the extrapolation of data from animal models to humans. Discussing the results of their pharmacokinetic studies of nalbuphine prodrugs, a narcotic agonist/antagonist, they stated that "Animal models should be similar to man with regard to: (i) nalbuphine disposition (bioavailability) after oral dosing; and (ii) prodrug hydrolysis rates". It is evident that due to great inter-species variability in xenobiotic metabolism (well known from a toxicological point of view) this goal is quite difficult to achieve. Accordingly, animal experiments should be analyzed with caution and experiments in humans must be performed. It could even be argued that for some active principles it would be wiser to conduct all experiments in man. As an example of inter-species differences, Vickers et al. [1978] found that the pivaloy-oxyethyl ester of methyldopa was essentially hydrolyzed presystematically to pivalic acid and methyldopa, while the succinimidoetyl derivative was readily hydrolyzed in the rat, but not in the dog.

2. CHEMICAL BOND

Prodrug design and synthesis is an important field of pharmaceutical chemistry. This review will, however, be centered around biopharmaceutical aspects of prodrugs. From a pharmacokinetic point of view it is important to understand the nature of the chemical bond linking the active moiety to its "carrier moiety", and the nature of the "carrier moiety". Knowledge of the nature of the chemical bond may help to explain the nature of the biotransformation process and its location in specific tissues or cells. The study of the fate in the body of the "carrier" moiety is particularly important from a safety point of view and should be investigated just as thoroughly as the active moiety. Clearly, in some cases, such as the esters of methanol or ethanol, the fate of the released "carrier moiety" is well known, and no extra study is needed during drug development. In other cases, additional pharmacokinetic investigations may be necessary.

In the following paragraphs, some typical kinds of chemical bonds will be discussed. It is not intended here to present an overview on this subject, but only to give some examples of how chemistry and pharmacokinetics interact in the field of prodrugs.

2.1. Esters

As stated above, rational design of biologically reversible drug derivatives is based on the ability of the host tissue to regenerate the active moiety. The manner in which this is frequently accomplished is through the mediation of an enzyme system. Enzymes considered important to orally administered prodrugs are found in the gut wall, liver and blood. In addition, enzyme systems present in the gut microflora may be important in metabolizing the prodrug before it reaches the intestinal cells [Sinkula & Yalkowsky, 1975]. Due to the wide variety of esterases present in the target tissues for oral prodrug regeneration, it is not surprising that esters are the most numerous prodrugs designed when gastro-intestinal absorption is considered.

Chloramphenicol as palmitate is an interesting study example with respect to polymorphism. This ester, synthesized to avoid the bitterness of the active moiety (see below), exists as four polymorphs, three cristalline forms and an amorphous one. Polymorphs A and B have been found in commercial preparations but only form B leads to satisfactory blood levels. This effect was first attributed to the lower solubility of the "inactive" form A [Aguiar & Zelmer, 1969], but it was later proved that the higher susceptibility of form B to esterases was, in fact, involved [Burger, 1977].

In some cases such as that of terbutaline [Olsson & Svensson, 1984], it was observed that improved absorption of the diester ibuterol was accompanied by shorter effect duration and a slightly increased extent of first-pass metabolism. Other prodrugs were thus synthesized in an attempt to improve resistance to first-pass hydrolysis. Preliminary results with bambuterol, an N-isostere of ibuterol, displayed improved hydrolytic stability. This might result from the introduction of the nitrogen atom reducing the reactivity of the acyl function, and/or dimethycarbamate being a cholinesterase inhibitor, bambuterol hydrolysis could be partially inhibited because of reversible inhibition of the esterases responsible for its own degradation.

2.2. Mannich bases

Mannich base prodrugs are said to enhance the delivery of their parent drugs through the skin because of enhanced water solubility, as well as enhanced lipid solubility. Those more polar prodrugs are also more effective in improving topical delivery than prodrugs that have been designed to incorporate only lipid solubilizing groups into the structure of the parent drug [Sloan & Bodor, 1982]. Mannich base prodrugs are regenerated by chemical hydrolysis without enzymatic catalysis [Johansen & Bundgaard, 1981].

2.3. Macromolecular prodrugs

Although the term macromolecular prodrug is recent, the interest of the temporary covalent binding of pharmacologically active compounds to macromolecules for sustained release has been recognized for many years as potential method for controlled drug delivery and has received increasing attention [Vert, 1986].

As early as 1975, Ringsdorf drew attention to the need for tailor-making macromolecular prodrugs with a view to attaching a homing device or directing units, solubilizing units, and using a biostable or biodegradable backbone. Besides monoclonal antibodies used in conjunction with anti-cancer drugs, other macromolecules such as polyvinylic or polyacrylic, polysaccharidic, and poly(α-amino acid) backbones have been mostly used. Theoretically, the main advantage of synthetic polymers over naturally occurring macromolecules is that they can be tailored to meet individual requirements. As a recent example, one may mention dextran, soluble starch or hydroxyethyl starch-based ester-prodrugs of naproxen [Larsen, 1989], but other polysaccharides have been tested for their usefulness as transport groups for therapeutic agents. Among them one can mention starch for acetyl salicylic acid [Kratzl & Kaufmann, 1961; Kratzl et al, 1961], inulin for procainamide [Schacht et al, 1984; Remon et al, 1984], agarose for mitomycin [Kojima et al, 1978] or for adriamycin [Tritton et al, 1983], soluble starch for nicotinic acid [Puglisi et al, 1976] or for salicylate [Havron et al, 1974], cellulose for insulin [Singh et al, 1981] and hydroxypropyl cellulose for estrone and testosterone [Yolles, 1987; Yolles et al, 1979]. However, clinical experience with such prodrugs is still too limited for an adequate judgment on this aspect of macromolecular prodrugs.

Recently drug-antibody conjugates have become realities in the context of drug transport, drug targeting, and cancer therapy. An aspect of macromolecular prodrug synthesis that seems still to present a technical problem is the development of selective antitumor drug-protein conjugates with a covalent and reversible linkage for stability in biological fluids, together with adequate release at the site of action.

3. GASTRO-INTESTINAL ABSORPTION

3.1 Improvement of gastro-intestinal tolerance

The gastro-intestinal lesions produced by the acid NSAI agents are generally believed to be caused by two different mechanisms: a direct contact mechanism on the gastro-intestinal mucosa and a generalized systemic action appearing after absorption, which can be demonstrated following intravenous administration. The relative importance of these mechanisms may vary from drug to drug. Acetylsalicylic acid and most newer NSAI drugs are carboxylic acids. Temporary masking of the acid function has been proposed as a promising means of reducing gastro-intestinal toxicity due to the direct mucosal contact mechanism [Bundgaard & Nielsen, 1988].

Several attempts have been made to develop bioreversible derivatives of acetylsalicylic acid [Bundgaard et al, 1988]. They nicely illustrate the potential problems encountered with prodrugs of prodrugs. A major difficulty in the design of aspirin prodrugs is the great enzymatic lability of the acetyl ester in aspirin derivatized at its carboxyl group. Therefore, a prerequisite for any true aspirin prodrug is that the masking group cleaves faster than the acetyl ester moiety. Otherwise, the derivatives will behave as prodrugs of salicylic acid.

3.2. Increase in systemic availability

Until recently it seemed that the most fruitful area of derivatization was the improvement of passive drug absorption through epithelial tissue. Accordingly, an immense number of prodrugs featuring the addition of a hydrophobic group have been prepared in order to improve their gastro-intestinal absorption. As early as 1975 Sinkula and Yalkowsky listed a number of drug derivatives utilized as modifiers of absorption, and the list has steadily increased since then. It is in the field of poorly absorbed penicillin derivatives that the most successful prodrugs have been produced and commercialized. They include esters of ampicillin (e.g. bacampicillin, pivampicillin and talampicillin), mecillinam (pivmecillinam), and carbenicillin (carfecillin: for urinary tract infections only). As an example, plasma concentrations of ampicillin attained with these esters are up to five times higher than those seen after oral ampicillin [Leight et al, 1976; Ehrnebo et al, 1979; Rozencweig et al, 1976]. Other antibiotics for which the prodrug approach has been tested include cefotiam [Yoshimura et al, 1987].

The concept of bioreversible chemical modification has been proposed in order to diminish the gastro-intestinal and/or hepatic first-pass metabolism of drugs with high extraction ratios. This approach has been tried for a number of drugs such as methyldopa [Vickers et al, 1978, 1984; Dobrinska et al, 1982], etilefrine [Wagner et al, 1980], L-DOPA [Bodor et al, 1977], terbutaline [Hrnblad et al, 1976] and salicylamide [D'Souza et al, 1986], 5-fluorouracil [Buur & Bundgaard, 1986, 1987; Sasaki et al, 1986], progesterone [Basu et al, 1988], naltrexone [Hussain & Shefter, 1988; Hussain et al, 1987] and peptides [Buur & Bundgaard, 1988].

Another type of problem may arise if presystemic isomerization to an inactive compound occurs. As an example, ester prodrugs of the delta-3-isomers of cephalosporins may isomerize to the microbiologically inactive delta-2-isomer [Saab et al, 1988] and it is important for clinical effectiveness that, after reaching the blood, hydrolysis of the ester group must proceed much more rapidly than isomerization of the delta-3 double bond.

3.3. Sustained release prodrug systems

Sustained release systems using polymer matrices rely usually on release of the drug by either diffusion through the matrix, erosion of the matrix, or a combination of both. The prodrug concept can also be applied for such systems, for example by covalently binding the drug to a biodegradable polymer. This approach has been tested for naltrexone [Negishi et al, 1987] but not intended for oral administration. In these systems, two physicochemical processes may influence drug release. First, drug is released from the backbone polymer by hydrolysis (enzyme and/or acid-base catalyzed). Second, free drug diffuses through the polymer matrix. If the rate of hydrolysis is lower than the rate of diffusion, it will govern the release rate. The fact that this rate limiting process occurs at the boundary between unaffected nonswollen and previously swollen, degraded, or permeated material implies particular pharmaceutical designs.

Another procedure has been followed by synthesizing a monomer with the pendant drug, and then by polymerizing this new molecule. Chloramphenicol has thus been attached to a methacrylic derivative by an acetal function, and then copolymerized with 2-hydroxyethyl-methacrylate [Meslard et al, 1986]. The copolymer can then, in addition, be dispersed into a biocompatible non-degradable polymer such as Eudragit playing the role of a polymer matrix [Chafi et al, 1988].

Ion-exchange systems have found many medicinal applications. As a controlled-release system, ion exchange resins offer the benefit that release of a complexed drug is initiated by an influx of competing ions from the gastro-intestinal tract. However, in general, release rates from unmodified resins are too great for adequate control of delivery and diffusional barriers to delay drug egress are employed in some preparations. Homologous series of O-n-acyl propranolol prodrugs have been used to study the influence of physicochemical properties of the drug on its release from resinates [Irwin and Belaid, 1988a,b].

4. DISTRIBUTION

4.1. Tissue targeting

An interesting approach of specific drug delivery to the liver has been attempted with glutathione as a dextran conjugate for the potential treatment of hepatic poisoning [Kaneo et al, 1988]. This prodrug was developed with the aim of overcoming both the poor hepatic uptake of glutathione and its rapid renal degradation into constituent amino acids.

Tissue targeting may also be directed at tumors using monoclonal antibodies, as briefly mentioned above.

It is possible that for such types of drugs, new pharmacokinetic approaches would have to be derived in order to correctly describe their behaviour in the body, in particular if one is interested to evaluate the degree of tissue targeting.

4.2. Activation at the site of action

Several examples of prodrugs are found in the purine and pyrimidine analogues which substitute for natural nucleotides and inhibit nucleic acid formation. For example, 5-fluorouracil is essentially harmless to mammalian host and tumor cells. Upon administration, the drug is subject to one of two opposing metabolic fates [Diasio & Harris, 1989]. Inactivation and elimination are accomplished by catabolism (about 80% of the dose) and by urinary excretion of unchanged drug (about 5 to 20%). On the other hand, cytotoxicity in host and tumor cells only occurs following anabolism in actively proliferating cells. Due to a marked first-pass metabolism, oral administration leads to highly variable systemic availability and is thus an unsuitable and unreliable mode of therapy. It seems thus possible to improve oral delivery of this drug using the prodrug approach [Buur & Bundgaard, 1986, 1987]. As a matter of fact, 5'-deoxy-5-fluorouridine is a sugar derivative presently in clinical trials for the oral route. It has a half-life of about 10 to 30 minutes in patients and is transformed mostly into 5-fluorouracil. It is possible that the therapeutic index of 5-fluorouracil might be increased by the prodrug approach. It seems that doxifluridine, is selectively activated to 5-fluorouracil in sensitive tumor cells as opposed to bone marrow cells [Schaaf et al, 1988]. Should this be conclusively demonstrated, it would be an interesting example of increased target organ availability due to prodrug derivatization.

The kinetics of 5-fluorouracil and its metabolites are essentially nonlinear. Therefore it is extremely difficult to build models that would correctly describe the cascade of nonlinear transformations that are observed, starting from drug absorption to its transformation into the active moiety.

Omeprazole, a potent anti-ulcer agent, is an effective inhibitor of gastric acid secretion, as it is an inhibitor of the gastric H^+, K^+-ATPase (unlike acid secretion inhibitors such as cimetidine). The enzyme H^+, K^+-ATPase is responsible for gastric acid production and is located in the secretory membranes of the parietal cell. Omeprazole itself is not an active inhibitor of this enzyme, but is transformed within the acid compartments of the parietal cell into the active inhibitor, close to the enzyme.

Orally administered amines do not cross the blood-brain barrier, but neutral amino acids such as α-methyldopa are transported into the brain by a specific carrier system. The α-methyldopa is subsequently concentrated in neuronal cells where it becomes a substrate in the catecholamine biosynthesis and is transformed into α-methylnoradrenaline [Pardridge, 1983].

Codeine activation to morphine is controlled by the activity of the polymorphic cytochrome P-450 db1/bufI [Dayer et al, 1988]. The clinical relevance of this observation derives from the fact that about 10% of Caucasians lack this metabolic pathway. Since poor metabolizers cannot activate this widely used drug, for them codeine is an ineffective analgesic [Desmeules et al, 1989].

Morphine crosses the blood-brain barrier relatively slowly. Diacetylmorphine (heroin) has a greater lipid solubility, and thus transfers into the brain where hydrolysis converts heroin to morphine. The latter is ultimately responsible for the pharmacological effect [Inturrisi et al, 1984]. This is a good example of a prodrug which has found an unintended use although it has also been used to treat chronic pain.

These substances metabolized within the central nervous system illustrate well one of the difficulties that may be encountered with prodrug kinetics. The prodrug may follow an ADME pattern perfectly well described by its systemic availability, volume of distribution, hepatic and renal clearances, but still have a pharmacological effect whose dependency on blood kinetics is only indirect. This is particularly true if the drug must diffuse through the blood-brain barrier, and is then metabolized by enzyme systems different from those found in the liver.

4.3. Reversible and irreversible conversion

The natural conversion of an inactive stereo-isomer to an active one is another case in which no a priori intention existed when the compound was synthesized. Ibuprofen is a good example. From a pharmacokinetic point of view, one may consider that the reversibly formed metabolite represents a compartment for the parent compound, and adequate provision must be taken when analyzing this type of data.

Sulindac is an interesting case of reversible metabolism to an active derivative by reduction to a sulfide. It is also metabolized irreversibly to an inactive sulfone [Duggan et al, 1977, 1978]. In vitro anti-inflammatory test systems considered relevant to in vivo efficacy and correlations between drug concentration and biological effect ex vivo suggest that sulindac is a prodrug. Re-oxidation of the sulfide to the parent form occurs before elimination, but the sulfide has a long half-life and prolonged duration of action. There is strong evidence that both liver [Tatsumi et al, 1983] and gut flora [Strong et al, 1985] contribute to sulphide generation.

Although many alkylating agents are administered in the active form, cyclophosphamide is a prodrug activated in the liver. Initial conversion is to 4-hydroxycyclophosphamide which is in spontaneous equilibrium with its tautomeric form, aldophosphamide. These function as inactive "transport" molecules [Colvin et al, 1976]. Subsequent spontaneous elimination of acrolein from aldophosphamide generates the active moiety, phosphoramide mustard.

It is evident, that modelling of such compounds is difficult, and that great care must be taken not to generate models that have too many degrees of freedom or that cannot be solved in a univocal way.

5. PHARMACOKINETIC AND BIOPHARMACEUTICAL ASPECTS

5.1. Compartmental approach

If one considers the one-compartment open body model with first-order input and output, the concentration-time curve for the parent compound can be described by the Bateman function. For most drugs the absorption rate constant is significantly larger than the elimination rate constant. As a consequence, the shape of the concentration versus time curve will, after some time, depend only on the elimination

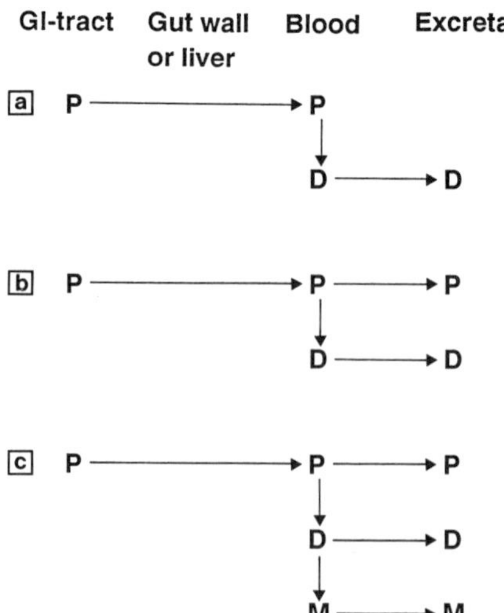

Figure 1. Pharmacokinetic models for a prodrug (P) biotransformed after reaching the systemic circulation into an active moiety (D) and, eventually (Scheme c), into a metabolite.

rate constant. During the postabsorptive phase therefore, it will be possible to estimate the "apparent half-life of elimination" since the latter is related to the first-order elimination rate constant. On the other hand, if elimination is faster than absorption, the rate constant obtained from the slope of the terminal portion of the curve will be the absorption rate constant and not the elimination rate constant. This is called a FLIP-FLOP situation, and may be observed with drugs that are very rapidly eliminated or with dosage forms that slowly release the drug in an apparent first-order fashion.

In contrast, the pharmacokinetic properties of metabolites are particularly interesting in this respect. At one time it was generally assumed that metabolite elimination is always faster than formation because metabolites were considered to be more easily eliminated from the body than the parent compound. This assumption may be true when polar compounds such as glucuronides, sulfates, or glycine conjugates are the major biotransformation products, but need not be true when drugs are, for example, acetylated or hydroxylated. Nevertheless, a basic difference usually exists between parent drug and metabolite kinetics : for metabolites, elimination is normally faster than formation, whereas for most drugs elimination is slower than gastro-intestinal absorption. The consequence of this reversal is that metabolites often show FLIP-FLOP kinetics.

For prodrugs, the situation is basically different since, in most cases, the chemical bond between transport moiety and active moiety is designed as labile in order to avoid, for example, urinary elimination of the prodrug which would diminish the amount of active compound available at the site of action. However, this is not an intangible rule, in particular if prodrug design is used in order to provide "slow formation" of the active moiety.

The pharmacokinetic models used to describe prodrug and active moiety kinetics are similar to those used for metabolites (Figures 1 and 2). As seen on scheme 2b or 2c, it may often be difficult to calculate the rate constants where prodrug transformation occurs both on first-pass and after reaching the systemic circulation. Even if these models are probably of little help for data analysis, they may be useful in order to determine the influence of relative magnitudes of the rate

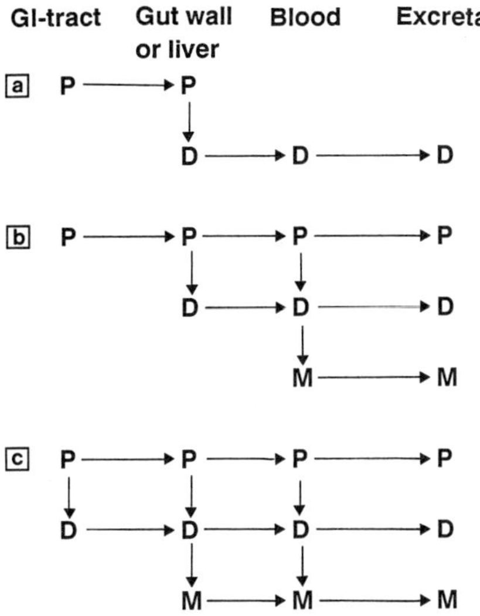

Figure 2. Possible first-order pharmacokinetic models for a prodrug submitted to first-pass and systemic activation.

constants on the behavior of the prodrug and the active moiety. Thus, comparing simulated data obtained with model 1a and 1b, it is possible to evaluate the impact of prodrug excretion on the AUC of the active moiety in function of the prodrug transformation rate (Figures 3,4,5,6).

5.2. Clearance approach

The clearance concept can successfully be applied to analyse metabolite kinetics in blood [Rowland & Tozer, 1989]. It can, by analogy be used for the analysis of the kinetics of the active moiety after prodrug administration. As with the compartmental approach, care must be taken when prodrug transformation occurs during first-pass and/or prodrug excretion is important after it reaches the systemic circulation. Here again, some differences with metabolite kinetics occur in function of two different aspects:

a) Metabolite kinetics are not always known after their administration to man, whereas the active moiety of a prodrug should be well investigated from a pharmacokinetic point of view. This means that the basic pharmacokinetic parameters [Balant et al, 1990b] are usually known, and that they can be used to extract relevant information when the prodrug is administered.

b) Metabolite formation is usually the rate limiting step, whereas this is not the case for the active moiety of a prodrug.

It must be stated that these two considerations also hold for the compartmental approach, although in the latter case, the danger of calculating artifactual parameters is more critical.

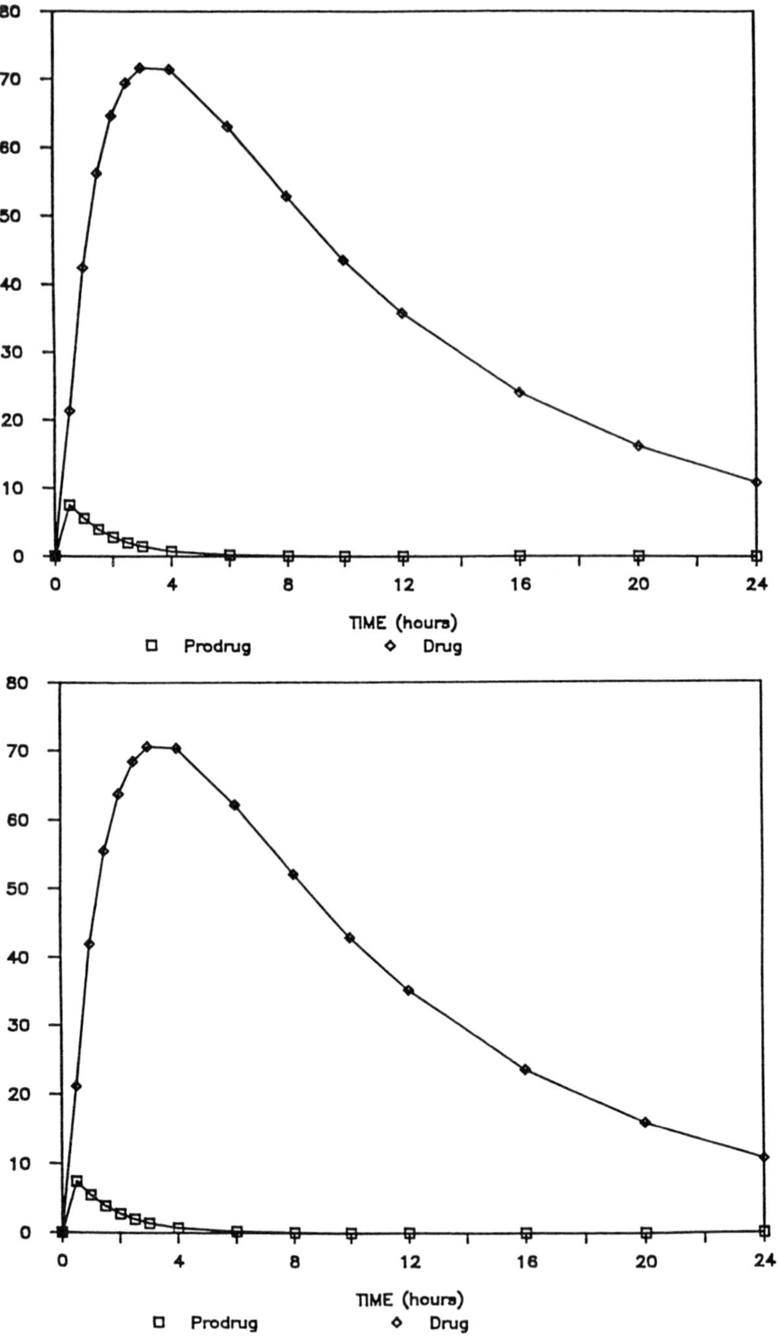

Figure 3. Top panel : Simulation of the behaviour of a prodrug and its active moiety after oral administration when there is no excretion of the prodrug (See scheme 1a). Activation is fast as compared to all other processes ($t_{1/2}$ = 0.1 h). Bottom panel : Same hypotheses as above with, in addition, elimination of the prodrug at the same rate as the active moiety (See scheme 1b). No measurable difference is observed between the two cases.

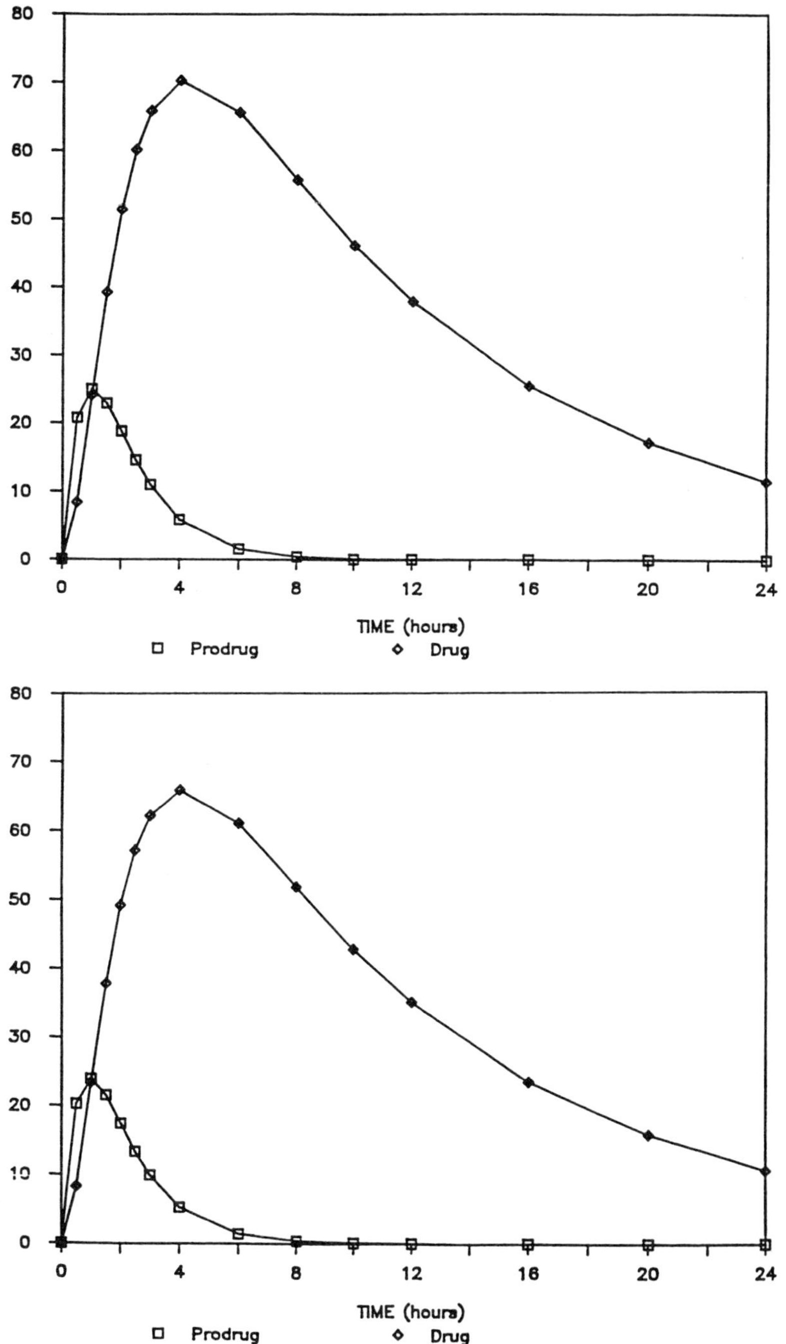

Figure 4. Same situation as in Figure 3. Activation half-life is 0.5 h, but still faster than the other processes. C_{max} for prodrug and active moiety are about 5% less when parallel elimination occurs (Bottom panel).

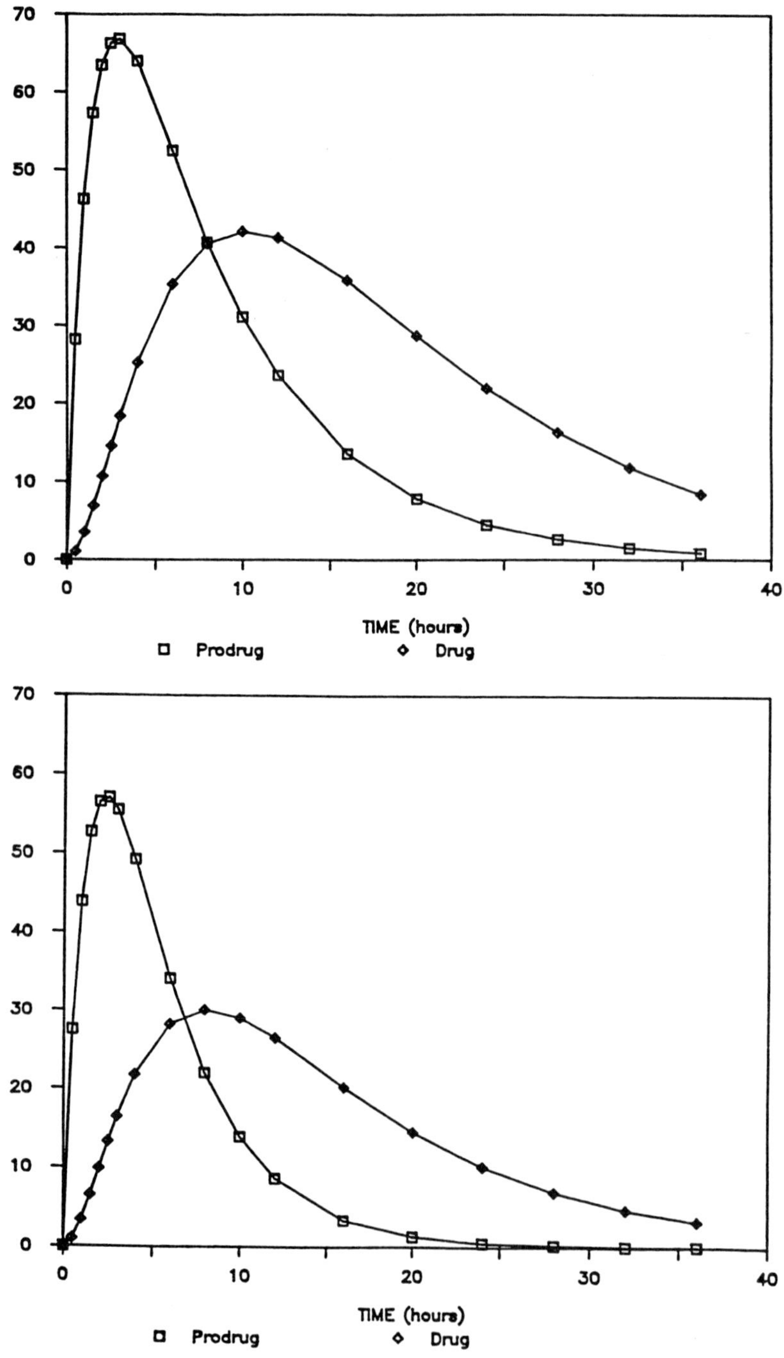

Figure 5. Same situation as in Figures 3 and 4. Activation into the active moiety is of the same order of magnitude but faster than active moiety elimination. C_{max} for the active moiety is 30% less when parallel elimination occurs (Bottom panel).

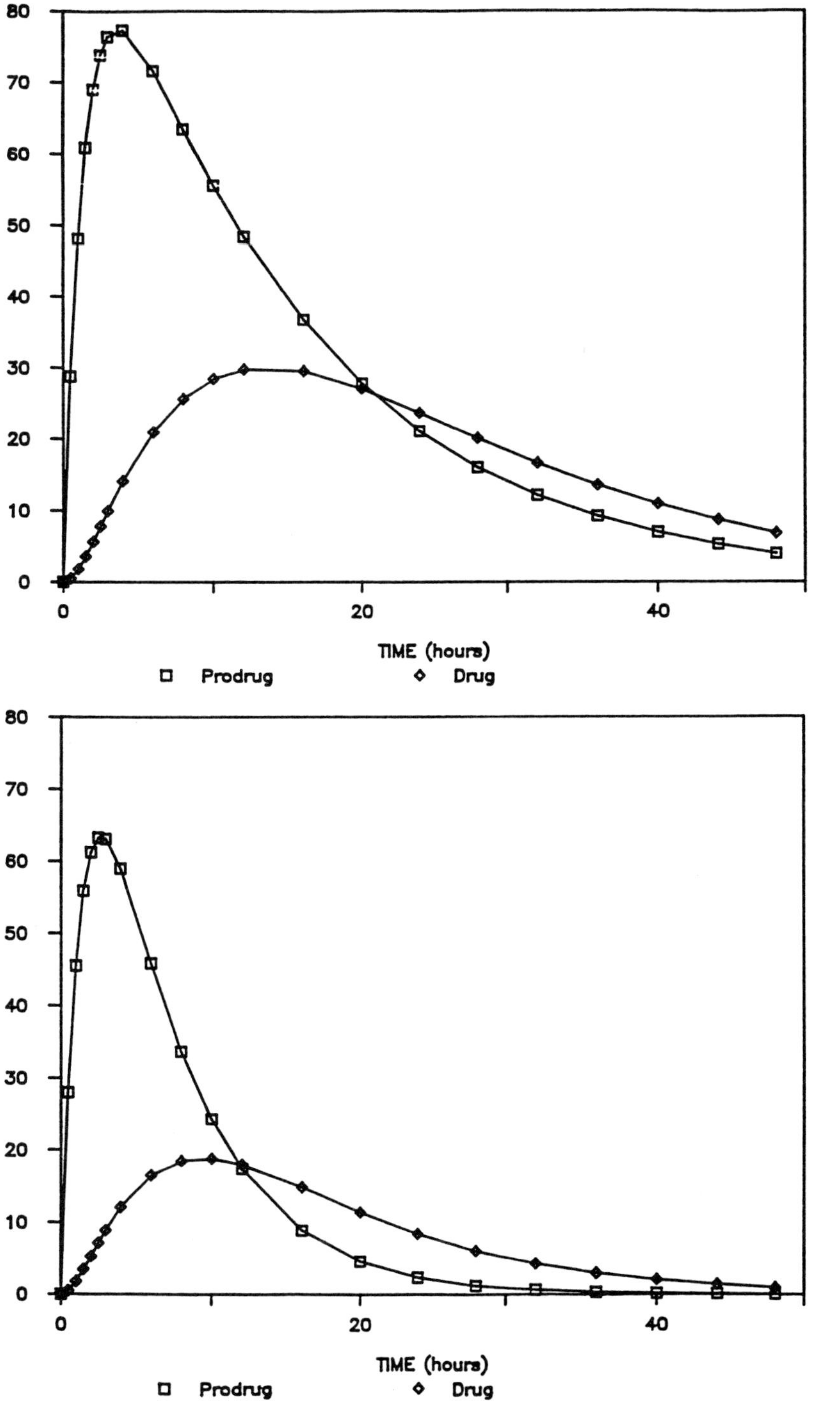

Figure 6. Same situation as in Figure 5. Activation rate is of the same order of magnitude but slower than active moiety elimination, C_{max} for the active moiety is 40% less when parallel elimination occurs (Bottom panel). In all 4 cases (Figures 3-6), elimination half-life of the active moiety is identical (i.e. 6.93 hours).

5.3. Study design for kinetic studies in man

Ideally, the kinetics and the prodrug and the active moiety should be studied in the same patients or healthy volunteers. It is not possible to give generally valid guidelines for study design since, as described in the previous paragraphs, many pharmacokinetic parameters may influence the analysis of the data, and consequently the way in which data must be collected. For example, it is certainly very different to study a prodrug such as bacampicillin with a very fast release of ampicillin after oral administration, or the decanoic ester of haloperidol given intramuscularly as a depot preparation in an oily vehicle.

Planning of pharmacokinetic studies for topically applied prodrugs is even more problematic, and it must be realized that even kinetic studies of topically applied drugs (i.e. dermal, ocular, rectal etc) are not yet clearly delineated. In particular, analytical sensitivity represents often a major obstacle for the conduct of properly designed pharmacokinetic investigations [Balant et al, 1990b].

Finally, it must be remembered that a prodrug intended for oral administration may show perfectly adequate release performances under normal conditions, but be unreliable (i.e. not transformed to the active moiety) in liver disease. This may be seen if regeneration occurs enzymatically in the liver. The prodrug should, accordingly, be tested in this pathological situation before marketing.

5.4. Bioavailability assessment

Case A: When a prodrug is pharmacologically inactive and is subject to an important pre-systemic transformation to the active moiety, it is correct to administer the active moiety by the intravenous route as the reference pharmaceutical form, as long as its behaviour (CL, V_d) is identical for both modes of administration. The ratio under the AUCs after oral and intravenous absorption is the absolute bioavailability of the active moiety. This is, for example, the procedure that was used with ampicillin and its prodrugs bacampicillin and pivampicillin [Ehrnebo et al, 1979].

Case B: When a prodrug is pharmacologically inactive, subject to an important pre-systemic and/or systemic biotransformation to the active moiety, and intended for both oral and intravenous administration, two practical situations may occur: the active moiety is available for parenteral administration or the active moiety cannot be administered intravenously to humans for practical, safety or ethical reasons.

In the first situation, the same procedure described in Case A is applicable. In the second situation, it is necessary to administer the parent compound intravenously as the reference drug. The ratio of the AUCs of the parent drug indicates its absolute bioavailability ; this parameter is often close to zero and of limited clinical relevance. The absolute bioavailability of the active moiety cannot usually be calculated, unless it is demonstrated that the prodrug is entirely biotransformed to the active moiety, or if the amount of parent drug transformed to other metabolites can be calculated for both routes of administration. Even if the absolute bioavailability of the active metabolite cannot be calculated in the strict sense of pharmacokinetic definitions, clinically relevant conclusions about the behaviour of the active moiety after i.v. and p.o. administration may nevertheless be obtained from analysis of its concentration-time curves. For example, if these curves are similar (AUC, t_{max}, C_{max}) after both modes of administration, it is probable that the therapeutic effects will be similar. In addition, if AUCs are equivalent, it can be stated that equivalent amounts of the active moiety are made available to the body following i.v. and p.o. administration.

6. CONCLUSIONS

Although prodrug design started more than 30 years ago and many reviews have been written on this subject, very little information is available in official guidelines or pharmacokinetic textbooks on the regulatory requirements or data analysis for this type of compounds. The present review is an attempt to gather and confront available information on this subject.

Some basic problems have, however, been left untouched. For example, the difficulty of extrapolating data from animal to man encountered during toxicokinetic and toxicologic studies with drugs is amplified with prodrugs since not only metabolism of the active moiety might differ, but also its availability from the prodrug. As a matter of fact, there is presently no published rationale for the conduct of animal and human pharmacokinetic programs during prodrug research and development.

We concluded a recent review on prodrugs [Balant et al, 1990a] quoting the question asked in 1985 by Stella et al.: "Do prodrugs have advantages in clinical practice ?". Our opinion on this matter was that "Today, the answer is certainly YES in some particular cases, but for many drugs this aspect of drug design has received no clear and satisfactory solution. The main reason for this situation is that most prodrugs have been synthesized starting from valuable and well known drugs. As a consequence, the potential advantage of the new chemical entity over its 'seasoned precursor' has often been only marginal. It is thus important that in the future, drug design of new chemical entities should incorporate 'delivery and/or targeting components' from the earliest stages of research and development. This strategy might help substances too toxic, or unable to show adequate pharmacologic effects in their basal form to go trough primary and secondary screening, before successfully reaching human testing. It is evident that if such an approach were to become an integral part of basic drug design and not just a hindsighted attempt to solve problems associated with older drugs, it would also be necessary to develop new biopharmaceutical and pharmacokinetic approaches to tackle the new challenges".

After this review, which focused more on pharmacokinetic aspects, we can conclude that, indeed, additional thinking on new ways to approach the pharmacokinetics of prodrugs and their active moiety is of paramount importance if prodrug design is to remain (or to become?) an important part for research and development of new therapeutic agents.

REFERENCES

Albert A., 1958. Chemical aspects of selective toxicity. Nature **182**:421.

Aguiar A., Zelmer J.E., 1969. Dissolution behavior of polymorphs of chloramphenicol palmitate and mefenamic acid. J. Pharm. Sci. **58**:983.

Aungst B.J., Myers M.J., Shefter E., Shami E.G., 1987. Prodrugs for improved oral nalbuphine bioavailability: Inter-species differences in the disposition of nalbuphine and its acetysalicylate and anthranilate esters. Int. J. Pharm. **38**:199.

Balant L.P., McAinsh J., 1980. Use of metabolite data in the evaluation of pharmacokinetics and drug action. In : "Concepts in Drug Metabolism" (P. Jenner and B. Testa, eds.), Part A, page 311. Marcel Dekker, New York.

Balant L.P., Doelker E., Buri P., 1990a. Prodrugs for the improvement of drug absorbtion via different routes of administration. Europ. J. Drug Metab. Pharmacokin. **15**:143.

Balant L.P., Roseboom H., Gundert-Remy U.A., 1990b. Pharmacokinetic criteria for drug research and development. In : "Advances in Drug Research" (B. Testa, ed.), Vol. 19, page 1. Academic Press Ltd, London.

Balant-Gorgia A.E., Balant L.P., 1987. Antipsychotic drugs. Clinical pharmacokinetics of potential candidates for plasma concentration monitoring. Clin. Pharmacokin. **13**:65.

Basu K., Kildsig D.O., Mitra A.K., 1988. Synthesis and kinetic stability studies of progesterone derivatives. Int. J. Pharm. **47**:195.

Bodor N., Sloan K.B., Higuchi T., Sasahara K., 1977. Improved delivery through biological membranes. 4. Prodrugs of L-Dopa. J. Med. Chem. **20**:1435.

Bundgaard H., Nielsen N.M., 1988. Glycolamide esters as a novel biolabile prodrug type for non-steroidal anti-inflammatory carboxylic acid drugs. Int. J. Pharm. **43**:101.

Bundgaard H., Nielsen N.M., Buur A., 1988. Aspirin prodrugs :synthesis and hydrolysis of 2-acetoxybenzoate esters of various N-(hydroxyalkyl)amides. Int. J. Pharm.**44**:151.

Burger A., 1977. Neue Untersuchungergebnisse von Chloramphenicolpalmitat. Sci. Pharm. **45**:269.

Buur A., Bundgaard H., 1986. Prodrugs of 5-fluorouracil. V. 1- Alkoxycarbonyl derivatives as potential prodrug forms for improved rectal or oral delivery of 5-fluorouracil. J. Pharm. Sci. **75**: 522.

Buur A., Bundgaard H., 1987. Prodrugs of 5-fluorouracil. VIII. Improved rectal and oral delivery of 5-fluorouracil via various prodrugs. Structure-rectal absorption relationships. Int. J. Pharm. **36**:41.

Buur A., Bundgaard H., 1988. Prodrugs of peptides. III. 5-Oxazolidinones as bioreversible derivatives for the α-amido carboxy moiety in peptides. Int. J. Pharm., **46**:159.

Chafi N., Montheard J.P., Vergnaud J.M., 1988. Dosage form with drug attached to polymer (polyanhydride) dispersed in a Eudragit matrix : Preparation and release of drug in gastric liquid. Int. J. Pharm. **45**:229.

Colvin M., Brundrett R.B., Kan M.N.N., et al., 1976. Alkylating properties of phosphoramide mustard. Cancer Res. **36**:1121.

Dayer P., Desmeules J., Leemann T., Striberni R., 1988. Bioactivation of the narcotic drug codeine in human liver is mediated by the polymorphic monooxygenase catalyzing debrisoquine 4- hydroxylation. Biochem. Biophys. Res. Commun. **152**:411.

Desmeules J, Dayer P., Gascon M.P., Magistris M., 1989. Impact of genetic and environmental factors on codeine analgesia. Clin. Pharmacol. **45**:122.

Diasio R.B., Harris B.E., 1989. Clinical pharmacology of 5-fluorouracil. Clin. Pharmacokin. **16**:215.

Dobrinska M.R., Kukovetz W., Beubler E., et al., 1982. Pharmacokinetics of the pivaloyloxyethyl (POE) ester of methyldopa, a new prodrug of methyldopa. J. Pharmacokin. Biopharm. **10**:587.

D'Souza M., Venkataramanan R., D'Mello A., Niphadkar P., 1986. An alternative prodrug approach for reducing presystemic metabolism of drugs. Int. J. Pharm. **31**:165.

Duggan D.E., Hare L.E., Ditzler C.A., et al., 1977. The disposition of sulindac in man. Clin. Pharmacol. Ther. **21**:326.

Duggan D.E., Hooke K.F., Noll R.M., et al., 1978. Comparative disposition of sulindac and metabolites in five species. Biochem. Pharmacol. **27**:2311.

Ehrnebo M., Nilsson S.O., Boreus L.O., 1979. Pharmacokinetics of ampicillin and its prodrugs bacampicillin and pivampicillin in man. J. Pharmacokin. Biopharm. **7**:429.

Gibaldi M., Perrier D., 1982. "Pharmacokinetics." Second Edition. Marcel Dekker, New York.

Havron A., Weiner B.Z., Zilkha A., 1974. Polymeric salicylate derivatives. J. Med. Chem. **17**:770.

Higuchi T., Stella V., 1975. "Pro-drugs as Novel Drug Delivery Systems." American Chemical Society, Washington DC.

Higuchi W.I., Kusai A., Fox J.L., et al., 1983. Controlled release of drugs: Prodrug performance in target tissues. In: "Controlled Release Delivery Systems" (T.J. Roseman and S.Z. Mansdorf, eds.), page 43. Marcel Dekker, New York.

Hrnblad Y., Ripe E., Magnusson P.O., Tegnes K., 1976. The metabolism and clinical activity of terbutaline and its prodrug ibuterol. Eur. J. Clin. Pharmacol. **10**:9.

Hussain M.A., Koval C.A., Myers M.J., et al., 1987. Improvement of the oral bioavailability of naltrexone in dogs: A prodrug approach. J. Pharm. Sci. **76**:356.

Hussain M.A., Shefter E., 1988. Naltrexone-3-salicylate (a prodrug of naltrexone): Synthesis and pharmacokinetics in dogs. Pharm. Res. **5**:113.

Inturrisi C.E., Mitchell B.M., Foley K.M., et al., 1984. The pharmacokinetics of heroin in patients with chronic pain. New Engl. J. Med. **310**:1213.

Irwin W.J., Belaid K.A., 1988a. Drug-delivery by ion-exchange. Hydrolysis and rearrangement of ester pro-drugs of propranolol. Int. J. Pharm. **46**:57.

Irwin W.J., Belaid K.A., 1988b. Drug-delivery by ion-exchange. Stability of ester prodrugs of propranolol in surfactant and enzymatic systems. Int. J. Pharm. **48**:159.

Johansen M., Bundgaard H., 1981. Decomposition of rotitetracycline and other N-Mannich bases and of N-hydroxy-methyl derivatives in the presence of plasma. Arch. Pharm. Chem. Sci. **9**:40.

Kaneo Y., Tanaka T., Fujihara Y., et al., 1988. Delivery of glutathione, as a dextran conjugate, into the liver. Int. J. Pharm. **44**:265.

Kojima T., Hashida M., Muranishi S., Sezaki H., 1978. Antitumor activity of timed-release derivative of mitomycin C, agarose bead conjugate. Chem. Pharm. Bull. **26**:1818.

Kratzl K., Kaufmann E., 1961. Versuche zur Darstellung hochmolekularer Pharmazeutika. 1. Mitt.: Synthese, Reaktionen und ^{14}C-Markierung eines Acetylsalicylsäurestärkeesters. Monatschr. Chem. **92**:371.

Kratzl K., Kaufmann E., Kraupp O., Stormann H., 1961. Versuche zur Darstellung hochmolekularer Pharmazeutika. 2. Mitt.: Stoffwechseluntersuchungen von Acetylsalicylsäurestärkeestern. Monatschr. Chem. **92**: 378.

Larsen C., 1989. Macromolecular prodrug. XII. Kinetics of release of naproxen from various polysaccharide ester prodrugs in neutral and alkaline solution. Int. J. Pharm. **51**:223.

Leight D.A., Reeves D.S., Simmons K., et al., 1976. Talampicillin, a new derivative of ampicillin. Brit. Med. J., **1**:1378.

Meslard J.C., Yean L., Subira F., Vairon J.P., 1986. Reversible immobilization of drugs on a hydrogen matrix. Makromol. Chem. **187**:787.

Negishi N., Bennett D.B., Cho C.S., et al., 1987. Coupling of natrexone to biodegradable poly(α-amino acids). Pharm. Res. **4**:305.

Olsson O.A.T., Svensson L.A., 1984. New lipophilic terbutaline ester prodrugs with long effect duration. Pharm. Res. **1**:19.

Pardridge W.M., 1983. Brain metabolism : a perspective from the blood-brain barrier. Physiol. Rev. **63**:1481.

Puglisi L., Caruso V., Paoletti R., et al., 1976. Macromolecular drugs. I : Long-lasting antilipolytic activtiy of nicotinic acid bound to a polymer. Pharmacol. Res. Commun. **8**:379.

Remon JP., Duncan R., Schacht E., 1984. Polymer-drug combinations: Pinocytic uptake of modified polysaccharides containing procainamide moieties by rat visceral yolk sacs cultured in vitro. J. Control. Rel. **1**:47.

Ringsdorf H., 1975. Structure and properties of pharmacologically active polymers. J. Polym. Sci. Polym. Symp. **51**:135.

Roche E.B., 1977. "Design of Biopharmaceutical Properties through Prodrugs and Analogs." American Pharmaceutical Association/ Academy of Pharmaceutical Sciences, Washington DC.

Rowland M., Tozer T.N., 1989. "Clinical Pharmacokinetics." Second Edition. Lea & Febiger, Philadelphia.

Rozencweig M., Staquet M., Klastersky J., 1976. Antibacterial activity and pharmacokinetics of bacampicillin and ampicillin. Clin. Pharmacol. Ther. **19**:592.

Saab A.N., Ditter L.W., Hussain A.A., 1988. Isomerization of cephalosporin esters: Implications for the prodrug ester approach to enhancing the oral bioavailability of cephalosporins. J. Pharm. Sci. **77**:906.

Sasaki H., Takahashi T., Nakamura J., et al., 1986. Intestinal absorption of 5-fluorouracil and its alkylcarbamoyl derivatives in the rat small intestine. J. Pharm. Sci. **75**:676.

Schaaf L.J., Dobbs B.R., Edwards I.R., Perrier D.G., 1988. The pharmacokinetics of doxifluridine and 5-fluorouracil after single intravenous infusions of doxifluridine to patients with colorectal cancer. Eur. J. Clin. Pharmacol. **34**:439.

Schacht E., Ruys E., Vermeersch J., Remon J.P., 1984. Polymer-drug combinations : Synthesis and characterization of modified polysaccharides containing procainamide moieties. J. Control. Rel. **1**:33.

Singh M., Vasudevan P., Sinha T.J.M., et al., 1981. An insulin delivery system from oxidized cellulose. J. Biomed. Mater. Res. **15**:655.

Sinkula A.A., 1977. Perspective on prodrugs and analogs in drug design. In "Design of Biopharmaceutical Properties through Prodrugs and Analogs" (E.B. Roche ed.), page 1. American Pharmaceutical Association/ Academy of Pharmaceutical Sciences, Washington DC.

Sinkula A.A., Yalkowsky S.H., 1975. Rationale for design of biologically reversible drug derivatives: Prodrugs. J. Pharm. Sci. **64**:181.

Sloan K.B., Bodor N., 1982. Hydroxymethyl and acycloxymethyl prodrugs of theophylline : Enhanced delivery of polar drugs through skin. Int. J. Pharm. **12**:299.

Stella V.J., Charman W.N.A., Naringrekar V.H., 1985. Prodrugs : Do they have advantages in clinical practice? Drugs **29**:445.

Strong H.A., Warner N.J., Renwick A.G., George C.F., 1985. Sulindac metabolism : The importance of an intact colon. Clin. Pharmacol. Ther. **38**:387.

Tatsumi K., Kitimura S. Yanada S., 1983. Sulfoxide reductase activity of liver aldehyde oxidase. Biochem. Biophys. Acta **747**:86.

Tritton T.R., Yee G., Wingaard L.B., 1983. Immobilized adriamycin : A tool for separating cell surface from intracellular mechanisms. Fed. Proc. **42**:284.

Vert M., 1986. "Polyvalent polymeric drug carriers." Critical Reviews in Theapeutic Drug Carrier Systems, Vol. 2. CRC Press, Boca Raton, FL.

Vickers S., Duncan C.A., White S.D., et al., 1978. Evaluation of succinimidoethyl and pivaloyloxyethyl esters as progenitors of methyldopa in man, rhesus monkey, dog and rat. Drug Metab. Dispos. **6**:646.

Vickers S., Duncan C.A.H., Ramjit H.G., et al., 1984. Metabolism of methyldopa in man after oral administration of the pivaloyloxyethyl ester. Drug Metab. Dispos. **12**:242.

Wagner J., Grill H., Henschler D., 1980. Prodrugs of etilefrine: Synthesis and evaluation of 3'-(O-acyl) derivatives. J. Pharm. Sci. **69**:1423.

Waller D.G., George C.F., 1989. Prodrugs. Brit. J. Clin. Pharmacol. **28**:497.

Yolles S., 1987. Time-release depot for anticancer drugs :Release of drugs covalently bonded to polymers. J. Parenter. Drug Assoc. **32**:188.

Yolles S., Morton J.F., Sartori M.F., 1979. Preparation of steroid esters of hydroxypropyl cellulose. J. Polym. Sci. Polym. Chem. Ed. **17**:4111.

Yoshimura Y., Hamaguchi N., Yashiki T., 1987. Synthesis and oral absorption of acyloxymethyl esters of cefotiam. Int. J. Pharm. **38**:179.

CARCINOGENIC RISK ASSESSMENT: SCIENCE OR FANTASY?

Ajit K. Thakur and Arvind Parthasarathi[1]

Hazleton Washington, Inc.
Vienna, VA 22182

INTRODUCTION

Important regulatory decisions about the fate and applicability of many chemicals are made from life-time rodent bioassays using mathematical models on tumor incidence data. Generally, these models are mostly based on empirical functions whose low-dose extrapolative behaviors are of questionable nature. In a few instances assumptions are made to generate these functions which cannot be experimentally established or rejected under given experimental details. Often investigators who use these models do not pay attention to the shapes and behaviors of the dose-response curves under consideration under regulatory presumptions.

The mechanisms of actions of different potential carcinogens as well as other chemicals are different. Some of them act through metabolic, some through endocrine, and yet some through a combination of the above two pathways. Furthermore, in many of the experiments, the animals are so overloaded with the chemicals in questions that the response is not brought about by any physiological actions; instead, they are caused by simple disruption of the system under question. In most cases, bioassay data do not provide adequate information toward the understanding of the inherent mechanism(s) involved. As a consequence, it is inconceivable that any single empirical function is adequate to describe all carcinogenic dose-response curves of different shapes. As a matter of fact, for many tumor incidence tables, statistical evidence may indeed indicate the inadequacy of many or all of the currently used empirical models unless one "doctors" the data. Yet, there are many firm believers who maintain that a particular model, viz. the multistage, should be used for low-dose extrapolation purposes for risk assessment with all tumor incidence data.

The purpose of this chapter is to outline the basic principles behind risk assessment. In the process, we will examine the low-dose behaviors of the commonly used mathematical models with some specific chemicals.

---

[1] Permanent Address: 9H Chamiers Road, Nandanam, Madras 600035, India.

New Trends in Pharmacokinetics, Edited by A. Rescigno and A.K. Thakur
Plenum Press, New York, 1991

EXPERIMENTAL DESIGN AND DATA

The general outline of chronic oncogenicity studies has been described previously [Thakur, 1988]. For our purpose here, a typical carcinogenic dose-response can be described as in Table 1. In this table, n_i, the number at risk at dose level l_i, is the actual number of animals examined for a particular tumor, and m_i is the number of animals having that tumor of interest. More complex situations arise when animals with tumor multiplicity are counted producing an $R \times C$ contingency tables. Such information is not generally dealt with in mathematical extrapolation of risk; instead, it is used qualitatively. Also, when onset time for a tumor is available, multiple $k \times 2$ tables as in Table 1 are formed and risk assessment becomes more complicated. In theory, these tables should initially be analyzed statistically to see whether there is any dose-response in the incidences.

Table 1

A Hypothetical $k \times 2$ Table of Lesions

Level	No. at Risk	No. with Lesions
l_0	n_0	m_0
l_1	n_1	m_1
l_2	n_2	m_2
l_3	n_3	m_3
...	...	...
l_k	n_k	m_k

In Table 1, the numbers at risk are the actual numbers of animals with complete follow-ups. More complex situations arise when there is differential mortality due to competing toxicity or when onset times for the tumors in question are available.

MODELS FOR LOW-DOSE EXTRAPOLATION

Most of these models fall under the following four categories:

(1) "Mechanistic" Models: These are based on certain assumptions regarding carcinogenic events in a tissue. In most cases, these assumptions either cannot be verified experimentally or are not well understood.

(a) Onehit Model. According to this model, a single critical molecular event ("hit") between a target cell and a proximate carcinogenic molecule produces cancer. The probability of a hit is proportional to the exposure level (d). The mathematical form of the model is:

$$P(d) = 1 - \exp[-(a + bd)] \tag{1}$$

where

$$a \geq 0 \text{ is tyhe background incidence rate,}$$
$$b \geq 0 \text{ is the empirical potency of a carcinogen.}$$

The model produces low-dose asymptotic linearity.

(b) Multistage Model. According to this model [Armitage and Doll, 1954; Crump

302

et al, 1976], a carcinogen may increase the rate(s) of event(s) or stage(s) a cell goes through before it becomes carcinogenic. This is the model of choice for the United States Environmental Protection Agency (EPA) and various other regulatory agencies.

In its simplest form, the multistage model is:

$$P(d) = 1 - \exp\left[-\sum_{i=0}^{k} b_i d^i\right]$$
(2)

where

$$b_i \geq 0 \text{ are the empirical potency parameters}$$
$$k \leq \text{Number of dose levels is the nomber of stages.}$$

At the low-dose region the model is asymptotically linear and is numerically indistinguishable from the onehit model.

(c) Multihit (Gamma) Model: According to this mode [Rai and Van Ryzin, 1981], k molecular events ("hits") are needed before cancer is induced in an organ. The distribution of these hits is assumed to be Poisson. In its mathematical form:

$$P(d) = \sum_{x=k}^{\infty} \frac{(bd)^x \cdot \exp(-bd)}{x!} = \int_0^{bd} \frac{x^{k-1} \cdot \exp(-x)}{(k-1)!} \cdot dx$$
(3)

where

$$b \geq 0 \text{ is the empirical potency, } k > 0 \text{ is the number of hits.}$$

The multihit is a generalization of the onehit model.

(2) Tolerance Distribution Models: According to these models, animals possess tolerance towards chemical insult and tumors are induced once the tolerance is exceeded. The tolerance is assumed to have certain distributions.

(a) Probit (Mantel-Bryan) Model: According to this model [Mantel and Bryan, 1961], the tolerance is lognormally distributed. Its mathematical form is:

$$P(d) = \frac{1}{\sqrt{2\pi}} \int_{-\infty}^{x} \exp(-\frac{u^2}{2}) \, du$$
(4)

with

$$x = a + b \ln d$$
$$a \geq 0 \text{ the intercept of the log-Probit plot (= background incidence)}$$
$$b \geq 0 \text{ the slope of the log-Probit plot (= empirical potency factor).}$$

(b) Logistic (Logit) Model: According to this model [Doll, 1971], the tolerance is binomially distributed. Mathematically:

$$P(d) = \frac{1}{1 + \exp\left(-(a + b{\cdot}\log d)\right)} \qquad (5)$$

where

$a \geq 0$ Intrercept of the log-Logit plot (= background risk)
$b \geq 0$ Slope of the log-Logit plot (= empirical potency factor)

(c) Weibull Model: According to this model [Carlborg, 1981], the tolerance is Weibull distributed. Its mathematical form is:

$$P(t,d) = 1 - \exp\left(g(d){\cdot}t^{k}\right) \qquad (6)$$

with

$g(d) = a + bd^{m}$ in its simplest version.

In Equation (6), a and b have the same meaning as in the previous models and m and k are two parameters of the model which do not have any specific biological meaning but determine the shape of the dose-response curve. The additional independent variable t here is time, with the assumption being that tumor formation is a function of both dose and time. This latter assumption is very difficult to establish in animal bioassays. To be able to use this additional information in the model, one has to assume that the first palpation time for a superficial tumor (as in skin or mammary tumors) or the time of death with a rapidly lethal tumor is the surrogate for time to tumor. Such an assumption introduces additional uncertainty in the model.
The multihit model, described in the previous section, also falls in this category. In that case the tolerance is assumed to be gamma distributed.

(3) Time-to-tumor Models: In these models, the probability of tumor formation in response to a chemical is a function of both dose of the chemical and time. Accordingly, a chemical shortens the latency of the carcinogenic processes in a dose-related fashion. The Weibull model, described previously, is an example of such models.

(a) Hartley-Sielken (General Product) Model: This model [Hartley and Sielken, 1977] is a generalization of both the multistage and Weibull models. The time to tumor distribution is assumed to be Weibull. Large numbers of animals with regular serial sacrifices are needed for the use of this model. Standard bioassay designs are mostly inadequate for application of this model. Mathematically, the model takes the form:

$$P(t,d) = 1 - \exp\left(-\sum_{j=0}^{k} a_{j}d^{j}h(t) \right) \qquad (7)$$

where

$k \geq 1$ is the number of stages of carcinogenesis
$h(t)$ is the time-to-tumor distribution, a positive non-decreasing function
of time (Weibull)
a_{j} is an empirical multistage type parameter.

Several other time-to-tumor models have been described in current literature [Prentice, Peterson, and Marek, 1982; Kalbfleisch, Krewski, and Van Ryzin, 1983].

(4) Pharmacokinetic Models:

(a) Cornfield's pharmacokinetic model [Cornfield, 1977] assumes that a carcinogen is in simultaneous reversible second order chemical reaction (activation and deactivation). The probability of a carcinogenic response is proportional to the activated complex concentration. The mathematical form of the model is:

$$P(d) = \frac{A}{A + K_a}, \ \text{if} \ d > T$$
$$= \frac{K_d d}{(S + K_a)K_d + K_a T}, \ \text{if} \ d \leq T \tag{8}$$

where

$$A = d - P(d)S - \alpha; \quad \alpha = \frac{K_a P(d) T}{P(d)K_a + [1 - P(d)]K_d}$$

Under various conditions on d, T, K_a, and K_d, the model is capable of simulating various forms of monotonic shapes [Thakur, 1988]. Some of the more recent works with this group of models will be discussed in the next three chapters of this book. At this time, these models are used mostly for research oriented low-dose extrapolations, not for regulatory purposes.

ARITHMETIC OF LOW-DOSE EXTRAPOLATION

Animal-to-Human Conversion: Human Equivalent Dose (HED)

Surface Area Conversion from animal to human dose is sometimes made using empirical scaling factors as follows:

$$\text{HED (mg/kg BW/day)} = \frac{\text{Animal Dose (mg/kg BW/day)}}{(\text{Human BW/Animal BW})^{0.667}}$$

The dose used in most of the life-time experiments is in terms of parts per million (ppm). The following conversion scheme is used for ppm to mg/kg BW/day by many:

Species	ppm	mg/kg BW/day
Mouse	1	1/6
Rat	1	1/15

Some people use Safety Factor (100) instead to avoid species and dose conversions:

$$Q^*_{\text{Human}} = Q^*_{\text{Animal}} \times 100.$$

Quantification of Risk

There are three ways of quantifying risk. One is to calculate the lower 95% confidence level (95% LCL) of the virtually safe dose (VSD), which is defined as the dose at which the cancer risk does not exceed one-in-a-million $(1/10^6)$ which is assumed to be an acceptable risk (AR) by most regulatory agencies. The second is to calculate the 95% upper confidence level (95%) of risk given a particular dose level. The third, which uses linear low-dose extrapolation, is to calculate the 95% UCL of the coefficient of the linear term in the model. This approach is widely used by individuals who specifically use the onehit and/or the multistage models to quantify risk. The EPA uses a variation of the third approach to calculate what is known as Q^* as follows:

$$Q^* = \frac{AR \text{ (usually } 1/10^6)}{95\% \text{ LCL of VSD}} \tag{9}$$

The term 'acceptable risk' is somewhat arbitrary particularly in the cases where the tumor has significant background rate, i.e. the incidence in the concurrent control group. People generally use either additional risk over background (AROB) or extra risk over background (EROB) to specify AR. These two terms are defined as follows:

$$AROB = P(d) - P(0) \quad \text{and} \quad EROB = \frac{P(d) - P(0)}{1 - P(0)} \, ,$$

where P(0) and P(d) are the proportions of animals with tumors in the untreated and treated (at dose level = d) groups respectively. To make things further complicated, the background could be assumed to be either independent or additive. Under independent background assumption, the effective dose is the same as the delivered dose and the background rate of tumor formation is independent from it. Mathematically it is expressed by Abbot's criterion:

$$P(d) = P(0) + [1 - P(0)]f(d),$$

where f(d) is the function describing the dose-response relationship. A direct corollary to independent background assumption is EROB. On the other hand, under the additive background assumption, the effective dose is the sum total of the delivered and any pre-existing dose.

Numerical Extrapolation

The common procedure is to linearly extrapolate using the UCL of the linear term in dose or from either the 1% or 10% EROB (or AROB). Some risk assessment software also use quadratic or higher order extrapolations or parametric or nonparametric bootstrap methods. Depending on the methods used, there may be several-fold differences in risk estimates for the same chemicals.

SOME EXAMPLES

Let us examine risk assessment of some well known chemicals. We will use the commonly used models such as the on-hit, multistage, multihit, logistic, probit, and the Weibull for this purpose. For comparisons, we will use linear extrapolation to

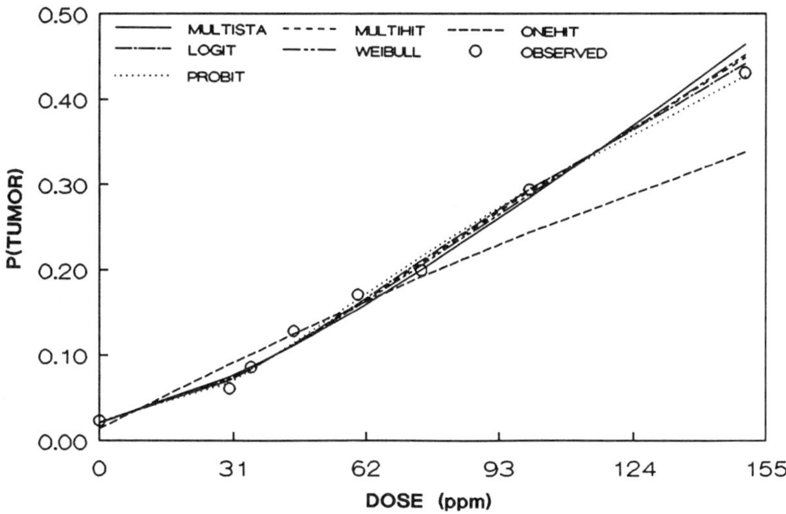

Figure 1. Liver Tumors in Female BALB/C Mice

compute risk in each case. As will be obvious, the shapes of the different dose-response curves to be examined here pose some interesting problems in the extrapolation procedure. In each case we will use the independent background (EROB) assumption with low-dose linearity assumption. The following software packages were used for risk extrapolation:

Model	Software
Multistage	GLOBAL 86 (Howe and Van Landingham, 1986)
Onehit and Multihit	MULTI80 (Rai and Van Ryzin, 1980)
Logistic, Probit and Weibull	DORES81 (Kovar and Krewski, 1981)

Example 1: 2-AAF Female Liver Tumors in BALB/C Mice

Dose (ppm)	No. at Risk	No. with Tumor	Proportion
0	383	9	0.0235
30	900	55	0.0611
35	639	55	0.0861
45	445	57	0.1281
60	415	71	0.1711
75	311	62	0.1994
100	160	47	0.2938
150	130	56	0.4308

The goodness of fit p-values and the 95% LCL of VSD for EROB of $1/10^6$ for different models for the above data set are shown below. The observed and expected values using different models are shown in Figure 1. Based on the goodness of fit p-values, only the onehit model shows serious lack of fit for the 2-AAF data set. Examination of Figure 1 reveals that the probit model fits the data the best followed by the multihit, logistic, Weibull, and finally the multistage models. Among the models that fit this data set well, the multistage seems to be the worst, although based on the goodness of fit p-value alone, it cannot be rejected. This issue will be further discussed in a later section.

Model	Goodness of Fit p-Value	95% LCL of VDS
Onehit	0.0055	$3.03 \cdot 10^{-4}$
Multistage	0.3580	$5.11 \cdot 10^{-4}$
Multihit	0.6577	$1.04 \cdot 10^{-2}$
Probit	0.8336	$5.50 \cdot 10^{-1}$
Logistic	0.7009	$1.35 \cdot 10^{-2}$
Weibull	0.5742	$5.53 \cdot 10^{-3}$

There is approximately 3 orders of magnitude difference in the VSD-values between the lowest (multistage) and highest (Probit) estimates among the models that presumably fit the data adequately. The 2-AAF dose-response curve is almost linear; yet there is such a discrepancy among these different models.

Example 2: DES Female Mammary Tumors in C3H Mice

Dose (ppm)	No. at Risk	No. with Tumor	Proportion
0	121	40	0.331
6.25	56	27	0.482
12.5	60	26	0.433
25	60	26	0.433
50	68	36	0.529
100	64	42	0.656
500	59	50	0.847
1000	58	50	0.862

The results of risk assessment of DES are shown in Figure 2 and the tables below. From a goodness of fit standpoint, the multistage model does not fit the data and the multihit barely makes it.

Graphically, from Figure 2, all models except the multistage seem to indicate approximately equivalent fit although, if one plots the residuals from the models,

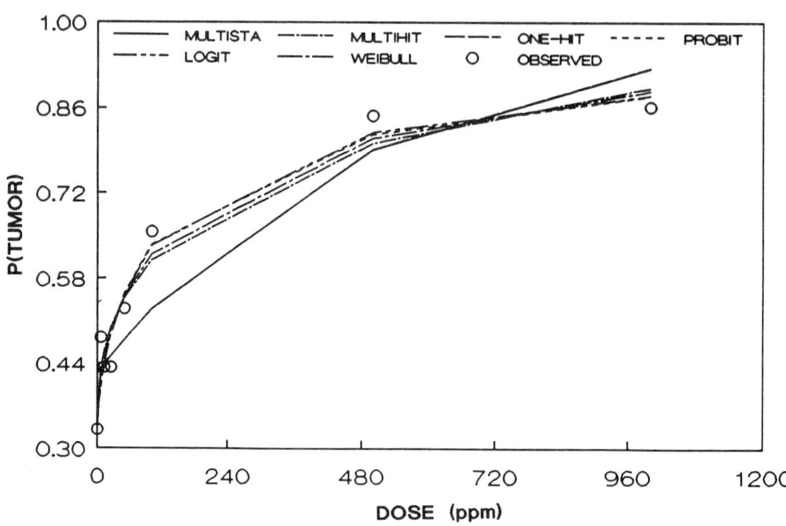

Figure 2. Mammary Tumors in Female C3H Mice

there is systematic deviations indicating poor choice of models. The VSD's for the DES data set for the models, once again, range between practically zero (multihit and Weibull) to about 10^{-3} (onehit). As can be seen in this example, the onehit and multistage models do not necessarily provide the most conservative estimates of risk. We will discuss it further under Discussion.

LOWER CONFIDENCE LIMIT OF VIRTUAL SAFE DOSE
(DES)

Model	Goodness of Fit P-Value	95% LCL of VDS
Onehit	0.1278	$1.19 \cdot 10^{-3}$
Multistage	0.0279	$6.28 \cdot 10^{-4}$
Multihit	0.0576	$1.67 \cdot 10^{-14}$
Probit	0.6418	$1.55 \cdot 10^{-4}$
Logistic	0.6949	$6.69 \cdot 10^{-9}$
Weibull	0.6496	$1.21 \cdot 10^{-13}$

Example 3: (a) Heptachlor: Male Liver Tumors in CD-1 Mice

TUMOR INCIDENCE RATES

Dose (ppm)	No. at Risk	No. with Tumor	Proportion
0	59	2	0.0339
1	58	1	0.0172
5	66	4	0.0606
10	73	37	0.5070

95% LOWER CONFIDENCE LIMIT OF VIRTUAL SAFE DOSE
(Heptachlor: Male)

Model	Goodness of Fit P-Value	95% LCL of VDS
Onehit	0.0002	$1.58 \cdot 10^{-5}$
Multistage	0.4428	$9.21 \cdot 10^{-5}$
Multihit	0.0569	$3.31 \cdot 10^{-1}$
Probit	0.5687	$7.69 \cdot 10^{-1}$
Logistic	0.5684	$1.28 \cdot 10^{-1}$
Weibull	0.5680	$7.60 \cdot 10^{-2}$

(b) Heptachlor: Female Liver Tumors in CD-1 Mice

TUMOR INCIDENCE RATES

Dose (ppm)	No. at Risk	No. with Tumor	Proportion
0	74	1	0.0135
1	71	0	0.0
5	65	3	0.0462
10	52	16	0.308

95% LOWER CONFIDENCE LIMIT OF VIRTUAL SAFE DOSE

(Heptachlor: Female)

Model	Goodness of Fit P-Value	95% LCL of VDS
Onehit	0.0016	$2.66 \cdot 10^{-5}$
Multistage	0.5938	$1.05 \cdot 10^{-4}$
Multihit	0.00001	$1.53 \cdot 10^{-8}$
Probit	0.3257	$3.22 \cdot 10^{-1}$
Logistic	0.3240	$3.10 \cdot 10^{-2}$
Weibull	0.3233	$2.04 \cdot 10^{-2}$

The observed and fitted values with the different models are shown in Figures 3(a) (Male) and 3(b) (Female).

In the cases of both male and female heptachlor incidence tables, the onehit does

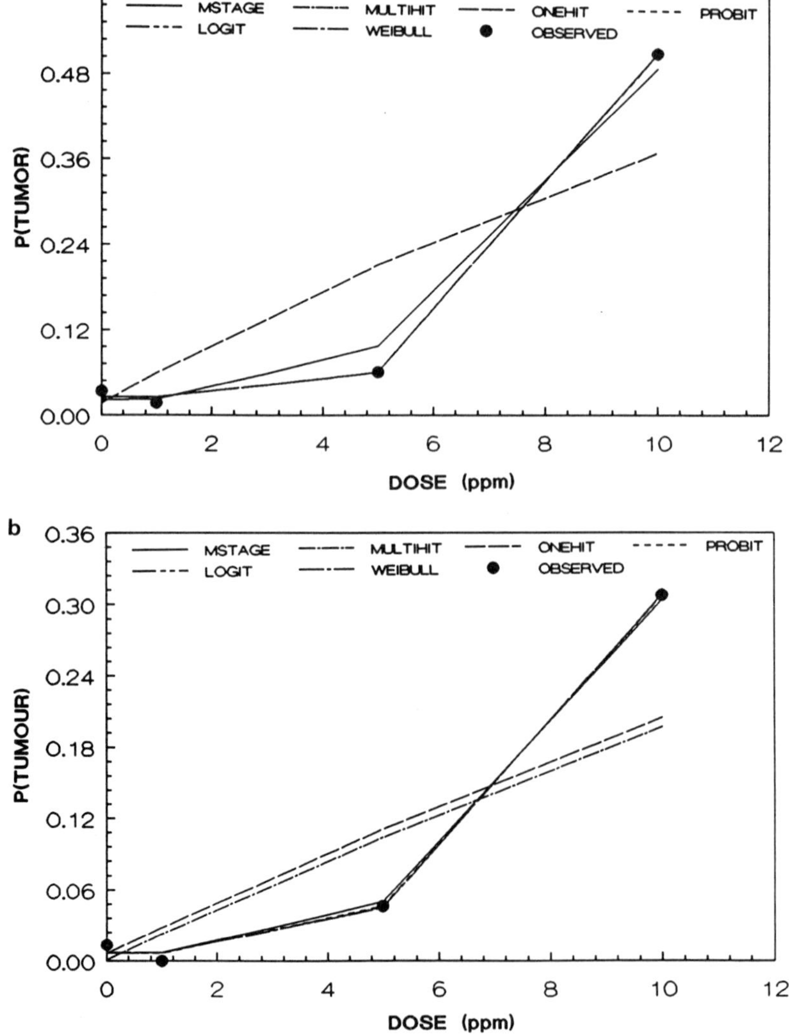

Figure 3. Liver Tumors in Male (top) and Female (bottom) CD-1 Mice

not fit the data; the multihit is the wrong choice for the female, whereas, in the case of the male, it barely makes it at 5% level of significance. Both these models can effectively be rejected for the heptachlor data. In either case, the multistage model allows for the smallest VSD differing from the other models by significant orders of magnitude. From that standpoint, the multistage is the most conservative for this data set.

DISCUSSION

There are many investigators who seem to be preferential toward one or more of the six empirical models discussed early. Many of them, including the U.S. EPA scientists, prefer the linearized multistage model since in many cases it produces the most conservative risk estimate (lowest VSD or highest risk). As can be seen below, depending on the model one uses, one may get million-fold or even higher disparity between models, for any particular response curve. The disparity is less when the dose level for which risk is being estimated is not very low or when the risk at which the VSD is being estimated is rather high. As the results indicate, even for a quite potent carcinogen such as 2-AAF, whose dose-response in this case is virtually linear, the discrepancy is rather phenomenal. When the shape of the response surface is not ideal, the discrepancy goes out of bounds. One then wonders about the futility of such endeavors.

95% LCL OF VSD AT DIFFERENT RISK LEVELS
(2-AAF Female Liver Tumor in BALB/C Mice)

MODEL	10^{-2}	10^{-3}	10^{-4}	10^{-6}	10^{-7}	10^{-8}
			RISK			
Onehit	3.06	0.305	$3.04 \cdot 10^{-2}$	$3.04 \cdot 10^{-4}$	$3.04 \cdot 10^{-5}$	$3.04 \cdot 10^{-6}$
Multistage	5.14	0.511	$5.11 \cdot 10^{-2}$	$5.11 \cdot 10^{-4}$	$5.11 \cdot 10^{-5}$	$5.11 \cdot 10^{-6}$
Multihit	7.55	1.46	$2.81 \cdot 10^{-1}$	$1.04 \cdot 10^{-2}$	$2.00 \cdot 10^{-3}$	$3.86 \cdot 10^{-4}$
Probit	1.16	4.46	$2.02 \cdot 10^{0}$	$5.50 \cdot 10^{-1}$	$3.14 \cdot 10^{-1}$	$1.87 \cdot 10^{-1}$
Logistic	8.21	1.65	$3.32 \cdot 10^{-1}$	$1.35 \cdot 10^{-2}$	$2.71 \cdot 10^{-3}$	$5.45 \cdot 10^{-4}$
Weibull	7.01	1.17	$1.97 \cdot 10^{-1}$	$5.53 \cdot 10^{-3}$	$9.26 \cdot 10^{-4}$	$1.55 \cdot 10^{-4}$

The shapes of most of the dose-response curves under the bioassay experiments are unique. In fact, in many cases, the same chemical does not even produce the same type of tumor in the same sex. In a very few instances, where the same tumor type is repeated in a second or third study, the shape may be quite different. As a consequence, with empirical models of the types described here, there seems to be hardly and scientific merit in attempting to perform curve fitting and risk assessment with one particular model.

Mechanisms of carcinogenesis are basically unknown in most cases. One generally sees the final outcome in the form of cancer and tries to correlate to some exposure or multiple exposures. It is inconceivable that such complex mechanisms (unknown) will produce similar shapes in dose-responses. There are certain biochemical, physiological, and pharmacological mechanisms which may be responsible for producing nonlinear, nonmonotone, and even irregular shapes in the dose-response. Let us briefly discuss a few of them.

(a) Threshold: Threshold is a physiological phenomenon under which a stimulus has to exceed a certain level before a response is exhibited by the system. As a consequence, there will be an initial flattening of the response. Once this level is

311

exceeded, there is response. Many physiological functions, chemical, electrical or others, exhibit threshold. Most scientists believe (Gehring and Blau, 1977) that many chemicals which act through enzymatic or hormonal pathways, may show threshold in their toxic and carcinogenic responses. The empirical models commonly used in carcinogenic risk assessment do not allow for such a phenomenon. Ignoring such a threshold in low-dose extrapolation produces scientifi"cally meaningless risk estimates from curves indicating such a behavior. Possibility for the presence of threshold should be investigated and models should be modified to account for it.

(b) Hormesis: Hormesis is a pharmacological mechanism by which a toxic chemical, at low dose levels, actually produces beneficial effects. For example, there may be decrease in natural mortality, increase in regulatory and immune response capabilities, decrease in cancer rate, etc. There is a growing amount of literature on this subject [Calabrese, McCarthy, and Kenyon, 1987, Boxenbaum, Neafsey, and Fournier, 1988, Neafsey, 1990]. In any case, as a consequence of hormesis, there will be an initial drop in the carcinogenic response curve. Many chemicals are showing such shapes. None of the models currently used in any risk assessment allows for such nonlinearity; they simply ignore them. Once again, the possibility of such a phenomenon for a chemical should be investigated, and if feasible, models should be modified or developed incorporating such mechanisms.

(c) Saturation: As in the case of enzyme kinetics, some carcinogenic dose-response curves may show flattening at the high-dose regions. This flattening may be due to occupation of all the target cells or receptors specific to the chemical in question. Failure to account for such behaviors produces incorrect risk estimates. Many investigators inappropriately exclude such points from the response curves, specially when they are using either the multistage or the onehit model.

(d) Competing Toxicity: Many chemicals produce concomitant cellular or target tissue toxicity while inducing carcinogenesis. As a result, at high dose levels, there may be cellular destruction or death of many experimental animals producing lesser numbers at risk. As a consequence, there are actually fewer numbers of animals with tumors at those dose levels which may produce a downward gradient at the high end of the curve. Once again, an appropriate model must incorporate such phenomena. None of the currently used models is capable of doing that.

From modeling standpoints, for many of the labeled and so-called potential carcinogens, the empirical models currently used are either inappropriate or applied without much statistical or biological considerations. Many of the regulatory agencies have adapted the linearized multistage model for risk assessment with the argument that it produces the most conservative risk estimate. For response curves with flattening or downward trend at the high end, the risk estimates are not conservative, contrary to the claims made. The DES example, discussed earlier is not an isolated example for such a behavior.

There are many physiological and pharmacological reasons why many scientists (the present author included), are skeptical about categorical risk assessment of any chemical disregarding the biology behind the process. Let us briefly mention some of them here.

(a) Toxic Dose: Because of regulatory requirements, investigators are forced to test chemicals at nonphysiological dose levels so that the study will be valid. The validity clause requires that there must be a high enough dose to induce some toxicity. These dose levels are generally based on acute and subchronic studies where animals are exposed to insult for much shorter durations. As a result, higher dose levels may actually exceed the maximum tolerated dose or MTD in the chronic study. In many such cases, one will not find any indication of carcinogenicity in the study until the highest dose level is reached. In other words, the dose response curve may be absolutely flat at the initial phase and then it suddenly rises at the highest dose level where all the regulatory and defense mechanisms in the animal's system may have been completely disrupted. Several recent works (Ames and Gold, 1990) address this issue.

(b) Species Differences: There are hormonal and enzymatic differences among various species. Even among the two popular strains of rats, the Sprague-Dawley (SD) and Fischer 344, there are such differences. A chemical which may seem to produce a particular type of tumor in an organ in one such species, may not be tumorigenic in human. In fact, there are differences between the two sexes within the same species as regards to chemicals which may induce tumors through enzymatic and/or endocrine mechanisms. Also, there are organs such as the harderian and the Zymbal glands in rats, but not in human. Tumor of these organs may not have any bearing to human.

The general design for toxicity/oncogenicity testing is not for investigating the underlying mechanisms or resolving the shapes of any dose-response curves which may arise from such testing. As a result, finding an appropriate empirical or mechanistic model to describe such mechanisms and shapes is a difficult, if not impossible task. Also, often a design which may be optimal to identify statistically significant toxicity or carcinogenicity, is non-optimal for quantitative risk assessment. Works need to be done in these fields. One of the later chapters will attempt to address some of these issues.

Dose and risk scaling from rodent to human either assumes that "man is a bigger rat" or uncertainty factors. A lot more pharmacokinetic and physiological investigation must be performed to justify such assumptions or to derive more appropriate factors. Physiologically based pharmacokinetic modeling (to be described in some of the later chapters) may have a future in these areas.

ACKNOWLEDGEMENT

Arvind Parthasarathi, a high school student from India, was funded with a summer research fellowship by Center for Excellence in Education's Research Science Institute program for him to work under the senior author. The authors thank Miss Terry Horner for her help in preparing this manuscript.

REFERENCES

Ames, B. and Gold, L.S., 1990. Too many rodent carcinogens: Mitogenesis increases mutagenesis. Science **249**:970.

Armitage, P. and Doll, R., 1954. The age distribution of cancer and a multistage theory of carcinogenesis. Brit. J. Cancer **8**:1.

Boxenbaum, H., Neafsky, P.J., and Fournier, D.J., 1988. Hormesis, Gompertz functions and risk assessment. Drug Metab. Rev. **19**:195.

Carlborg, F.W., 1981. Dose-response functions in carcinogenesis and the Weibull model. Food Cosmet. Toxicol. **19**:255.

Calabrese, E.J., McCarthy, M.E., and Kenyon, E., 1987. The occurrence of chemically induced hormesis. Health Phys. **52**:531.

Cornfield, J., 1977. Carcinogenic risk assessment, Science **198**:693.

Crump, K.S., Hoel, D.G., Langley, C.H., and Peto, R., 1976. Fundamental carcinogenic processes and their implications for low dose risk assessment, Cancer Res. **36**:2973.

Doll, R., 1971 .Age distribution of cancer. J. Roy. Stat. Soc., Series A **134**:133.

Gehring, P.J. and Blau, G.E., 1977. Mechanisms of carcinogenesis: Dose-response. J. Env. Path. Toxicol. **1**:163.

Hartley, H.O. and Sielken, R.L., 1977. Estimation of 'safe doses' in carcinogenic experiments. Biometrics **33**:1.

Howe, R.B. and Van Landingham, C., 1986. Global 86. Clement and Associates, Inc., Ruston, Louisiana.

Kovar, J.G. and Krewski, D.R. "DORES81: A Computer Program for Low Dose Extrapolation of Quantal Response Toxicity Data" Biostatistics Section, Environmental Health Directorate, Health and Welfare Canada, Ontario, Canada.

Mantel, N. and Bryan, W.R., 1961. "Safety" testing of carcinogenic agents. J. Nat. Cancer Inst. **27**:455.

Neafsey, P.J., 1990. Longevity hormesis: A review. Mech. Age Develop. **51**:1.

Rai, K. and Van Ryzin, J., 1981. A generalized multi-hit dose-response model for low-dose extrapolation. Biometrics **37**:341.

Rai, K. and Van Ryzin, J., 1980. MULTI80: A Computer Program for Risk Assessment of Toxic Substances. A Rand Note, N-1512-NIEHS, Rand, Santa Monica, California.

Thakur, A.K., 1988. Modeling and Risk Assessment of Carcinogenic Dose-response. In: "Pharmacokinetics: Mathematical and Statistical Approaches to Metabolism and Distribution of Chemicals and Drugs" (A. Pecile and A. Rescigno, ed.), page 227. Plenum Press, New York.

THE VALUE OF BIOKINETIC DATA
IN HAZARD AND RISK ASSESSMENT

James T. Stevens and Darrell D. Sumner

CIBA-GEIGY Corporation
Greensboro, N.C. 27419

INTRODUCTION

Paracelsus, 16th century Swiss physician and recognized father of Toxicology, is quoted as stating [Klaassen and Doull, 1980] "All substances are poisons; there is none which is not a poison. The right dose differentiates between a poison and a remedy". Therefore, a functional definition for toxicity would state that virtually every substance possesses the potential to produce injury or death (hazard) if available in a sufficient amount (exposure). This idea encompasses not only the response to the substance but also the delivery of the dose to the target tissue(s), and hence implies hazard and exposure or risk. In the biological model, risk is determined by a dynamic sequence of events; biokinetics, whether restricted to a pharmacological response or an overdose expressing itself as a toxicological response, is central to understanding risk.

Biokinetics plays a critical role in dose response. Any thorough toxicity evaluation (Hazard Assessment) for a substance, and subsequently Risk Assessment, must consider biokinetics and biotransformation. Studies to examine these processes serve at least two general functions in the Risk Assessment process [Barnes and Dourson, 1988]. First, the data collected can be used to help elucidate the mechanism of toxicity, define target organs, and estimate critical exposure levels such as the Maximum Tolerated Dose (MTD) and the No-Observable-Effect Level (NOEL). The second is to help select the appropriate experimental animal model and procedures to extrapolate from animal toxicity data to man.

Throughout the course, the focus has been on pharmacokinetic and statistical techniques that can be used to elucidate our understanding of pharmaceutical agents. The discussions entailed a spectrum of subjects ranging from an appreciation of compartmental modeling to stereospecific and receptor kinetics. This chapter considers the value of such approaches in understanding chemical toxicity in the hazard and risk assessment processes. Examples of the pragmatic use of biokinetics ranging from the setting of dose levels to an understanding of the nontransparent alteration of complex systems, and the events of the interaction of a chemical with cellular constituents will be delineated.

New Trends in Pharmacokinetics, Edited by A. Rescigno and A.K Thakur
Plenum Press, New York, 1991

THE CONCEPT OF AN INTEGRATED APPROACH

Toxicology is the study of the nature, effects and detection of chemicals in living systems [Klaassen and Doull, 1980]. It is a multidisciplinary subject that includes the study of chemistry, biochemistry, biology, pharmacology, pathology, and other related medical sciences.

Chemical toxicity can manifest itself in many ways, the most serious of which is, of course, death. Also of major concern are cancer; reproductive effects (interference with fertility, pregnancy, fetal development, and development of the newborn); mutagenicity; debilitating effects (damage to organs and tissues); and irritation or allergic reactions.

Every chemical has the inherent ability to produce harmful effects in biological systems. The toxicity of any chemical is governed by its biological activity at a particular dose. It may be assumed that, in large enough doses, all chemicals are poisonous; conversely, there is likely to be a sufficiently small dose at which no toxic effect occurs.

To determine the "right dose", it is necessary to consider the duration of exposure, routes of exposure (ingestion, inhalation, skin contact), rates of absorption through the skin, lungs, or intestinal tract, distribution of the chemical to various tissues, chemical modification by bodily processes (metabolism and excretion), and the possibility of accumulation in tissues. In other words, the biokinetics and biodynamics of the substance in the body.

Interpretation of Toxicology Information

Risk is defined as a measure of the possibility of loss, injury, or harm, whereas safety is defined as a judgment of the acceptability of risk. Since the taking of risks, both personal and societal, is inherent in all human activity, eliminating all levels of risk is not possible.

Risk (R) is often described as a function (&) of the product of hazard (H) and exposure (E) or R = &(HE); where Hazard is the inherent potential to produce injury or death. Toxicity occurs if the amount of exposure is sufficient. This idea encompasses not only the event or response to the substance but also the delivery of the dose to the target tissue(s) and hence implies hazard and exposure or risk. The integrated process of assessing risk in chemical exposures involves four major components [Barnes and Dourson, 1988]: a) dose-response relationship, b) hazard identification, c) exposure assessment, and d) risk evaluation.

Hazard Identification

Hazard identification involves the overall evaluation of data from acute, subchronic, chronic, reproduction, teratogenicity, and mutagenicity studies. This evaluation begins with a careful examination of the validity and quality of each study, and comprehensive interpretation of results.

Hazard identification may involve defining the mechanism of toxicity, target organs, and estimated critical exposure levels such as the Maximum Tolerated Dose (MTD) and No-Observable-Effect Level (NOEL). The MTD is defined as the highest dose that can be administered to a test animal without significantly altering the accuracy and interpretability of the data obtained by simultaneously eliciting excessive toxicological effects, including mortality or by overwhelming normal metabolic processes [Paynter, 1985]. The NOEL is a level or dose at which no discernible response is noted. The driving force for much of the toxicity testing required under regulatory mandate is the completion of studies in an attempt to establish these levels.

To ensure minimal risk to man, man is assumed to be more sensitive than the most sensitive species tested. Therefore, the NOEL from the most sensitive species evaluated is divided by a "safety factor" of 10, 100, or 1000. This safety factor generally considers a 10-fold species sensitivity factor and an additional uncertain factor which may be 1 for cholinesterase inhibition, 10 for chronic effects, and 100 for more severe effects such as cancer. However, this approach is without a true biological foundation, but has a more perceptive or empirical basis which has been established as a tradition [Lehman and Fitzhugh, 1954; Barnes and Dourson, 1988].

The one major exception is the hazard of cancer. Unlike the other hazards, carcinogens are considered not to have thresholds or initiate a response in a dose-dependent manner. Instead, cancer-producing agents are either banned or regulated using a nonthreshold criterion.

In order to carefully evaluate the hazard of cancer, the International Agency for Research on Cancer (IARC) in 1978 established a more formalized approach often described as the weight-of-evidence criteria, which might be used to determine whether a chemical poses a carcinogenic (tumor-producing) hazard to humans [IARC, 1978]. This system has been adopted and/or adapted by several regulatory agencies including those in the United States [United States Environmental Protection Agency (USEPA), 1986].

Instead of using exclusively the animal cancer study results or limited data from human epidemiology studies, the following considerations are made based on the available evidence as described in Table 1.

Table 1

Weight-of-Evidence Approach for Classifying Carcinogens

o Carcinogenicity data (primary)

o Lifetime animal bioassays (verified in two species)

— Species and target specificity

— Type of response (benign versus malignant, common versus rare)

o Epidemiology (studies of exposed populations of humans)

Supplemental Data

o Mutagenicity and other short-term tests (for genetic effects)

o Structure-activity relationships

o Metabolic and biokinetic studies

Using such an approach, carcinogens can be classified and risk management criteria defined . This is shown in Table 2 for the USEPA [1986].

Dose-Response Relationship

The dose-response relationship usually demonstrates that the severity and/or incidence of a toxic effect increases in relation to increasing dose levels and exposure duration. Biokinetics, whether with a prefix of "pharmaco," originally used to describe the action of a drug, or with the prefix "toxico," presumably used to signify the biokinetics of a toxicant, is the study of the time course of all processes which determine the overall fate of a chemical in the body [Watanabe, Young, and Gehring, 1977]. However, often the dose-response factor which captures the kinetic feature is ignored for expediency.

Understanding the biokinetics of an agent at various dose levels is essential to assessing the hazard of the chemical [Van Der Heuden, 1988]. This is particularly important as regulatory guidelines frequently require testing at extremely high dose levels in the quest for the MTD discussed in Section 1.2. The current approach to setting an MTD estimate without using biokinetics and considering the overload phenomena is another example where regulation continues to collide with science and knowledge becomes the casualty [Stevens, 1986].

Table 2
USEPA Classification of Carcinogens

Carcinogen category	Criteria for classification	Evaluation	
		Safety factor	mathematical modeling
A - Human	- Sufficient evidence in man		X
B- Probable human	B1 - Limited evidence in man - Sufficient evidence in animal (two species with tumors)		X
	B2 - Inadequate human evidence - Sufficient animal evidence		X
C - Possible human	- No evidence in man - Limited evidence in animals	X	or X
D- Not classified	- Inadequate animal or human data	X	
E - No evidence of potential	- Sufficient animal testing and human experience	X	

Exposure Assessment

Exposure assessment involves determining the source and level of exposure to a chemical. Exposure may occur in the workplace, under conditions of use, and/or through residues in food or water. The general public is more likely to encounter chemical exposure through the intentional ingestion, absorption, and/or inhalation of pharmaceuticals or the unintentional ingestion, absorption, and/or inhalation of trace residues in food, water, or air of naturally occurring and man-made chemicals in our homes and environment .

Risk Assessment

The fifth component in evaluating risk is called Risk Assessment. This is the actual integration of the hazard, dose-response, and exposure information into a useful tool to limit hazard and exposure to a particular chemical so that the risk to man and his environment will be controlled to an acceptable level. Most often this involves defining a level of exposure at which the risk is deemed insignificant. In the workplace this might mean setting a threshold limit value for an industrial chemical. Establishing a tolerance or acceptable level in food or water might be the case for a naturally occurring material such as aflatoxin, a food or water additive such as fluoride or chloride, or a contaminant such as a pesticide. Usually these acceptable limits are based on a careful consideration of hazard, most often using data collected from animal models. Generally the hazard(s) is(are) assumed to be threshold(s), below which no effects are observed; this is the embodiment of the dose-response concept. To further integrate the idea of minimizing risk to man, the NOEL from the most sensitive species tested is used in combination with a safety factor as previously considered (see Section 1.2). Further as discussed, the safety factor tradition has been accepted for chronic toxicity, teratogenicity, and reproductive toxicity, but not carcinogenicity.

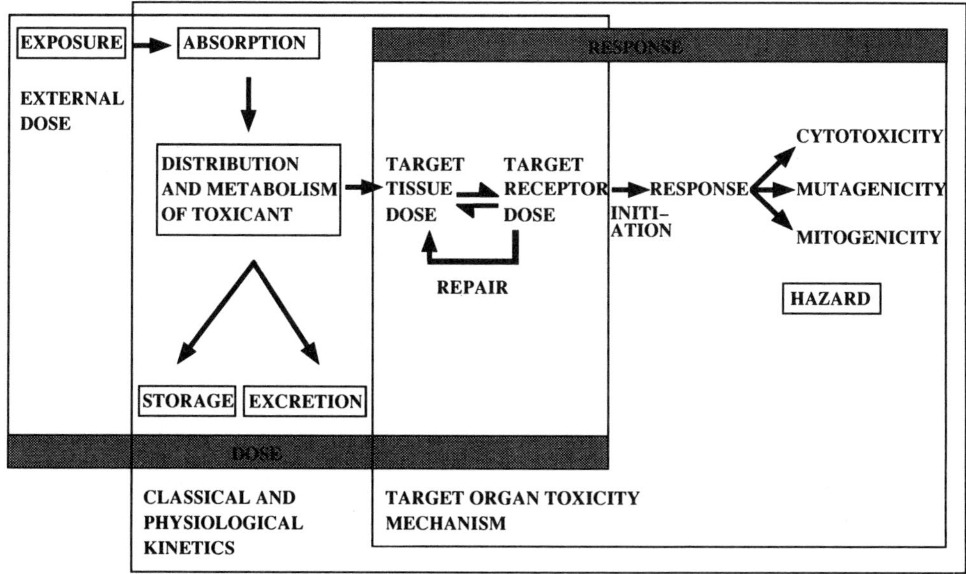

Figure 1. A Holistic Diagram of the Toxicological Process

Quantitative Risk Assessment, which attempts to mathematically model the relationship between dose and oncogenic response in animal studies and to extrapolate that data to determine the risk of developing cancer in humans, will not be discussed. A discussion of this approach can be found in Stevens and Sumner [1982]. Although this approach can add merit to the risk assessment process, the current approach taken by some regulatory agencies diminishes the potential value-added factor [USEPA, 1986].

An attempt to combine the kinetics, risk assessment, dose-response hazard evaluation, and toxicological process into a package is presented in Figure 1.

CHARACTERIZATION OF PROPERTIES OF ABSORPTION, DISTRIBUTION, METABOLISM, AND ELIMINATION (ADME) — DOSE-DEPENDENT BIOKINETICS

Understanding of the pharmacokinetics of a chemical at various dose levels is important in assessing hazard, particularly in light of the fact that chemicals are tested frequently for toxicity in animals at extremely high doses in order to achieve an MTD [Watanabe, Young, and Gehring, 1977; WHO, 1986; Klaassen, 1980]. This becomes very important when it is realized that ADME for many chemicals are saturable, that is, following non linear dose-dependent kinetics [Ramsey and Gehring, 1979]. Therefore, ADME should be pursued for each chemical as a standard feature in safety assessment seeking indication of nonlinear biokinetics over a broad dose range such as described in Table 3.

When one or more of the ADME processes are capacity-limited, that is, saturable, the internal concentration of the chemical and/or its metabolites may not be directly related to the administered dose. When this happens the resulting toxicity (if any) may or may not be relevant to the test animal or man. Therefore, the use of saturation points for capacity-limited processes in dose selection may distinguish between a rational approach to safety evaluation and a flirtation with overdose phenomena, i.e., toxicokinetics.

Table 3
Features Indicating Nonlinear Biokinetics

o Elimination cannot be described by a single exponential process

o The elimination half-life increases with dose

o The area under the plasma concentration versus time curve (avg.) changes with increasing dose

o The composition of excretion products changes qualitatively or quantitatively with increasing doses

This saturation dose established by biokinetics may be lower than a toxicologically defined MTD [Munro, 1977; Apostolou, 1990].

If one reflects on the kinetic rate equation [Ramsey and Gehring, 1979]:

$$R = \frac{V_{max}C}{K_{max} + C}$$

where R equals the rate
    V_{max} equals the maximum velocity
    C equals the concentration of chemical; and
    K_m equals the concentration of chemical at which the rate is
      one half that maximally possible (Michaelis-Menten constant).

• When C is much greater than its K_m value for a biological process, the rate will be essentially fixed as C changes; hence, R is considered zero order (constant).

• When C is much less than its K_m value, then the rate of the process is nearly directly proportional to the changing C; R is considered first order.

Biologically, enzymatic reactions are often saturable, zero order processes. Saturable processes can be capacity-limited in vivo [Gillette, 1976; Clewell and Anderson, 1987]. Examples of saturable processes are delineated in Table 4.

Table 4
Zero Order Processes That Can Be Capacity-Limited

o Active transport systems

o Enzymatically mediated biotransformation reactions

- Enzymatic-metabolic overload (alternative pathways)

- Cofactor depletion (glutathione, etc.)

- Inhibition of metabolism (suicide inhibitors)

o Repair of damaged DNA

o Protein and/or macromolecular binding

o Resynthesis of cellular components

o Biliary excretion

o Tubular reabsorption

Since the behavior of a chemical and/or its metabolites in the body is dependent on the administered dose, with repeated administration the concentration of chemical (C) in the body will generally increase until a steady state is achieved. Again, when

320

C is handled in the body by first order processes or C is well below the K_m values for capacity-limited processes, the distribution as well as the metabolite levels of the chemical will remain directly proportional to the dose level. In such a system, if a toxic response is elicited, a linear association between C and the toxicity would be anticipated. When the zero order process is saturated, i.e., C in the body approaches or exceeds the K_m value for any saturable process, then linearity for that process is lost; C at which this occurs in relationship to the toxicity manifested will fix the utility of this value in the dose selection process.

DEFINITION OF CRITICAL EXPOSURE VALUES SUCH AS THE MTD AS WELL AS TARGET ORGAN AND MECHANISM OF TOXICITY

The discussion to follow will utilize several examples which demonstrate how biokinetic features, in particular, zero order, dose-dependent saturation kinetics, are of value in performing risk assessments and understanding the toxicological process.

SATURATION KINETICS

Saturation of absorption kinetics at high dosage levels

CHEMICAL: Simazine

DESCRIPTION: Simazine is a symmetrical chlorotriazine herbicide used in agriculture for broadleaf-weed control. The mechanism for its herbicidal activity is inhibition of photosynthesis.

CHEMICAL PROPERTIES: Simazine is relatively insoluble in water (3.5 ppm) and has a Log P of 4.18. Its molecular weight is 201.7.

HAZARD PROFILE: The hazard profile for simazine has been carefully evaluated and it is devoid of mutagenic, teratogenic, reproductive, and oncogenic toxicity except in the female Sprague-Dawley rat. The response noted in the female rat is an increase in the incidence of highly spontaneously occurring mammary tumors. Evidence has been presented [Wetzel , et al., 1990] that supports that simazine at high levels binds weakly, but competitively to estrogen receptors and endocrine tissue in the rat, altering normal hormonal balance and resulting in an apparent precocious senescence of the reproductive system in this strain of rat. This response was noted after lifetime administration of simazine at 5 mg/kg/day and 50 mg/kg/day; both levels exceeded the MTD as evidenced by an elevation in mortality.

ADME: Simazine represents an example of saturation of absorption at the high levels of administration. Simazine is rapidly metabolized in the rat with the majority of the administered dose excreted within the first 24 hours. The major pathways of metabolism are dechlorination, N-dealkylation, mercapturic acid formation, glutathione conjugation, and oxidation of amino group. However, these metabolic responses occur only if the compound is absorbed from the gastrointestinal tract. There is experimental evidence [Orr and Simoneaux , 1986; Orr, 1985] that there is a dose-dependent elimination process as the dose is increased. This phenomenon is presented in Figure 2 for female rats.

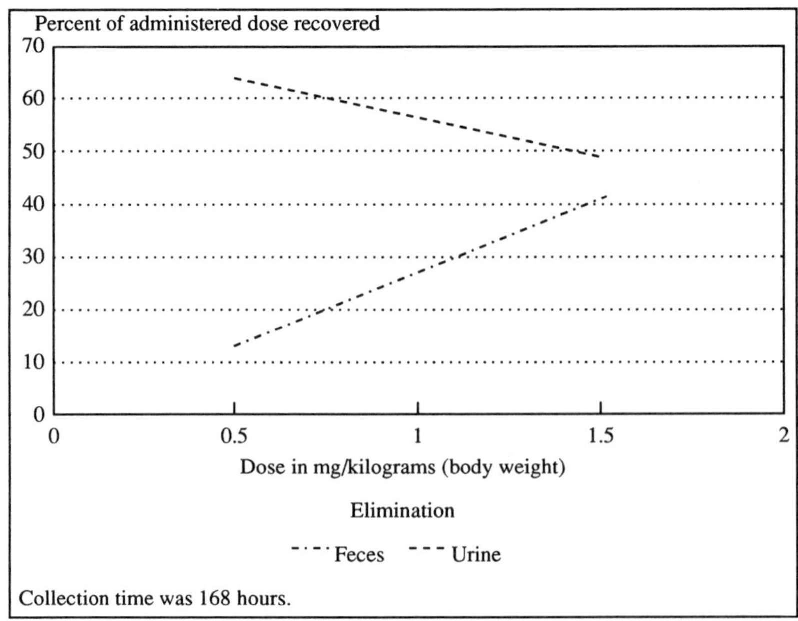

Figure 2. Dose-Dependent Elimination of Simazine Single Dose (po)
to Female Rats. Collection time was 168 hours.

It can be seen that with a shift of the C^{14}-dose of simazine from 0.5 mg/kg to
1.5 mg/kg the percentage of the dose recovered in the urine decreased from 62% to
49%. On the other hand, the percentage found in the feces increased from 13% to
42%.

This simple biokinetic assessment enables the toxicologist to realize that the
administration of high doses of simazine may not reach target tissues and is of value
in selecting levels of administration for repeat-dose studies.

SATURATION OR INHIBITION OF ACTIVE TRANSPORT

CHEMICAL: Sulfinpyrazole [Brazeau,
1970]

DESCRIPTION: Sulfinpyrazole is a pharmaceutical agent which is indicated for the
treatment of gout; it is a uricosuric agent.

CHEMICAL PROPERTIES: Sulfinpyrazole
is a strong organic acid with a pK_a of 2.8. It is rather soluble in water and has a
molecular weight of 404.7.

HAZARD PROFILE: Sulfinpyrazole belongs to the pyrazole class of compounds.
There are inconclusive results in regard to teratogenic potential of this chemical in
animals. Sulfinpyrazone has been shown to cause gastrointestinal irritation and
depression of hematopoiesis.

ADME: Sulfinpyrazole is well absorbed after oral administration. It is oxidized to
form the sulfoxide sulfinpyrazone. It is strongly bound to plasma albumin; it is

322

excreted by kidney glomerules, secreted by proximal tubules, and is a potent inhibitor of renal tubular reabsorption.

Sulfinpyrazole is example of an inhibitor of active transport and zero order kinetics.

ALTERED METABOLISM AT HIGH DOSAGE LEVELS

Formation of a new metabolite

CHEMICAL: Primisulfuron [USEPA, 1990]

DESCRIPTION: Primisulfuron is a recently introduced sulfonylurea herbicide.

CHEMICAL PROPERTIES: Primisul-furon is moderately soluble in water (23 ppm). It has a Log P value of 1.15 and a pK_a of 5.1. Primisulfuron has a molecular weight of 468.3.

HAZARD PROFILE: Primisulfuron is not acutely toxic and is not genotoxic or mutagenic in standard assays. Primisulfuron is not a developmental toxicant, it is not teratogenic in the rat or rabbit, and it does not have any adverse effect on reproduction in the rat. The liver, kidneys, testes, and bone have been identified as target organs in subchronic and chronic toxicity studies conducted in the rat, mouse, and dog. The No-Observable-Effect Level (NOEL) in the most sensitive species was established at 300 ppm (15 mg/kg) in a chronic rat study. An increased incidence of liver tumors was observed in male and female mice at feeding levels that were near to

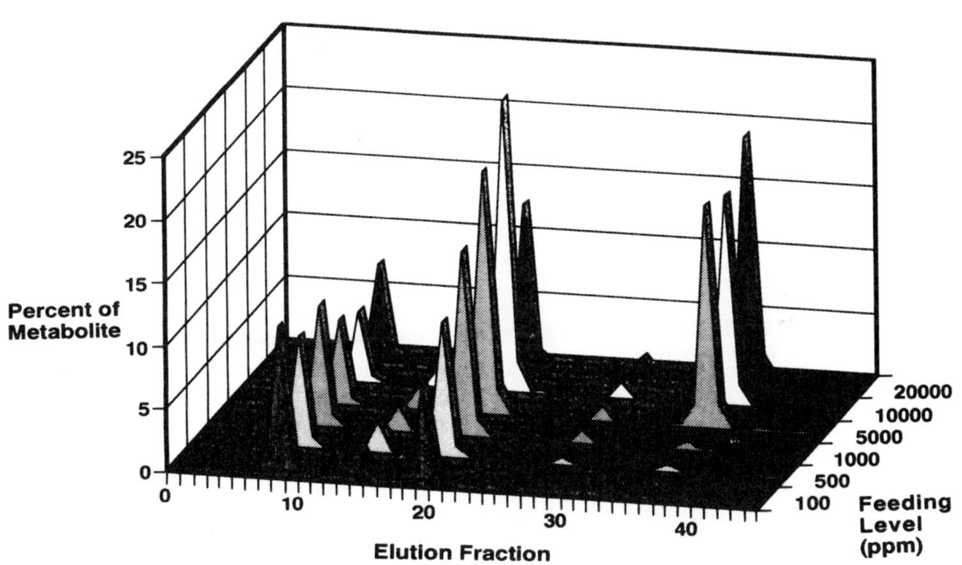

Figure 3. Radiochromatograms Depicting the Metabolite Profile of Primisulfuron in the Feces of Male Mice

or that exceeded the Maximum Tolerated Dose (MTD). This response was discounted as the MTD was exceeded. There was no evidence of an oncogenic response in the rat at the maximum tolerated dose of 8000 ppm.

BIOKINETICS: Metabolism studies suggest that different metabolic pathways may be involved at high feeding levels in the mouse which apparently are not operative at low feeding levels.

Figure 3 depicts a series of radiochromatograms for primisulfuron metabolites in feces of male mice. It can be seen that as the dose increases above 5000 ppm a new metabolite is formed.

Primisulfuron provides an example of nonlinear biokinetics characterized by a qualitative change in excretion products with increasing doses.

COFACTOR DEPLETION

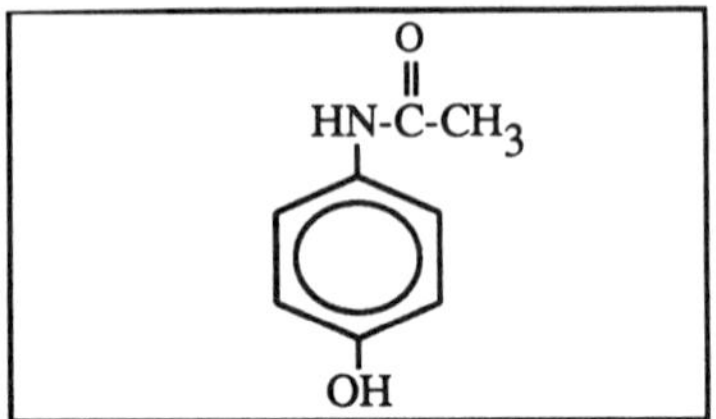

CHEMICAL: Acetaminophen [Insel, 1990]

DESCRIPTION: Acetaminophen is an analgesic agent that is sold without a prescription.

CHEMICAL PROPERTIES: Acetaminophen is a weak acid. Its molecular weight is 151.2. It is only slightly soluble in water.

HAZARD PROFILE: Doses of acetaminophen above 25 g are potentially fatal to adults. The severity of the initial symptoms may belie the potential hazard. Initial reactions are varied but usually include nausea, vomiting, anorexia, and a general feeling of malaise, not unlike the common flu. Delayed hepatotoxicity may occur in three to five days with potentially fatal results if not treated. The predominant toxic effect is centrilobular hepatic necrosis, but renal tubular necrosis may also occur.

ADME: Acetaminophen is metabolized primarily by the formation of glucuronide and sulfate conjugates. A small amount may be excreted unchanged and about 4% is biotransformed by the oxidative metabolism system to a reactive quinoid-type structure. When high exposure levels occur, the hepatic glutathione is consumed by the protective processes and the oxidation product begins to accumulate and is available to react within the cell. Animal data suggest that this occurs when approximately 70% of the hepatic glutathione is consumed.

Since it appears that the toxic effect is so intrinsically connected to the cellular level of glutathione, the treatment for intoxication utilizes replacement of this cofactor.

N-Acetylcystiene is administered to provide a sulfhydryl-containing amino acid to at least partially supplement the depleted cofactor.

In the evaluation of risk it is important to recognize situations where cofactor depletion is responsible for toxicity. From a public health standpoint, better understanding leads to better policies. From an individual standpoint, care can be taken to ensure those individuals with altered metabolic states are properly protected.

The depletion of protective cofactors is a mechanism by which toxins can present an unusual or exaggerated dose-response relationship. It is by such a mechanism that acetaminophen provides adequate margins of safety at therapeutic doses but substantial toxicity at exaggerated exposures.

CHEMICAL: Bromobenzene [Plaa, 1990; Levi, 1987]

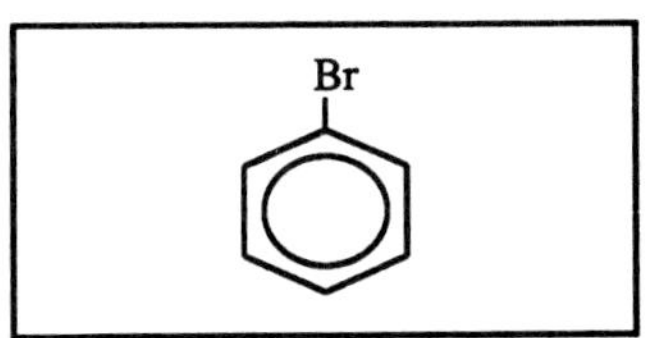

DESCRIPTION: Bromobenzene is used as a solvent and an additive to motor oils.

CHEMICAL PROPERTIES: The molecular weight of bromobenzene is 157. It is practically insoluble in water (0.45 ppm).

HAZARD PROFILE: Bromobenzene produces centrilobular hepatic necrosis at high doses but not at low levels of exposure. Its toxicity is enhanced by pretreatment with phenobarbital and decreased by co-administration of inhibitors of oxidative metabolism. This suggests the formation of reactive metabolites but that conclusion is complicated by the observation that 3-methylcholanthrene pretreatment also decreases toxicity. Glutathione is implicated because depletion of glutathione increases toxicity.

ADME: The basic metabolism of bromobenzene is shown in Figure 4.

Bromobenzene is metabolized by epoxidation to form a 2,3 epoxide and a 3,4 epoxide. The 2,3 epoxide degrades to form 2-bromophenol. The 3,4 epoxide degrades to 4-bromophenol, the corresponding diol, and glutathione conjugates. In addition, covalent binding to cellular macromolecules is observed at toxic doses.

Induction of cytochrome P-448 by pretreatment with 3-methylcholanthrene decreases toxicity and increases the formation of the 2,3 epoxide and its daughter product, 2-bromophenol. Induction of cytochrome P-450 by pretreatment with phenobarbital increases toxicity as well as the formation of the 3,4 epoxide and its daughter products, including the material covalently bound to the cellular macromolecules.

From these observations several conclusions can be made: 1) toxicity can be correlated with the formation of a minor but reactive intermediate, 2) a threshold tissue concentration of the reactive intermediate must be attained before injury occurs, 3) glutathione protects the cell from injury by removing the reactive intermediate, 4) other pathways such as the 2,3 epoxidation and epoxide hydrolase play an important role in protecting the cell, and 5) agents that selectively induce metabolic pathways may decrease or enhance toxicity.

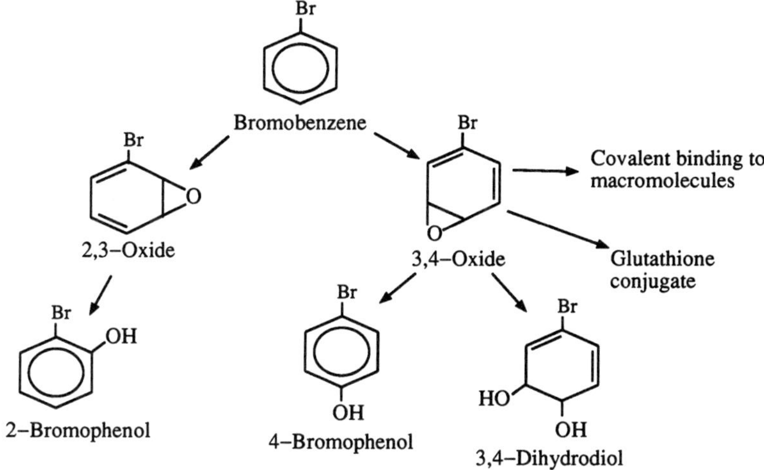

Figure 4. Metabolism of Bromobenzene.

Saturation of several protective mechanisms in the cell accounts for the toxicity of bromobenzene. Only by understanding the biokinetics of these processes can an accurate assessment of the risks be determined.

FORMATION OF ACTIVE METABOLITES

Reactive metabolites

CHEMICAL: Vinyl Chloride [Torkelson and Rowe, 1981; Andrews and Snyder, 1986]

DESCRIPTION: Vinyl chloride is used as a chemical intermediate in the plastics industry, as a refrigerant, and in organic synthesis.

CHEMICAL PROPERTIES:HCHHCClVinyl chloride is a colorless gas with a molecular weight of 62.5. It is sparsely soluble in water (2.5 ppm at 250C).

HAZARD PROFILE: Vinyl chloride has been shown to produce liver damage and tumors in laboratory animals. It has been shown to be genotoxic in mutagenicity tests. Most importantly, the empirical finding has been observed in man where hepatic angiosarcomas have been noted in those occupationally exposed to high concentrations.

ADME: The basic metabolism of vinyl chloride is shown in Figure 5.

The chloroacetaldehyde is a known mutagen. The toxic effects are thought to develop after liver glutathione levels have been sufficiently depleted that protective conjugation no longer occurs.

Vinyl chloride addresses the importance of understanding metabolic activation in toxicology and risk assessment. The toxicity more closely follows epoxide formation than the dose of vinyl chloride.

SELECTION OF THE APPROPRIATE ANIMAL MODEL AND EXTRAPOLATION

As discussed in the introduction [Barnes and Dourson, 1988], biokinetics can serve at least two functions in risk assessment. The first function, to elucidate the mechanism of toxicity , define target organs, and estimate critical exposure levels, has been discussed briefly with some examples of utilizing this approach provided. The second function, the selection of the appropriate animal model, will now be

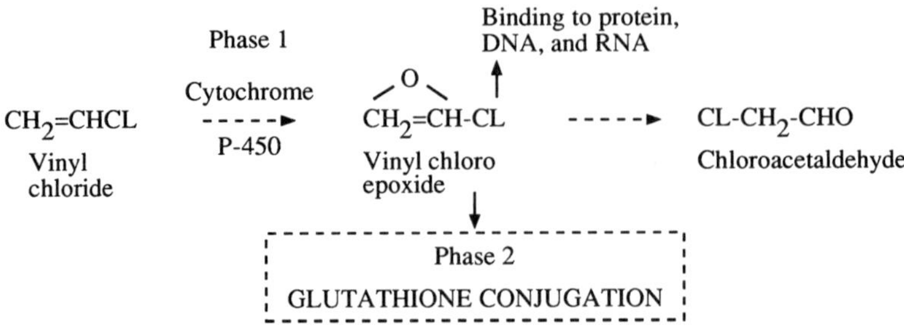

Figure 5. Metabolism of Vinyl Chloride.

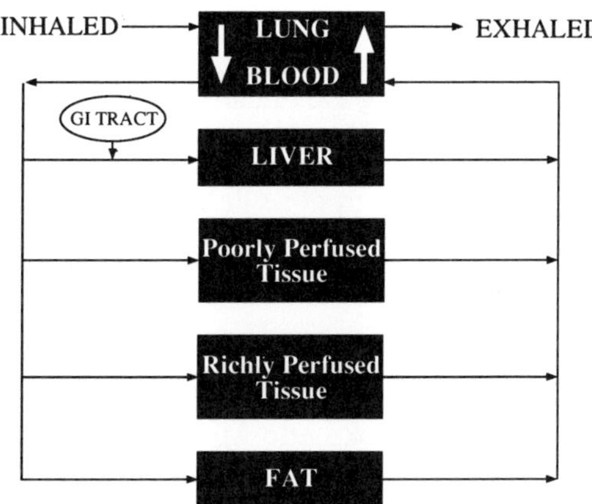

Figure 6. Schematic Representation of Different Regions of the Body Arranged in a Flow Diagram Which Constitutes a Physiological Pharmacokinetic Model

considered. Excellent reviews of this subject include Vocci and Farber [1988] and Clewell and Anderson [1987].

Use of Physiological Kinetics.

With increased availability of faster, more powerful computers it has become possible to develop biokinetic models which more closely approximate the physiology of the organism. That is, it is possible to model multiple organs, multiple perfusion rates, varying partition coefficients, organ-specific metabolism, and maximum rates for saturable processes. These models are commonly described as physiologically based pharmacokinetic models. An example is depicted in Figure 6.

This diagram illustrates the principles of the physiologically based pharmacokinetic models. Each equilibrium can be independently estimated and Michaelis-Menten type considerations incorporated for each saturable process [after Gibaldi and Perrier, 1982] . Of course, the determination of more variables is required than when using the more conventional one- or two-compartment models.

CHEMICAL: 1,4-Dioxane [Leung and Paustenbach, 1990]

DESCRIPTION: 1,4-Dioxane has been widely used as a solvent, preservative, fumigant, and deodorant.

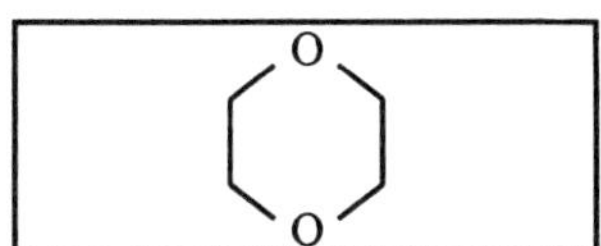

CHEMICAL PROPERTIES: 1,4-Dioxane has a molecular weight of 88.1. It is soluble in water.

HAZARD PROFILE: 1,4-Dioxane may cause CNS depression and necrosis of the liver and kidney. Prolonged or repeated exposure can cause death resulting from renal failure. It has been shown to produce hepatomas, hepatocellular carcinomas, and nasal cavity tumors in Sprague-Dawley rats.

ADME: 1,4-Dioxane is readily absorbed and is metabolized in the liver by cytochrome P-450 to b-hydroxy-ethoxyacetic acid, which is excreted. The metabolism of dioxane is dose-dependent due to saturation of the metabolic enzymes at high levels.

Physiologically based pharmacokinetic models have been used by Leung and Paustenbach [1990] to describe tissue concentrations and risks of dioxane exposure.

Estimation of human risks using the linearized multistage extrapolation does not consider that metabolism of dioxane is saturable. Leung and Paustenbach [1990] present an alternative approach to risk assessment incorporating a quantitative relationship between animal dose and dose delivered to the target organ as well as consideration of the saturable metabolic processes. Parameters used in the model are shown in Table 5.

Table 5

Parameters Used in the Physiologically
Based Pharmacokinetic Model for Dioxane
[Leung and Paustenbach, 1990]

Parameter	Species Rat	Human
Weight		
Body (kg)	0.25	84.1
Liver (%)	4	4
Fat (%)	7	20
Richly perfused (%)	5	5
Slowly perfused (%)	75	62
Blood flow		
Cardiac output (l/hr)	5.4	399
Liver (%)	25	25
Fat (%)	5	5
Richly perfused (%)	51	51
Slowly perfused (%)	19	19
Alveolar ventilation (l/hr)	5.4	399
Partition coefficient		
Liver/blood	0.85	0.85
Fat/blood	0.4	0.4
Richly perfused/blood	0.85	0.85
Slowly perfused/blood	0.54	0.2
Metabolic constant		
V_{max}	1.9	300
K_{max}	7.5	15

Organ weights and perfusion rates are standard values. Tissue/air partition coefficients were determined experimentally. Tissue/blood partition coefficients were calculated from the tissue/air and blood air partition values. Richly perfused tissues were considered equivalent to liver. For comparison of human and rodents the tissue/blood partition coefficients were considered equal while alveolar ventilation and cardiac output were considered proportional to body surface area.

Metabolism constants V_{max} and K_m were obtained by optimization of the model with experimental data. Notably the model appeared to demonstrate sufficient sensitivity that at doses >300 mg/kg the V_{max} for rats was increased. The authors proposed this could be a response to enzyme induction.

Their model applied to the data of Watanabe, Young, and Gehring [1977] showed reasonably good correlation between predicted and observed values in the curve of urinary excretion of b-hydroxy-ethoxyacetic acid as a metabolite of dioxane.

The physiologically based pharmacokinetic model showed that conventional approaches to risk assessment as used in the U.S. overestimated risks by as much as 80-fold when compared to the time-weighted tissue concentration predicted by the model.

USE OF CLASSICAL KINETICS

CHEMICAL: Atrazine [Orr, et al., 1987; Davidson, 1988]

DESCRIPTION: Atrazine is a symmetrical chlorotriazine herbicide used for control of broadleaf weeds. It is the largest-selling chemical of this type in the world. It exerts its phytotoxic effects through inhibition of the Hill Reaction in photosynthesis.

CHEMICAL PROPERTIES: Atrazine is relatively soluble in water (33 ppm at 220C) and has a Log P value of 2.75. Its molecular weight is 215.7.

HAZARD PROFILE: Atrazine is not acutely toxic, not a mutagen, not a teratogen, and not a reproductive hazard . Atrazine is not oncogenic in mice of either sex or male rats; however at high feeding levels, female rats exhibit an increase in the incidence of mammary tumors.

The mechanism of this enhancement in the appearance of this highly spontaneous tumor is endocrine-mediated [Wetzel, et al., 1990]. Atrazine appears to be both agonistic and antagonistic to estrogen in the Sprague-Dawley rat.

ADME: Atrazine metabolism was evaluated in Sprague-Dawley rats [Orr, et al., 1987]. Both 1 and 100 mg/kg of atrazine were administered acutely and 1 mg/kg was administered daily for 14 days. Approximately 74% of the administered atrazine was eliminated by the kidneys with 19% being found in the feces; the remainder was found as tissue residues. No significant differences were found in the pattern of urinary or fecal elimination for low and high dose, single or repeat administration, in male or female. The primary chlorotriazine metabolites of atrazine found in the urine were the two monodealkylated forms (% of the administered dose) and the didealkylated material. Kinetically the didealkylated data best fit a linear two-compartment open model. The alpha (distributive and excretory phase) and the beta (whole body renal elimination phase) were 6.9 hours + 0.2 (SE) and 31.1 hours + 2.8 (SE), respectively. The half-life of renal elimination from the central compartment (K_{10}) was 12.4 hours + 1.3 (SE). The equation for the kinetic model for atrazine elimination of the didealkylated metabolite in the rat is:

$$A(t) = A \exp(-\alpha t) + B \exp(-\beta t)$$

where $A(t)$ is the amount in the body at any time. No differences were found between males and females. A dominant factor in the residue was the concentration found in the red blood cells. This apparent species-specific response is likely due to covalent binding of the triazine ring to exposed cysteine sulfhydryls in the rat hemoglobin chain [Hamboeck, et al., 1981] and explains the peripheral compartment seen in the rat. This second compartment would certainly lengthen the duration of exposure to atrazine metabolites in the rat.

Indeed, the calculated volume of distribution (V_d), as well as the half-time and predicted mean steady-state concentration with repeated administration of atrazine, supports the red blood cell as the key component of the peripheral compartment [Thede and Kahrs, 1987].

This differs remarkably from what has been found when atrazine was administered orally [Davidson, 1988] to six human males at 0.1 mg/kg. The overall kinetics for urinary elimination of the didealkylated metabolite for each of the six subjects is best described by a single-compartment first order model where:

$$A(t) = A_0 \exp(-k_e t)$$

The monodealkylated metabolites were best described biokinetically as two-compartment first order models.

Unlike the Sprague-Dawley rat, man appears to have only a central compartment based on urinary excretion; the half-life for renal elimination of atrazine or its metabolites averages 11.5 ± 0.4 hour (SE) in man.

In human blood, atrazine is primarily found as the deethylated and didealkylated metabolites; no deisopropylated metabolism is seen [Davidson, 1988]. The deethylated metabolite is best described as a two-compartment model (most likely its formation and later conversion to the didealkylated material; the didealkylated metabolite is a single-compartment model.

The first order nature of didealkylated biokinetics as described for blood disappearance and urine elimination indicates P-450 oxidative dealkylation of atrazine is first order and not saturated [Davidson, 1988].

These metabolism data illustrate basic differences between the way atrazine is distributed, metabolized, and excreted in the Sprague-Dawley rat and man. Considering these findings and the toxicologic observations, the results in the Sprague-Dawley rat do not appear relevant to humans.

Classical kinetics can be used to uncover species differences of toxicological significance when first order kinetics predominate.

CONCLUSION

Despite the arduous road that has been traversed, it should be apparent that the use of biokinetics in risk assessment is critical and mandatory if we are to make progress. To ignore biokinetics is to ignore a basic tenet upon which the foundation of toxicology is based, i.e., Dose-Response. As toxicologists, it is imperative that we incorporate these techniques into our procedures for analysis and interpretation. As risk assessors and modelers, it is vital that we do not let ourselves be dulled by overly simplistic approaches such as the linearized multistage model [USEPA, 1986], but insist that good science prevail and that knowledge is based on wisdom and not ignorance.

REFERENCES

Adams, N. H., Levi, P. E., and Hodgson, E., 1990. In vitro Studies of the Metabolism of Atrazine, Simazine, and Terbutryn in Several Vertebrate Species. J. Agri. Fd. Chem. **38**:6, 1411.

Andrews, and Snyder, R., 1980. Toxic Effects of Solvents and Vapors. Chapter 20. In: "Casarett and Doull's Toxicology: The Basic Science of Poisons" (C. D. Klaassen, M. O. Amdur, and J. Doull, eds.), Chapter 20, page 636. Third Edition. Macmillan Publishing Co., Inc., New York.

Apostolou, A., 1990. Revelence of Maximum Tolerated Dose to Human Carcinogen Risk. Regul. Toxicol. Pharmacol. **11**:1, 68.

Barnes, D. G., and Dourson, M., 1988. Reference Dose (RfD): Description and Use in Health Risk Assessments. Regul. Toxicol. Pharmacol. **8**, 471.

Bogen, K. T., 1988. Pharmacokinetics for Regulatory Risk Analysis: The Case of Trichloroethylene. Regul. Toxicol. Pharmacol.**8**:447.

Brazeau, P. Inhibitors of Tubular Transport of Organic Compounds. In: "The Pharmacological Basis of Therapeutics" (L. S. Goodman and A. Gilman, eds.), Chapter 41, page 886. 4th edition. The MacMillan Company, New York, Toronto.

Cheung, M. W., and Kahrs, R. A., 1990. Analysis of Human Urine to Determine Residues of Atrazine, G-28273, G-28276, and G-30033 Resulting from Oral Ingestion of Atrazine, Including Storage Stability Results. CIBA-GEIGY Report ABR-90034 (Unpublished).

Clewell, H. J. III, and Anderson, M. E., 1987. Dose, Species, and Route Extrapolation Using Physiologically Based Pharmacokinetic Models. In: "Pharmacokinetics in Risk Assessment: Drinking Water and Health", Vol. 8, page 159. National Academy Press, Washington, D.C.

Conolly, R. B., 1990. Biologically-Based Models for Toxic Effects. Tools for Hypothesis Testing and Improving Health Risk Assessments. CIIT Activities. Chemical Industry Institute of Toxicology. Research Triangle Park, North Carolina **10**:5, 1.

Davidson, I. W. F., 1988. Metabolism and Kinetics of Atrazine in Man. Project 101947. Bowman Gray School of Medicine (Unpublished).

Dietz, R. A., Reitz, R. H., Watanabe, P. G., and Gehring, P. J., 1982. "Translation of pharmacokinetics/biochemical data in risk assessment. In: "Biological Active Intermediates. II. Chemical Mechanisms and Biological Effects" (R. Synder, D. V. Parke, J. J. Kocsis, D. J. Jollow, C. G. Gibson, and C. M. Witmer, eds.), page 1399. Plenum, New York.

Di Simplicio, P., Dolara, P., and Lodovici, M., 1984. Blood Glutathione as a Measure of Exposure to Toxic Compounds. J. Appl. Toxicol.**4**:227.

Doull, J., 1980. Factors Influencing Toxicology. In: "Casarett and Doull's Toxicology: The Basic Science of Poisons" (J. Doull, C. D. Klaassen, and M. O. Amdur, eds.), Chapter 5, page 70. Second Edition, Macmillan Publishing Co., Inc., New York.

Gehring, P. J., Watanabe, P. G., and Park, C. N., 1976. Risk of Angiosarcomas in Workers Exposed to Vinyl Chloride from Studies in Rats. Toxicol. Appl. Pharmacol.**37**:49.

Gibaldi, M., and Perrier, D., 1982. "Pharmacokinetics". In: "Drugs and the Pharmaceutical Sciences (J. Swarbrick, ed.), Chapter 7, page 271. 2nd Ed., Marcel Dekker, New York.

Gibaldi, M., and Perrier, D., 1982. "Pharmacokinetics". In: "Drugs and the Pharmaceutical Sciences (J. Swarbrick, ed.), Chapter 9, page 355. 2nd Ed., Marcel Dekker, New York.

Gillette, J. R., 1976. Application of Pharmacokinetic Principles in the Extrapolation of Animal Data to Humans. Clin. Toxicol. **9**:709.

Grushenka, H. I., Wolfgang, A., Gandolfi, J., Stevens , J. L., and Brendel, K., 1989. N-Acetyl S-(1,2-dichlorovinyl)-L-Cysteine in Rabbit Renal Slices: Differential Transport and Metabolism. Toxicol. Appl. Pharmacol.**101**:2, 205.

Hamboeck, H., Fischer, R. W., DiIorio, E. E., and Winterhalter, K. H., 1981. The Binding of s-Triazine Metabolites to Rodent Hemoglobins Appears Irrelevant to Other Species. Molecular Pharmacology **20**:579.

Insel, P., 1990. Analgesic-Antipyretic and Anti-inflammatory Agents: Drugs Employed in the Treatment of Rheumatoid Arthritis and Gout. In: "The Pharmacological Basic of Therapeutics (L. S. Goodman and A. Gilman, eds.), page 638. 8th Edition. Pergamon Press, New York.

International Agency for Research on Cancer (IARC), 1978. International Agency for Research on Cancer Monographs on the Evaluation of the Carcinogenic Risk of Chemicals to Humans **17**:1.

Jollow, D. J., Thorgeirsson, S. S., Potter, W. Z., Hashimoto , M., and Gillette, J.R., 1974. Acetaminophen-induced hepatic necrosis. VI. Metabolic disposition of toxic and nontoxic doses of acetaminophen. Pharmacology (Basel) **12**:251.

Jollow, D. J., Roberts, S., Price, V., and Smith, C., 1982. Biochemical basis for dose response relationships in reactive intermediate toxicity. In: "Biological Active Intermediates II. Chemical Mechanisms and Biological Effects (R. Synder, D. V. Parke, J. J. Kocsis, D. J. Jollow, C. G. Gibson, and C. M. Witmer, eds.), page 99. Plenum, New York.

Klaassen, C. D., and Doull, J., 1980. Evaluation of Safety: Toxicologic Evaluation In: "Casarett and Doull's Toxicology: The Basic Science of Poisons (J. Doull, C. D. Klaassen, and M. O. Amdur, eds.), Chapter 2, page 12. Second Edition, Macmillan Publishing Co., Inc., New York.

Klaassen, C. D.1980. Absorption, Distribution, and Excretion of Toxicants. In: "Casarett and Doull's Toxicology: The Basic Science of Poisons (J. Doull, C. D. Klaassen, and M. O. Amdur, eds.), Chapter 3, page 28. Second Edition, Macmillan Publishing Co., Inc., New York.

Lehman, A. J., and Fitzhugh, O. G., 1954. 100-Fold Margin of Safety. Assoc. Food Drug Off., U.S.Q. Bulletin **18**:33.

Leung, H. W., and Paustenbach, D. J., 1990. Cancer risk assessment for dioxane based upon a physiologically-based pharmacokinetic approach. Toxicol. Letters **51**:147.

Levi, P., 1987. Toxic Action. In: "A Textbook of Modern Toxicology (E. Hodgson and P. Levi, eds.), page 133. Elsevier, New York.

Munro, I. C.,1977. Considerations in Chronic Toxicity Testing: The Chemical, The Dose, The Design. J. Environ. Pathol. Toxicol. **1**:183.

Orr, G. R., Simoneaux, B. J., and Davidson, I. W. F., 1987. Disposition of Atrazine in the Rat. CIBA-GEIGY Report ABR-87048 (Unpublished).

Orr, G. R., 1985. Review of Simazine Metabolism in the Rat. CIBA-GEIGY Report ABR-85052 (Unpublished).

Orr, G. R., and Simoneaux, B. J., 1986. Deposition of Simazine in the Rat. CIBA-GEIGY Report ABR-86032 (Unpublished).

Paynter, O. E., 1985. "Hazard Evaluation Division, Standard Evaluation Procedure. Oncogenicity Potential (Guidance for Analysis and Evaluation of Long-Term Rodent Studies)." U.S. Environmental Protection Agency, Washington.

Plaa, G. L., 1980. Toxic Responses of the Liver. In: "Casarett and Doull's Toxicology: The Basic Science of Poisons (J. Doull, C. D. Klaassen, and M. O. Amdur, eds.), Chapter 9, page 286. Second Edition. Macmillan Publishing Co., Inc., New York.

Ramsey, J. C., and Gehring, P. J., 1979. Application of Pharmacokinetic Principles in Practice . In: "Extrapolation of Laboratory Toxicity Data to Man: Factors Influencing the Dose-Toxic Response Relationship". An ASPET Symposium. 63rd Annual Meeting of the Federation of American Societies for Experimental Biology, Dallas, Texas, April 2, 1979.

Reitz, R. H., Ramsey, J. C., Anderson, M. E., and Gehring, P. J., 1988. Integration of Pharmacokinetics and Pathological Data in Dose Selection for Chronic Bioassays. In: "Carcinogenicity: The Design and Interpretation of Long-Term Animal Studies (H. C. Grice and J. L. Ciminera, eds.), page 54. Springer-Verlag, New York.

Stevens, J. T., 1986. The Maximum Tolerated Dose Concept: The Key to Overdose Phenomena . Toxicol. Letters. **31**:P1-1.

Stevens, J. T., and Sumner, D. D., 1982. The importance of Metabolite Identification in Quantitative Risk Estimation. J. Toxicol. Clin. Toxicol. **19**:6-7, 781.

Thede, B., and Kahrs, R. A., 1987. Study of 14C -Atrazine Dose/Response Relationship in the Rat. CIBA-GEIGY Report ABR-87087 (Unpublished).

Torkelson, T. R., and Rowe, V. K., 1981. Halogenated Aliphatic Hydrocarbns Containing Chloride, Bromine and Iodine. In: "Patty's Industrial Hygiene and Toxicology" (G. D. Clayton and F. E. Clayton, eds.), Chapter 48, page 3537. Third Edition. John Wiley and Sons, New York.

United States Environmental Protection Agency (USEPA), 1986. Risk assessment guidelines for carcinogenicity, mutagenicity, complex mixtures, suspect developmental toxicants, and estimating exposures. Fed. Regist. **51**:33,992.

United States Environmental Protection Agency (USEPA), 1990. Pesticide Tolerances for Primisulfuron-Methyl. Fed. Regist. **55**:(102) 21547.

Van der Heuden, C. A., 1988. Introduction on Comparative Toxicology. Regul. Toxicol. Pharmacol. **8**, 385.

Vocci, F., and Farber, T., 1988. Extrapolation of Animal Toxicology Data to Man. Regul. Toxicol. Pharmacol. **8**:389.

Watanabe, P. G., Schumann, A. M,, and Reitz, R. H., 1988. Toxicokinetics in the Evaluation of Toxicity Data. Regul. Toxicol. Pharmacol. **8**:408.

Watanabe, P. G., Young, J. D., and Gehring, P. J., 1977. The importance of Non-Linear (Dose-Dependent) Pharmacokinetics in Hazard Assessment. J. Environ. Pathol. Toxicol. **1**:147.

Wetzel, L. T., Breckenridge, C. B., Eldridge, J. C., Tisdel, M. O. and Stevens, J. T., 1991. Possible Mechanism of Mammary Tumor Formation in Sprague-Dawley Female Rats Following the Administration of Chloro-triazine Herbicides. J. Am. Coll. Toxicol. (Submitted as Abstract).

World Health Organization (WHO), 1986. Assessment of Toxicokinetic Studies. In: "Principles of Toxicokinetic Studies. Environmental Health Criteria 57", Chapter 10, page 126. IPCS International Programme on Chemical Safety.

PHYSIOLOGICAL MODELING AND CANCER RISK ASSESSMENT

Kannan Krishnan and Melvin E. Andersen

Chemical Industry Institute of Toxicology
Research Triangle Park, NC 27709

INTRODUCTION

Quantitative risk assessment for carcinogens is frequently performed based on evidence of their tumorigenicity in animals. In animal bioassay studies, the test chemicals are administered at high doses and by routes often different from anticipated human exposures. A particularly challenging aspect in cancer risk assessment is the extrapolation of tumor responses in animals to humans exposed at concentrations much below those tested under the range of the animal bioassay conditions. Since high doses of chemicals are usually administered during these studies, a metabolic saturation may lead to nonlinearities of target tissue dose and exposure concentrations. Tumor response in such cases may not be directly proportional to the external or exposure concentration of the chemical, but rather will reflect the complex nonlinear pharmacokinetic processes occurring in the organism at high concentrations. Therefore, a quantitative understanding of the critical biological determinants of pharmacokinetics and cancer response is necessary to confidently predict the carcinogenic risks associated with human exposure to chemicals. This chapter presents a review of the conventional cancer risk assessment methodology, and discusses the utility of biologically-based tissue dosimetry and cancer response modeling approaches in reducing some of the uncertainties associated with it.

RISK ASSESSMENT

Risk assessment is the process of estimating the probability of potential effects in individuals and populations of biota arising from exposure to chemical, physical or biological agents. The assessment of health risks associated with chemical exposure is performed in four steps, namely, hazard identification, exposure assessment, dose-response assessment, and risk characterization [National Academy of Sciences, 1983].

Hazard identification involves the determination of whether or not a particular chemical is causally linked to the incidence of adverse health effects. It involves characterizing the nature and strength of the evidence for causation. Essentially this process serves to identify chemicals that can cause an increase in the incidence of

New Trends in Pharmacokinetics, Edited by A. Rescigno and A.K. Thakur
Plenum Press, New York, 1991

adverse health conditions in humans. The hazard potential of a substance is defined by its toxic properties (e.g., acute toxicity, carcinogenicity, etc.). Characteristics of the source and of the environmental compartment(s) to which it is discharged also determine the degree and incidence of adverse effects resulting from exposure. Typically, the identification of carcinogens is performed with information obtained from animal bioassay studies, human epidemiological studies, results from predictive assays such as mutagenicity, and comparisons to the molecular structure of known carcinogens.

Exposure assessment is the process of measuring or estimating the intensity, frequency, and duration of the occurrence of contact between the chemical and the population of interest. In calculating total exposure one needs to take into account exposure by various routes such as inhalation, oral ingestion, dermal absorption etc. In conventional exposure assessment, humans are presumed to be exposed to the chemical of interest thoughout their lifetime. Daily water consumption at the rate of 2 liters, and daily inhalation of 12 meter^3 of air are assumed in order to calculate the total exposure to a chemical. Unless information to the contrary is available, 100% of the ingested/inhaled chemical is assumed to be absorbed, i.e., all of the chemical is considered bioavailable.

Dose response assessment involves the determination of the nature and extent of adverse health effects associated with human exposure to a particular chemical. A dose-response assessment requires (1) the extrapolation of response incidence rates observed at high doses in test animal species to low exposure doses and (2) the extrapolation of toxicologically-equivalent doses from animals to humans.

(1) High dose — Low dose extrapolation: The relationship between the administered dose of a chemical and the incidence of an adverse effect is evaluated by exposing laboratory animals to variable but usually high doses via inhalation, ingestion or injection (subcutaneous, intramuscular, intravenous, intraperitoneal), thus warranting an extrapolation of the toxicity observations (e.g., tumor incidence) to low exposure doses. It is impractical and uneconomical to ascertain experimentally small increases in tumor frequency over background exposure levels, especially at very low levels of chemical exposure. Even in experiments in which 24,000 mice were used, calculating the doses that yielded response frequencies below 1% of the test group was extremely difficult [Gaylor, 1979; Fishbein, 1980]. Therefore, the dose level corresponding to one in a million tumor response is calculated by extrapolating from the dose causing observable responses in test animals. Since there usually are not enough data points to yield a convincing dose-response relationship at very low doses of carcinogens, mathematical models have been used for extrapolation. The modeling approach most commonly used by the regulatory agencies is the linearized multistage procedure [Moolenaar, 1989].

The linearized multistage (LMS) procedure is a mathematical approach to low dose extrapolation of cancer prevalence that is consistent with a no-threshold concept for the action of carcinogens [Crump et al., 1976]. In this model, the probability of tumor incidence, P(d), is related to the administered dose, d, through an exponential function which contains a polynomial, the degree of which is chosen to fit the data, and is not based on any knowledge of the carcinogenic process.

The multistage model is represented mathematically by the equation,

$$P(d) = 1 - \exp(-(q_0 + q_1 d + q_2 d^2 + \ldots + q_k d^k))$$

where d = life-time dose rate,
    k = number of stages, determined by fitting to actual data,
    q_0 = background incidence in control groups,

q_k = non-negative constants used to fit the data (only two parameters are usually used for this model because of the limited number of available experimental points; as a general rule, the number of estimated constants should not exceed $k - 1$, where k is the number of dose groups for which cancer prevalence is available).

The extra risk over the background level is given by $\Sigma q_i d^i$. At low doses, this expression essentially becomes a one-hit or a linear extrapolation model, because the equation simplifies to

$$P(d) = 1 - \exp(-(q_0 + q_1 d)).$$

Typically, the behavior predicted by this model is linear when $q_1 d > 0$, and sublinear (quadratic, upward trend) if $q_1 = 0$ and $q_2 d^2 > 0$, with a behavior for very low doses defined by $P(d) = q_1 d$ and $P(d) = q_2 d^2$, respectively. The model parameters are estimated by computer optimization to fit animal bioassay data through the maximum likelihood method [Howe and Crump, 1982; Howe, 1983], by introducing the treatment doses and the observed frequencies of response into the model.

Conventionally, all carcinogens are treated as if they caused mutations at all dose levels, which assumes that there is no threshold. The dose-response for non-carcinogenic and non-mutagenic agents, on the other hand, is believed to have an identifiable threshold, a dose level below which adverse effects are not observed (NOAEL: no observable adverse effect level) in either the individual or in the population.

(2) Animal — Human extrapolation: Responses associated with low exposure concentrations in experimental animals are then extrapolated to humans. There are several issues to be considered in scaling the toxicologically equivalent doses across species. These include, but are not limited to, relative body size, life span, qualitative and quantitative differences in tissue response and pharmacokinetics. The extrapolation of toxicologically-equivalent doses across species is performed by a safety factor approach for most non-carcinogenic end points (systemic toxicants, as referred to by US EPA)[US EPA, 1989]. Accordingly, a reference dose (RfD), is an exposure dose level for a chemical below which a significant risk of adverse effects is not expected. RfDs are calculated by dividing the NOAEL by a safety factor, the magnitude of which ranges from 10 to 1000 according to the exposure conditions of the experiment. When the NOAEL is determined in rodents as a result of chronic exposure to a chemical, a safety factor of 100 is used to derive the RfD for humans. This factor of 100 is the result of two multiplicative components: a factor of 10 to account for interindividual differences, and another factor of 10 to account for interspecies differences [Dourson and Stara, 1983]. An additional factor of 10 is included if the NOAEL was determined in an experiment involving short term, high level exposure to the chemical. The major criticism of this approach is that the choice of safety factor is arbitrary. Further, the NOAEL is dependent upon the sample size and statistical test chosen by the investigator. Currently, a modifying factor (1 to 10) is also included in the determination of RfDs, depending upon the scientific merit of the study under consideration [U.S. Environmental Protection Agency, 1989].

For carcinogens, when toxicologically equivalent exposure doses for humans are to be derived, a dose level that is expected to increase the background level of tumors in some fraction of the population (e.g., one in a million) of experimental animals is first determined. The exposure dose for such an acceptable risk level in humans is then derived from that of the rodents using a body surface or a body weight scaling factor. The assumption here is that human exposure to 1 mg of chemical per unit body weight (or body surface) per day is pharmacokinetically and toxicologically equal to 1 mg chemical per unit body weight (or body surface) per

337

day in the rat. The US Food and Drug Administration uses body weight scaling, while the US Environmental Protection Agency has adopted body surface scaling to calculate equivalent doses across species.

Risk characterization is the process of estimating the incidence and severity of a health effect expected from exposure to the chemical in question. It is accomplished by combining the outcome of exposure and dose-response assessments. Risk associated with exposure to a chemical is characterized as the probability of an undesired event (e.g., cancer) for a given daily dose. Thus, the maximum allowable exposure level of a chemical for humans is derived as the daily dose, which on daily exposure for a lifetime, will not produce any significant adverse effects (for non-carcinogens) or that will not increase the incidence of tumors in a fraction of population not exceeding an "acceptable" level (e.g., one in a million). This value is then adjusted for the rate and extent of food and water consumption and of breathing to derive exposure limits for common media which may contain the chemical.

The process of risk assessment, despite the uncertainties associated with it, [Cothern and Schnare, 1986], it useful in regulating the use and environmental levels of chemicals because (1) it provides a rationale for regulation and (2) explicitly quantifies the probabilities and frequencies of occurrence of adverse effects on exposure to any level of a chemical. It should, however, be emphasized that the projected risk level (e.g., 1 in 1,000,000) does not necessarily imply the occurrence of such number of deaths in a population, but it is only a probability estimate resulting from the particular assumptions used in the process of risk assessment. The actual risks are given as a range extending from the upper bound of the LMS for a $1 \cdot 10^{-6}$ risk to as low as zero.

The current regulatory approaches have implicit in them a number of assumptions on the disposition of chemicals and mechanisms of carcinogenesis [Houk, 1989; Moolenaar, 1989; Park, 1989]: (1) the uptake, disposition, and target tissue dosimetry of a chemical are similar for all routes of exposure in all species and are linearly related to the administered dose; (2) all chemical carcinogens are genotoxic and act by a similar mechanism of tumorigenesis; (3) humans are as sensitive as the most sensitive test species; and (4) the tissue response is linearly related to exposure dose in all species. The physiologically-based dosimetry and response modeling approaches can not only be useful in verifying the validity of these assumptions but also be of use to elucidate the mechanistic bases of chemical disposition and action, and thus help to improve the surrent risk assessment procedure.

CANCER RISK ASSESSMENT: ISSUES RELATING TO PHARMACOKINETICS

The conventional cancer risk assessment approach followed by US EPA is based upon the perceived relationship between the external or exposure concentration and the incidence of tumors in test species (Figure 1). Quantitative information on neither tissue dosimetry nor tissue response is incorporated at the mechanistic level. Considering the issues related to cancer risk asssessment from the standpoint of tissue dosimetry, the following can be pointed out.

High dose — low dose extrapolation

In the rodent cancer bioassay studies, very high doses of chemicals are administered to assess their carcinogenic potential. The tumor prevalence in such cases is frequently not proportional to the dose throughout the range of exposure doses but is more complex due to the non-linearities in pharmacokinetic processes

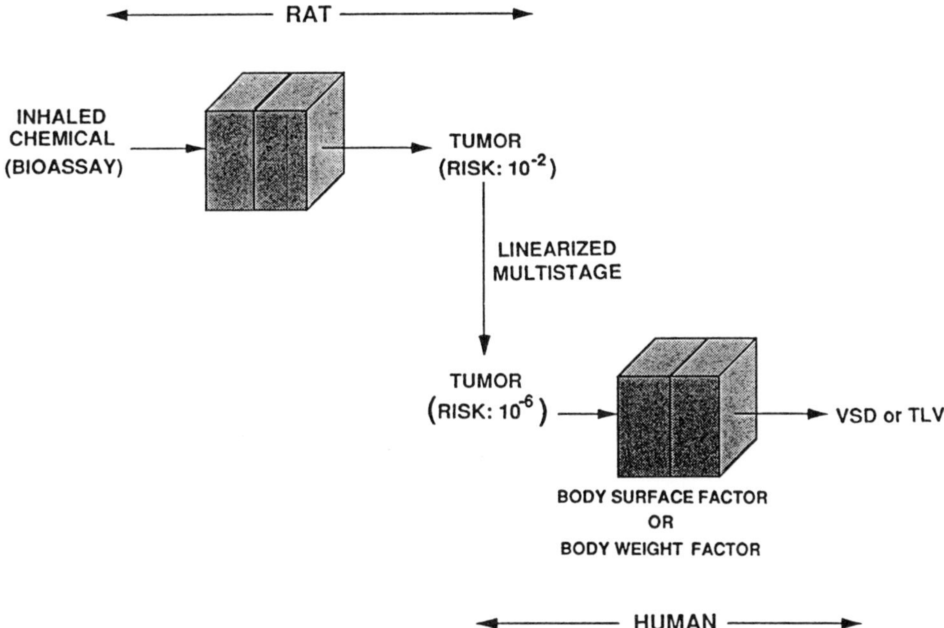

Fig. 1. Schematic representation of the conventional approach to cancer risk assessment of chemicals (VDS: virtually safe dose; TLV: Threshold limit value).

(i.e., absorption, metabolism, distribution and excretion) which occur at high doses [Levy, 1968; Gehring et al., 1976; van Ginneken and Russel, 1989]. The relative concentration of the parent chemical or reactive metabolite(s) at the target site is frequently disproportional to the administered dose used in animal bioassay studies.

Route — Route extrapolation

In the conventional analysis, uncertainties also arise because the routes of administration in rodent toxicity studies are often different from anticipated human exposures. Route-to-route extrapolations are performed by assuming that the relationship between administered dose and the internal/target tissue dose is independent of the route of exposure and the species. There is no obvious toxicological or pharmacokinetic basis for such an assumption.

Interspecies extrapolation

The equivalent safe exposure concentration or dose of a chemical for humans is predicted from that of rodents using a conversion factor, usually body weight to the power of 2/3 or 1 [Krishnan and Andersen, this volume]. The former power ($BW^{2/3}$) is considered scaling on the basis of body surface area, the latter (BW^1) is direct body weight scaling. Based on the fact that several physiological and biochemical processes are more closely related to body surface area rather than body weight, the 2/3 scaling has been adopted by the US EPA. The problem, however, is that this approach is used for all chemicals regardless of the toxic moiety (parent or metabolite) and mechanism of toxicity [National Academy of Sciences, 1987].

Thus, in the conventional quantitative risk assessment approach, the potential carcinogenic risk associated with human exposure to chemicals is estimated without accounting for dose-, exposure route- or species-dependent pharmacokinetic differences. Extrapolation of the kinetic behavior for a given chemical from high dose to low dose, from inhalation to other routes of exposure, or from rodents to people can only be done with a mechanistic understanding of its uptake and disposition. This can effectively be accomplished with a physiological modeling approach.

PHYSIOLOGICAL MODELING

Physiological modeling involves computer simulation of the uptake and disposition of chemicals based upon their tissue solubility characteristics, metabolism/binding rates and physiology of the test species. The concept of physiological modeling dates back to the research of Haggard [1924], who mathematically described the uptake of inhaled diethylether from a physiological perspective. Further developments in physiological modeling using vapors were contributed by Kety [1951] and Riggs [1963], who provided mathematical descriptions of the pharmacokinetics of chemicals based on their tissue solubility and physiology of the test species, and by Mapleson [1963], who developed an electric analog approach of the physiological system to describe the pharmacokinetics of inert gases. This electric analog approach was expanded by Fiserova-Bergerova [1975] to describe the pharmacokinetic behavior of metabolized vapors and gases. Beginning in the early 1970's, scientists trained in chemical engineering became involved in developing physiological models to describe the pharmacokinetics of antineoplastics such as methotrexate and 5-fluorouracil [Bischoff et al., 1970; Collins et al., 1982]. Subsequently, Ramsey and Andersen [1984] used the physiological modeling approach to describe the pharmacokinetics of an industrial chemical (styrene) and then successfully utilized the model to extrapolate the kinetic behavior of styrene from high to low exposure doses, from inhalation to other routes of exposure, and from rodents to people. In the styrene work, the critical biological determinants of chemical disposition (i.e., physiological parameters, biochemical rate constants, tissue:blood partition coefficients etc.) were derived from previously published blood time course data. The critical determinants required to develop physiological models to describe the pharmacokientics of xenobiotics can be determined independently and are briefly discussed below.

Physiological parameters such as alveolar ventilation rate, cardiac output, blood flow rates and tissue volumes can be obtained from biomedical literature [Caster et al., 1956; International Commission on Radiation Protection, 1975]. A compilation of this information from various sources for a variety of animal species is also available [Arms and Travis, 1988].

Partition coefficients, indicative of the relative partitioning of a chemical between two media (e.g., blood and air, tissue and blood) at equilibrium are instrumental in describing the pulmonary uptake and tissue distribution of volatile organic chemicals as well as the relative tissue distribution of most uncharged xenobiotics. The blood:air and tissue:air partition coefficients of volatile organic chemicals can be determined *in vitro* using the vial equilibration technique of Sato and Nakajima [1979a] as modified by Gargas et al. [1989]. This technique involves determining the headspace concentration of the chemical in test vials containing the tissues of interest and comparing the concentration measured to appropriate reference vials. Partition coefficients for nonvolatile xenobiotics can be determined *in vitro* using equilibrium dialysis or ultrafiltration techniques [Lin et al., 1982; Igari et al., 1983; Jepson, 1986; Fisher et al., 1990 .

Biochemical parameters required for modeling are mainly metabolic rate constants, which can be determined either *in vivo* or *in vitro*. Two innovative methods have been devised for estimation of the metabolic rate constants of volatile organic chemicals. These are the (1) closed chamber or gas uptake method and (2) the exhaled breath chamber method. The gas uptake system involves continuous monitoring of the decline in the chamber concentration of a chemical in a closed desiccator-type chamber, with recirculating atmosphere, containing two or three animals [Andersen et al., 1980; Filser and Bolt, 1981]. Highly soluble chemicals are unsuitable for use in the gas uptake system. For these, the exhaled breath technique is more useful [Gargas and Andersen, 1989]. Here an animal previously exposed to a chemical, is placed in a closed recirculated chamber, and the increase in the chamber concentration of the chemical is monitored [Gargas and Andersen, 1989]. Alterations in the the chamber concentration of chemicals, resulting from either of these techniques, are analysed with a physiological description to derive the metabolic rate constants [Gargas et al., 1986a; Gargas, 1990]. The *in vivo* metabolic rate constants for the model can also be estimated based on *in vitro* determinations [Dedrick et al., 1972; Sato and Nakajima, 1979b; Hilderbrand et al., 1981]. These *in vitro* approaches need to consider that the rate-limiting factor for metabolism of drugs and environmental chemicals *in vivo* might be blood flow rather than enzyme activity [Rane et al., 1977; Pang et al., 1978; Andersen, 1981].

Once formulated by integrating information on animal physiology, rate constants for kinetic processes, and tissue partition coefficients of a chemical, the physiological model can be used to simulate the kinetic behavior of the chemical in the test species. Model simulations of per cent dose exhaled, amount of metabolites produced, level of hepatic and extrahepatic glutathione depletion, tissue and blood concentrations of parent chemical or its metabolite(s), etc., can be generated for a variety of exposure scenarios in the test species [Clewell and Andersen, 1987]. The model predictions can be compared with experimental observations; when the model accurately predicts the experimental data obtained independently of the model development process, it is considered to be "validated". Failure of a model to accurately predict the pharmacokinetic behavior of a chemical indicates incomplete understanding on the part of the investigator of the critical processes involved in its uptake and disposition. In such cases, further experimentation to obtain information of a specific nature, to refine and validate the model might be required [Clewell and Andersen, 1987]. Validated models can then be used for dose-, exposure route-, and interspecies-extrapolation of pharmacokinetic behavior of chemicals.

High dose to low dose extrapolation of pharmacokinetic behavior is accomplished with physiological modeling by accounting for the nonlinear kinetic behavior of chemicals. The description of metabolism may include linear processes or a capacity-limited oxidative process which becomes saturated at high exposure doses, or both [Ramsey and Andersen, 1984]. Nonlinearities in tissue exposure arising from mechanisms other than saturable metabolism — including enzyme induction [Andersen et al., 1984], enzyme inactivation [Andersen et al., 1986b], depletion of glutathione reserves [Andersen et al., 1986a; D'Souza et al., 1988], and protein binding [Leung et al., 1990] — have also been described with physiologic pharmacokinetic modeling.

Exposure route extrapolation of chemical disposition can be performed by adding appropriate equations to the basic inhalation model [Ramsey and Andersen, 1984]. For instance, intravenous injection can be described by including a term representing the rate of infusion into the mixed venous blood. Oral gavage can be modeled by introducing either a first-order or a zero-order uptake rate constant, with the chemical assumed to appear in the liver after gastric absorption [Ramsey and Andersen, 1984; Fisher et al., 1990]. Dermal absorption can be modeled by including a diffusion-limited compartment to represent skin as a portal of entry [McDougal et al., 1986].

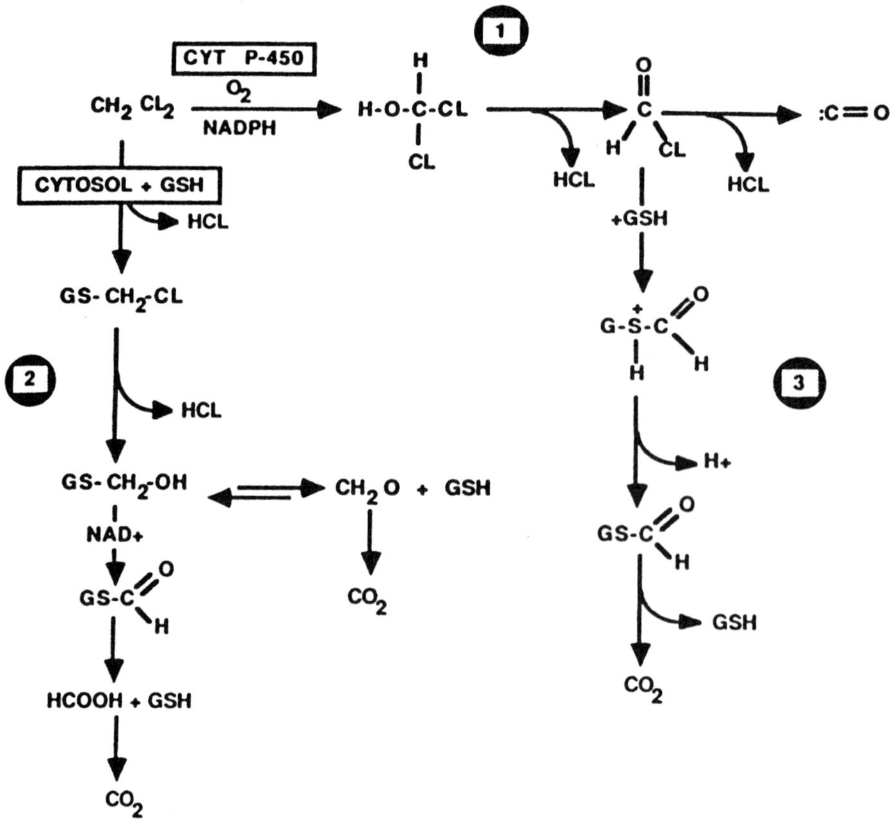

Fig. 2. Oxidative (CYT P-450) and conjugation (CYTOSOL + GSH) pathways of dichloromethane metabolism. (From Andersen et al., 1987.)

Interspecies extrapolation of the pharmacokinetic behavior of chemicals can be performed by scaling the physiological parameters in the rodent model to the species of interest (e.g., humans), and by detemining the chemical-specific parameters for the species of interest. Whereas the blood:air partition coefficients need be determined, the human tissue:air partition coefficients have been assumed to be similar to rodents [Ramsey and Andersen, 1984], an assertion which needs experimental verification. The *in vivo* metabolic rate constants for the human model can be estimated from the *in vitro* rate constants determined with liver preparations obtained from autopsy, by comparing them to the relationship between the *in vivo* and *in vitro* metabolic rate constants seen in rodents [Reitz et al., 1988]. In the absence of such information, the rate constants for metabolism have sometimes been scaled across species according to body weight or basal metabolic rate [Tuey and Matthews, 1980; Ramsey and Andersen, 1984; Sawada et al., 1985]. This latter procedure for estimating rate constants for metabolism should be regarded skeptically in risk assessment calculations.

The principal application of physiological pharmacokinetic models is to predict tissue dosimetry of the toxic moiety (i.e., the parent chemical or its active metabolite(s)). Quantitative information on the internal/target dose of the active form of a chemical can undoubtfully provide a better basis of relating to the latter's toxicity. Because physiological models allow prediction of target tissue dosimetry in people based on physiological and mechanistic considerations, they can also help reduce the uncertainty of extrapolation procedures adopted in conventional risk assessment approaches. Such an application has already been demonstrated with dichloromethane [Andersen et al., 1987].

PHYSIOLOGICAL MODELING IN CANCER RISK ASSESSMENT: THE DICHLOROMETHANE EXAMPLE

Methylene chloride (CH_2Cl_2, dichloromethane, DCM) caused dose-related increases in the incidence of liver and lung tumors in mice after chronic inhalation

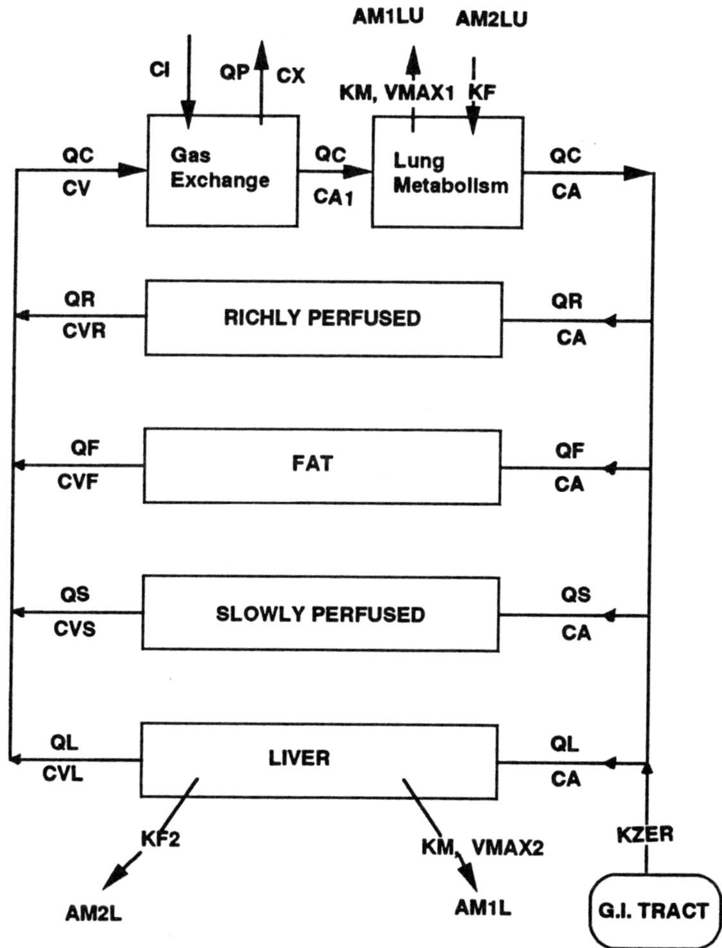

Fig. 3. The physiologically based pharmacokinetic model used to describe the kinetics of dichloromethane. Q terms are air and blood flow; C terms are concentrations. These are followed by letters F, M, R, L, V, A or A1 which represent fat, muscle, richly perfused tissues, liver, venous blood, arterial blood and the blood leaving the gas exchange compartment. CI, QP and QC are concentration of DCM in inhaled air, alveolar ventilation rate, and cardiac output. Kinetic constants for MFO-oxidation are Vmax (maximum velocity) and Km (Michaelis constant), and for GSH-conjugation is KF. From Andersen et al. (1987).

343

exposures of up to 4000 ppm [National Toxicology Program, 1985]. The toxic moiety responsible for the cellular interaction leading to the tumorigenicity of DCM has not been identified; however, it is known that potentially reactive intermediates are produced by the two major metabolic pathways [Kubic et al., 1974; Gargas et al., 1986b; Reitz et al., 1986]. DCM is metabolized in both target organs, by a cytochrome P-450-mediated oxidative pathway that yields carbon monoxide and carbon dioxide , as well as by a glutathione (GSH)-dependent conjugation pathway that yields only carbon dioxide and not carbon monoxide (Figure 2). Using the physiological modeling approach, the pharmacokinetics of DCM was studied in rodents and humans, and information on tissue dosimetry of parent chemical and metabolites in the most sensitive test species (i.e., mouse) was obtained. Target tissue exposure to the appropriate dose surrogate was then related to the tumor levels seen in the NTP carcinogenicity bioassay to derive the acceptable internal/target and external/exposure levels, and these were then compared with those obtained with the conventional risk assessment approach adopted by the US EPA.

The physiological pharmacokinetic model for DCM consisted of several tissue compartments connected by arterial and venous blood pools. The model structure was similar to that of styrene, described previously by Ramsey and Andersen [1984], except that the DCM model contained a metabolically-active lung compartment placed between the pulmonary gas exchange and systemic arterial blood compartments (Figure 3). The metabolically-active lung compartment was included because this tissue was a target for DCM-induced cancer. The physiological parameters for the rodent model were obtained from the literature; breathing rates for mice were also determined from the inhalation kinetics using a closed chamber system. The tissue:air and blood:air partition coefficients for the rodent model were determined by vial equilibration, and the metabolic rate constants were estimated in gas uptake studies [Gargas et al., 1986b]. The whole body metabolism for these two pathways (P-450 oxidation, GSH conjugation) was then apportioned between lung and liver by assuming that the distribution of enzyme activities metabolizing DCM was the same as the distribution of enzyme activities acting upon two model substrates, 7-ethoxycoumarin for oxidation and 2,4-dinitrochlorobenzene for GSH conjugation [Lorenz, 1984].

The rate of change in the amount of DCM in metabolizing tissues (i.e., liver, lung) was represented by the following equations:

$$\frac{dA_i}{dt} = Q_i(C_a - C_{vi}) - \frac{dAmet_i}{dt}$$

$$\frac{dAmet_i}{dt} = \frac{Vmax_i \cdot C_{vi}}{K_m + C_{vi}} + K_f V_i \cdot C_{vi}$$

where dA_i/dt = rate of change in the amount of DCM in tissue "i" (mg/hr),
    Q_i = blood flow to "i" (L/hr),
    C_a = concentration of DCM in arterial blood (mg/L)
    C_{vi} = concentration of DCM in the venous blood leaving "i" (mg/L),
    $dAmet_i/dt$ = rate of the amount of DCM metabolized in "i" (mg/hr),
    $Vmax_i$ = maximum enzymatic reaction rate in "i" (mg/hr),
    K_m = Michaelis constant for enzymatic reaction in "i" (mg/L)
    K_f = first order rate constant (hr^{-1}),
    V_i = volume of tissue "i" (L).

In the equation representing the rate of metabolism, the first term accounts for loss of DCM by microsomal oxidation, a saturable metabolic process. The second term represents metabolism by GSH conjugation, which is a first order process at all exposure concentrations examined.

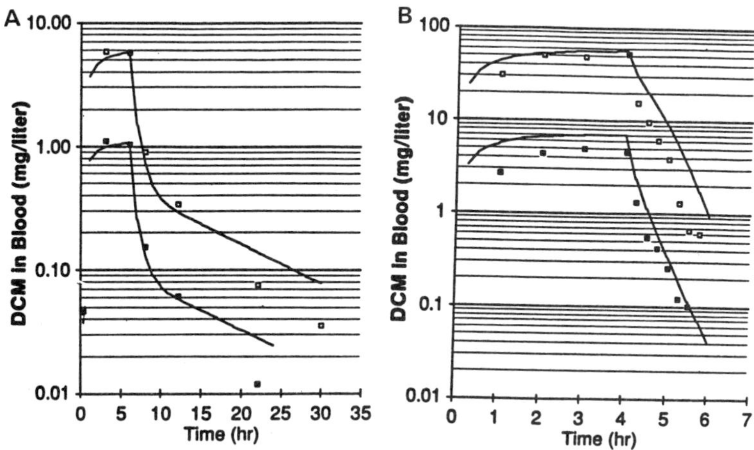

Fig. 4. Validation of the physiological pharmacokinetic model for dichloromethane with experimental data. Model simulations (solid lines) are compared with observed blood DCM levels (closed or open symbols) in Fischer-344 rats (A) and humans (B) following inhalation exposure. From Andersen et al. [1987].

The rodent physiological model for DCM, formulated by integrating information on its solubility characteristics, metabolic rate constants and animal physiology was utilized to describe the disposition of DCM in rodents at various dose levels delivered by inhalation and intravenous administration (Figure 4a). The physiological model which successfully described the rodent data was then scaled to predict human tissue dosimetry after scaling/determining the model parameters for humans. The physiological parameters were scaled allometrically; the tissue:blood partition coefficients were calculated by dividing rodent tissue:air partitions by human blood:air partition values. The metabolic rate constants for the oxidative pathway for the human model were estimated from studies conducted at Dow Chemical Co. in which levels of DCM and carboxyhemoglobin in blood during and after 6-hr exposures to 100 and 350 ppm DCM were determined [Andersen et al., 1991]. The level of glutathione S-transferase activity in humans was set equal to the highest activity found in rodents. The scaled human physiological model accurately simulated the blood levels of DCM observed in humans after a 6-hr inhalation exposure to 100 or 350 ppm DCM (Figure 4b).

The mouse physiological pharmacokinetic model was then used to calculate internal dose of metabolite(s) and parent chemical arising from exposure scenarios comparable to those in the NTP bioassay studies and their relationship to predicted tumor incidence was examined (Table 1). The mouse DCM model was used to estimate the time-course concentrations of the parent chemical and the amount of metabolites arising from both pathways in lung and liver. Since DCM is very unreactive, it is unlikely to be involved directly in tumor initiation; therefore, the relationship of the tumorigenicity of DCM to its metabolite levels was considered. The dose surrogates based on oxidative pathway did not vary between DCM

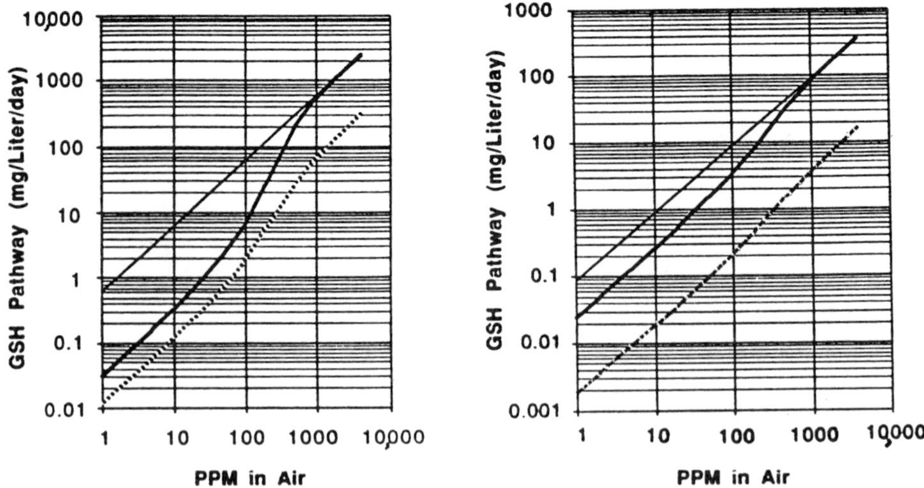

Fig. 5. Relationship between tissue dose and external exposure concentration for humans and mice for the glutathione pathway. The physiological modeling approach was used to determine the target tissue dose from the GSH pathway in both liver (a) and lung (b) at various exposure concentrations in mice (——) and humans (---). The lighter solid line depicts linear back extrapolation to 1 ppm. From Andersen et al. [1987].

exposure concentrations despite a clear dose-dependent increase in tumor incidence. The increase in the amount of metabolism via GSH pathway at high doses corresponded well with the degree of DCM tumorigenicity (Table 1). This was consistent with a role for the metabolite(s) resulting from GSH pathway in DCM tumorigenicity. Very recently, Casanova et al. [1991] have reported DNA-formaldehyde-protein crosslinks from DCM exposure, further strengthening the case for the GSH conjugation as the pathway leading to carcinogenic metabolites.

Table 1.
Methylene Chloride-Dose Response in Female Mice

Exposure Concentr.	Microsomal Pathway Dose*		Gluthathione Pathway Dose*		Tumor Prevalence	
ppm	Liver	(Lung)	Liver	(Lung)	Liver	(Lung)
0	---	---	---	---	6	(6)
2000	3575	(1531)	851	(123)	33	(63)
4000	3701	(1583)	1811	(256)	83	(85)

*Tissue dose is cumulative daily exposure (mg metabolized/volume tissue/day)

The mouse and human physiological pharmacokinetic models for DCM were used to predict target tissue exposures of the GSH conjugate resulting from 6-hr inhalations of 1-4000 ppm of DCM. The model predictions of the target tissue doses calculated for both mouse and human (Figure 5) were also compared with those predicted by the linear back-extrapolation from the concentrations used in the mouse carcinogenicity bioassay. Thus, the models indicate that the target tissue exposure to DCM-GSH reactive products in humans is expected to be lower than those in mice at all concentrations. Furthermore, the target tissue dose in mice exposed to low

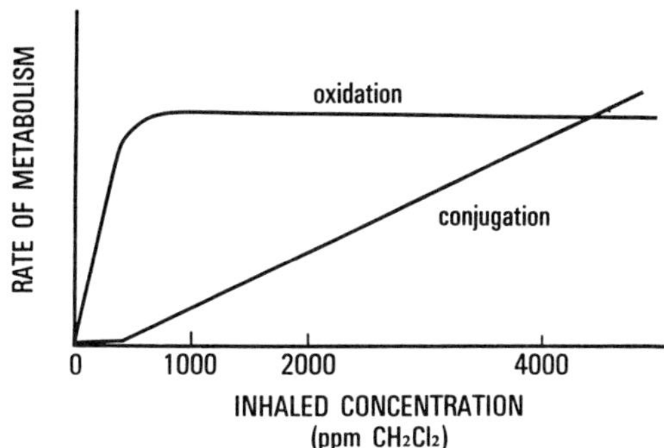

Fig. 6. Representation of rate of metabolism of dichloromethane at various inhaled concentrations. Dichloromethane is metabolized by oxidation and by conjugation with glutathione. These two metabolic pathways have different characteristics. At low inhaled concentrations oxidation is the preferred manner of metabolism, but at higher concentrations conjugation becomes more favoured.

concentrations is less than that expected by a linear back-extrapolation of the high concentration behavior. This nonlinear behavior at high exposure concentrations (>100 ppm) is a consequence of the saturated cytochrome-P-450-mediated oxidation pathway and a corresponding disproportionate rise in the flux through the GST pathway (Figure 6) at high inhaled concentrations.

The linear extrapolation procedure thus gives rise to 21-fold higher estimate for the tissue dose of DCM-GSH conjugate when compared to that obtained by physiological modeling. In addition, using the physiological pharmacokinetic approach, the target tissue dose for humans was estimated to be some 2.7 times

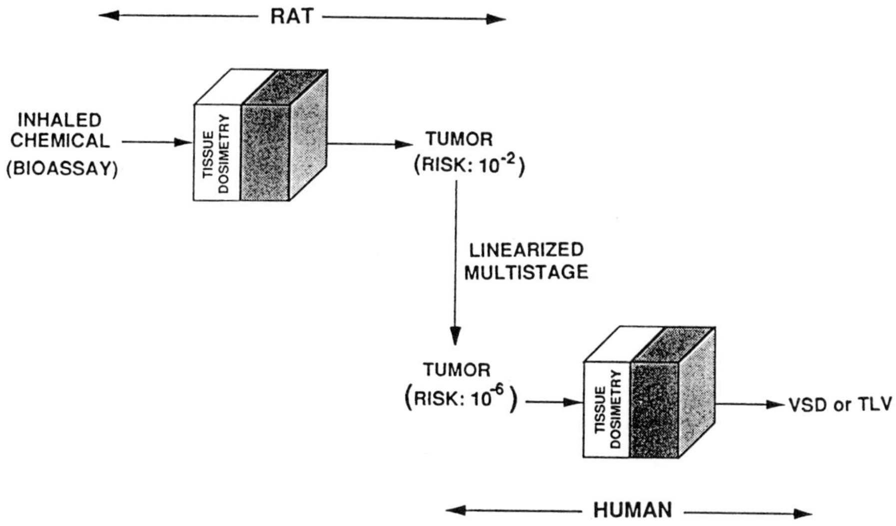

Fig. 7. Schematic representation of the physiologically based pharmacokinetic modeling (PBPK)-based approach to cancer risk assessment (VSD: virtually safe dose; TLV: threshold limit value).

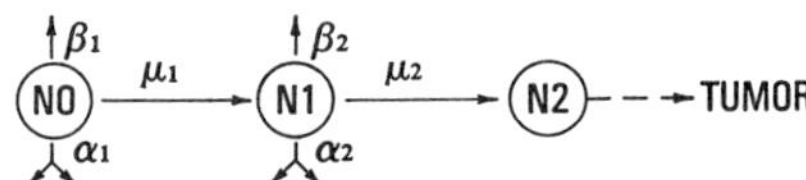

Fig. 8. Schematic of the two-stage, Moolgavkar-Venzon-Knudson (MVK) model for cancer. N_0, N_1, and N_2 are, respectively, normal cells, intermediate cells with a single mutation at one allele of a critical gene locus, and the cancer cell with mutations in both alleles of the critical gene locus. Birth rates of normal and intermediate cells are α_1 and α_2, death rates are β_1 and β_2, and mutation frequencies between cell types are μ_1 and μ_2. Whereas genotoxicants increase mutation rates (μ_1 and μ_2), the cytotoxicants alter cell death and birth rates (α_1, β_1, α_2, β_2), and the promoters convey growth advantages on the intermediate cell population (α_2, β_2).

lower than that for the mouse. Considered together then, the human tissue dose of the DCM-GSH metabolite for a 6-hr exposure to 1 ppm DCM is expected to be some 57 times lower than that expected by linear extrapolation of the results from the mouse bioassay. In contrast, the US EPA's approach to DCM risk assessment resulted in an increase, by a factor of 3, of the dose calculated with the LMS procedure, in order to account for differences in surface area and breathing rate between mice and humans [Singh et al., 1985]. The net result was that liver tissue dose and tumorigenic risk associated with DCM exposure as estimated by US EPA were 168 times that of the values determined by the PBPK studies. Similar analysis of predicted dose to lung tissue indicated that the LMS procedure estimated a target dose 143-times higher than that calculated with the physiological modeling approach. The US EPA has amended its original risk assessment for DCM in order to reduce the over-estimation by incorporating some, but not all of the concepts used in the physiological model-based risk assessment approach [U.S. Environmental Protection Agency, 1987 .

The use of physiologically-based pharmacokinetic modeling to estimate internal/target dose of the toxic moiety in test species and humans, can help illuminate the dosimetry part of the "black box" approach to carcinogenic risk assessment (Figure 7). Thus, relating tumor response to internal or target dose provides greater confidence than exposure or external concentration for low dose and interspecies extrapolation. The use of physiological modeling in risk analysis does not rely on the use of body surface/body weight factors for dose conversion across species; neither is it dependent upon the use of any mathematical models for low dose extrapolation of chemical disposition behavior. However, it can be used in conjunction with the LMS procedure, as done with DCM by Reitx et al. [1988], to relate the levels of the dose surrogate in the target tissues with tumor incidence. The optimal approach would be to illuminate the other half of the "black box" as well, with the understanding of the mechanistic and biological basis of tissue response.

BIOLOGICALLY BASED CARCINOGENIC RESPONSE MODELING

A biologically based cancer model has been developed by Moolgavkar, Venzon, and Knudson (MVK) [Moolgavkar and Venzon, 1979]. It is a two-stage model (Figure 8) that describes carcinogenesis as the end result of two mutagenic events (μ_1 and μ_2), that correspond to mutations at a single, critical gene locus. Accordingly, the normal cells (No) first progress to an intermediate cell type (N1) by mutation at one allele of the gene locus. This intermediate cell may have different growth characteristics from the normal cells but is not itself malignant. The second event (u2), which alters the second allele of the critical gene locus, produces cells of a malignant genotype (N2) which produce a tumor by clonal expansion. The MVK description includes explicitly the critical determinants of the rate at which mutations accumulate in a tissue. These include (1) the number of normal and initiated cells, (2) the rates at which the normal and intermediate cells die and replicate, and (3) the probabilities with which critical mutations occur. Using this model, the mechanism of carcinogenic action of chemicals can be expressed in a quantitative manner by accounting for their effect on one or more of the critical determinants of carcinogenesis. Whereas genotoxic chemicals primarily cause an increase in the mutation rates, the cytotoxicants would be expected to alter death and birth rates of the normal and/or malignant cells; promoters could convey growth advantages to the intermediate cell population.

The use of this biologically-motivated cancer model should be a significant advance compared to the conventional linearized multistage procedure. The following are some of the limitations arising from the use of LMS in the risk assessment process [Thorslund et al., 1987]:

1. The dose-response relationship used in the LMS procedure often is not derived from any underlying biological theory of cancer induction; it is obtained by fitting data to the multistage polynomial,
2. the LMS procedure does not account for agent-induced stimulation of cell proliferation in an explicit manner, and
3. it is difficult to incorporate important experimental observations other than carcinogenesis bioassay data into the risk assessment procedure.

In contrast, the parameters in the MVK model are interpretable in biological terms and therefore can potentially be estimated using data obtained from sources other than carcinogenesis bioassays. For example, the parameters related to dose-reponse of cellular transformation estimated from initiation-promotion experiments can be included [Cohen and Ellwin, 1990; Moolgavkar et al., 1990]. Thus, using this biologically-based cancer reponse model, relevant information obtained from experiments supplementary to the chronic animal bioassays can be incorporated into the risk assessment process. The relevance of incorporating mechanistic data in the risk assessment process is especially evident in case of non-genotoxic carcinogens [Butterworth, 1990]. The US EPA has begun incorporating such information, as it becomes available, into its risk assessment procedure. This has already been done for formaldehyde [US EPA, 1990] and unleaded gasoline [US EPA 1991]; but these examples are not germane to use of a biologically-based tumor induction model.

FUTURE DIRECTIONS

An improved strategy for cancer risk assessment would be to develop and adopt a biologically-realistic and mechanistically-based approach. This strategy would involve the development of (1) physiological models to describe the pharmacokinetic behavior of chemicals, (2) quantitative descriptions of the mechanistic link between target tissue dose and cellular responses, and (3) biologically-based response models to account for cancer induction based on cellular responses [Conolly et al., 1988]. This integrated approach is comprised of four essential components (Figure 9): an animal physiological dosimetry element, a mechanistic link, an animal biological response element and a human physiological dosimetry element. The animal dosimetry model serves to identify critical determinants of chemical disposition and the appropriate dose surrogate of target tissue exposure. The toxic moiety in the

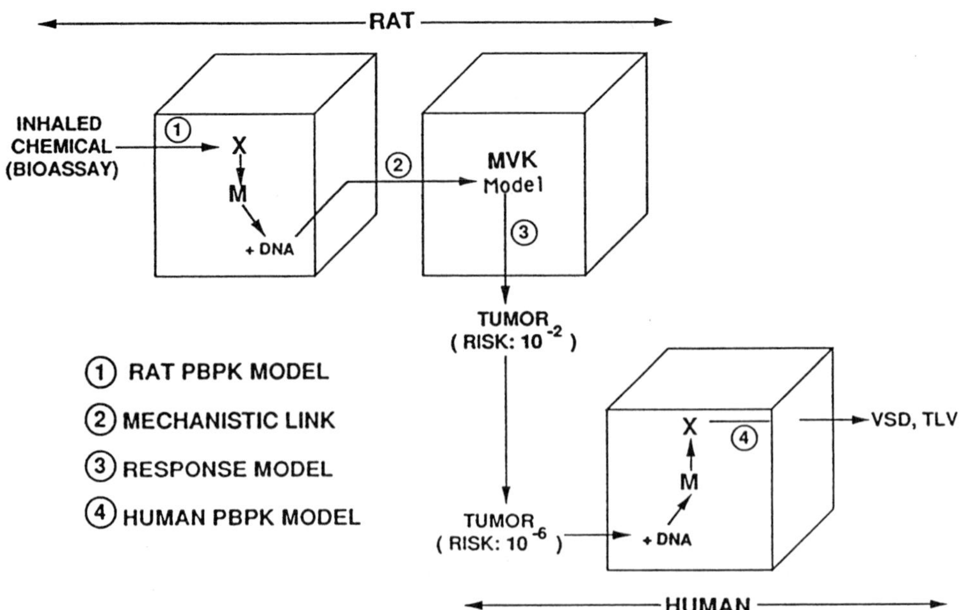

Fig. 9. Schematic representation of a biologically- and mechanistically-based approach to cancer risk assessment. It consists of four components: (1) a rodent physiologically based pharmacokinetic element, (2) a mechanistic link, (3) a rodent biologically based response element and (4) a human physiologically based pharmacokinetic element (VSD: virtually safe dose; TLV: threshold limit value; X: parent chemical; M: reactive metabolite; MVK: Moolgavkar-Venzon-Knudson).

target tissue(s) is then related to the response model by a definition of its mechanism of action (e.g., DNA reactivity, cytotoxicity, mitogenic action). The response model then specifies the number of mutations (stages) required for malignancy and the role of target cell death and birth processes in the accumulation of these mutations. The internal/target dose of the potential toxic moiety of a chemical with respect to a predetermined tumor incidence level can be calculated using the animal dosimetry model. Then, assuming that humans would be as sensitive as the test species, equal internal/target doses integrated over time are expected to produce equal tumor responses in both species. Finally, allowable exposure concentration of the chemical corresponding to the allowable internal dose of its potential toxic moiety can be determined with the human physiological pharmacokinetic model.

In summary, physiological modeling allows the prediction of tissue concentrations of toxic moieties, either the parent chemical or its metabolite(s), thus providing a better basis for relating exposure to the carcinogenic response seen in chronic bioassay studies than simply using external or exposure concentration. Whereas physiological pharmacokinetic models provide the link between exposure and internal dose of chemicals, the pharmacodynamic models describe the mechanistic link between tissue dosimetry and tumor response. Development of combined dosimetry-response models would allow computer simulation of the "exposure concentration-internal dose-tissue response" relationships in experimental animals. Scaling the critical biological determinants to people will considerably enhance our ability to better predict human health risks associated with exposure to carcinogenic and otherwise toxic chemicals.

ACKNOWLEDGEMENTS

The authors would like to thank Drs. R. Conolly, F. Miller and L. Recio for valuable discussions during the preparation of this chapter.

REFERENCES

Andersen M.E., 1982. Recent advances in methodology and concepts for characterizing inhalation pharmacokinetic parameters in animals and man. Drug Metab. Rev. **13**:799.

Andersen M.E., Clewell III H.J., Gargas M.L. and Conolly R.B., 1986a. A physiological pharmacokinetic model for hepatic glutathione depletion by inhaled halogenated hydrocarbons. Toxicologist **6**:148.

Andersen M.E., Clewell III H.J., Gargas M.L., MacNaughton M.J., Reitz R.H., Nolan R. and McKenna M., 1991. Physiologically based pharmacokinetic modeling with dichloromethane, its metabolite carbon monoxide and blood carboxyhaemoglobin in rats and humans. Toxicol. Appl. Pharmacol. (in press).

Andersen M.E., Clewell III H.J., Gargas M.L., Smith F.A. and Reitz R.H., 1987 Physiologically based pharmacokinetics and the risk assessment process for methylene chloride. Toxicol. Appl. Pharmacol. **87**:185.

Andersen M.E., Gargas M.L., Jones R.A. and Jenkins L.J.Jr, 1980. Determination of the kinetic constants for metabolism of inhaled toxicants in vivo using gas uptake measurements. Toxicol. Appl. Pharmacol. **54**:100.

Andersen M.E., Gargas M.L. and Clewell III H.J., 1984. Inhalation pharmacokinetics: evaluating systemic extraction, total metabolism and the time course of induction for inhaled styrene based on steady state blood:gas concentration ratios. Toxicol. Appl. Pharmacol. **73**:176.

Andersen M.E., Gargas M.L. and Clewell III H.J., 1986b. Suicide inactivation of microsomal oxidation by cis- and trans-dichloroethylene in male Fischer rats. Toxicologist **6**:148.

Arms A.D. and Travis C.C., 1988. Reference physiological parameters in pharmacokinetic modeling. Office of Health and Environmental Assessment, U.S. EPA, Washington, DC. NTIS PB 88-1906019.

Bischoff K.B., Dedrick R.L., Zakharo D.S. and Longsreth J.A., 1971. Methotrexate pharmacokinetics. J. Pharm. Sci. **60**:1128.

Butterworth B. E., 1990. Consideration of both genotoxic and nongenotoxic mechanisms in predicting carcinogenic potential. Mutat. Res. **239**:117.

Casanova M., d'Heck H. and Deyo D.F., 1991. Dichloromethane: metabolism to formaldehyde and formation of DNA-protein crosslinks in mice and hamsters. Toxicologist **11**:180.

Caster W.O., Poncelet J., Simon A.B. and Armstrong W.D., 1956. Tissue weights of the rat. I. Normal values determined by dissection and chemical methods. Proc. Soc. Exp. Biol. Med. **91**:122.

Clewell III H.J. and Andersen M.E., 1987. Dose, species and route extrapolations with a physiologically based model. Drinking Water and Health **8**:159.

Cohen S.M. and Ellwin L.B., 1990. Cell proliferation and cancer. Science **249**:1007.

Collins J.M., Dedrick R.L., Flessner M.F. and Longstreth J.A., 1982. Concentration dependent disappearance of fluorouracil from peritoneal fluid in the rat: experimental observations and distributed modeling. J. Pharm. Sci. **71**:735.

Conolly R.B., Reitz R.H., Clewell III H.J. and Andersen M.E., 1988. Pharmacokinetics, biochemical mechanism and mutation accumulation: a comprehensive model for chemical carcinogenesis. Toxicol. Lett. **43**:189.

Cothern C.R. and Schnare D.W., 1986. The limitation of summary risk management data. Drug Metab. Dispos. **17**:145.

Crump K.S., Hoel D.G., Langley C.H. and Peto R., 1976. Fundamental carcinogenic processes and their implications for low dose risk assessment. Cancer Res. **36**:2973.

Dedrick R.L., Forrester D.D. and Ho D.H.W., 1972. *In vitro* correlation of drug metabolism — deamination of 1,-B-D-arabinofuransyl cytosine. Biochem. Pharmacol. **21**:1.

Dourson M.L. and Stara J.F., 1983. History and experimental support of uncertainty (safety) factors. Regul. Toxicol. Pharmacol. **8**:45.

D'Souza R. W., Francis W. R. and Andersen M. E., 1988. Physiological model for tissue glutathione depletion and increased resynthesis after ethylene dichloride exposure. J. Pharmacol. Exp. Ther. **245**:563.

Filser J.G. and Bolt H.M., 1981. Inhalation pharmacokinetics based on gas uptake studies. Arch. Toxicol. **47**:279.

Fiserova-Bergerova V., 1975. Mathematical modeling of inhalation exposure. J. Combust. Toxicol. **32**:201.

Fishbein L., 1980. New concepts of design and utility of large-scale carcinogenicity studies. J. Toxicol. Environ. Health **6**:1275.

Fisher J.W., Whittaker T.A., Taylor D.H., Clewell III J.H. and Andersen M.E., 1990. Physiologically based pharmacokinetic modeling of the pregnant rat: a multiroute exposure model for trichloroethylene and trichloroacetic acid. Toxicol. Appl. Pharmacol. **99**:395.

Gargas M.L., 1990. An exhaled breath chamber system for assessing rates of metabolism and rates of gastrointestinal absorption with volatile chemicals. J. Amer. Coll. Toxicol. **9**:477.

Gargas M.L. and Andersen M.E., 1989. Determining kinetic constants for chlorinated ethylene metabolism in the rat from rates of exhalation. Toxicol. Appl. Pharmacol. **99**:344.

Gargas M.L., Andersen M.E. and Clewell III H.J., 1986a. A physiologically based simulation approach for determining metabolic constants from gas uptake data. Toxicol. Appl. Pharmacol. **86**:341.

Gargas M.L., Andersen M.E. and Clewell III H.J., 1986b. Metabolism of inhaled dihalomethanes : differentiation of kinetic constants for two independent pathways. Toxicol. Appl. Pharmacol. **87**:211.

Gargas M.L., Burgess R.J., Voisard D.E., Cason G.H. and Andersen M.E., 1989. Partition coefficients of low molecular weight volatile liquids and tissues. Toxicol. Appl. Pharmacol. **98**:87.

Gaylor D. W., 1979. The EDOL study: Summary and conclusions. J. Environ. Pathol. Toxicol. **3**:179.

Gehring P.J., Watanabe P.G. and Blau G.E., 1976. Pharmacokinetic studies in evaluation of the toxicological and environmental hazard of chemicals. In: "New Concepts in Safety Evaluation" (M.A. Mehlman, R.E. Shapiro and H. Blumenthal, eds.), page 193. Hemisphere, New York.

Haggard H.W., 1924. The absorption, distribution and elimination of ethyl ether. II. Analysis of the mechanism of the absorption and elimination of such a gas or vapor as ethyl ether. J. Biol. Chem. **59**:753.

Hilderbrand R.L., Andersen M.E. and Jensen jr L.J., 1981. Prediction of in vivo kinetic constants for metabolism of inhaled vapors from kinetic constants measured in vitro. Fundam. Appl. Pharmacol. **1**:403.

Houk V.N., 1989. The risk of risk assessment. Regul. Toxicol. Pharmacol. **9**:257.

Howe R.B., 1983. GLOBAL83: An experimental program developed for the U.S. Environmental Protection Agency as an update to GLOBAL82. OSHA Contract No. 41USC252C3.

Howe R.B. and Crump K.S., 1982. GLOBAL83: a computer program to extrapolate quantal animal toxicity data to low doses (May 1982). OSHA Contract No. 41USC252C3.

Igari Y., Sugiyama Y., Sawada Y., Iga Y. and Hanano M., 1983. Prediction of diazepam disposition in rat and man by a physiologically based pharmacokinetic model. J. Pharmacokinet. Biopharmaceut. 11:577.

International Commission on Radiation Protection, 1975. Report of the task group on reference man. ICRP Publication No 23, Pergamon, New York.

Jepson G.W., 1986. A kinetic model for acetylcholinesterase inhibition by diisopropylfluorophosphate in crude rat brain homogenate. Master's thesis, Wright State University, Dayton, OH.

Kety S.S., 1951. The theory and application of the exchange of the inert gas at the lungs. Pharmacol. Rev. 3:1.

Kubic V.L., Anders M.W., Engel R.R., Barlow C.H. and Caughey W.S., 1974. Metabolism of dihalomethanes to carbon monoxide. I. Studies. Drug Metab. Dispos. 2:53.

Leung H.W., Paustenbach D.J., Murray F.J. and Andersen M.E., 1990. A physiologically based pharmacokinetic description and enzyme inducing properties of 2,3,7,8-tetrachlorodibenzo-p-dioxin in the rat. Toxicol. Appl. Pharmacol. 103:411.

Levy G., 1968. Dose dependent effects in pharmacokinetics. In: "Importance of Fundamental Principles in Drug Evaluation" (D.H. Tedeschi and R.E. Tedeschi, eds), page 141. Raven, New York.

Lin J.H., Sugiyama Y., Awaza S. and Hanano M., 1982. In vitro and evaluation of the tissue to blood partition coefficient for physiological pharmacokinetic models. J. Pharmacokinet. Biopharmaceut. 10:637.

Lorenz J., Glatt H.R., Fleischmann R., Ferlinz R. and Oesch F., 1984. Drug metabolism in man and its relationship to that in three rodent species. Monooxygenase, epoxide hydrolase and glutathione S-transferase activities in subcellular fractions of lung and liver. Biochem. Med. 32:43.

Mapleson W.W., 1963. An electric analog for uptake and elimination in man. Acta Pharmacol. Toxicol. 14:265.

McDougal J.N., Jepson G.W., Clewell III H.J., MacNaughton M.G. and Andersen M.E., 1986. A physiological pharmacokinetic model for dermal absorption of vapors in the rat. Toxicol. Appl. Pharmacol. 85:286.

Moolenaar R.J., 1989. Commentary on EPA carcinogenic risk assessment guidelines. Regul. Toxicol. Pharmacol 9:230.

Moolgavkar S.H., Venzon D.J., 1979. Two event models for carcinogenesis: incidence curves for childhood and adult tumors. Math. Biosci. 47:55.

Moolgavkar S.H., Leubeck E.G., de Gunst M., Port R.E. and Schwarz M., 1990. Quantitative analysis of enzyme-altered foci in rat carcinogenesis experiments. I. Single agent regimen. Carcinogenesis 11:1271.

National Academy of Sciences, 1983. Risk assessment in federal government: managing the process. National Academy Press, Washington, DC.

National Academy of Sciences, 1987. Pharmacokinetics in risk assessment. National Academy Press, Washington, DC.

National Toxicology Program, 1985. NTP Technical Report on the toxicology and carcinogenesis studies of dichloromethane in F-344 rats and B6C3F1 mice (inhalation studies). NTP TR TR 306.

Pang K.S., Rowland M. and Tozer T.N., 1978. Evaluation of Michaelis-Menton constants of hepatic drug elimination system. Drug Meta. Dispos. 6:197.

Park C.N., 1989. Mathematical models in quantitative assessment of carcinogenic risk. Regul. Toxicol. Pharmacol. **9**:236.

Ramsey J.C. and Andersen M.E., 1984. A physiologically based description of the inhalation pharmacokinetics of styrene in rats and humans. Toxicol. Appl. Pharmacol. **73**:159.

Rane A., Wilkinson G.R. and Shand D.G., 1977. Prediction of hepatic extraction ratio from in vitro measurements of intrinsic clearance. J. Pharmacol. Exp. Ther. **200**:420.

Reitz R.H., Mendrela A.L., Park C.N., Andersen M.E. and Guengerich F.P., 1988. Incorporation of in vitro enzyme data into the physiologically based pharmacokinetic model for methylene chloride: implications for risk assessment. Toxicol. Lett. **43**:97.

Reitz R.H., Smith F.A. and Andersen M.E., 1986. Metabolism of ^{14}C-methylene chloride. Toxicologist **6**:260.

Riggs D.S., 1963. The mathematical approach to physiological problems: a critical primer. MIT Press, Cambridge, MA.

Sato A. and Nakajima T., 1979a. Partition coefficients of some aromatic hydrocarbons and ketones in water, blood and oil. Br. J. Ind. Med. **36**:231

Sato A. and Nakajima T., 1979b. A vial equilibration method to evaluate the drug metabolizing enzyme activity for volatile hydrocarbons. Toxicol. Appl. Pharmacol. **47**:41.

Sawada Y., Harashima H., Hanano M., Sugiyama Y. and Iga T., 1985. Prediction of the plasma concentration of time courses of various drugs in humans based on data from rats. J. Pharmacobiodyn. **8**:757.

Singh D.V., Spitzer H.L. and White P.D., 1985. Addendum to the health risk assessment for dichloromethane. Updated carcinogenicity assessment for dichloromethane. EPA/600/8-82/004F.

Thorslund T.W., Brown C.C. and Charnley G., 1987. Biologically motivated cancer risk models. Risk Anal. **7**:109.

Tuey D.B. and Matthews D.H., 1980. Use of physiological compartmental model for the rat to describe the pharmacokinetics of several chlorinated biphenyls in the mouse. Drug Metab. Dispos. **8**:397.

United States Environmental Protection Agency, 1987. Update to the health assessment document and addendum for dichloromethane: pharmacokinetics, mechanism of action and epidemiology. EPA 600/8-87/030A.

United States Environmental Protection Agency, 1989. Interim methods for development of inhaled reference doses. (Office of Research and Development, Office of Health and Environmental Assessment, Environmental Criteria and Assessment Office, EPA Report # EPA/600/8-88/066F, Research Triangel Park, NC.

United States Environmental Protection Agency, 1990. Formaldehyde risk assessment update (External review draft). Office of Toxic Substances, US EPA, Washington, DC.

United States Environmental Protection Agency, 1991. Alpha-2µ-globulin: association with chemically-induced renal toxicity and neoplasia in the mate rat. EPA 62%/3-91. Office of Research and Development, OHEA-ECAO, Cincinnati, OH.

van Ginneken C.A.M. and Russel F.G.M., 1989. Saturable pharmacokinetics in the renal excretion of drugs. Clin. Pharmacokinet. **16**:38.

PHYSIOLOGICALLY BASED PHARMACOKINETIC MODELS: APPLICATIONS IN CARCINOGENIC RISK ASSESSMENT

D. Krewski, J.R. Withey, L.F. Ku

Health Protection Branch
Health & Welfare Canada
Ottawa, Ontario, Canada

C. C. Travis

Office of Risk Analysis
Oak Ridge National Laboratory
Oak Ridge, TN 37831

INTRODUCTION

Pharmacokinetics is the study of the absorption, distribution, metabolism and elimination of chemicals in biological systems. As such, it provides a means of resolving some of the ambiguities in exposure assessment and of evaluating the scientific assumptions upon which risk assessment is based.

Protocols for the conduct of pharmacokinetic studies and biostatistical methods used to curve fit blood, plasma, serum, fecal or urinary temporal data have recently been reviewed by Withey [1990] and Collins [1990] respectively. These techniques facilitate the mathematical description of the rates of absorption, distribution, metabolism, elimination and excretion of the substance of interest. This, in turn, may provide information on biological parameters that are not measured directly in such studies, such as the apparent volume of distribution, bioavailability, and steady-state levels achieved after repeated dosing. Pharmacokinetic studies conducted by different routes of administration, in different species, and at a range of doses (including those where certain rate limiting pathways may be saturated) have illuminated some critical aspects involved in route, species and dose extrapolation [Watanabe et al., 1988; National Research Council, 1987].

An important development in the area of pharmacokinetics has been the advent of physiologically based pharmacokinetic (PBPK) models. In this approach, each compartment represents a well-defined physiological entity. Relying on actual physiological parameters such as body weight, cardiac output, breathing rates, blood flow rates and tissue volumes to describe the metabolic process, PBPK models relate exposure concentrations to organ and tissue concentrations over a range of exposure

New Trends in Pharmacokinetics, Edited by A. Rescigno and A.K. Thakur
Plenum Press, New York, 1991

levels [Ramsey & Andersen, 1984; Andersen et al., 1987b; Paustenbach et al., 1988; Ward et al., 1988]. The physiological parameters are combined with chemical-specific parameters and metabolic parameters to predict the dynamics of a compound's movement through mammalian systems. By selecting the proper values for the parameters, the model can predict the transport and metabolism of the substance of interest via several routes of administration, across species, and over temporal variations in exposure [Travis, 1987]. Continual development and refinement of these models will reduce the uncertainties involved in exposure assessment, and subsequently in risk assessment.

In this chapter we explore the use of PBPK models as a tool for determining doses of the substance of interest and its metabolites to critical organs and tissues. The use of measures of dose delivered to the target tissues as a dose metameter is also examined.

MATHEMATICAL DESCRIPTION OF PBPK MODELS

A PBPK model is comprised of a series of compartments representing organs or tissue groups with realistic weights determined from the literature [Arms & Travis, 1988]. These models require physiological information on tissue volumes, blood flow rates to tissues, cardiac output, alveolar ventilation rates, and possibly membrane permeabilities. The models also utilize biochemical information such as blood/air partition coefficients and metabolic parameters. The uniqueness of the physiologically-based approach is its reliance on measured physiological and biochemical parameters. An appealing aspect of these physiological models is that they allow ready extrapolation from one animal species to another untested species simply by placing the appropriate physiological and biochemical parameters in the model [Dedrick, 1973b; Travis, 1987; Ward et al., 1988]. Similarly, the effect of route of administration can be investigated by allowing for several different input functions.

Physiological Compartments

A general representation of an individual physiological compartment within a PBPK model is shown in Figure 1. Here, $X(t)$ denotes the mass of the parent compound or one of its metabolites present in the compartment at time t. The concentration of the substance of interest in the compartment is then $C(t) = X(t)/U$, where U denotes the tissue volume. The arterial and venous blood concentrations, denoted by $A(t)$ and $V(t)$ respectively, reflect entry and exit of the substance to and from the compartment via the blood supply. To maintain continuity of blood flow, the flow rates of arterial and venous blood are usually assumed to be equal. In some cases, the substance may enter and be removed from the compartment of interest by pathways other than blood. The amounts of the substance entering and leaving the compartment by such an alternative pathway are denoted by $Y(t)$ and $Z(t)$ respectively. All of these parameters may depend on the compartment under consideration, which will be indexed by the subscript i.

The concentration of the substance of interest in each compartment may be described by a dynamic equation depending on the nature of the compartment. We now discuss dynamic equations which may be used with organ, blood and combined lung-blood compartments.

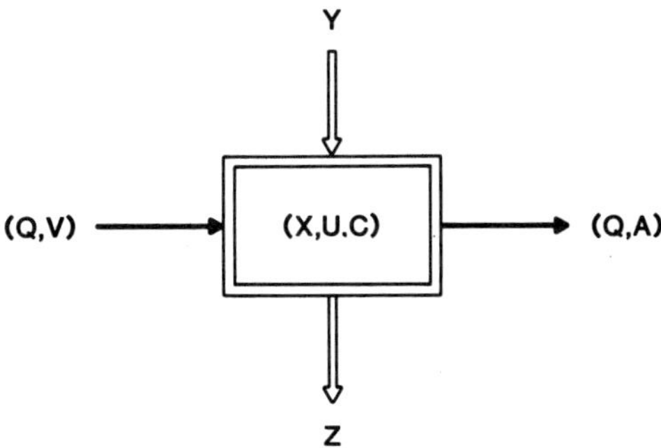

Figure 1. Schematic Representation of a Basic Physiologic Pharmacokinetic Compartment.

Legend: $A(t)$ = arterial blood concentration
$V(t)$ = venous blood concentration
Q = blood flow rate
$X(t)$ = mass of substance
$C(t)$ = compartmental concentration
U = compartmental volume
$Y(t)$ = input
$Z(t)$ = removal

Organ Compartments

The dynamic equation for most organ compartments is the mass balance equation

$$dX_i(t)/dt = Q_i(A_i(t) - V_i(t)) + Y_i(t) - Z_i(t), \tag{1}$$

where the arterial blood concentration $A_i(t) \equiv A(t)$, the concentration in the blood compartment. In general, concentration of the substance in the ith compartment is simply

$$C_i(t) = X_i(t)/U_i, \tag{2}$$

where $X_i(t)$ is the solution to the mass balance equation (1).

If the solubility of the substance in the tissue in the ith compartment is the same as in blood, we have

$$C_i(t) = V_i(t). \tag{3}$$

In many cases, however, the solubility of the substance in blood is different from that in tissues; their relationship is then defined as

$$C_i(t) = V_i(t)P_i, \tag{4}$$

where P_i is the tissue to blood partition coefficient. The determination of the value of P_i, which may be greater or less than unity, is discussed later in this chapter.

If solubility changes with concentration due to binding of the substance to proteins in the liver, we have

$$C_l(t) = V_l(t)P_l + B_{m1}V_l(t)/([K_{b1} + V_l(t)]U_l), \tag{5}$$

where B_{m1} is the binding capacity of the substance to a protein, and K_{b1} is the binding constant to the same protein. The second term on the right hand side of (5) permits a saturation mechanism to slow down the increase in the tissue concentration when the concentration approaches K_{b1}. Note that when $V_l(t) \ll K_{b1}$,

$$C_l(t) \cong V_l(t)(P_l + B_{m1}/[K_{b1}U_l]),$$

and when $V_l(t) \gg K_{b1}$,

$$C_l(t) \cong V_l(t)P_l + B_{m1}/U_l.$$

For TCDD in liver, it is necessary to add a more complex secondary binding mechanism described by [Leung et al., 1990a],

$$C_l(t) = V_l(t)P_l + B_{ml}V_l(t)/[(K_{b1} + V_l(t)U_l] + B_{m2}V_l(t)/[K_{b2} + V_l(t)U_l], \tag{6}$$

where $B_{m2} = B_{m20} + B_{m2i}V_l(t)/[K_{b1} + V_l(t)]$, and B_{m20}, B_{m2i} are constant. Note that equation (1) coupled with an appropriate equation for $V_i(t)$ such as (3) or (4) provides the two equations needed to find the solution for the two dependent variables $X_i(t)$ and $V_i(t)$.

Blood Compartment

The blood compartment supplies arterial blood to all other compartments and receives venous blood from other compartments. In general, the dynamic equation for the blood compartment is given by

$$dX_b(t)/dt = Q_b V_b(t) - A_b(t) + Y_b(t), \tag{7}$$

where $V_b(t) = \sum_{i \neq b} Q_i V_i(t)/Q_b$, $Q_b = \sum_{i \neq b} Q_i$, and $Y_b(t)$ is the rate of input of the substance of interest into the blood compartment.

In most applications, the arterial blood concentration $A_b(t) \equiv A(t)$ is simply

$$A(t) = C_b(t) = X_b(t)/U_b. \tag{8}$$

However, to allow for TCDD binding with blood cells, we may take

$$A(t) = C_b(t)/(1 + K_{ab}), \tag{9}$$

where K_{ab} is a constant set equal to 2 by Leung et al. [1990a]. Therefore, the free arterial concentration delivered to other compartments is only a fraction $1/(1 + K_{ab})$ of the blood concentration.

Equation (7) and either one of (8) or (9) provide the two equations needed to find the solution for the two dependent variables $X_b(t)$ and $A(t)$.

Lung-Blood Compartment

In the modeling of volatile substances, it is essential that lung function be incorporated. This is usually done by representing the lung by an alveolar chamber which is attached to the blood compartment as shown in Figure 2a [Ramsey & Andersen, 1984]. Equation (7) for blood is then modified to

$$dX_b(t)/dt = Q_{alv}[C_{inh}(t) - C_{alv}(t)] + Q_b V_b(t) - Q_b A_b(t) - Z_b(t),$$

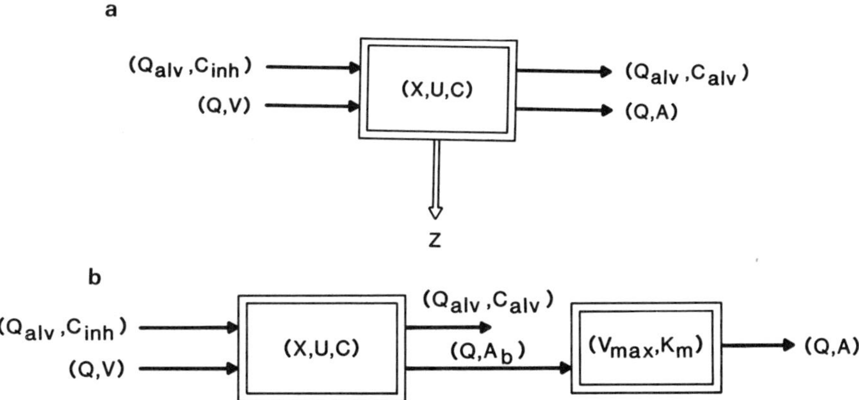

Figure 2. Lung-Blood Compartment with Metabolism

Legend:
$A(t)$ = arterial blood concentration
$V(t)$ = venous blood concentration
Q = blood flow rate
$X(t)$ = mass of substance
$C(t)$ = compartmental concentration
U = compartmental volume
$Z(t)$ = removal
C_{inh} = inhaled concentration
C_{alv} = alveolar concentration
Q_{alv} = alveolar ventilation
V_{max} = maximum reaction rate
K_m = Michaelis constant

[Whalen et al., 1989], where Q_{alv} is the alveolar ventilation rate, $C_{alv}(t)$ is the concentration in the alveoli, $C_{inh}(t)$ is the inhaled concentration, and $Z_b(t)$ denotes removal rate in the lung-blood compartment.

Note that Q_{alv} is another physiological parameter used in PBPK models for volatile substances, that can be readily determined by direct observation of respiratory function. However, two new variables $C_{inh}(t)$ and $C_{alv}(t)$ have been introduced into the dynamic equation, so that two more equations are needed to find a solution. Studies of volatile substances are usually carried out by placing the test animals in an inhalation chamber. In many cases, $C_{inh}(t)$ serves as an input which can be closely regulated by equipment. Otherwise, it can be related to $X_{air}(t)$ by

$$C_{inh}(t) = X_{air}(t)/U_{air},$$

where U_{air} is the volume of the exposure chamber and X_{air} is the substance in the air. Ramsey & Andersen [1984] assumed that

$$A_b(t) = C_{alv}(t)P_a,$$

where P_a is the blood to air partition coefficient.

There are several variations on the representation of the lung-blood compartment. Ramsey & Andersen [1984] ignored the terms $dX_b(t)/dt$ and $Z_b(t)$; others ignored the term $dX_b(t)/dt$ and considered removal $Z_b(t)$ to take place in a separate compartment as shown in Figure 2b [Andersen et al, 1987a].

Input and Removal

Chemical substances may enter the body in different ways, including intravenous and intragastric administration, oral uptake, inhalation, and dermal absorption. Chemicals may be removed from the body in expired air, by discharge in urine or fesces, and by transformation to metabolites. All of these uptake and elimination pathways can be incorporated into PBPK models to accommodate input and removal of substances from the compartmental system.

Input

Consider, for example, the case of intravenous injection of a dose y_k of a specific compound at time t_k. This input directly into the blood compartment can be expressed as

$$Y_b(t) = y_k\, \delta(t - t_k), \qquad\qquad (10)$$

where $\delta(t)$ is the Dirac deta function with $\delta(t) = 1$ for $t = 0$ and $\delta(t) = 0$ otherwise. For continuous intravenous infusion between times $t_1 \le t < t_2$, the input function is

$$Y_b(t) = y_k I_{[t_1,t_2]}(t), \q\qquad\qquad (11)$$

where

$$I_{[t_1,t_2]} = 1 \text{ for } t \in [t_1,t_2]$$
$$= 0 \text{ for } t \notin [t_1,t_2].$$

Both (10) and (11) can be used for repeated exposures by summing the input functions for individual exposures. The input function for inhalation exposure is analogous to that in (11), except that entry into the system is via the lung.

Oral and dermal uptake can be more complicated since absorption by either route may follow first order kinetics. For example, letting Y_g denote uptake by the gastrointestinal tract, we have

$$Y_g(t) = I_{[t_g,\infty]} k_g y_g e^{-\alpha_g(t-t_g)}$$

for a single and oral dose y_g administered at time t_g.

Removal

Removal of substances from the body is accommodated in a similar fashion. For water soluble substances eliminated in urine, we have

$$Z_k(t) = K_k\, U_k\, V_k(t)$$

where $Z_k(t)$ denotes the rate at which the substance is removed from the kidney, K_k is the removal coefficient, U_k is the volume of kidney tissue, and $V_k(t)$ is the venous concentration in the kidney. Implementation of this particular removal process is accomplished by isolating the kidney as a separate compartment attached to the liver [Paustenbach et al., 1988].

Xenobiotics and their metabolites may also be removed from the body directly by liver metabolism. For first-order metabolic processes,

$$dM_l(t)/dt = K_f \, U_l \, V_l(t), \tag{12}$$

where $M_l(t)$ is the amount of the metabolite produced at time t, K_f is the rate coefficient for elimination, U_l is the volume of the liver, and $V_l(t)$ is the venous concentration of the parent substance in the liver at time t.

Many metabolic processes are enzymatically mediated, and saturable upon enzyme depletion. Saturable metabolic processes are often described by the Michaelis-Menten equation

$$dM_l(t)/dt = V_{max} \, V_l(t)/[K_m + V_l(t)], \tag{13}$$

where V_{max} is the maximum reaction rate, and K_m is the Michaelis constant.

For some substances, more than one elimination pathway may exist. This occurs with methylene chloride, which is metabolized in the liver both by the linear glutathione-s-transferase (GST) pathway and by the saturable mixed function oxidose (MFO) pathway [Andersen et al., 1987a]. In this case, the total metabolic product is obtained by summing the GST and MFO pathways. It is also possible that an intermediate metabolite is produced which is then transformed into several further metabolites. For example, benzene is first metabolized into benzene oxide, and then converted to hydroquinone glucuronide, or hydroquinone sulfate, phenol glucuronide or phenal sulfate, muconic acid, and prephenyl mercapturic acid or phenyl mercapturic acid [Medinski et al., 1989]. Different metabolic products may also be produced by different pathways [Paustenbach et al., 1988].

Selection of Compartments

The number of compartments to be included in a PBPK model depends on the objective of the study. In general, however, the number of compartments should be kept as small as possible to simplify the analysis and interpretation of the results. In this regard, Fiserova-Bergerova [1983a] suggests that tissues in which chemical concentrations increase or decrease at the same rate can be treated as a single compartmental unit.

If the solubility of the chemical in tissue cells is the same as in blood, the rate of change of the chemical can be represented by rate of change of the blood in the tissue Q_i/U_i, where Q_i is the blood perfusion rate to the tissue, and U_i is the volume of the tissue. The values of Q_i, U_i and $Q_i/U_i i$ are given in Table 1 for a 70 kg man. Although the ratios span a wide range, it has been suggested that tissues with rate constants differing by less than a factor of five can be grouped into one compartment [Fiserova-Bergerova, 1983a]. Since all tissues in the fat and the muscle groups and the tissues in the liver and organ groups have this property, all body organs could, in principle, be pooled into two compartments for pharmacokinetic purposes. If all body organs were pooled into one compartment, the volume of this compartment (6 liters) would comprise about 8.6% of the total body volume, and be perfused by 75% of the cardiac output (4 liter/min).

In most applications, some refinement of the above grouping is needed. For example, the red marrow, muscle, lean subcutaneous tissue and skin are usually grouped into a muscle group which comprises of 54% of the total body volume and is perfused by 18% of the cardiac output (1.1 liter/min). Since metabolism occurs primarily in the liver, this organ is generally treated as a unique compartment separate from other organ tissues. Furthermore, the solubility of many chemicals in certain tissues is different from that in the blood. This difference is represented by the tissue to blood partition coefficient of a specific chemical P_i, and the rate of change of the chemical is modified as $Q_i/(P_iU_i)$. Since the partition coefficient for fatty marrow and adipose tissue is generally much larger than that of other tissues, they are generally grouped into a fat group which comprises 20% of the total body volume and is perfused by 5% of the cardiac output (0.26 liter/min.). Therefore, most PBPK models include the following four compartments: liver, organs, fat, and muscle.

A general five compartment PBPK model for non-volatile substance is shown in Figure 3. This is comprised of the four compartments identified previously, with organs and muscle identified as richly and poorly perfused compartments respectively, along with blood as an additional compartment. Input into the model occurs by intravenous administration; elimination takes place by metabolism in the liver. Additional compartments and entry or exit pathways may be added to this model depending on the properties of the compound being studied.

Solution of a Five Compartment PBPK Model

In this section, we will show how a PBPK model can be characterized mathematically by a system of equations. Each compartment is represented by two or more equations: one is a differential mass balance equation describing compartmental dynamics; the other specifies certain relationships among the variables involved in the mass balance equation. We seek a solution of these equations expressed in terms of the amount of the substance of interest $X_i(t)$ in the ith compartment at time t, or the concentration of that substance $C_i(t)$. If a reactive metabolite is produced, a solution for $M_i(t)$ is also required.

If all of the equations involved in defining the PBPK model are linear, then a closed form solution may be obtained by analytical means. To illustrate, consider the

Table 1

Volumes of Distribution and Blood Perfusion Rates for Selected Body Tissues[a,b]

Tissue	Volume of Distribution U (liter)	Blood Perfusion rate Q (liter/min)	Ratio Q/U (min^{-1})
Liver	3.9	1.58	0.41
Fat			
Fatty marrow	2.2	0.06	0.27
Adipose tissue	10.0	0.20	0.020
Organs			
Adrenals	0.02	0.1	5
Kidneys	0.30	1.24	4.1
Thyroids	0.02	0.08	4.0
Gray matter	0.75	0.60	0.8
Heart	0.30	0.24	0.8
White matter	0.75	0.16	0.21
Muscle			
Red marrow	1.4	0.12	0.086
Muscle	30.0	0.85	0.028
Lean subcutaneous	4.8	0.07	0.015
Skin	3.6	0.045	0.012

[a] For a 70 kg man in resting condition.

[b] Source: Fiserova-Bergerova [1983a].

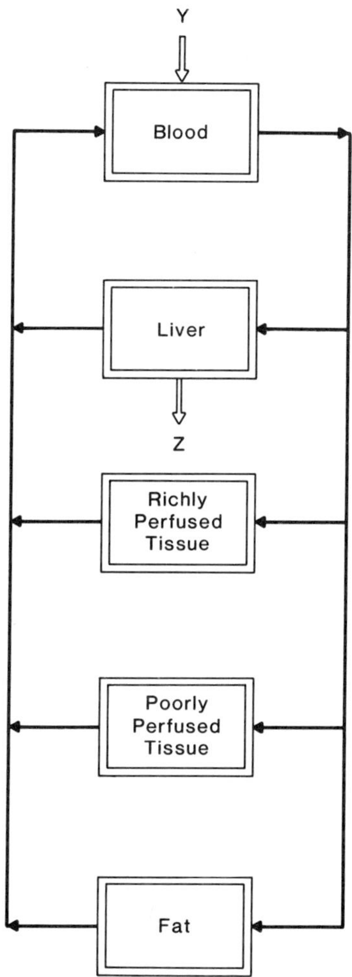

Figure 3. A Five Compartment PBPK Model

five compartment model in Figure 3 which is described by the following equations.

Blood: $\qquad dX_b(t)/dt = \sum_{i \neq b}(Q_i X_i(t)/U_i) - Q_b X_b(t)/U_b + Y(t) \qquad$ (14)

Liver: $\qquad dX_l(t)/dt = Q_l[X_b(t)/U_b - X_l(t)/(U_l P_l)] \qquad$ (15)

Richly Perfused Tissue: $\quad dX_r(t)/dt = Q_r[X_b(t)/U_b - X_r(t)/(U_r P_r)] \qquad$ (16)

Slowly Perfused Tissue: $\quad dX_s(t)/dt = Q_s[X_b(t)/U_b - X_s(t)/(U_s P_s)] \qquad$ (17)

Fat: $\qquad dX_f(t)/dt = Q_f[X_b(t)/U_b - X_f(t)/(U_f P_f)] \qquad$ (18)

Here, the subscripts b, l, r, s and f represent the blood, liver, richly perfused tissue, slowly perfused tissue and the fat compartment respectively.

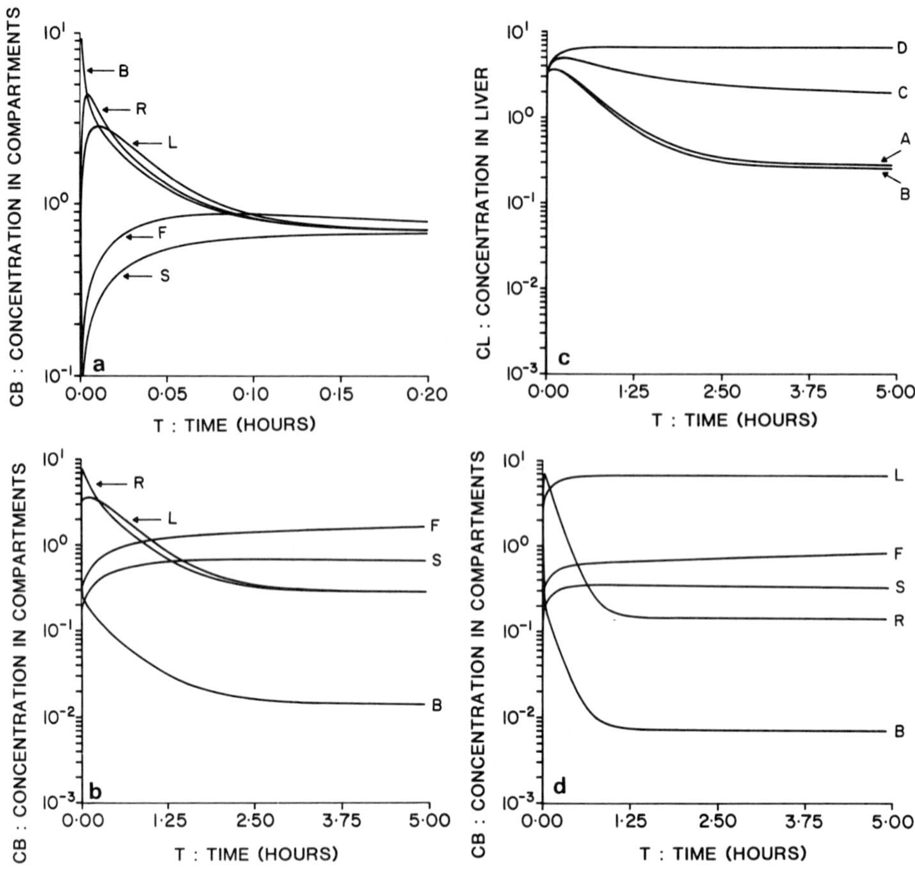

Figure 4.
Illustrative Solutions to the Five Compartment PBPK Model
a. Simple circulatory model
b. Partition coefficients added
c. Accumulation in liver with partition coefficient [A],
   with linear removal [B], with primary binding [C]
   and with secondary binding [D]
d. With binding in plasma

The coefficient Q_i/U_i describes the rate at which the ith compartment is replenished with new blood, and can be thought of as the natural frequency of that compartment. The term $Q_i/(U_iP_i)$ represents the rate of replacement of the substance in the compartment, and can be considered as the effective natural frequency of the compartment for a specific substance.

The solution to the above equations can be expressed as

$$X_i(t) = \sum_{j=1}^{5} D_{ij}e^{\gamma_j t} + B_j \int_0^t Y(t)\,dt \quad (i=b,l,r,s,f), \tag{19}$$

where the γ_j are the eigenvalues of the system associated with the solution of all compartments, D_{ij} is a constant determined by the initial concentration in each compartment, and B_i is another constant determined by the input function Y.

The steady state solution for an initial value of $X_b(0)$ in the blood compartment can be computed quickly by

$$X_b(\infty) = X_b(0)/(1 + \Sigma_{i \neq b} U_i P_i/U_b) \quad (i = l,r,s,f)$$

and

$$X_i(\infty) = X_b(\infty)/(P_i U_i/U_b).$$

The solution of a model with linear removal processes will have the same form as (19), and all eigenvalues will be non-zero.

The solution of a PBPK model is more complicated if any of the equations which characterize the model are nonlinear. In this event, a numerical solution is usually obtained.

Illustrative solutions to the five compartment model in Figure 3 are shown in Figure 4. These solutions were obtained under different conditions using the parameter values given by Leung et al. [1990a] for TCDD, and represent the concentration of the chemical of interest in each of the compartments at a given time following the administration of a single intravenous injection.

Consider first the simple circulatory model in Figure 4a, in which the response in compartment i is determined by the rate constant Q_i/U_i. Since the blood and richly perfused tissue have the highest rate constants (383 and 244 liter/hour respectively), the concentration in these two compartments is higher than in liver, fat, and slowly perfused compartments (with rate constants of 96, 9 and 5 liter/hour respectively) immediately post-dosing. Since no tissue binding is involved, however, the concentration of the chemical reaches the same steady state level in all five compartments.

Suppose now that tissue binding does occur (Figure 4b). The partition coefficients P_i used by Leung et al. [1990a] are 20, 20, 40, and 350 for liver, richly and slowly perfused tissue, and fat respectively. Because of the large partition coefficient for fat, the highest steady state concentrations of the test chemical occur in this tissue. Note that because the partition coefficients for liver and richly perfused tissue are identical, the same steady state concentration is reached in these two tissues.

The case in which the test chemical is removed from the liver according to first order kinetics ($K_f = 2.87$ hour^{-1}) is illustrated in Figure 4c, along with the case of primary and secondary binding in this tissue $B_{m1} = 0.054$, $B_{m20} = 25$, $B_{m2i} = 175$ nmol; $K_{b1} = .015$, $K_{b2} = 7$ nM. While the steady state concentration of the test chemical in the liver is lower with linear elimination than without, primary and secondary binding in the liver leads to increased steady state concentrations. Although not shown in this display, the increased concentrations of the test chemical in the liver resulting from either primary or secondary binding are accompanied by decreased concentrations in other compartments.

If plasma binding occurs ($K_{ab} = 2$ hour^{-1}), the free arterial concentration increases (Figure 4d). As with binding in the liver, steady state concentrations in other tissues are decreased.

DETERMINATION OF PHYSIOLOGICAL, BIOCHEMICAL AND METABOLIC PARAMETERS

Physiological Parameters

The physiological parameters used in pharmacokinetic modeling include tissue volumes, blood flow rates, cardiac output, and alveolar ventilation. The range of values used for cardiac output in mice, rats, and humans used in previous applications of PBPK models is summarized in Table 2. Cardiac output is defined as the flow of blood pumped by each ventricle of the heart. Cardiac output can be determined by multiplying the heart rate and the volume of blood ejected by each ventricle during each beat [Vander et al., 1975]. The major determinant of cardiac output in all species is the oxygen demand of the tissues. During physical activity, cardiac output increases, but to a smaller extent than alveolar ventilation [Astrand, 1983]. As a result, the perfusion-ventilation ratio is decreased by exercise. Thus, the ratio of cardiac output to alveolar ventilation decreases from a value of 1.3 at rest to a value of 0.4 to 0.5 with mild activity and a value of 0.4 with strenuous activity [Astrand, 1983]. To achieve consistency in risk assessment, it is desirable to have a set of reference physiological parameters for popular laboratory animals like mice, rats, and for humans. Arms & Travis [1988] have determined reference physiological parameters for use in pharmacokinetic modeling based on a generic 0.025 kg mouse, 0.25 kg rat, and a 70 kg man, without regard to sex, strain or age (Table 3). Weight-dependent differences in physiological parameters are accounted for by scaling within species on the basis of body weight as described later.

Table 2
Cardiac Output in Mice, Rats, and Humans (litre/min)

Source	Species		
	Mouse	Rat	Human
Mapleson [1963]	-	-	6.48
Munson and Bowers [1967]	-	-	5.00
Dedrick and Bischoff [1968]	-	-	6.48
Bischoff et al. [1971]	0.0044	0.03	4.04
Dedrick et al. [1972]	-	-	4.04
Dedrick et al. [1973]	0.0044	-	4.04
Harrison and Gibaldi [1977]	-	0.0673	-
Lutz et al. [1977]	-	0.0244	-
Fernandez et al. [1977]	-	-	6.5
Tuey and Matthews [1980]	0.0047	-	-
Ramsey and Andersen [1984]	0.015	0.0940	4.83
Fiserova-Bergerova [1985]	-	-	6.54
Andersen et al. [1987]	0.039	0.0850	5.80

Partition Coefficients

Partition coefficients are used to measure the affinity of a test chemical to tissues in the different compartments of the PBPK model. The partition coefficient P_i of a given chemical between two media is defined as the ratio of the chemical concentration in the first medium to the chemical concentration in the second medium when the media are in equilibrium. Various methods of measuring partition coefficients have been devised [Fiserova-Bergerova, 1983b], including estimation from thermodynamic properties and estimation from uptake or washout of a chemical from a given sample when a carrier gas is passed through the sample. For volatile substances, the vial-equilibration technique described by Sato & Nakajima [1979] is currently the most commonly used method. With this approach a known amount of the test chemical is sealed in an airtight vial and allowed to equilibrate. The concentration of the chemical in air and in tissue or blood is then measured. The ratio of the concentration in blood or in tissue against that in air is called the blood to air or the tissue to air partition coefficient respectively. The tissue:blood partition coefficient is then obtained by taking the ratio of the tissue:air to the blood:air partition coefficients. Under normal conditions, partition coefficients are independent of the dose level.

Table 3
Reference Physiological Parameters for Three Species[a]

	Species		
Parameter	Mouse	Rat	Human
Body weight (kg)	0.025	0.25	70.0
Tissue volume(fractions)			
Liver	0.055	0.04	0.026
Fat	0.10	0.07	0.19
Organs	0.05	0.05	0.05
Muscle	0.70	0.75	0.62
Cardiac output(liter/min)	0.017	0.083	6.2
Tissue perfusion(fractions)			
Liver	0.25	0.25	0.26
Fat	0.09	0.09	0.05
Organs	0.51	0.51	0.44
Muscle	0.15	0.15	0.25
Minute volume(liter/min)	0.037	0.174	7.5
Alveolar ventilation(liter/min)	0.025	0.117	5.0

[a]Source: Arms & Travis [1988]

It is possible that the vehicle of administration of a chemical can affect the determination of its partition coefficient. Angelo & Pritchard [1984] showed significant differences in the distribution, metabolism, and elimination between oral doses of methylene chloride administered in water and corn oil. They concluded that changes in the value of tissue to blood partition coefficient between water and corn oil dosing were necessary to match tissue and blood concentrations for these two vehicles, even after accounting for the differences in gastrointestinal (GI) tract absorption. Angelo & Pritchard [1984] also noted that the balance between exhalation and metabolism was different for methylene chloride administered in water and corn oil.

Metabolic Parameters

If metabolism depends on an enzyme whose supply is limited with respect to the time of the reaction, the rate of the metabolism will saturate as a function of time. This is referred to as Michaelis-Menten metabolism [Riggs, 1963], with, by equation (13),

$$dM_m(t)/dt = V_{max}V(t)/(K_m + V(t)), \qquad (20)$$

where M_m is the amount of metabolite, $V(t)$ is the concentration of the parent compound in venous blood, V_{max} is the maximum metabolic rate and K_m, the Michaelis constant, is the venous blood concentration at which the metabolic rate is half of V_{max}.

When the rate of metabolite formation is nonsaturable and depends linearly on the venous blood concentration, it is characterized by equation (12), with

$$dM_m(t)/dt = K_f U V(t), \tag{21}$$

where U is the volume of the tissue compartment in which the metabolism occurs and K_f is the linear metabolic rate constant. Linear (nonsaturable) metabolism will occur whenever there is a limitless supply of the compounds involved in the metabolic reaction. In the case of the chlorohydrocarbons, where metabolism occurs in the liver, U is the volume of the liver.

In liver metabolism of the chlorohydrocarbons, the formation of epoxide intermediates requires the enzyme nicotine adenine diphosphate (NADPH). Since the quantity of NADPH is limited during the time of metabolism of the parent compound, chlorohydrocarbon metabolism is described by the Michaelis-Menten formula. The metabolism of the haloethylenes is thought to involve linear metabolism due to the interaction of the parent compound with the sulfur of reduced glutathione (GSH). Thus, both linear and nonlinear metabolic models will be required in specific applications. The total metabolic product is the sum of the linear and nonlinear components.

The metabolic parameters V_{max}, K_m, and K_f can be estimated by fitting urinary metabolite data and blood and alveolar concentrations of the parent compound for mice and rats [Medinski et al., 1989; Richard et al., 1988]. Since chlorohydrocarbons show very little metabolism in humans [Reynolds & Moslen, 1980], determining the optimum metabolic parameters for humans is difficult.

Scaling of Physiologic and Metabolic Parameters

The most desirable method of obtaining the physiological parameters used in a pharmacokinetic model is direct measurement. When such values are not available for an untested species, or for an animal within the same species but different body weight, they can be obtained through scaling. The concept of using body weight scaling in pharmacokinetic modeling was introduced by Dedrick [1973b] and Dedrick & Bischoff [1980]. This is possible because both large and small animals are physiologically similar, and demonstrate an orderly variation in anatomical and physiological parameters with body weight [Adolph, 1949]. With some exceptions, the mechanisms of absorption, renal excretion, distribution, and storage are relatively similar in different species of animals, including humans [Fiserova-Bergerova & Hughes, 1983].

Many of the physiological parameters used in PBPK models vary with body weight according to the power function

$$Y = a\,W^b, \tag{21}$$

where Y is the physiological parameter of interest, W denotes body weight, and a and b are constants [Fiserova-Bergerova & Hughes, 1983]. If $b=1$, the physiological parameter y correlates directly with body weight. If $b=2/3$, Y is said to correlate roughly with surface area. The general power function in (21) is discussed in the now classical paper of Adolph [1949]. Tissue volumes, for example, scale across species in proportion to body weight raised to a power $0.70 \le b \le 0.99$, depending on the tissue considered (Table 4).

Table 4
Tissue Volumes as Functions of Body Weights[a]

Tissue	Power Function[b]	
	Constant a	Exponent b
Kidney	0.212	0.85
Brain	0.081	0.07
Heart	0.0066	0.98
Lungs	0.0124	0.99
Liver	0.082	0.87
Stomach and intestines	0.112	0.94
Blood	0.055	0.99

[a]Source: Adolph [1949]

[b]Based on $U = aW^b$, where U denotes tissue volume and W denotes body weight.

Other physiologic parameters may scale differently. Interspecies scaling of cardiac output is known to depend on the 3/4 power of body weight [Stahl, 1967; Holt et al., 1968; White et al., 1968; Takezawa et al., 1980]. There is also evidence that minute volume, and hence alveolar ventilation, scales across species in a similar fashion [Kleiber, 1961; Holt et al., 1968; Schmidt-Nielsen, 1970].

For modeling purposes, it is common to assume that tissue volume fractions and blood tissue perfusion fractions are constant and independent of body weight.

Ramsey & Anderson [1984] have suggested that the metabolic parameters V_{max} and K_f can be correlated with the body weight of the particular organism. They proposed power functions of the following form:

$$V_{max} = V_0 W^{0.7}$$

$$K_f = K_0 W^{-0.3},$$

where V_0 and K_0 are constant across species.

Table 5
Metabolic Parameters for Methylene Chloride in Four Species[a]

Parameter	Mice	Rats	Hamsters	Humans
Body weight (kg)	0.0345	0.233	0.140	70.0
Metabolic constants				
V_{max} (mg/hr)	1.054	1.50	2.047	118.9
K_m (mg/l)	0.396	0.771	0.649	0.580
K_f (hr^{-1})	4.017	2.21	1.513	0.53

[a] Source: Andersen et al. [1987a]

Andersen et al. [1987a] constructed a PBPK model of methylene chloride (DCM) disposition in four mammalian species: mice, rats, hamsters, and humans. The metabolic parameters were found by minimizing the difference between actual data and model predictions (Table 5). (Note that K_m is nearly constant across species.) A least-squares analysis of these data shows that V_{max} and K_f scale as follows:

$$V_{max} = 6.69 \ W^{0.75} \quad (r = 0.985);$$

$$K_f = 1.40 \ W^{-0.24} \quad (r = 0.946).$$

These latter observations are consistent with observations that other metabolic parameters are related to the 3/4 power of body weight [Kleiber, 1961; Holt et al., 1968; Schmidt-Nielsen, 1970].

Model Validation and Parameter Optimization

A PBPK model is characterized by the physiological compartments and metabolic processes selected to represent the body and by the physiological, biochemical, and metabolic parameters which govern the kinetics of tissue dosimetry. As discussed previously, values for these parameters may be obtained from published sources or from *in vitro* and *in vivo* studies of chemical partitioning, chemical binding, and metabolic rates. It is then desirable to validate the fully specified PBPK model by comparing model predictions with independently gathered experimental observations. This approach has been successfully followed with styrene [Ramsey and Andersen, 1984].

Another approach to reconciling model prediction with experimentally observed data is to choose values for critical model parameters which lead to reasonable agreement between observed and predicted results [Richard et al., 1988]. This may be done using statistical procedures such as least squares which minimizes the sum of squared differences between observed and predicted values [Andersen et al., 1987a]. Optimization of parameter values in this way works best when there are only a few unknown constants to be estimated. When there are many unknowns, the optimization problem becomes more difficult. As discussed later, applications of optimization techniques generally involve only one or two unknown metabolic parameters.

Those interested in the construction of physiologically relevant pharmacokinetic models may prefer, at least initially, to develop a complete PBPK model and then subject the model to some form of experimental validation. This approach is reasonable when all of the unknown parameters are readily estimated and when the model validation process proves successful. In some cases, however, it may not be possible to estimate all of the unknown parameters directly. This can occur, for example, with human metabolic constants which cannot be determined by experimental means for procedural or ethical reasons. It may also happen that modification to the original parameter values may be necessary to achieve adequate predictivity. In such cases, statistical methods for parameter optimization represent a useful tool for model development.

APPLICATIONS OF PBPK MODELS

Although sophisticated PBPK models have been developed only within the last two decades, sufficient experience with PBPK modelling has accumulated to demonstrate the utility of this technique in studying the pharmacokinetics and metabolism of xenobiotics. Following the pioneering work of Bischoff & Brown [1966] [see also Bischoff & Dedrick, 1968; Bischoff et al., 1970, 1971] in developing PBPK models for drugs, other investigators have developed PBPK models for volatile organic compounds such as methylene chloride and benzene, as well as nonvolatile chemicals like the polychlorinated biphenyls (PCBs) and 2,3,7,8 tetrachlorodibenzo-p-dioxin (TCDD). The term *physiological pharmacokinetics* was not used until about 1973 by Dedrick [1973a] [cf. Bischoff, 1987]. In this section, we review PBPK models developed for ten chemicals or chemical classes. Our purpose is to examine how PBPK models have been applied in practice, and to evaluate the overall utility of PBPK modeling as a tool for tissue dosimetry.

Styrene

Ramsey & Andersen [1984] proposed a relatively simple flow limited physiological model to describe the disposition, metabolism and tissue kinetics of volatile compounds. This model included fat, muscle and richly perfused compartments with the liver identified separately as the principal site of metabolism. The lung, blood and alveolar space were also identified separately. With relatively minor modification, this model has been applied to numerous other volatile compounds, some of which are discussed below.

The first compound to be addressed using this model was styrene, for which animal kinetic data, following administration by several routes, and human inhalation data was available [Withey, 1976; Withey & Collins, 1977; Young et al., 1979; Ramsey et al., 1980]. Metabolism was considered to be a saturable process that takes place principally in the liver. Previous studies on the hepatic metabolism allowed the determination of the Michaelis-Menten metabolic parameters [Andersen et al., 1984]. Tissue to blood partition coefficients were determined using a vial equilibration technique which involves first measuring the blood to air and homogenized tissue to air partition coefficients. Values for individual tissues to blood were then obtained as ratios of the air coefficients [Sato & Nakajima, 1979]. Model based predictions of styrene in blood were particularly sensitive to changes in the parameters that described its partitioning to fat and to the maximum metabolic rate V_{max}.

The metabolism of styrene was found to be saturable at inhalation exposures above 200 ppm in mice, rats and humans. Below this level, the concentration of styrene in blood relative to that in the inhaled air was controlled by perfusion limited metabolism. Above 200 ppm, the same ratio was controlled by the blood to air partition coefficient. This model adequately described the blood and tissue concentrations in several species, and accommodated the observed non-linearity over a range of exposure concentrations.

Methylene Chloride

One of the most comprehensive applications of PBPK modeling in carcinogenic risk assessment involved the carcinogenic response of the liver and the lungs of mice and rats exposed to inhaled doses of methylene chloride [Andersen et al., 1987a; Portier & Kaplan, 1989]. Mice were found to be much more sensitive to tumour induction in the lung and liver than either rats or hamsters. Andersen and his colleagues were able to calculate, using allometric scaling and measured parameters in their model, the actual delivered dose of methylene chloride and its principle metabolic products in these organs for each species.

The metabolism of methylene chloride occurs via two pathways. One of these, utilizing mixed function oxidases (MFO), follows Michaelis-Menten kinetics and was shown to be saturable over the range of exposures used in the cancer studies [Gargas et al., 1987]. The other pathway, involving cytosolic glutathione-s-transferase (GST), was not saturable over the same range of exposure concentrations. Andersen et al. [1987a] calculated the daily average of the areas under the concentration curves for both types of metabolite in the liver and the lung of each species. They found that the induction of tumours in the liver and lung was related to the delivered dose of GST metabolite in these organs. Subsequent studies on the relative metabolic conversion by GST pathway showed that it was 10 to 12 times more active in the mice than in rats, and its activity in hamsters and man was even lower [Green et al., 1986, 1987; European Chemical Industry Ecology and Toxicology Centre, 1989]. Thus, it was possible in this case to obtain a measure of the delivered dose of the active metabolite to the target organs which produced the same measure of effect in all species.

Perchloroethylene

The pharmacokinetics of perchloroethylene have been described for the rat, mouse and humans by Ward et al. [1988], using the same PBPK model as was used for methylene chloride. Metabolism was considered to occur only in the liver, and to involve principally the hepatic microsomal cytochrome P-450 system [Farber & Fisher, 1980], which is saturable. The secondary metabolic pathways for perchloroethylene have not been clearly identified, but were considered, in part, to involve glutathione conjugation similar to that observed for trichloroethylene [Dekant et al., 1986].

Kinetic data for mice, rats and humans, following inhalation and gavage dosing, were obtained from the literature. A good fit was obtained for the experimental data for mice by assuming that urinary metabolites accounted for 65% of total metabolism, independent of the administered dose. At low gavage doses of about 100 mg/kg, the fraction of the total metabolites obtained by the linear pathway was only 65%. At high doses of about 2000 mg/kg where the non-linear pathway is nearly saturated, the linear pathway contributed 45% of the total metabolites. This is consistent with the data of Buben & O'Flaherty [1985]. For experimental data on rats [Pegg et al., 1979], best fits were obtained only after the fat to air partition coefficient was increased by some 40%. Satisfactory fits to the experimental data for humans [Stewart et al., 1970; Monster et al., 1979; Fernandez et al., 1976] were obtained by using the partition coefficient determined for rats. It was found that improved fits for the data could be obtained if the muscle to air partition coefficient determined for the rat was increased.

Metabolic parameters for humans are not available, largely because human exposure studies with perchloroethylene have been conducted only at low exposure levels. Values for the metabolic parameters in humans were obtained by the use of scaling parameters using a power function of body weight as discussed previously. Although adequate fits were obtained by scaling rat metabolic parameters in this way, scaling of the metabolic parameters of mice appeared to overestimate human metabolism [Ward et al., 1988].

Trichloroethylene/Dichloroethylene

A gas-uptake system consisting of a closed 9 liter desiccator jar and a means of monitoring the atmospheric concentration of the vapour in the jar during animal exposure [Gargas et al., 1986] was used to determine kinetic constants for trichloroethylene and 1,1-dichloroethylene, alone and in combination, in rats [Andersen et al., 1987b]. Vial equilibration methods were used to obtain partition coefficients, and blood flows through the four compartments of the physiological model (fat, liver, muscle and viscera) were obtained by allometric scaling [Dedrick, 1973b].

The metabolic constants were obtained by fitting gas uptake curves for each chloroethylene [Gargas et al., 1986], and were found to be adequately described by a model which consisted of a single saturable pathway involving the microsomal mixed function oxidases with a high binding affinity for the substrates. Each of these compounds was considered to be an inhibitor of the other's metabolism; 1,1-

dichloroethylene was found to be a slightly better substrate for microsomal oxidation than trichloroethylene.

Rats were exposed to 0, 100, 200, 300 and 400 ppm of 1,1-dichloroethylene for six hours to examine hepatoxicity. Measurements of hepatic enzymes following these exposures revealed a dramatic increase in serum-glutamic-oxaloacetic transaminase (SGOT) at exposures above 100 ppm. In similar co-exposure studies, rats were exposed to 300, 713 and 1718 ppm of 1,1-dichloroethylene along with 500 ppm of trichloroethylene. When the increase in serum SGOT was plotted against the exposure concentration of 1,1-dichloroethylene, the curve was significantly displaced relative to that obtained for 1,1-dichloroethylene alone. However, no displacement was observed when the response was plotted against the expected amount of substrate metabolized. The metabolic data, together with the successful prediction of the uptake curves using the PBPK model, clearly demonstrated that these two structurally similar compounds were strictly competitive in their kinetic interactions within their common metabolic pathway.

Carbon Tetrachloride

Paustenbach et al. [1986] exposed rats to ^{14}C-labelled carbon tetrachloride for 8 or 11.5 hours per day for periods of one to two weeks to obtain temporal data on exhaled carbon tetrachloride, carbon dioxide and related metabolites. A PBPK model was developed which was able to predict the behavior of carbon tetrachloride in rats during repeated exposures, and to predict its pharmacokinetic properties in humans and monkeys [Paustenbach et al., 1988]. This model was also used to predict the potential for accumulation in adipose tissue on repeated inhalation exposure in rats and humans.

Live metabolism was adequately described by a single saturable pathway. Fecal and urinary metabolites were non volatile. The amounts of metabolite excreted via expired air, feces and urine were apportioned to three separate compartments. The simplest explanation for the kinetics of metabolism was that 4% of the initially metabolized carbon tetrachloride was directly converted to carbon dioxide, and the remainder was bound to biological substrates. These bound adducts then degraded slowly with an apparent half-life of approximately 24 hours. Repeated exposure did not change the values derived for the Michaelis-Menten constants, indicating that there was no significant enzyme inhibition or destruction at the exposure levels used.

There did not appear to be significant accumulation of carbon tetrachloride in adipose tissue of the rat following repeated 8 or 11.5 hours per day exposures to 100 ppm. However, this model predicted that humans exposed to the current threshold limit value of 5 ppm over an 8 hour working day would accumulate carbon tetrachloride in fatty tissues.

Methylchloroform

Reitz et al. [1988] developed a PBPK model for rats and mice subjected to intravenous, oral gavage, and drinking water exposure to methylchloroform. The object of this study was to demonstrate the ability of the model to extrapolate from high to low doses, between exposure routes, and between species. In addition, the model was used to predict changes that occurred in methylchloroform disposition in older animals [Schumann et al., 1982], and to compare chronic animal inhalation studies and human exposures. Metabolism was assumed to involve hepatic mixed function oxidases and to be saturable.

Data obtained from older rats (aged 18.5 months) was satisfactorily described by the PBPK model after increasing the size of the fat compartment in the model from 7 to 18 % of body weight. However, a similar increase in the size of the fat compartment did not account for all of the observed differences between young and old mice. The authors speculated that this may have been due to other factors such as an increased metabolic capacity in older mice.

The PBPK model developed for rodents predicted that humans exposed to methylchloroform in drinking water would achieve steady-state levels in the liver within a few days. It was further assumed that human liver would experience mild and reversible toxic changes, without cytotoxicity or necrosis, as were observed in chronic studies with rats and mice. [Rampy et al., 1978; Quast et al., 1984; Mc Nutt et al., 1975]. Although it is uncertain whether methylchloroform or its metabolites cause liver damage, methylchloroform *per se* was used to represent hepatic exposure. Target tissue doses calculated in this way for humans consuming two litres of water per day containing from 1 to 10 ppb of methylchloroform were predicted to be 4 to 6 orders of magnitudes lower than the tissue doses in rodents exposed to 875 to 1500 ppm by inhalation 6 hours per day, five days per week. These doses failed to produce hepatotoxicity in rats and mice. On the other hand, if tissue doses were expressed in the form of metabolites rather than the parent compound, the estimated risk in humans would be considerably lower since allometric calculations showed that the concentration of the metabolizing enzyme was 13-fold lower in the human as compared to mouse liver.

Benzene

Benzene has been known to induce hematopoietic effects in humans that range from pancytopenia to leukemia [Goldstein, 1977; Aksoy & Erdem 1978]. Studies on the chronic toxicity of benzene revealed that mice were much more sensitive than rats when dosed by gavage [National Toxicology Program, 1986a, b].

The PBPK model developed by Ramsey & Andersen [1984] was modified by Medinsky et al. [1989] to determine whether these species differences could be explained by tissue doses. The major pathways for the metabolism of benzene yield mixed metabolic products which include hydroquinone glucuronide or sulfate, phenyl glucuronide or sulfate, muconic acid and prephenyl mercapturic acid and phenyl mercapturic acid [Kalf, 1987; Rusch et al., 1977; Bechtold et al., 1988]. Metabolism is assumed to take place only in the liver and to follow Michaelis-Menten kinetics.

At inhalation levels of up to 1000 ppm, mice metabolized two to three times more benzene per unit body weight than rats, and produced 10-fold more hydroquinone conjugates. Rats produced phenyl sulfate as the primary metabolite, whereas mice formed substantial amounts of hydroquinone glucuronide and muconic acid in addition to phenyl sulfate. These metabolites of benzene may be translocated to other organs, including the principal target organ [Eastmond et al,. 1987]. It has been suggested that the metabolic pathways leading to the formation of hydroquinone and muconic acid are also associated with the formation of the putative toxic metabolites of benzene. This toxification pathway was found to be characterized by high affinity and low capacity. These differences in the metabolism of benzene by rats and mice are consistent with the increased susceptibility to benzene exhibited by mice.

Travis et al. [1990a] developed a PBPK model to describe the pharmacokinetics of benzene in mice, rats, and humans. Metabolism was assumed to take place in both liver and bone marrow and to follow Michaelis-Menten kinetics. For mice and rats, the model produces results similar to those of Medinsky et al. [1989]. For human inhalation exposures, the model was able to reproduce blood and expired air concentrations of benzene, as well as phenol concentrations in urine.

Polychlorinated Biphenyls and Halogenated Hydrocarbons

All of compounds considered to this point are volatile and, for the most part, have been analyzed using the PBPK model proposed by Ramsey & Andersen [1984]. Tissue to blood partition coefficients for these compounds were obtained from air to blood or tissue values obtained by vial equilibration techniques. Nonvolatile compounds such as polychlorinated biphenyls and halogenated hydrocarbons usually persist in the body for longer durations, and may be stored in body depots and redistributed. Tissue to blood partition coefficients for these involatile substances were obtained from actual tissue and blood concentrations measured following exposure. Bungay et al. [1979] described the characteristics of a suitable physiological model for these compounds using hexachlorobiphenyl and Kepone as illustrative examples.

Hexacholorobiphenyl, a nonpolar lipid soluble compound, was considered to be metabolized by the liver to its glucuronide conjugate, a water soluble polar metabolite which is readily excreted via the kidneys. Distribution of these moieties was considered to be limited by the blood flow through the various organs and tissues embodied in the model. The disposition of hexachlorobiphenyl to the fat depots was facilitated by its very high fat to blood partition coefficient compared to the low value for the polar metabolite [Lutz et al., 1977]. The preferential distribution of hexachlorobiphenyl to the fat, coupled with the poorly perfused nature of this tissue by the capillary system, indicated that it would accumulate in the fat depot. Furthermore, its redistribution to other tissues would be slow. In this regard, experimental data indicated that concentrations of hexachlorobiphenyl in liver and muscle tissue following the administration of intravenous doses peaked within the first two hours, while levels in the fat were still increasing four days post dosing.

This experiment, involving the sampling of blood following a single intravenous dose of hexachlorobiphenyl, was continued for 42 days, thereby allowing the increase in body weight of the animals and the consequent increase in the proportion of body weight represented by the fat depot to be incorporated into the model. The importance of the gut compartment excretion of the glucuronide metabolite via the bile and its subsequent reabsorption through the gut wall was also recognized. This enterohepatic recycling was also important to the elimination pathways considered in the model, and supported the conclusion that the accumulation of metabolites was negligible relative to that of the parent compound.

The disposition of Kepone (chlordecone) following intravenous adminstration was markedly different to hexachlorobiphenyl. This was considered to be principally due to the relatively low partition coefficient [Egle et al., 1978] of Kepone in fat (a range of 8 to 31) and its much higher value for the liver (a range of 28 to 126). It has been suggested that the high liver and low fat levels for Kepone may be a consequence of the formation of chlordecone hydrate or its strong binding to plasma proteins [Cohn et al., 1978].

Only a small percentage of the body burden of Kepone was excreted via the kidney after intravenous adminstration, the bulk being excreted via the bile and ultimately the feces. However, since actual measurements of biliary excretion were insufficient to account for the total amount excreted by the fecal route, the bile was suggested as not the only pathway that was available for elimination from the systemic circulation to the feces. Bungay et al.,[1979] suggested that Kepone was able to diffuse into the contents of the gut from the blood by perfusing the gut wall, as well as being absorbed in the reverse direction.

Tetrachlorodibenzofurans and Tetrachlorodibenzodioxin

Nonvolatile chlorinated benzofurans and dioxins have attracted considerable attention as persistent environmental contaminants because they are highly toxic in some animal species [Moore et al., 1979; Schwetz et al., 1973]. The acute toxicity of 2, 3, 7, 8,-tetrachlorodibenzofuran (TCDF) has been studied in several species including weight loss, thymic atrophy and hepatoxicity [Goldstein et al., 1978; Moore et al., 1979]. The single dose LD50 was found to be between 5 and 10 µg/kg in guinea pigs, and about 1000 ug/kg in monkeys. While LD50 was not reached in studies with mice and rats, toxic signs were found at about 1000 and 10 000 µg/kg, respectively, in these species. In a study of the disposition of TCDF in the rat, Yoshihara et al. [1981] found no relationship between tissue distribution and biological response. Thus, disposition alone may not account for the large interspecies difference in toxicity.

The longitudinal pattern of tissue distribution and excretion of TCDF and its metabolites has been described by a PBPK model [King et al., 1983] similar to that used for chlorinated biphenyl and Kepone [Bungay et al., 1979] based on data generated in previous reports [Birnbaum et al., 1980; 1981; Decad et al., 1981 a,b]. The flow-limited model consisted of blood, liver (considered to be the only site of metabolism), fat, skin and muscle. The model also incorporated excretion of metabolites via the urine, bile, gut lumen, and feces. Metabolite clearance rates from the liver were available for the mouse and rat; clearance rates for the monkey were obtained by scaling up rates for the rat, in proportion to body weight raised to the power 0.7. The compound was concentrated in the liver with tissue to blood partition coefficients of 130, 100, 100 and 30 for this organ in two strains of mice (C57 and DBA), the rat, and the monkey, respectively. Metabolism by the liver was modeled using first order linear kinetics. Urinary clearance of metabolites was a minor route of excretion in all species, the principal route being via the bile and feces. TCDF concentrated in the fat, with the tissue to blood partition coefficients being similar (between 25 and 40) for these species.

A similar PBPK model has been used to describe the tissue distribution of 2, 3, 7, 8,-tetrachlorodibenzo-p-dioxin in the rat [Leung et al., 1990a] and mouse [Leung et al., 1988]. The binding of TCDD to blood components was described by a first order linear process. Two classes of liver proteins were considered as binding sites in the liver, one being a high affinity, low capacity receptor [Poland et al., 1976], and the other an inducible low affinity, high capacity microsomal protein [Voorman & Aust, 1987]. Data from several single and multiple dose studies were used to analyze disposition and enzyme induction (TCDD is known to be a potent inducer of hepatic microsomal enzymes) following oral exposure [McConnell et al., 1989; Rose et al., 1976; Kociba et al., 1976; 1978].

Using the enzyme induction data of McConnell et al. [1989], the dissociation constant of the cytosolic arylhydrocarbon hydroxylase was determined to be 15 picomoles. Liver to fat concentration was about 4:1. This ratio was principally determined by the dissociation constant of the microsomal binding protein. Using these values, the two primary sites of accumulation were accurately described by the model under both single and repeated dosing. The kinetic behaviour was found to be very sensitive to binding capacities over the dose range for which data was available, and induction of microsomal binding proteins was included in the model in order to account for differences in the disposition at low (0.01 µg/kg) and high (1.0 µg/kg) doses. Since the carcinogenic mechanism is thought to involve the anylhydrocarbon hydroxylase complex, these data were considered to be useful for the development of pharmacodynamic models for cancer risk assessment.

In a follow-up investigation, Leung et al. [1990b] examined the effect of an inducing dose of TCDD administered i.p. 3 days earlier on the pharmacokinetics of a

TCDD analogue, radiolabelled 2-iodo-3,7,8-trichlorodibenzo-p-doxin. When injected with TCDD alone, the mice had the highest concentration of TCDD in fatty tissues; those pretreated with TCDD had the highest concentration in the liver. Pretreated mice had elimination rates for TCDD that were almost twice those observed when TCDD was administered alone. The inducing dose increased the amount of microsomal TCDD binding protein in the liver by about 12-fold and increased the rate coefficient for TCDD metabolism 3-fold. Again, the principal factor that influenced the liver to fat concentration ratio was the affinity and capacity of microsomal TCDD-binding proteins. The physiological model developed for these compounds was considered to work well in interpreting principal organ concentration, enzyme activity and induction, as well as in the interpretation of single and repeated dose data. This model should, therefore, be useful in the extrapolation of high dose rodent data to low dose human environmental exposure levels.

Methotrexate

A membrane-limited PBPK model for methotrexate has been developed by Bischoff et al., [1970, 1971] and Dedrick et al. [1973]. Methotrexate, used in treating leukemia, lymphoma, and other cancers, is a nonvolatile liquid with a high molecular weight, readily soluble in water Because of methotrexate's nonvolatility and its usual intravenous administration, there is no need for a lung compartment. Tissue compartments chosen in this model are those tissues known to be active sites for the chemical as an anticancer agent: spleen, bone marrow, kidney, and liver. Skin is also included as it appears to play a role in storing the chemical, thus providing elevated plasma concentrations over prolonged time periods following exposure. Finally, the gastrointestinal tract is considered as a separate compartment, and the rest of the poorly perfused tissue, besides skin, is combined into a muscle compartment.

Three of the tissue compartments in the methotrexate PBPK model (spleen, bone marrow, and the GI tract) demonstrate membrane-limited transport. This transport can be modeled by a series of compartments which mimic the diffusion process by slowing down the progress of the drug in steps. The physiological parameters required to model membrane-limited transport include the membrane permeabilities governing transfer of the chemical from the plasma to the interstitial fluid, and the permeabilities controlling passage from the interstitial fluid to the intracellular fluid.

The methotrexate model uses both linear and nonlinear representations of tissue binding. Nonlinear tissue binding occurs in the liver and kidney because of methotrexate's strong inhibition of the enzyme dihydrofolate reductase, which catalyses the formation to tetrahydrofolate [Dedrick et al., 1973]. Methotrexate also binds linearly to other proteins in those tissues.

CARCINOGENIC RISK ASSESSMENT

Up to this point, only the development and use of PBPK models for tissue dosimetry has been discussed. In order to address the pharmacodynamic properties of reactive metabolites, analogous models are required to describe the process of neoplastic transformation of cells in target tissue. Coupled with the PBPK models used for phamacokinetic purposes, such pharmacodynamic models will embody the important features of chemical carcinogenesis.

Models of Carcinogenesis

Moolgavkar & Venzon [1979] and Moolgavkar & Knudsen [1981] describe a biologically based two stage birth-death mutation model of carcinogenesis [Moolgavkar & Luebeck, 1990]. This model assumes that two mutations are required to convert a normal stem cell in the target tissue to a malignant cancer cell. The rates of occurrence of cancer of these two mutations are assumed to be related to the dose of the proximate carcinogen delivered to the target tissue. The kinetics of initiated cells which have sustained the first mutation is described by a birth-death process. Agents which increases the net birth of initiated cells are called promotors [Thorslund et al., 1987]. Such selective clonal expansion of the initiated cell population increases the pool of cells available for neoplastic, transformation thereby increasing the overall level of risk. Tissue growth is accommodated in the two-stage model by allowing the number of cells in the tissue of interest to vary with age.

In contrast to the two-stage model, the Armitage-Doll multi-stage model assumes that k genotoxic events are required to convert a normal stem cell to a malignant cell [Armitage & Doll, 1961]. As described by Armitage [1985], the multi-stage model explicitly incorporates neither tissue growth nor cell kinetics as does the two-stage model.

The formulation of an appropriate biologically based multi-stage model to describe the effects of a particular carcinogen is a complex process, analogous to the development of a PBPK model for tissue dosimetry. While there are clearly more than two genetic events involved in neoplastic transformation induced by a given carcinogen, not all of these will be rate-limiting. By allowing for tissue growth and cell kinetics, an adequate description of most toxicological and epidemiological data on carcinogenesis can be achieved by postulating only two critical genotoxic events. Without these non-genetic factors, more than two stages are often required to describe the available data.

Low Dose Risk Assessment

Since human exposure to most environmental carcinogens is low, it is often of interest to estimate the level of risk associated with very low doses. One approach to low dose risk estimation is to fit a particular dose response model to carcinogen bioassay data obtained at moderate to high doses inducing observable tumour occurrence rates, and then extrapolate to the low dose region of interest using the fitted model [Krewski et al., 1989]. The problem with this approach is that different models can lead to markedly different estimates of risk upon extrapolation outside the experimentally observable response range [Krewski & Van Ryzin, 1981]. Because of this, upper bounds on low dose risks calculated under the assumption that the dose response curve is linear at low doses are often used [Office of Science and Technology Policy, 1985; U.S. Environmental Protection Agency, 1986]. Such upper bounds may be calculated using the linearized multi-stage model discussed by Crump [1984] or using the model-free procedure described by Krewski et al. [1991]. Although both methods of linear extrapolation generally lead to similar estimates of risk, the model-free method avoids unnecessary assumptions concerning the mechanisms of carcinogenesis other than that of linearity at low doses. As experience accumulates with biologically based models of carcinogenesis such as the two-stage clonal expansion model, it is possible that more refined estimates of risk will be obtained in the future.

Administered and Delivered Doses

In the past, the dose administered to test animals, or the environmental level of exposure to human populations, has been used to describe the relationship between dose and risk. PBPK models provide an opportunity to improve estimates of carcinogenic risk through improved tissue dosimetry.

When the dose of the reactive metabolite is directly proportional to the administered dose of the parent compound, estimates of risk will be identical using either dose metric. In the presence of saturable activation or detoxification pathways, the relationship between administered and delivered dose will be nonlinear. This was demonstrated by Hoel et al. [1983] using a theoretical paradigm for metabolic activation developed by Ghering & Blau [1977]. Saturation of an activation pathway at high doses results in a decrease in the amount of active metabolite reaching the target tissue; saturation of a detoxification pathway results in an increase in the amount of metabolite formed at high doses.

Krewski et al. [1987] investigated the impact of saturation kinetics on estimates of low dose risk obtained using model-free extrapolation (MFX) by computer simulation. In this study, risk was essentially assumed to be directly proportional to delivered dose. In the absence of saturation effects, the MFX procedure produced accurate estimates of risk using administered doses. However, serious overestimation of risk occurred when upon saturation of an elimination pathway. Saturation of an activation pathway, on the other hand, resulted in underestimation of risk. Similar findings have been reported by Hoel et al. [1983] and Whittemore et al. [1986].

Taken as a whole, these results suggest that improved tissue dosimetry can lead to improved estimates of low dose cancer risk, at least in the case where risk is proportional to the dose delivered to the target tissue. Thus, the use of PBPK models for tissue dosimetry offers the promise more accurate estimates of carcinogenic risk.

Tissue Dosimetry

A long standing problem in toxicology is the extrapolation of observed experimental results between animal species and man. It has historically been assumed that experimental results can be extrapolated between species if administered dose is standardized using one of two metrics: mg/kg body weight/day (body weight scaling) or mg/m^2/day (surface area scaling). It is understood that neither of these two metrics will be exactly correct for all chemical compounds and that interspecies scaling should depend upon the kinetic behavior of the particular compound and its mechanism of toxicity [National Research Council, 1987].

With the advent of PBPK models, it has become possible to provide a reasonably accurate description of the pharmacokinetics of parent compounds and metabolites in mice, rats, and humans. In cases where such models exist, it is no longer necessary to extrapolate experimental results on the basis of administered dose. The effective dose to the target tissue can be estimated using the pharmacokinetic model and interspecies extrapolation done on this basis. It is generally agreed that the most appropriate measure of dose to target tissue is the time course of the concentration of the toxic moiety in the target tissue. However, since it is inconvenient to compare tissue concentration curves at all time points, the area under the tissue curve in chronological time (AUC_{ct}) of the toxic moiety is often used as a convenient surrogate.

Travis et al. [1990b] have proposed an alternative definition of dose to target tissue. They suggest that individuals from different species will receive the same dose to target tissue if, an only if, the time profiles in physiological time of the concentration of toxic moiety in the target tissue are the same. Physiological time is an allometric scale which relates biological time functions to body weight and renders the rate of all biological events approximately constant across species when compared per unit of physiological time. Physiological time (t') can be defined in terms of chronological time (t) and body weight (W) by:

$$t' = t \cdot W^{-0.25}.$$

Support for this concept has been provided by many investigators who have shown that breath duration, heartbeat duration, longevity, pulse time, breathing rates, and blood flow rates are approximately constant when expressed in these internal time units. Thus, Travis et al. [1990b] proposed the surrogate measure for dose to target tissue to be aea under the tissue concentration curve in physiological time of the toxic moiety (AUC_{pt}).

To clarify this concept, consider the definition of AUC_{ct}. Using a change of variable from chronological time t to physiological time t',

$$AUC_{ct} = \int_0^\infty C(t)\,dt$$

$$= \int_0^\infty C(t')\,dt' \cdot W^{0.25}$$

$$= AUC_{pt} \cdot W^{0.25}$$

where C(t) denotes the concentration of the reactive metabolite in the target tissue at time t. It follows that $AUC_{pt} = AUC_{ct}/W^{0.25}$. Within a species, no difficulty arises from assuming that effect is proportional to AUC_{ct} rather than AUC_{pt} since the two measures of dose differ by a constant. For interspecies extrapolation, however, the two different dose measures will produce different results.

Travis et al. [1990b] used PBPK models for mice, rats and humans to demonstrate that if toxic response is a function of the time profile in physiological time of the concentration of the toxic moiety in the target tissue, then the appropriate interspecies scaling law for toxic compounds which are metabolically deactivated is mg/kg per unit of physiological time. At low dose rates, this metric is approximately equivalent to $mg/kg^{0.75}/day$.

Impact of Uncertainty in Model Parameters

In the previous section, we discussed examples in which the dose of the reactive metabolite reaching the target tissue was used as a dose metameter for pharmacodynamic effects. In each of these examples, tissue dose was evaluated on the basis of an appropriate PBPK model. Since PBPK models depend on a number of physiological, biochemical, and metabolic parameters, uncertainty in the PBPK model parameters will impart uncertainty in the estimates of tissue dose based on this model. The extent to which this occurs has recently been examined in sensitivity analyses conducted using PBPK models for methylene chloride and tetrachloroethylene.

Methylene Chloride

The effects of uncertainty in the parameters of the PBPK model in Figure 2b developed by Andersen et al. [1987a] for methylene chloride was addressed by Portier & Kaplan [1989]. This PBPK model includes a total of 23 distinct parameters consisting of tissue weights, blood flow rates, partition coefficients, and metabolic constants. Estimates of the variability of these parameters were based on literature sources whenever possible; otherwise, the coefficient of the variation was arbitrarily assigned a value of 20%, 50%, 100% or 200%. This information was used to develop a probability distribution of plausible values for each model parameter, which formed the basis for a sensitivity analysis of tissue dose to variation in the model parameters.

Rather than express the results of their analyses in terms of tissue dose *per se*, Portier & Kaplan [1989] focused on the variability in the 10^{-6} risk specific dose (RSD), which is defined as the dose estimated to induce a lifetime cancer risk of 10^{-6}. This was estimated by fitting an essentially linear one-stage model of carcinogenesis to simulated experimental outcomes, using tissue doses derived from the PBPK model. Following Andersen et al. [1987a], tissue dose was expressed in terms of the area under the concentration-time curve in lung and liver tissue for metabolites of methylene chloride produced via the GST pathway. The reported distributions of RSDs obtained by computer simulation thus allowed for variability due to uncertainty in the PBPK model parameters and for variability in the experimental data on tumourigenesis.

Portier & Kaplan [1989] found that the distribution of RSDs obtained allowing for uncertainty in the PBPK parameter value exhibited substantially greater dispersion than when these parameters were treated as known constants. Under maximal uncertainty, for example, standard deviation of the distribution of RSDs was increased by a factor of approximately 10-fold.

Trichloroethylene

A similar analysis of the effects of uncertainty in the values of the parameters of the PBPK model for trichlorethylene was conducted by Farrar et al. [1989]. Parameter uncertainty was evaluated on the basis of variation in published values, sometimes in conjunction with subjective evaluations based on expert opinion. Since model parameters are not necessarily independent of one another, multivariate probability distributions were used to characterize uncertainty.

In this study, three measures of tissue dose were considered. These included the average daily area under the concentration time curve for trichloroethylene in the liver (AUCL), the average daily area under the concentration time curve trichloroethylene in arterial blood (AUCA), and the average daily amount of trichlorethylene metabolites per volume of liver tissue (CML).

In addition to evaluating the variability in these three measure of delivered dose imparted by uncertainty in the value of the PBPK model parameters, Farrar et al. [1989] also considered the variability in the corresponding estimates of extra risk at an administered dose of 50 ppm for 8 hours/day 5 days/week (the current OSHA standard). Only a single neoplastic response (hepatocellular carcinoma) was considered, with best estimates and upper confidence limits on the risk obtained using the multi-stage model with the number of stages fixed at two, corresponding to the number of nonzero doses in the available bioassay data.

This investigation produced the following results. First, the medians of the distributions of the three dose surrogates (AUCL, AUCA and CML) were close to the values obtained ignoring the uncertainty in the PBPK model parameters. There was, however, appreciable variations about the median, with the range from the 2.5th to 97.5th percentile spanning a range of about 4 to 100-fold, depending on the dose surrogate. The best estimates of risk based on these three dose surrogates demonstrated even greater variability.

Although risk estimates assuming no uncertainty in the parameters of the PBPK model for trichloroethylene differed markedly depending on which dose surrogate was used, Farrar et al. [1989] gave no indication as to which measure of dose (AUCL, AUCA, CML, or the administered dose) was considered most appropriate.

The authors concluded that "for tetrachloroethylene, it appears that the structural uncertainty associated with the selection of an appropriate dose metric for cross-species extrapolation is of relatively greater importance than is the uncertainty associated with the values of the PBPK model parameters".

SUMMARY AND CONCLUSIONS

The classical method of estimating risk from exposure to a volatile organic substance assumes that the biological effect of a compound is directly related to the exposure dose absorbed via the lungs. Pharmacokinetics allows for estimation of risk to be based on the quantity of toxic substance which actually reaches the target tissue. The dose to target tissue can be determined through computer solution of a set of equations describing the distribution and metabolism of the compound in the human body. This set of equations is termed a pharmacokinetic model.

PBPK models utilize three types of data: the anatomical and physiological structure of the animal, the partitioning of chemicals in various tissues, and metabolic parameters in various organs. For animals, all of these parameters can be obtained from the literature or from laboratory experiments. For humans, the physiological and partitioning parameters are available, but the metabolic parameters generally must be indirectly inferred.

The PBPK model is a tool with diverse and wide-ranging applications. Used appropriately, it can aid in understanding species differences in xenobiotic metabolism and response, provide a clearer means of measuring effective or target tissue dose, and aid in estimating risk due to exposure to potentially hazardous materials.

Perhaps the most significant aspect of this kind of pharmacokinetic modeling, with respect to either the design or interpretation of toxicity studies, has been the identification of non-linear kinetic processes that are indicated by variation in kinetic parameters with the administered dose [McKenna, 1987]. This concept has been especially important in the interpretation of several studies involving animal exposure to volatile aliphatic halogenated hydrocarbons and their extrapolation to man. If any rate-limiting pharmacokinetic process becomes saturated, then the delivered internal dose of either the parent compound or its metabolites to the target tissue will not be directly proportional to the administered dose, and disproportionate increases or decreases in the delivered dose will occur. Levy [1968] has given some guidelines for determining whether there is an indication of the saturation of a kinetic process:

(i) The elimination of a xenobiotic in the body is not described by an exponential curve.
(ii) The terminal half-life of elimination increases with increasing administered dose.
(iii) Areas under the blood concentration-time curve are not proportional to increasing doses.
(iv) Excreted metabolites will be quantitatively changed with increasing dose.
(v) Competitive inhibition of enzymes by other chemicals, metabolized by the same metabolic pathway, will be likely.
(vi) Dose-toxic response curves will show increases or decreases in response that are disproportionate at dose levels higher than those where saturation occurs.

Recent advances in the application of physiological parameters to evaluate blood and excretion data have allowed a much more rational approach to the characterization and extrapolation of kinetic data between species [Gerlowski & Jain, 1983;

Andersen et al., 1987a; Himmelstein & Lutz, 1979; National Research Council, 1987]. The physiological model groups realistic physiological compartments, such as poorly perfused or well perfused tissues, and metabolic or excretory organs, connected together as a series of parallel shunts by the arterial and venous blood flow. When available, actual physiological parameters, like the breathing rate, blood flow rates through individual organs, tissue volumes, blood to tissue partition coefficients and metabolic rates are used to evaluate the kinetic behaviour and distribution of an administered dose. These physiological parameters may be measured either by *in vitro* or *in vivo* methods, or, if these are not available they can be obtained from allometric relationships [Travis et al., 1990b; Boxenbaum & D'Souza, 1990]. By using the appropriate parameters specific to the species, the pharmacokinetic behaviour of any xenobiotic can be extrapolated to any species or to any route of application.

PBPK models for volatile organics such as chlorohydrocarbons are flow-limited, have linear tissue binding, and involve several sites of chemical metabolism. A PBPK model which has found frequent application in the description of the inhalation of volatile organics includes compartments based on (1) organs such as brain, heart, kidney, and viscera, (2) muscle, (3) fat, and (4) metabolic organs (principally liver). A separate compartment has been assigned to fat because chlorohydrocarbons are lipophilic and deposit mostly in fatty tissue. Bone is ignored as volatile organics are not expected to reside there. Since volatile chemicals are cleared to a considerable extent through exhalation, the model is completed by connecting all tissue compartments via the blood flow pathway with a lung compartment. While a lung compartment is necessary when modeling inhalation data, it is also very important for volatile chemicals when the dose route is ingestion. Finally, the metabolism of the chemical in the liver is modeled by metabolic rate equations which can be linear or nonlinear.

It should be pointed out that no single PBPK model can be used to determine the distribution of all chemicals in the body. Like any other tool, the primary rule in model selection is that the model should provide the simplest description consistent with its intended purpose. The number of compartments and the way they are connected will vary from chemical to chemical depending upon the chemical's metabolic behaviour and the nature of the carcinogen bioassay data available for the chemical.

Substantial practical experience with PBPK modeling has now accumulated to demonstrate their utility as tools for improved tissue dosimetry. Successful application of PBPK models with ten chemical carcinogens have been described in this chapter alone. This does not imply that predictions of the dose of the reactive metabolite delivered to target tissues based on such pharmacokinetic models are always superior to less refined measure of exposure such as the administered dose level. Care needs to be taken to validate predictions of tissue dose based on PBPK models, and to ensure that the error associated with such predictions is within reasonable limits of uncertainty.

In addition to the pharmacokinetic dimensions of carcinogenic risk assessment, consideration also needs to be given to the pharmacodynamic processes governing the interaction between critical metabolic products and target tissues. Biologically based models of carcinogenesis such as the two-stage clonal expansion model which allow for tissue growth and cell kinetics as well as genotoxic interactions with DNA offer considerable promise as tools for obtaining accurate estimates of risk. Ideally, risk assessment methods incorporating realistic assumptions about both the pharmacokinetic and pharmacodynamic processes involved in carcinogenesis will ultimately result in improvements over current risk assessment methodologies based on crude measures of exposure and linearized upper bounds on low dose risks.

ACKNOWLEDGEMENTS

We are grateful to Dr. Ajit Thakur for helpful comments on the original version of this article.

REFERENCES

Adolph, E.F., 1949. Quantitative Relations in the Physiological Constitutions of Mammmals. Science **109**:579.

Aksoy, M. and S. Erdem, 1978. Follow-up Study on the Development of Leukemia in 44 Pancytopenic Patients with Chronic Exposure to Benzene. Blood **52**:285.

Andersen, M.E., R.L. Archer and M.G. MacNaughton, 1984. A Physiological Model of the Intravenous and Inhalation Pharmacokinetics of Three Dihalomethanes-CH_2Cl_2, CH_2BrCl, CH_2Br_2 in the Rat. Toxicologist **4**:111.

Andersen, M.E., H.J.III. Clewell, M.L. Gargas, F.A. Smith and R.H. Reitz, 1987a. Physiologically Based Pharmacokinetics and the Risk Assessment Process for Methylene Chloride. Toxicol. Appl. Pharmacol. **87**:185.

Andersen, M.E., M.L. Gargas, H.J. Clewell III, and K.M. Severyn, 1987b. Quantive Evaluation of the Metabolic Interactions between Trichloroethylene and 1,1-Dichloroethylene in Vivo Using Gas Uptake Methods. Toxicol. Appl. Pharmacol. **89**:149.

Angelo, M. and A. Pritchard, 1984. Simulations of Methylene Chloride Pharmacokinetics Using a Physiologically-Based Model. Toxicol. Appl. Pharmacol. **4**:329.

Angelo, M.J. and A.B. Pritchard, 1987. Route to Route Extrapolation of Dichloromethane Exposure Using a Physiological Pharmacokinetic Procedure. Pharmacokinetics in Risk Assessment: Drinking Water and Health. Vol.8:254.

Armitage, P., 1985. Multistage Models of Carcinogenesis. Environ. Health Perspect. **63**:195.

Armitage, P. and R. Doll, 1961. Stochastic Models for Carcinogenesis. In: "Proceedings of the Fourth Berkeley Symposium" (J. Neyman, ed.), page 19. University of Calfironia Press, Berkeley, California

Arms, A.D. and C.C. Travis, 1988. Reference Physiological Parmeters in Pharmacokinetic Modelling. U. S. EPA Final Report. EPA 600/6-88/004.

Astrand, I., 1983. Effect of Physical Exercise on Uptake. Distribution and Elimination of Vapors in Man. In: "Modeling of Inhalation Exposure to Vapor: Uptake, Distribution and Elimination" (V. Fiserova-Bergerova, ed.), Volume 2, page 107. CRC Press, Boca Raton, Florida.

Bechtold, W.E., P.J. Sabourin and R.F. Henderson, 1988. A Reverse Isotope Dilution Method for Determining Benzene and Metabolites in tissues. J. Anal. Toxicol. **12**:176.

Birnbaum, L.S., D.M. Decad and H.B. Matthews, 1980. Disposition and Excretion of 2,3,7,8 Tetrachlorodibenzofuran in the Rat. Toxicol. Appl. Pharmacol. **55**:342.

Birnbaum, L.S., G.M. Decad, H.B. Matthews and E.E. McConnell, 1981. Fate of 2,3,7,8-tetraclorodibenzofuran in the Monkey. Toxicol. Appl. Pharmacol. **57**:189.

Bischoff, K.B., 1987. Physiologically Based Pharmacokinetic Modeling. In: "Pharmacokinetics in Risk Assessment: Drinking Water and Health", Vol. 8, page 36. National Academy Press, Washington, D.C.

Bischoff, K.B. and R.G. Brown, 1966. Drug Distribution in Mammals. Chemical Engineering Progress Symposium Series **66**:33.

Bischoff, K.B., R.L. Dedrick and D.S. Zaharko, 1970. Preliminary Model for Methotrexate Pharmacokinetics. J. Pharm. Sci. **59**:149.

Bischoff, K.B., R.L. Dedrick, D.S. Zaharko and J.A. Longstreth, 1971. Methotrexate Pharmacokinetics. J. Pharm. Sci. **60**:1128.

Bischoff, R.L., and R.L. Dedrick, 1968. Thiopental Pharmacokinetics. J. Pharm. Sci. **57**:1346.

Boxenbaum, H., 1982. Interspecies Scaling, Allometry, Physiological Time, and the Ground Plan of Pharmacokinetics. J. Pharmacokin. Biopharm. **10**:201.

Boxenbaum, H., 1984. Interspecies Pharmacokinetic Scaling and the Evolutionary-comparative Paradigm. Drug Metab. Rev. **15**:1071.

Boxenbaum, H. and R. D'Souza, 1990. Interspecies Pharmacokinetic Scaling, Biological Design and Neoteny. Adv. Drug Res. **19**:139.

Buben, J.A., and E.J. O'Flaherty, 1985. Delineation of the Role of Metabolism in the Hepatoxicity of Trichloroethylene and Perchloroethylene. A Dose-Effect study. Toxicol. Appl. Pharmacol. **78**:105.

Bungay, P.M., R.L. Dedrick and H.B. Matthews, 1979. Pharmacokinetics of Halogenated Hydrocarbons. Ann. N. Y. Acad. Sci. **7**:257.

Cohn, W.J., J.J Boylan, R.V. Blanke, M.W Fariss, J.R. Howell and P.S. Guzelian, 1978. Treatment of Chlodecone (Kepone) Toxicity with Cholestyramine. Results of a Controlled Clinical study. New Engl. J. Med. **293**:243.

Collins, B.T., 1990. Pharmacokinetic Models. In: "Handbook of In Vivo Toxicity Testing" (D.L. Arnold, H.C. Grice, and D.R. Krewski, eds.), page 339. Academic Press, New York.

Crump, K.S., 1984. An Improved Procedure for Low-Dose Carcinogenic Risk Assessment from Animal Data. J. Environ. Pathol. Toxicol. Oncol. **6**:339.

Decad, G.M., L.S. Birnbaum and H.B. Matthews, 1981a. 2,3,7,8-tretachlorodibenzofuran Tissue Distribution and Excretion in Guinea Pigs. Toxicol. Appl. Pharmacol. **57**:321.

Decad, G.M., L.S. Birnbaum and H.B. Matthews 1981b. Distribution and Excretion of 2,3,7,8-tetrachlorodibenzofuran in C57BL/6J and DBA/2J Mice. Toxicol. Appl. Pharmacol. **59**:564.

Dedrick, R.L., 1973a. Physiological Pharmacokinetics. J. Dyn. Syst. Mess. Cant. Trans. ASME, September 1973, 255.

Dedrick, R.L., 1973b. Animal scale-up. J. Pharmacokin. Biopharm. **1**:435.

Dedrick, R.L. and K.B. Bischoff, 1968. Pharmacokinetics in Applications of the Artificial Kidney. Chem. Eng. Prog. Symp. Series **84** (64):32.

Dedrick, R.L. and K.B. Bischoff, 1980. Species Similarities in Pharmacokinetics. Fed. Proc. **39**:54.

Dedrick, R.L., D.D. Forrester and H.W. Ho., 1972. In Vitro-In Vivo Correlation of Drug Metabolism-Determination of 1-B-arabinofuransylcytosine. Biochem. Pharmacol. **21**:1.

Dedrick, R.L., D.D. Forrester, J.N. Cannon, S.M. El Dareer and L.B. Mellett, 1973. Pharmacokinetics of 1-B-arabinofuranosylcytosine. Biochem. Pharmacol. **22**:2405.

Dekant, W., M. Metzler and D. Henschler, 1986. Identification of S-1, 2-dichlorovinyl-N-acetyl Cystine as a Uninary Metabolite of Trichloroethylene. A Possible Explanation for its Nephrocarcinogencity in Male Rats. Biochem. Pharmacol. **35**:2455.

Eastmond, D.A., M.T. Smith and R.D. Irons, 1987. An Interaction of Benzene Metabolites Reproduces the Myelotoxicity Observed with Benzene Exposure. Toxicol. Appl. Pharmacol. **91**:85.

Egle, J.L., S.B. Fernandez, P.S. Guzellian and J.F. Borzelleca, 1978. Distribution and Excretion of Chlodecone (Kepone) in the Rat. Drug Metab. Dispos. **6**:91.

European Chemical Industry, Ecology and Toxicology Centre, 1989. Methylene Chloride (Dichloromethane): An Overview of Experimental Work Investigating Specie Differences in Carcinogenicity and their Relevance to Man. Technical Report No 34.

Farber, E. and M.M. Fisher, 1980. "Toxic Injury of the Liver", page 569. Marcel Dekker. New York.

Farrar, D., B. Allen, K. Crump and A. Shipp, 1989. Evaluation of Uncertainty in Input Parameters to Pharmacokinetic Models and the Resulting Uncertainy in Output. Toxicol. Lett. 49:371.

Fernandez, J.G., P.O. Droz, B.E. Humbert and J.R. Caperos, 1977. Trichloroethylene Exposure: Simulation of Uptake, Excreation, and Metabolism Using a Mathematical Model. Br. J. Ind. Med. 34:43.

Fernandez, J.G., J. Guberan and J. Caperos, 1976. Experimental Human Exposures to Tetrachloroethylene Vapor and Elimination in Breath after Inhalation. Am. Ind. Hyg. Assoc. J. 37:143.

Fiserova-Bergerova, V., 1983a. Physiological Models for Pulmonary Administration and Elimination Inert Vapors And Gases. In: "Modelling of Inhalation Exposure or Vapors: Uptake, Distribution, and Elimination" (V. Fiserova-Bergerova, ed.), Vol. 1, page 73. CRC Press., Boca Raton, Florida.

Fiserova-Bergerova, V., 1983b. Gases and their Solubility: A Review of Fundamentals. In: "Modeling of Inhalation Exposure to Vapors: Uptake, Distribution, and Elimination" (V. Fiserova-Bergerova, ed.), Vol. 1, page 3. CRC Press, Boca Raton, Florida.

Fiserova-Bergerova, V., 1985. Toxicokinetics of Organic Solvents. Scand. J. Work Environ. Health. 11:7.

Fiserova-Bergerova, V. and H.C. Hughes, 1983. Species Differences in Bioavailability of Inhaled Vapors and Gases. In: "Modeling of Inhalation Exposure to Vaports: Uptake, Distribution, and Elimination" (V. Fiserova-Bergerova, ed.), Vol. 2, page 97. CRC Press., Boca Raton, Florida.

Gargas, M., H. Clewell and M. Andersen, 1987. Metabolism of Inhaled Dihalomethanes In Vitro: Differentiation of Kinetic Constants for Two Independent Pathways. Toxicol. Appl. Pharmacol. 87:211.

Gargas, M.L., H.J.III. Clewell and M.E. Andersen, 1986. A Physiologically-Based Approach for Determining Metabolic Constants from Gas Uptake Data. Toxicol. Appl. Pharmacol. 86:341.

Gerlowski, L.E. and R.K Jain, 1983. Physiologically Based Pharmacokinetic Modeling: Principles and Applications. J. Pharm. Sci. 72:1103.

Ghering, P.J. and G.E. Blau, 1977. Mechanism of Carcinogenisis: Dose Response. J. Environ. Pathol. Toxicol.1:63.

Goldstein, J.A., J.R. Hass, P. Linko and D. Harvan, 1978. 2,3,7,8-Tetrachlorodibenzofuran in a Commercially Available 99% Pure Polychlorinated Biphenyl Isomer Identified as the Inducer of Hepatic Cytochrome p-448 and Aryl Hydrocarbon Hydroxylase in the Rat. Drug Metab. Dispos. 6:258.

Golstein, B.D., 1977. Hematoxicity in Humans. J. Toxicol. Environ. Health 2:69.

Green, T., J.A. Nash and G. Mainwaring, 1986. Methylene Chloride (Dichloromethane): In Vitro Metabolism in Rat, Mouse, and Hamster Liver and Lung Fractions and in Human Liver Fractions. ICI Technical Report CTL/R/879. Imperial Chemical Industries, Maccelsfield, England.

Green, T., J.A. Nash and S.J. Hill, 1987. Methylene Chloride (Dichloromethane) Glutathione-S-Transferase Metabolism in Vitro in Rat, Mouse, Hammster and Human Cytosol Fractions. ICI Technical Report, CTR/R/934. Imperial Chemical Industries, Maccelsfield, England.

Harrison, L.I. and M. Gibaldi, 1977. Physiologically Based Pharmacokinetic Model for Digoxin Distribution and Elimination in the Rat. J. Pharm. Sci. 66:1138.

Himmelstein, K.J. and R.J Lutz, 1979. A Review of the Applications of Physiological Based Pharmacokinetic Modeling. J. Pharmacokin. Biopharm. 7:127.

Hoel, D.G., N.L. Kaplan and M.W. Anderson, 1983. Implication of Nonlinear Kinetics on Risk Estimation in Carcinogenisis. Science 219:1032.

Holt, J.P., E.A. Phodes and H. Kines, 1968. Ventricular Volumes and Body Weight in Mamals. Am. J. Physiol. 215:704.

Kalf, G.F., 1987. Recent Advances in the Metabolism and Toxicology of Benzene. CRC Crit. Rev. Toxicol. 18:141.

King, F.G., R.L. Derick, J.M. Collins and L.S. Birnbaum, 1983. Physiological Model for the Pharmacokinetics of 2,3,7,8-tetrachlorodibenzofurn in Several Species. Toxicol. Appl. Pharmacol. 67:390.

Kleiber, M., 1961. "The Fire of Life". Wiley, New York.

Kociba, R.J., P.A. Keeler, C.N. Park and P.J. Gerhing, 1976. 2,3,7,8-tetrachlorodibenzo-p-dioxin (TCDD): Results of a 13-Week Oral Toxicity Study in Rats. Toxicol. Appl. Pharmacol. 35:553.

Kociba, R.J., D.G. Keyes, J.E. Beyer, R.M. Carreon, C.E. Wade, D.A. Dittenber, R.P. Kalnins, L.E Franson, C.N. Park, S.D. Barnard, R.A. Hummel and C.G. Humiston, 1978. Results of a Two Year Chronic Oncogenicity Study and Oncogenicity Study of 2,3,7,8-tetrachlorodibenzo-p-dioxin in Rats. Toxicol. Appl. Pharmacol. 46:279.

Krewski, D., D.J. Murdoch and J.R. Withey, 1987. The Application of Pharmacokinetic Data in Carcinogenic Risk Assessment. In: "Pharmacokinetics in Risk Assessment". Drinking Water and Health, Volume 8, page 441. National Academy Press, Washington, D.C.

Krewski, D., Gaylor, D.W. and M. Szyszkowicz, 1991. A Model-Free Approach to Low Dose Extrapolation. Environ. Health Perspect. 90:279.

Krewski, D. and J. Van Ryzin, 1981. Dose Response Models for Quantal Response Toxicity Data. In: "Statistics and Related Topics" (M. Csorgo, D. Dawson, J. N. K. Rao, and E. Saleh, eds.), page 201. North Holland, Amsterdam.

Krewski, D., Murdoch, D. and J.R. Withey, 1989. Recent Developments in Carcinogenic Risk Assessment. Health Phys. 57(Supplement 1):313.

Leung, H.W., D.J. Paustenbach, F.J. Murray and M.E. Andersen, 1988. A Physiologically-Based Pharmacokinetic Model for 2,3,7,8-tetrachlorobenzo-p-dioxin in C57BL/2J Mice. Toxicol. Lett. 42:15.

Leung, H.W., D.J. Paustenbach, F.J. Murray and M.E. Andersen, 1990a. A Physiologic Pharmacokinetic Description of the Tissue Distributon and Enzyme-Inducing Properties of 2,3,4,8-tetrachlorodibenzo-p-dioxin in Mice. Toxicol. Appl. Pharmacol. 103:399.

Leung, H.W., A. Poland, D.J. Paustenbach, F.J. Murray and M.E. Andersen, 1990b. Pharmacokinetics of [125I]-2-iode-3,7,8-trichlorodibenzo-p-dioxin in Mice: Analysis with a Physiological Modeling Approach. Toxicol. Appl. Pharmacol. 103:411.

Levy, G., 1968. Dose Dependent Effects in Pharmacokinetics. In: "Importance of Fundamental Principles in Drug Evaluation" (D.H. Tedeschi and R.E. Tedeschi eds.), page 141. Raven Press, New York.

Lutz, R.J., R.L. Dedrick, H.B. Matthews, T.E. Eling and M.W. Anderson, 1977. A Preliminary Pharmacokinetic Model for Several Chlorinated Byphenyls in the Rat. Drug. Metab. Dispos. 5:386.

Mapleson, W.W., 1963. An Electric Analogue for Uptake and Exchange of Inert Gases and Other Agents. J. Appl. Physiol. 18:197.

McKenna, M.J., 1987. The Role of Studies of Absorption, Metabolism, Distribution and Elimination in Animal Selection and Extrapolation. In: "Human Risk Assessment: The Role of Animal Selection and Extrapolation" (M.V. Roloff, ed.), page 113. Taylor & Francis , Philadelphia.

McNutt, N.S., R.L. Amster, E.E. McConnell and F. Morris, 1975. Hepatic Lesions in Mice after Continuous Exposure to 1,1,1-Trichloroethane. Lab. Invest. **32**:642.

McConnell, E.E., G.W. Lucier, R.C. Rumbaugh, P.W. Albo, D.J Harvan, J.R. Hass and M.W. Harris, 1989. Dioxin in Soil: Bioavailability after Ingestion by Rats and guinea pigs. Science **223**:1077.

Medinsky, M.A., P.J. Sabourin, G. Lucier, L.S. Birnbaum and R.F. Henderson, 1989. A Physiological Model for Simulation of Benzene Metabolism by Rats and Mice. Toxicol. Appl. Pharmacol. **99**:193.

Monster, A.C, G. Boersma and H. Steenweg, 1979. Kinetics of Tetrachloroethylene in Volunteers: Influence of Exposure Concentrations and Work Load. Int. Arch. Occup. Environ. Health. **42**:303.

Moolgavkar, S.H. and A.G. Knudson, 1981. Mutation and Cancer: A Model for Human Carcinogenesis. J. Natl. Cancer Inst. **66**:1037.

Moolgavkar, S.G., E.G. Luebeck, 1990. Two-event Model for Carcinogenesis: Biological, Mathematical and Statistical Considerations. Risk Anal. **10**:323.

Moolgavkar, S.H. and D.J. Venzon, 1979. Two Event Models for Carcinogenesis. Incident Cancer in Childhood and Adult Tumors. Math. Biosci. **47**:45.

Moore, J.A., E.E. McConnell, D.W. Dalgard and M.W. Harris, 1979. Comparative Toxicity and Halogenated Dibenzofurans in Guinea Pigs, Mice and Rhesus Monkeys. Ann. N. Y. Acad. Sci. **320**:151.

Munson, E.S. and D.L. Bowlers, 1967. Effects of Hyperventilation on the Rate of Cerbral Anestetic Equilibration. Anesthesiology **2**:371.

National Research Council, 1987. "Pharmacokinetics in Risk Assessment". Drinking Water and Health Vol. 8. National Academy Press, Washington, D.C.

National Toxiciology Program, 1986a. Toxicology and Carcinogenesis Styudies of Tetrachloroethylene (Perchloroethylene) in F344/N Rats and B6C3F$_1$ Mice. NTP Technical Report N.311. U. S. Department of Health and Human Services, Washington, D.C.

National Toxicology Program, 1986b. Toxicology and Carcinogenesis Studies of Benzene in F334/N Rats and B6C3F$_1$ Mice. NTP Technical Report N.289. U. S. Department of Health and Human Services, Washington, D.C.

Office of Science and Technology Policy, 1985. Chemical Carcinogens: A Review of the Science and Associated Principles. Fed. Regist. **50** :10372.

Paustenbach, D.J., G.P. Carlson, J.E. Christan and G.S. Born, 1986. A Comparative Study of the Study of the Pharmacokinetics of Carbon Tetrachloride on the Rat Following Repeated Inhalation Exposures of 8 and 11.5/day. Fundam. Appl. Toxicol. **6**:484.

Paustenbach, D.J., H.J.III. Clewell, M.L. Gargas and M.E. Andersen, 1988. A Physiologically-based Pharmacokinetic Model for Inhaled Carbon Tetrachloride. Toxicol. Appl. Pharmacol. **96**:191.

Pegg, D.G., J.A. Zemple, W.H. Braun and P.G. Watanabe, 1979. Disposition of (14C) Tetrachloroethylene Following Oral and Inhalation Exposure in Rats. Toxicol. Appl. Pharmacol. **51** :465.

Poland, A., E. Glover, and A.S. Glende, 1976. Stereospecific High Affinity Binding of 2,3,7,8-tetrachlorodibenzo-p-dioxin by Hepatic Cytosol. J. Biol. Chem. **251** :4936.

Portier, C.J. and N.L. Kaplan, 1989. Variability of Safe Dose Estimates when using Complicated Models of the Carcinogenic Process. Fundam. Appl. Toxicol. **13**:533.

Quast, J.F., L.L. Calhoun and M.J. McKenna, 1984. 1,1,1-Trichloroethane: A Chronic Inhalation Toxicology and Oncogenic Study in Rats and Mice. Part I. Results of Findings in Mice. Toxicologist **5**:14.

Rampy, L.W., J.F. Quast, B.K.J. Leong and P.J. Gehring, 1978. Results of Long-term Inhalaton Studies on Rats of 1,1,1-trichloroethylene and Perchloroethlene

Formulation. In: "Proceedings of the 1st International congress on Toxicology" (G.L. Plaa and W.A.M. Ducan, eds.), page 562. Academic Press, New York.

Ramsey, J.C., and M.E. Andersen, 1984. A Physiologically Based Description of the Inhalation Pharmacokinetics of Styrene in Rats. Toxicol. Appl. Pharmacol. 73:159.

Ramsey, J.C., J.D. Young, R. Karbowski, M.B. Chenoweth, L.P. McCarty and W.H. Braun, 1980. Pharmacokinetics of Inhaled Styrene in Human Volunteers. Toxicol. Appl. Pharmacol 53:54.

Reitz, R.H., J.N. McDougal, M.W. Himmelstein, R.J. Nolan and A.M. Schumann, 1988. Physiologically-based Pharmacokinetic Modeling with Methylchloroform: Implications for Interspecies, High Dose/Low Dose, and Dose Route Extrapolations. Toxicol. Appl. Pharmacol. 95:185.

Reynolds, E.S., and M.T. Molsen, 1980. Enviromental Liver Injury. In: "Toxic Injury of the Liver, Part B". (E. Farber and M. Fisher, eds.), page 541. Marcel Dekker, New York.

Richard, C. V., C. C. Travis, D. M. Hetrick, M. E. Andersen, and M. L. Gargas, 1988. Pharmacokinetics of Tetrachloroethylene. Toxicol. Appl. Pharmacol. 93:108

Riggs, D.S., 1963. "The Mathematical Approach to Physiological Problems". M.I.T. Press., Cambridge, Massachusetts.

Rose, J.Q., J.C. Ramsey, T.H. Wentzler, R.A. Hummel and P.J. Gehring, 1976. The Fate of 2,3,7,8-Tetrachlorodibenzodioxin Following Single and Repeated Oral Doses to the Rat. Toxicol. Appl. Pharmacol. 36:209.

Rusch, G.M., B.K.J. Leony and S. Laskin, 1977. Benzene Metabolism. J. Toxicol. Environ. Health 2:23.

Sato, A. and T. Nukajima, 1979. Partition Coefficients of Some Aromatic Hydrocarbons and Ketones in Water, Blood and Oil. Br. J. Ind. Med. 36 :231.

Schmidt-Neilsen, K., 1970. Energy Metabolism, Body Size, and the Problem of Scaling. Fed. Proc. 29:1524.

Schumann, A.M., T.R. Fox, and P.G. Watanabe, 1982. C-Methyl Choroform (1,1,1-Trichloroethane): Pharmacokinetics in Rats and Mice Following Inhalation Exposure. Toxicol. Appl. Pharmacol. 62:390.

Schwetz, B.A., J.M. Noris, G.L. Sparschu, V.K. Rowe, P.J. Gerhing, J.L. Emerson and C.G. Gervig, 1973. Toxicology of Chlorinated Dibenzo-p-dioxins. Environ. Health. Perspect. 5:87.

Stahl, W.L., 1967. Scaling of Respiratory Variables in Mamals. J. Appl. Physiol. 22:453.

Stewart, R.D., E.D. Baretta, H.C. Dodd and T.R. Torkelson, 1970. Experimental Human Exposure to Tetrachloroetylene. Arch. Environ. Health. 20:224.

Takezawa, J., F.J. Miller and J.J. O'Neil, 1980. Single-breath Diffusing Capacity and Lung Volumes in Small Laboratory Animals. J. Appl. Physiol. 48:1052.

Thorslund, T.W., Brown, C.C. and G. Charnley, 1987. Biologically Motivated Cancer Risk Models. Risk Anal. 7:109.

Travis, C.C., 1987. Interspecies and Dose-route Extrapolations. In: "Pharmacokinetics in Risk Assessment". Drinking Water and Health, Vol.8, page 208. National Academy Press, Washington, D.C.

Travis, C.C., Quillen, J.L. and A.D. Arms, 1990a. Pharmacokinetics of Benzene Toxicol. Appl. Pharmacol. 102:400.

Travis, C.C., R.K. White, and R.C. Ward 1990b. Interspecies Extrapolation of Pharmacokinetics. J. Theor. Biol. 142:285.

Tuey, D.B., and H.B. Matthews, 1980. Use of Physiological Compartmental Model for the Rat to Describe the Pharmacokinetics of Several Chlorinated Biphenyls in the Mouse. Drug Metab. Dispos. 8:397.

U.S. Environmental Protection Agency, 1986. Guidelines for Carcinogen Risk Assessment. Fed. Regist. 51:33992.

Vander, A.J, J.H. Sherman and D.S. Luciano, 1975. "Human Physiology: The Mechanisms of Body Function". McGraw-Hill, New York.

Voorman, R., and S.D. Aust., 1987. Specific Binding of Polyhalogenated Aromatic Hydrocarbon Inducers of Cytochrone P-450d to the Cytochrome and Inhibition of its Estraliol 2-hydroxylase Activity. Toxicol. Appl. Pharmacol. **90**:69.

Ward, R.C., C.C. Travis, D.M. Hetrick, M.E. Andersen and M.L. Gargas, 1988. Pharmacokinetics of Tetrachloroehylene. Toxicol. Appl. Pharmacol. **93**:108.

Watanabe, P.G., A.M. Schumann and R.H. Reitz, 1988. Toxicokinetics in the Evaluation of Toxicity Data. Regul. Toxicol. Pharmacol. **8**:408.

Whalen, C.L., and B.E. Kirstein, 1989. A Pharmacokinetic Model for Trichlorethylene. Report submitted to the U.S. Environmental Protection Agency under SAIC Contract #68-02-4402. Science Applications International, San Diego, California.

White, L., Haines, H. and T. Adams, 1968. Cardiac Output Related to Body Weight in Small Mammals. Comp. Biochem. Physiol. **27**:559.

Whittemore, A.S., S.C. Grosser and A. Silvers, 1986. Pharmacokinetics in Low Dose Extrapolation using Animal Cancer Data. Fundam. Appl. Toxicol. **7**:183.

Withey, J.R., 1976. Quantitative Analysis of Styrene Monomer in Polystryene and Foods including some Preliminary Studies of the Uptake and Pharmacodynamics of the Monomer in Rats. Environ. Health Perspect. **17**:125.

Withey, J.R., 1990. Pharmacokinetics: Principles, Mechanisms and Methods. In: "Handbook of Vivo Toxicity Testing" (D. L. Arnold, H. C. Grice and D. R. Krewski, ed.), page 303. Academic Press, New York.

Withey, J.R. and P.G. Collins, 1977. Pharmacokinetics and Distribution of Styrene Monomer in Rats after Intravenous Adminstration. J. Toxicol. Environ. Health **3**:1011.

Yoshihara, S., K. Nagato, H. Yoshimura and Y. Masuda, 1981. Inductive Effect of Hepatic Enzymes and Acute Toxicity of Individual Polychlorinated Dibenzofuran Cogeners in Rats. Toxicol. Appl. Pharmacol. **59**:580.

Young, J.D., J.C. Ramsey, G.E. Blau, R.J. Karbowski, K.D. Nitschke, R.W. Slauter and W.H. Braun, 1979. Pharmacokinetics of Inhaled or Intraperitoneally Administered Styrene in Rats. In: "Toxicology and Occupational Medicine" (W.B. Diechmann, ed.), page 287. Elsevier/North Holland, New York.

GENETIC POLYMORPHISMS IN HUMAN DRUG METABOLISM

L.P. Balant,[1,2] A.E. Balant-Gorgia,[1] M. Gex-Fabry,[1] and M. Eichelbaum[3]

[1]Psychiatric University Institutions of Geneva
[2]School of Pharmacy, University of Geneva, Switzerland
[3]Fischer-Bosch Institut für Klinische Pharmakologie,
  Stuttgart, Germany

INTRODUCTION AND DEFINITIONS

Basically, the behaviour of all drugs is determined, at least to some extent, by genetic factors. It is thus possible to state that "pharmacogenetics" is a basic component of the interindividual variability of all drugs. It is, however, customary to restrict the term "pharmacogenetics" when one (or more) distinct subpopulations can be identified as far as their metabolic capacities towards drugs or group of drugs are concerned.

Genetic factors influencing drug metabolism can be mono- or polygenic. If they are of monogenic type, they present themselves either as polymorphisms or as rare phenotypes. A genetic polymorphism is a monogenic trait that exists in the population in at least two phenotypes and at least two genotypes, neither of which is rare, i.e. less frequent than 1% [Meyer et al., 1990].

Genetic polymorphisms of drug metabolism usually segregate the population into two groups differing in their ability to metabolize certain drugs. Individuals with a deficient metabolism are termed "poor metabolizers" (PM), as compared to normal or "extensive metabolizers" (EM).

In the present review, only N-acetylation (briefly) and C-oxidation polymorphisms will be discussed since, at the present time, it seems that other polymorphisms are either little investigated or have little clinical relevance and pharmacokinetic interest. The description of the molecular aspects of polymorphisms is treated in details in recent reviews [see Meyer et al., 1990].

New Trends in Pharmacokinetics, Edited by A. Rescigno and A.K Thakur
Plenum Press, New York, 1991

HISTORY

N-Acetyltransferase

The classical example of a monogenic defect in drug metabolism is the N-acetylation polymorphism, first observed by Evans et al. in 1960. Urinary recovery data and plasma concentrations allowed subjects to be classified as "rapid" or "slow" acetylators of the drug.

N-acetylation of isoniazid, hydralazine, sulfasalazine or procainamide is under the control of N-acetyltransferase. It is generally accepted that slow acetylators have a higher risk to suffer from drug concentration dependent adverse drug reactions if standard daily doses are used.

This genetic polymorphism will not be further discussed in this review which concentrates on the oxidation polymorphism of the debrisoquine/sparteine-type.

Debrisoquine/sparteine-type Oxidation Polymorphism

Between 1975 and 1977, three groups independently discovered this polymorphism. A dramatic happening in a pharmacokinetic study with a new beta-blocking agent, bufuralol, suggested that the metabolism of this drug might be under polymorphic genetic control [Balant et al., 1976]. A volunteer suffered from marked orthostatic hypotension with simultaneous very high concentrations of the parent drug, and very low concentrations of the hydroxylated metabolite. Further work by the same group confirmed this hypothesis [Dayer et al., 1982, 1983]. In Great Britain, investigators of the behaviour of debrisoquine, an antihypertensive drug, found that marked hypotensive response in one volunteer was due to impaired 4-hydroxylation of the drug [Mahgoub et al., 1977]. At about the same time, a group of physicians in Bonn observed increased side effects associated with decreased oxidative metabolism of sparteine, an anti-arrhythmic and oxytocic drug [Eichelbaum et al., 1975, 1979a,b].

Since these early observations, numerous investigations have followed indicating that the proportion of "poor" metabolizers is about 7% in Caucasian populations. Clinical studies have demonstrated that poor metabolizers of debrisoquine are at high risk to develop adverse reactions due to excessively high drug concentrations [Brøsen and Gram, 1989a]. This is particularly true for tricyclic antidepressants [Balant-Gorgia et al., 1986, 1989] and for anti-arrhytmics. For other classes of drugs such as β-blockers, the genetic defect can also be observed [Lennard et al., 1986], but the clinical consequences are less dramatic [Brøsen and Gram, 1989a]. Finally, it must be stated that it is now widely accepted that this aspect of drug metabolism should be taken into account during new drug development as discussed below in this review [Balant et al., 1989].

Molecular Biology of the DQ/SP-Polymorphism

The gene coding for the synthesis of cytochrome P-450IID6 is located on the long arm of chromosome 22 [Eichelbaum et al., 1987]. The gene locus (CYP2D) consists of two pseudogenes (D8 and D7) and one functional gene (D6) [Kimura et al., 1989]. The CYP2D6 gene locus is highly polymorphic and several mutations causing the PM phenotype have been discovered [Skoda et al., 1988; Gonzalez et al., 1988a; Kagimoto et al., 1990; Heim and Meyer, 1990].

The most common mutant allele 29B has multiple mutations with base changes in exons 1, 2 and 9 and a point mutation at the consensus sequence of the splice site of the 3rd intron. The 29A mutant allele is characterized by a simple nucleotide deletion in the 5th exon with consequent frameshift [Kagimoto et al., 1990; Heim and Meyer, 1990].

Table 1
Listing of some drugs known to be metabolized via the sparteine/debrisoquine
cytochrome P450db1 system

β-blockers	
Bufuralol, tolamolol	Balant et al. 1976
	Dayer et al. 1982, 1983
Metoprolol	Lennard et al. 1983
Timolol	Lewis et al. 1985
Cardioactive Drugs	
Sparteine	Eichelbaum et al. 1975, 1979a,b
Encainide	Wang et al. 1984
N-propylajmaline	Zekorn et al. 1985
Indoramin	Pierce et al. 1987
Perhexiline	Cooper et al. 1987
Propafenone	Siddoway et al. 1987
Flecainide	Beckmann et al. 1988
	Mikus et al. 1989
Tricyclic Antidepressants	
Nortriptyline	Bertilsson et al. 1980
Amitriptyline	Balant-Gorgia et al. 1982
	Mellstrøm et al. 1983
Desipramine	Bertilsson & Åsberg-Wistedt, 1983
Clomipramine	Balant-Gorgia et al. 1986
Imipramine	Brøsen et al. 1986b
Neuroleptics	
Perphenazine	Dahl-Puustinen et al. 1989
Thioridazine	von Bahr et al. 1989
Miscellaneous	
Debrisoquine	Mahgoub et al. 1977
Phenformin	Oates et al. 1983
Amiflamine	Alvan et al. 1984
Dextromethorphan	Schmid et al. 1985
Codeine	Dayer et al. 1988

Another mutant allele of 44 kb size after XbaI digestion has the same mutation as mutant allele 29B. The mutant allele of 11.5 kb after XbaI digestion is due to the deletion of the entire functional IID6 gene [Gaedigk et al., 1991]. In addition another rare mutation which results in a 16 + 9 kb XbaI fragment is associated with the PM phenotype [Evans and Relling, 1990].

Thus there are at least 12 genotypes responsible for the PM phenotype. Despite the diversity of genotypes causing the PM phenotype all these mutations result in the absence of P450IID6 in the liver of PMs [Gonzalez et al., 1988b; Zanger et al., 1988].

As work proceeds, new drugs are added to the list of compounds shown to be under the metabolic control of cytochrome P450IID6 [Brøsen and Gram, 1989a]. Accordingly, the information of Table 1 must be considered as only temporarily valid.

PHARMACOKINETIC ASPECTS OF PHENOTYPING

Two subpopulations of a polymorphic system are different at the gene level, showing a "genotype" that can, for example, be determined from lymphocyte DNA. The expression of this difference can also be determined by the capacity of a person to metabolize a substance usually called a "probe drug". In this way, the "phenotype" of this person is determined. Clearly, the genotype is an invariant characteristic of a given individual, whereas the phenotype may be influenced by the concomittant intake of other drugs susceptible to cause enzyme induction or inhibition.

Usually, phenotyping is performed after a single administration of the probe drug. It is theoretically possible to determine the phenotype based on blood concentrations of parent compound and metabolite. For practical reasons, most of these studies have been based on urinary excretion analysis. As discussed below there are good theoretical reasons to believe that in many instances this is a correct approach.

There are, however, cases for which the measurement of parent drug and metabolite concentrations in blood at a given time may allow an adequate separation of EM and PM subjects. This is for example the procedure chosen for bufuralol [Dayer et al., 1982, 1983]. Another example is mephenytoin for which both genetic and stereospecific factors may be used for the development of a simple phenotyping procedure. The S-, but not the R-enantiomer of mephenytoin, is hydroxylated in a polymorphic manner [Küpfer and Preisig, 1984]. Since EM subjects naturally eliminate the S-enantiomer much faster than the R-enantiomer, the ratio is usually much lower than 1 in these subjects. As the 4-hydroxylation of S-mephenytoin is impaired in PM subjects, the elimination of the two enantiomers becomes comparable after a single dose of the racemate leading to a S/R ratio of about 1. This observation is particularly clear some time after drug intake [Wedlund et al., 1984]. Such a property has been used to determine the incidence of PM of S-mephenytoin in a Swedish population [Sanz et al., 1989].

The Simple Case

If a drug is metabolized to only one metabolite, which is then excreted into urine, the following model may be derived

$$
\begin{array}{ccc}
G \longrightarrow & B & \longrightarrow M \\
& \downarrow & \quad\downarrow \\
& E & \quad P
\end{array}
$$

where G represents the parent drug in the gastro-intestinal system, B in the central compartment, E in the urine, and M is the metabolite in the central compartment, P in the urine. The areas under the curves of B and M will be proportional to their clearances. From such a model, ratios of areas under the curves or amounts excreted into urine can easily be derived. They represent, however, little interest since in reality the metabolic patterns of drugs are never as simple as depicted here. In particular, most drugs known to be under polymorphic control display more than one metabolic pathway and undergo more or less extensive first-pass metabolism. This is the reason why Jackson et al. [1986] have proposed a more appropriate model which was further discussed more recently [Tucker et al., 1990].

The "Metoprolol" Model

A model with the following properties has been developed:

- Complete absorption of intact drug across the gastro-intestinal mucosa.
- Linear, time-invariant kinetics.
- Production of two metabolites, only one of which exhibits genetic polymorphism in its formation(metabolite 1).
- Metabolites are end products and are excreted unchanged into urine together with the parent drug.
- Metabolites exhibit formation rate limited kinetics.
- Elimination of unchanged drug and its metabolites is described by mono-exponential functions.
- The "well-stirred" model of hepatic drug clearance applies.

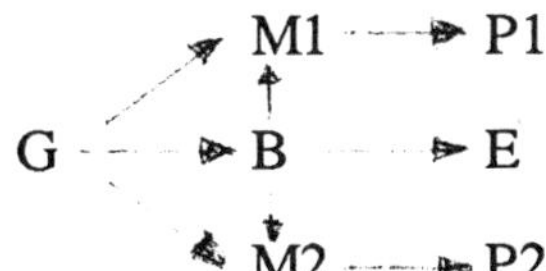

The symbols are similar to those in the previous paragraph.

The area under the blood concentration curve of the parent drug (AUC) has been proposed as an index for polymorphism [Jack et al., 1983]. The expression of AUC contains the total clearance of the drug, and it is thus impossible to extract, from this data only, the information about a possible polymorphism if the frequency of the gene defect is low. As a matter of fact it can be shown that experimental indices based upon measurement of both drug and metabolite are more discriminatory than those based on measurement of either drug or metabolite alone.

The value of AUC in this model is:

$$AUC = F_H \cdot Dose/(CL_M + CL_R) \tag{1}$$

where F_H is the systemic availability of the parent drug, CL_M the total metabolic clearance of the drug, and CL_R its renal clearance. Despite its limited value, AUC is an adequate parameter if none of the two subpopulations has a low frequency, and it has been used, for example, for isoniazid data analysis in Caucasians.

The urinary drug to metabolite ratio ($R_{D/M1}$) has been extensively used for the phenotyping of populations with debrisoquine or sparteine. Today, poor metabolizers are defined (by statistical analysis of the antimode in a bimodal distribution) as individuals with a metabolic ratio greater than 12.6 and 20 respectively, for the ratio of debrisoquine to 4-hydroxy-debrisoquine, and of sparteine to 2-and 5-dehydrosparteine [Evans et al., 1980; Eichelbaum et al., 1986a]. The metabolic ratio can be defined as:

$$R_{D/M1} = \frac{Ae}{Ae(M1)} \tag{2}$$

where Ae is the amount of parent drug recovered in urine at infinity, and Ae(M1) the amount of metabolite. Knowing that:

$$Ae = AUC \cdot CL_R \tag{3}$$

and
$$Ae(M1) = \frac{E_H \cdot Dose \cdot CL_{M1}}{CL_M} + CL_{M1} \cdot AUC \tag{4}$$

where E_H is the hepatic extraction ratio of the drug, equation 2 becomes:

$$R_{D/M1} = \frac{CL_R}{CL_{M1}} \frac{1}{1 + E_H/F_H(1 + CL_R/CL_M)} \tag{5}$$

It can also be shown that the intrinsic clearance for the formation of M1 $[CL_{int}(M1)]$ is:

$$CL_{int}(M1) = \frac{Ae(M1) \cdot Q_H}{AUC \, (Q_H + CL_R)} \tag{6}$$

where Q_H is the hepatic blood flow.

Combining these equations and assuming $CL_R \ll Q_H$, the following expression is obtained:

$$R_{D/M1} = \frac{CL_R}{CL_{int}(M1)} \tag{7}$$

In practice, urinary drug to metabolite ratios are rarely determined from complete recoveries and time truncated ratios are measured. They have, nevertheless, been found to represent powerful tools for the determination of the phenotype of debrisoquine and sparteine.

It must be remembered that renal drug clearance, plasma drug binding, hepatic blood flow and metabolic clearance down non-polymorphic routes may all be confounding variables, depending upon the index chosen. These considerations are of obvious importance in separating the contribution of genetic, environmental and pathophysiological factors to variability in drug metabolism. They are probably especially relevant to studies of ethnic differences in oxidative polymorphism [Jackson et al., 1986].

The "Clomipramine" Model

The metabolic ratio approach can probably be used for drugs such as nortriptyline or desipramine for which hydroxylation is a major route of elimination. For substances such as clomipramine, the ratio of demethyl-clomipramine to hydroxy-demethyl-clomipramine is a better index of hydroxylation than the clomipramine to hydroxy-clomipramine ratio. A new alternative approach may be considered [Gex-Fabry et al., 1990].

For substances such as clomipramine, the following model applies:

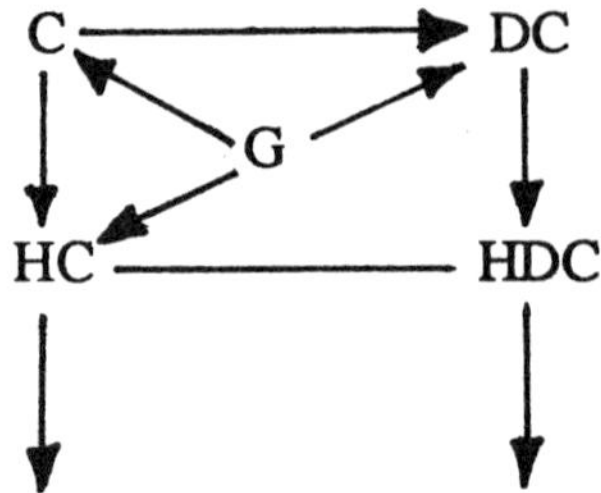

where C, DC, HC, and HDC are clomipramine, demethyl-clomipramine, hydroxy-clomipramine and hydroxy-demethyl-clomipramine.

For a fixed dose administered at a constant dosing interval τ, the average blood concentration over the interval at steady-state can be written as [Gibaldi & Perrier, 1982]:

$$C_{SS} = \frac{F_H \cdot Dose}{\tau} \frac{1}{CL_{OH} + CL_{DE}} \tag{8}$$

$$DC_{SS} = \frac{Dose}{\tau} \frac{1}{CL_{OH}} \left(E_{OH} + F_H \frac{CL_{DE}}{CL_{OH} + CL_{DE}} \right) \tag{9}$$

$$HC_{SS} = \frac{Dose}{\tau} \frac{1}{CL_{DE} + CL_{E1}} \left(E_{OH} + F_H \frac{CL_{OH}}{CL_{OH} + CL_{DE}} \right) \tag{10}$$

$$HDC_{SS} = \frac{Dose}{\tau} \frac{1}{CL_{E2}} \left[F_H \left(\frac{CL_{DE}}{CL_{OH} + CL_{DE}} + \frac{CL_{OH}}{CL_{OH} + CL_{DE}} \cdot \frac{CL_{DE}}{CL_{DE} + CL_{E1}} \right) \right.$$
$$\left. + E_{DE} + E_{OH} \frac{CL_{DE}}{CL_{DE} + CL_{E1}} \right] \tag{11}$$

CL_{DE}, CL_{OH}, CL_{E1}, and CL_{E2} are demethylation clearance, hydroxylation clearance and elimination clearances of hydroxy-clomipramine and hydroxy-demethyl-clomipramine respectively. F_H is the systemic availability or fraction of the dose of clomipramine escaping first-pass metabolism, i.e. $F_H = 1 - E_{OH} - E_{DE}$, where E_{OH} and E_{DE} are the liver extraction ratios for hydroxylation and demethylation. These parameters are related to the clearances according to $CL_{OH} = E_{OH} \cdot Q_H$ and $CL_{DE} = E_{DE} \cdot Q_H$, where Q_H is the hepatic blood flow. It should be pointed out that the equations above are strictly valid only when whole blood concentrations are measured, as opposed to plasma.

Equations (8) to (11) can be solved in order to write clearances as functions of concentrations. The following expressions are obtained for hydroxylation and demethylation clearances:

$$CL_{OH} = \left(\frac{C_{SS} + DC_{SS}}{Dose/\tau} + \frac{1}{Q_H} \right)^{-1} \tag{12}$$

$$CL_{DE} = \left(\frac{C_{SS}}{Dose/\tau} + \frac{1}{Q_H} \right)^{-1} - CL_{OH} \tag{13}$$

Similar equations may be derived for CL_{E1} and CL_{E2}. Assuming a single elimination rate for the two hydroxy-metabolites (i.e. $CL_{E1} = CL_{E2}$), they simplify to:

$$CL_E = \frac{Dose/\tau}{HC_{SS} + HDC_{SS}} \tag{14}$$

The equations given above hold under the following hypotheses. (a) linear processes are assumed; (b) renal clearances of C and DC are neglected; (c) direct conjugation of C and DC is also negligeable; (d) the hydroxylation clearances for C and DC are identical; (e) HC and C are demethylated at the same rate; (f) the elimination rate for HC and HDC is expected to reflect the combined effect of conjugation with glucuronic acid and direct renal excretion; (g) a hepatic first-pass effect is taken into account for demethylation and hydroxylation, but the presystemic sequential action of the two isozymes is supposed to be very unlikely; (h) in addition, a constant hepatic blood flow of 1.5 liters/minutes is assumed in order to allow the calculation of clearances from C and DC concentrations.

Support to these assumptions from the literature and sensitivity analysis have been presented [Gex-Fabry et al., 1990], indicating that equations (12) and (13) may provide fairly good estimates of hydroxylation and demethylation clearances of clomipramine. The validity of equation (14) to estimate the elimination rate of hydroxy-metabolites is probably more questionable [Gex-Fabry et al., 1990].

Practical Considerations

From the above considerations it appears that before starting a phenotyping program, it is necessary to understand the relative importance of different metabolic routes and to adequately choose the measured species and pharmacokinetic model in order to avoid artifactual conclusions.

In principle, it should be possible to phenotype with any of the known substrates of cytochrome P450db1 or acetyl-transferase. However, it is obviously preferable to choose a probe drug which is: a) fully characterized for its metabolism in vitro ; b) devoid of any tocixity or pharmacological action; c) available worldwide ; d) easy to assay in biological fluids; and e) accurate in determining the phenotype [Dayer, 1990].

For cytochrome P450IID6, the following substances are presently available and recommended: debrisoquine, sparteine and dextromethorphan. None of them has all the 5 basic qualities described in the previous paragraph, but they represent the best possible probe drugs at the present time [Dayer, 1990].

For N-acetyl-transferase 4 test compounds have been advocated: isoniazid, sulphamethazine, dapsone and caffeine. The latter substance is so universally used, totally safe at the doses needed for phenotyping (even in children) that it can be proposed as the presently best available probe drug [Bechtel and Bechtel, 1990].

STATISTICAL CONSIDERATIONS

As underlined by Jackson et al. [1990], in classical genetic analysis phenotypes are categorical (i.g. blue or dark eyes) whereas in drug metabolism they show continuous variability either for genetic reasons or because of the influence of environmental factors or because other similar enzymes are able to biotransform the drug to the same metabolite. Accordingly, it is often difficult to separate two peaks in the population distribution of a given index such as a metabolic ratio.

Distributions and Index Transformation

The detection of bimodality is facilitated if the distribution of the index within each phenotype is normal (Gaussian). This is, however, not always the case, and there are many reasons to explain such deviations [Jackson et al., 1990]. For example, it has been discussed whether metabolic ratios should be expressed as drug/metabolite or metabolite/drug, and whether it should or should not be logtransformed. Theoretically, the distribution of the ratio is undefined, but when variance in the denominator is small it will approximate the distribution of the numerator. As a consequence, it would be advantageous to have the polymorphic element in the numerator. Unfortunately, in the majority of cases this is not known when large scale studies are undertaken. Log transformation may overcome some of these uncertainties, but it may introduce other problems. As an example, log transformation of an unimodal skewed distribution may lead to bimodality.

Data Analysis

As usual, the simplest method of data analysis is a graphical representation such as a histogram of the chosen index. Such methods are, however, not very robust since on occasions, simple shifts of cell position may alter the apparent modality. Another, often used, method is the probit plot. There are also a number of situations in which such plots may lead to erroneous conclusions. In particular, the application of such approaches to small sample sizes combined with inadequate automated methods may lead to catastrophic conclusions.

Statistical methods such as the chi-squared test, the Kolmogorov-Smirnov test and others have been used to test for bimodality either directly or on transformed data. Some methods have the disadvantage of assuming a normal distribution or a superposition of normal distributions. Accordingly, other approaches have been derived. They have been reviewed by Jackson et al. [1990]. The conclusion of this review is that: "Graphical methods to suggest and statistical methods to test the presence of bimodality have many potentiel problems. Making assumptions about the form of distribution of important indices may alleviate some, but not all, of the difficulties". We believe that such assumptions can only be made in a rational way if the underlying pharmacokinetic basis has been thoroughly analyzed.

NONLINEAR KINETICS

One of the functional characteristics of P450IID6 (the isozyme of cytochrome P-450 responsible for the debrisoquine/sparteine polymorphism) appears to be nonlinear kinetics for some drugs in EM, but linear kinetics in PM with oral dosing, probably reflecting saturation of P450IID6 during first-pass through the liver [Brøsen and Gram, 1988]. On the other hand, Dayer et al. [1985] reported dose-dependent first-pass metabolism of bufuralol in PM subjects. It is possible that these differences are to be found in substrate affinity and metabolizing capacity of the different metabolic pathways involved in the biotransformation of these two drugs.

Such nonlinearities have not only potential clinical relevance, [Brøsen et al., 1986a] but may confuse bioavailability issues. As an example, the absolute bioavailability of bufuralol was found to be greater than one in a PM volunteer having received 30 mg orally and 20 mg as a 2 hour infusion [Dayer et al., 1985]. In addition, the presence of undetected PM may increase interindividual variability [Kübli et al., 1982]. Accordingly, for drugs known to be under polymorphic genetic control, it is wise to phenotype the subjects recruited for bioavailability studies.

The nonlinearity may, on rare occasions, manifest itself in a particularly striking mode. As an example, a poor metabolizer of debrisoquine, treated with clomipramine showed clear nonlinear kinetics of desmethyl-clomipramine, and had measurable blood concentrations of this active metabolite up to three months after therapy with this antidepressant was suspended [Balant-Gorgia et al., 1987].

STEREOSELECTIVE METABOLISM

When metoprolol is administered as a racemate, the inactive R-enantiomer is metabolized faster than the active S-enantiomer, whereas metabolism is not stereoselective in PM [Lennard et al., 1983]. As a consequence, the difference in plasma concentrations between phenotypes is smaller for the active S-enantiomer than for the inactive R-enantiomer.

On the other hand, debrisoquine has no chiral center, but 4-hydroxylation introduces a chiral center into the molecule, giving rise to two possible enantiomers. In EM subjects, 4-hydroxylation leads almost exclusively to the formation of S(+)-4-hydroxy-debrisoquine [Eichelbaum et al., 1988]. In contrast, PM subjects are not only characterized by a decreased total 4-hydroxy-debrisoquine formation, but also by a marked loss of enantio-selectivity in product formation.

These two examples illustrate the importance of the combined study of enantioselectivity in drug metabolism and metabolite formation and genetic particularities during the development of new drugs. This subject is discussed in more details in the contribution of Eichelbaum to be found in this volume.

PHARMACOKINETIC DRUG INTERACTIONS

The phenotypic difference between PM and EM subjects of cytochrome P450IID6-dependent probe drugs can be reduced or abolished by functional impairment of the enzyme system. For example, quinidine is a very potent competitive inhibitor of sparteine and debrisoquine. Hence, genetically determined EM subjects may change to phenotypical PM when taking quinidine. Brøsen and Gram [1989b] found that during quinidine co-administration, the total oral clearance of imipramine was reduced by 35%, and that of desipramine by 85%. The clearance of imipramine via demethylation was not significantly reduced, whereas its clearance via 2-hydroxylation was strongly affected. As a consequence, 2-hydroxy-imipramine and 2-hydroxy-desipramine were detected in plasma before, but not during quinidine treatment. These results confirm that quinidine is a potent inhibitor of cytochrome P450db1, but not of the P450 isozyme responsible for demethylation of imipramine. Similar findings have been found for patients treated with thioridazine, levomepromazine or propafenone. As a consequence, phenotyping during treatment with such drugs acting as potent inhibitors is uninformative [Brøsen and Gram, 1989a].

The activity of P450IID6 is also impaired by tricyclic antidepressants and other neuroleptics than those of the previous paragraph, but not to such an extent that the EM phenotype appears to change to PM. Nevertheless, it is not recommended to phenotype patients under treatment with these drugs.

When inhibitors are added to the treatment with a polymorphically oxidized drug, the pharmacokinetics may change and major increases in blood concentrations may be observed [Balant-Gorgia et al., 1986, 1989; Zhou et al., 1990]. The clinical significance of such interactions depends on: a) the frequency of the drug combination; b) the role of P450IID6 in overall elimination of the drug ; c) the potency of the inhibition ; d) the practical consequence of a raised drug concentration and the therapeutic margin of the active principle ; and e) the feasibility of therapeutic drug monitoring [Brøsen and Gram, 1989a].

INTERPLAY BETWEEN GENETIC AND ENVIRONMENTAL FACTORS: POPULATION APPROACHES

It is important to realize that there are interactions between genetic and environmental factors. Genetically controlled enzymatic systems often require participation of environmental factors ; conversely, the latter require a genetic apparatus to exert their effects on enzyme activity [Breimer, 1990]. As an example, a patient may start or stop taking other drugs and this will possibly change metabolizing activities to a greater or lesser extent as discussed in the previous Paragraph.

To investigate this type of interplay, a relatively large number of subjects must receive the drug under investigation and kinetic data or steady state concentrations of parent drug and metabolite(s) must be obtained. In order to analyse this type of data, classical pharmacokinetic methods soon become very cumbersome. Accordingly, it is probably more useful to use "population approaches" for this type of investigations.

INTERETHNIC DIFFERENCES

As stated above, the metabolism of drugs is under the control of both environmental and genetic factors. As these factors vary between different populations, it is not surprising to find pharmacokinetic and pharmacodynamic differences between ethnic groups [Bertilsson, 1990].

N-Acetyl-transferase

Interethnic differences are well established. Generally in individuals of Caucasian origin, the percentage of rapid acetylators varies from 30% to 67%. The figures are not very different in Africans. Slow acetylators are predominant in some sub-populations of Mediterranean origin with only 18% of rapid acetylators in Egyptians. The reverse is observed in Asian populations with about 90% of rapid acetylators in Japanese subjects [Bechtel and Bechtel, 1990].

Debrisoquine 4-hydroxylation

Among the Caucasians, between 5 to 10% of subjects are classified as poor metabolizers with a metabolic ratio of debrisoquine higher than 12.6. This proportion drops to less than one percent in native Chinese both from the Han and Mongolian origin [Lou et al., 1987]. Similar results were obtained with mainland Chinese from Hunan [Horai et al., 1989]. Three studies in native Japanese [Nakamura et al., 1985; Ishizaki et al. 1987; Horai et al. 1989] have also shown that there is a low proportion of PM in this population (from 0 to 2.5%). These studies involved phenotyping procedures, and no antimode could be found when debrisoquine was used with an antimode for the metabolic ratio of 12.6. However, using genomic leucocyte DNA from some Chinese, and analysis of restriction fragment length polymorphisms (RFLPs), it was shown that a number of Chinese with low metabolic ratios (i.e. EM phenotype) are genotypically PM [Yue et al., 1989]. This study indicates that the antimode for a probe drug determined in one population might not be appropriate for another one, and that in the Chinese the antimode for the debrisoquine-metabolic ratio is probably lower than 12.6. The mechanism behind the dissociation between debrisoquine hydroxylation phenotype and genotype is not known. It is possible that the Chinese, in addition to the debrisoquine hydroxylase (P450IID6) have another enzyme able to hydroxylate debrisoquine. Another possibility is that the gene which is inoperational in the Caucasian PM is operational in the Chinese, producing the debrisoquine hydroxylase [Bertilsson, 1990].

Studies in the Chinese and Japanese [Nakamura et al., 1985; Horai et al., 1989] have shown that 15% to 22% of these individuals are PM of S-mephenytoin, whereas the proportion is lower than 5% among the Caucasians.

CLINICAL CONSEQUENCES

The clinical impact of genetic polymorphism in drug metabolism has been extensively reviewed by Brøsen and Gram [1989a], Gram and Brøsen [1990]. Two main aspects are to be considered: a) the pharmacokinetic importance of the metabolic pathway under polymorphic control, and b) the therapeutic margin of the drug and its active metabolites. It is only when these two aspects combine unfavorably that clinical relevance is achieved. This is essentially the case for tricyclic antidepressants for which drug monitoring represents an important contribution for therapeutic success.

There are a few observations related to inter-ethnic differences which warrant a short discussion, even if no final conclusions can be drawn at the present time.

First, it is most probable that genetic polymorphism in drug metabolism does not explain all differences in drug response found between different ethnic groups since environmental factors certainly also play an important role in this context. In addition, it is possible that phamacodynamic differences in response exist in different groups. As an example, despite the fact that propranolol is metabolized faster in Chinese subjects than in the Caucasians, the Chinese are more sensitive to the beta-blocking effect. Stereoselective metabolism of the drug does not explain these differences suggesting a pharmacodynamic cause for the increased sensitivity [Zhou and Wood, 1990].

Another interesting observation is related to the S-mephenytoin 4-hydroxylation polymorphism [Küpfer and Preisig, 1984]. Although this polymorphism has been found to be of little clinical relevance up to now, the situation might change. For example, the finding that the metabolism of diazepam (mainly N-demethylation) and desmethyl-diazepam (mainly C-hydroxylation to oxazepam) is related to S-mephenytoin hydroxylation, but not to debrisoquine hydroxylation [Bertilsson et al., 1989], together with the fact that there are more PM of S-mephenytoin in Orientals than in the Caucasians could be of great interest [Sanz et al., 1989]. Ghoneim et al. [1981] had indeed found that the Orientals had a lower metabolic clearance than the Caucasians. Interestingly, in another study it was noted that "many Hong-Kong physicians routinely prescribe smaller diazepam doses for the Chinese than for the white Caucasians" [Kumana et al., 1987].

REGULATORY CONSEQUENCES

The consequences of genetic polymorphism in drug metabolism have been discussed in the frame of a concerted action of the European Communities about the "Criteria for the choice and definition of healthy volunteers and/or patients for Phase I and II studies" [Balant et al., 1989; Cano et al., 1990; von Bahr, 1990; Case, 1990; Gundert-Remy, 1990; Rane, 1990]. It was concluded that "the fact that the metabolism of a new drug is under substancial polymorphic genetic control, resulting in considerable pharmacokinetic variability, should not be an a priori reason for stopping its development. If compounds had been abandoned on this basis, valuable drugs, such as beta-adrenoceptor antagonists and tricyclic antidepressants, would not be available today, or they would have encountered additional difficulties in reaching the market" [Balant et al., 1989]. However "the issue becomes a relevant influence on the decision for marketing authorization in cases where the therapeutic range is narrow and side effects or ineffectiveness may be related to variability. If the applicant does not provide measures which are supported by experimental and clinical data, which allow the safe handling of the drug in the therapeutic situation, the reviewer has to take into consideration not only the situation as presented in the application file, but also further implications regarding the use of drugs in the "therapeutic wilderness". If, on the other hand, the applicant provides evidence for a clear concept in the development, relevant data and detailed sound advice for the handling of the drug, polymorphic drug metabolism, even in drugs with a narrow therapeutic margin, will not represent an additional difficulty in reaching the market" [Gundert-Remy, 1990].

Since the Caucasians and the Asians may differ with respect to their metabolizing capacities, some precautions must be taken when licensing in products. For example, it should be remembered that drugs developed in the western world have been tested mostly on the Caucasians, and that different doses may have to be used on the Orientals. The converse is naturally also true. Experience shows that volume of distribution and renal clearance are relatively constant in different ethnic groups. Accordingly, the main efforts should be laid on determining the metabolic capacities in new ethnic groups. Thus, if a drug is to be administered in the same pharmaceutical form in Asians and Caucasians, it is usually sufficient to perform a pharmacokinetic study in healthy volunteers and to compare the results to historical controls in the other ethnic group. Emphasis should be laid on the measurement of metabolites, even if they are inactive. The situation is more complex if dosage regimens and/or pharmaceutical forms are going to be different. Then clinical trials must be performed in the new ethnic group. "Pharmacokinetic equivalence" may simplify the problem of dose-finding, while non-equivalence may necessitate a full development program as far as Phase I to III clinical trials are concerned.

IMPLICATIONS FOR DRUG DEVELOPMENT

An important point to consider is the impact of polymorphic drug oxidation on the development of new drugs. So far volunteers participating in Phase I clinical trials have been selected on the basis of some normal biochemical characterististics and normal liver and renal function. Less attention has been paid to characterise them with respect to their drug metabolising capacity. The use of so called model drugs such as antipyrine for this purpose has been disappointing since it is of predictive value for only a limited number of drugs. Furthermore it is known which cytochrome P450 isozymes are catalysing the biotransformation of most probe drugs. The knowledge gained from the isolation of certain isoenzymes exhibiting a genetic polymorphism, and their characterisation with regard to substrate specificity, should be incorporated into drug development at an early stage to determine whether or not the drug is metabolised by a particular isozyme and hence subject to a genetic polymorphism. Since Phase I clinical trials are carried out at a rather late time during drug development, usually five to seven years after the initial discovery, a strategy which allows for an earlier recognition of this phenomenon would be desirable. If it were possible to predict that the metabolism of a drug cosegregates with a known polymorphism at the preclinical stage, the decision on whether or not to pursue development of the drug would be facilitated. Provided the drug is of great therapeutic potential the fact that its metabolism is polymorphic should not lead to its abandonment, but rather may be considered an advantage since prior knowledge of its variability in metabolism could be incorporated in Phase I and II clinical trials with regard to dose finding and the selection of volunteers and patients.

Several in vitro approaches have been developed or will become available in the near future which will allow a prediction to be made during preclinical testing if the metabolism of a new drug is subject to genetic polymorphism. The common structural features present in substrates of cytochrome P450IID6 are a basic nitrogen atom, a lipophilic domain, a type I binding spectrum and a distance between the basic nitrogen and the site of oxidation between 5 and 7 A [Meyer et al., 1986; Guengerich et al., 1986]. At the present time this information is insufficient to predict on the basis of structure alone whether a given compound is a substrate for P450IID6. However as more knowledge concerning the structure of the isozyme is gained, this may become possible in the future.

Cytochrome P450IID6 and other P450 enzymes have been expressed in cell lines and this system offers a convenient tool to study the metabolism of drugs by specific isozymes and to identify the metabolic pathways that are catalyzed. These studies can be supplemented and extended by studying the metabolism of the substrate in human liver microsomes of known phenotype/genotype in comparison to a known polymorphic substrate, such as bufuralol. This approach allows to elucidate how many pathways are involved in the biotransformation of the compound and which of the pathways are controlled by this polymorphism. If a drug is a substrate for cytochrome P450IID6 it can be anticipated that the rate of metabolite formation should correlate with the formation of 1'-hydroxybufuralol. In addition the drug should be a competitive inhibitor of bufuralol 1'-hydroxylation. The use of antibodies directed against P450IID6 will allow the polymorphic metabolic pathways to be elucidated and to establish whether these metabolites are formed solely by P450IID6 or whether other isozymes are involved. Whereas metabolite formation by P450IID6 is inhibited by antibodies, the formation of metabolites not catalysed by P450IID6 is not inhibited. These approaches have been successfully applied in investigating the metabolism of propafenone [Kroemer et al., 1989]. The same test systems can be used to predict already in vitro the potential for interactions of the candidate compound with other substrates of P450IID6. In addition, these in vitro techniques can also be used to identify whether the metabolism of drugs already used therapeutically is controlled by a known polymorphism. Cytochrome P450IID6 is not induced in both phenotypes by the classical antipyrine inducers phenobarbital or rifampicin [Eichelbaum et al., 1986b]. Thus in contrast to other cytochrome P450 isozymes, the isozyme involved in polymorphic sparteine/debrisoquine oxidation is predominantly under genetic control and environmental factors such as treatment with enzyme inducing drugs have a negligible influence on its activity. Therefore, drug interactions due to enzyme induction are very unlikely to occur.

Since all these studies can be carried at an early stage of drug development a great deal of information about the drug can be gained prior to Phase I clinical trials. If these data indicate that the metabolism of the new drug is subject to the debrisoquine/sparteine or another genetic polymorphism, Phase I clinical trials should be conducted in subjects of both EM and PM phenotype in order to assess the genetic variability in drug metabolism and response in relation to phenotype. Provided the drug is of therapeutic potential and it is decided to pursue its development, this knowledge should be incorporated into the design of the future studies. Patients and volunteers participating in Phase II and III trials should be characterised with respect to their oxidative metabolising capacity and particular emphasis should be placed on determining the different doses required for both phenotypes. In addition such studies should investigate whether a relationship exists between the metabolic ratio of a patient and the steady state plasma concentration attained. If such a relationship is established, then this will certainly facilitate the individualisation of drug dosage and hence the safe and effective use of the drug in the future.

These in vitro techniques also allow to evaluate the interaction potential of a new drug. If the drug is a substrate for cytochrome II D6 it can be anticipated that all drugs which inhibit P450IID6 activity can impair the metabolism of the drug and vice versa. Whether or not the new drug will cause a drug interaction or will itself be affected depends on the affinity of the particular drug for P450IID6. Thus by carrying out mutual in vitro inhibition studies it is possible to predict already in vitro the interaction potential prior to any clinical studies.

CONCLUSIONS

Interindividual differences in the response to drugs are of major concern to health professionals in private practice, academia, the pharmaceutical industry and regulatory agencies. A major source of variability is in the metabolism of drugs. Interest in the pharmacogenetic aspect of the problem was much stimulated by reports 10 to 15 years ago [Eichelbaum, 1975; Balant et al., 1976; Mahgoub et al., 1977] showing that the metabolism of drugs by oxidation, could be subject to genetic polymorphism as previously shown for acetylation. The number of publications in this research area is increasing almost exponentially. In Europe, a coordinated action under the assistance of the Organisation for European Cooperation Action in the Field of Scientific and Technical Research (COST) was started in late 1984 [Bechtel and Alvan, 1989] under the title "COST B1: Criteria for the Choice and Definition of Healthy Volunteers and/or Patients for Phase I and II studies in Drug Development". A Consensus Conference on Pharmacogenetics was organized in Besançon in October 1989, and the interest raised in the scientific community by this issue indicates that new achievements will certainly occur in this area in a near future.

REFERENCES

Alvan G., Grind M., Graffner C., Sjöqvist F., 1984. Relationship of N-demethylation of amiflamine and its metabolite to debrisoquine hydroxylation polymorphism. Clin. Pharmacol. Ther. 36:515.

Balant LP., Gundert-Remy U., Boobis AR., von Bahr Ch., 1989. Relevance of genetic polymorphism in drug metabolism in the development of new drugs. Europ. J. Clin. Pharmacol. 36:551.

Balant LP., Gorgia A., Tschopp JM., Revillard C., Fabre J., 1976. Pharmacocinétique de deux médicaments bêta-bloquants: Détection d'une anomalie pharmacogénétique ? Schweiz. Med. Wschr. 106:1403.

Balant-Gorgia AE., Balant LP., Garrone G., 1989. High blood concentrations of imipramine or clomipramine and therapeutic failure: A case report study using drug monitoring data. Ther. Drug Monitoring, 11:415.

Balant-Gorgia AE., Balant L., Zysset Th., 1987. High plasma concentrations of desmethylclomipramine after chronic administration of clomipramine to a poor metabolizer. Europ. J. Clin. Pharmacol. 32:101.

Balant-Gorgia AE., Balant L., Genet Ch., Dayer P., Aeschlimann JM., Garrone G., 1986. Importance of oxidative polymorphism and levomepromazine treatment on the steady-state blood concentrations of clomipramine and its major metabolites. Europ. J. Clin. Pharmacol. 31:449

Balant-Gorgia AE., Schulz P., Dayer P., Balant L., Kübli A., et al., 1982. Role of oxidation polymorphism on blood and urine concentrations of amitriptyline and its metabolites in man. Archiv für Psychiatrie und Nervenkrankenheiten 232:215.

Bechtel PR., Alvan G., 1989. Criteria for the choice and definition of healthy volunteers and or patients for phase I and phase II studies in drug development:

Commentary on an action for cooperative research. Europ. J. Clin. Pharmacol. **36**:549.

Bechtel PR., Bechtel Y., 1990. N-acetyltransferase. In: "European Consensus Conference on Pharmacogenetics", (G. Alvan, LP. Balant, PR. Bechtel, AR. Boobis, LF. Gram and K. Pithan, eds.), page 161. Commission of the European Communities, Luxembourg.

Beckmann J., Hertrampf R., Gundert-Remy U., Mikus G., Gross AS., Eichelbaum M., 1988. Is there a genetic factor in flecainide toxicity ? Br. Med. J. **297**:1316.

Bertilsson L., 1990. Interethnic differences in drug oxidation polymorphism. In: "European Consensus Conference on Pharmacogenetics" (G. Alvan, LP. Balant, PR. Bechtel, AR. Boobis, LF. Gram and K. Pithan, eds.), page 171. Commission of the European Communities, Luxembourg.

Bertilsson L., Åsberg-Wistedt A., 1983. The debrisoquine hydroxylation test predicts steady-state plasma levels of desipramine. Br. J. Clin. Pharmacol. **15**:388.

Bertilsson L., Eichelbaum M., Mellström B., Säwe J., Schulz HU., Sjöqvist F., 1980. Nortriptyline and antipyrine clearance in relation to debrisoquine hydroxylation in man. Life Sci. **27**:1673.

Bertilsson L., Henthorn TK., Sanz E., Tybring G., Säwe J., Villén T., 1989. Importance of genetic factors in the regulation of diazepam metabolism: Relationship to S-mephenytoin, but not debrisoquine, hydroxylation phenotype. Clin. Pharmacol. Ther. **45**:348.

Breimer DD., 1990. Potential clinical relevance of the interplay between genetic and environmental factors. In: "European Consensus Conference on Pharmacogenetics" (G. Alvan, LP. Balant, PR. Bechtel, AR. Boobis, LF. Gram and K. Pithan, eds.), page 69. Commission of the European Communities, Luxembourg.

Brøsen K., Gram LF., 1988. First-pass metabolism of imipramine and desipramine: Impact of the sparteine oxidation phenotype. Clin. Pharmacol. Ther. **43**:400.

Brøsen K., Gram LF., 1989a. Clinical significance of the sparteine/debrisoquine oxidation polymorphism. Europ. J. Clin. Pharmacol. **36**:537.

Brøsen K., Gram LF., 1989b. Quinidine inhibits the 2-hydroxylation of imipramine and desipramine but not the demethylation of imipramine. Europ. J. Clin. Pharmacol. **37**:155.

Brøsen K., Gram LF., Klysner R., Bech P., 1986a. Steady-state levels of imipramine and its metabolites: Significance of dose-dependent kinetics. Europ. J. Clin. Pharmacol. **30**:43.

Brøsen K., Klysner R., Gram LF., Otton SV., Bech P., Bertilsson L., 1986b. Steady-state concentrations of imipramine and its metabolites in relation to the sparteine/debrisoquine polymorphism. Europ. J. Clin. Pharmacol. **30**:679.

Cano JP., Berger Y., Fabre G., 1990. Relevance of genetic polymorphism in metabolism for drug development: Industrial point of view. In: "European Consensus Conference on Pharmacogenetics" (G. Alvan, LP. Balant, PR. Bechtel, AR. Boobis, LF. Gram and K. Pithan, eds.), page 113. Commission of the European Communities, Luxembourg.

Case DE., 1990. Genetic polymorphism in drug development: One industrial viewpoint. In: "European Consensus Conference on Pharmacogenetics" (G. Alvan, LP. Balant, PR. Bechtel, AR. Boobis, LF. Gram and K. Pithan, eds.), page 123. Commission of the European Communities, Luxembourg.

Cooper RG., Evans DAP., Price AH., 1987. Studies on the metabolism of perhexiline in man. Europ. J. Clin. Pharmacol. **32**:569.

Dahl-Puustinen ML., Lidén A., Alm Ch., Nordin C., Bertilsson L., 1989. Disposition of perphenazine is related to polymorphic debrisoquin hydroxylation in human beings. Clin. Pharmacol. Ther. **46**:78.

Dayer P., 1990. Advantages and drawbacks of probe drugs for the assessment of phenotypic expression of cytochrome P450 db1 (P450 IID6). In: "European Consensus Conference on Pharmacogenetics" (G. Alvan, LP. Balant, PR. Bechtel, AR. Boobis, LF. Gram and K. Pithan, eds.), page 33. Commission of the European Communities, Luxembourg.

Dayer P., Desmeules J., Leemann T., Striberni R., 1988. Bioactivation of the narcotic drug codeine in human liver is mediated by the polymorphic monooxygenase catalyzing debrisoquine 4-hydroxylation. Biochem. Biophys. Res. Commun. 152:411.

Dayer P., Balant L., Küpfer A., Courvoisier F., Fabre J., 1983. Contribution of the genetic status of oxidative metabolism to variability in the plasma concentrations of beta-adrenoceptor blocking agents. Europ. J. Clin. Pharmacol. 24:797.

Dayer P., Balant L., Küpfer A., Striberni R., Leemann T., 1985. Effect of oxidative polymorphism (debrisoquine/sparteine type) on hepatic first-pass metabolism of bufuralol. Europ. J. Clin. Pharmacol. 28:317.

Dayer P., Balant L., Courvoisier F., Küpfer A. Kübli A., Gorgia A., Fabre J., 1982 The genetic control of bufuralol metabolism in man. Europ. J. Drug. Metab. Pharmacokin. 7:73.

Eichelbaum M., 1975. Ein neuentdeckter Defekt im Arzneimittel-stoffwechsel des Menschen: Die fehlende N-Oxidation des Spartein. Habilitationsschrift, Medizinische Fakultät Rheinischen Friedrich-Wilhelms-Universität, Bonn.

Eichelbaum M., Spannbrucker N., Dengler HJ., 1979a. Influence of the defective metabolism of sparteine on its pharmacokinetics. Europ. J. Clin. Pharmacol. 16:189.

Eichelbaum M., Spannbrucker N., Steincke B., Dengler HJ., 1979b. Defective N-oxidation of sparteine in man: A new pharmacogenetic defect. Europ. J. Clin. Pharmacol. 16:183.

Eichelbaum M., Reetz KP., Schmidt EK., Zekorn C., 1986a. The genetic polymorphism of sparteine metabolism. Xenobiotica 16:465.

Eichelbaum M., Mineshita S., Ohnhaus E.E., Zekorn C., 1986b. The influence of enzyme induction on polymorphic spartein oxidation. Br. J. Clin. Pharmacol. 22:49.

Eichelbaum M., Baur MP., Dengler HJ., Osikowska-Evers BO., Tieves G., Rittner C., 1987. Chromosomal assignement of human cytochrome P450 (debrisoquine/sparteine type) to chromosome 22. Brit. J. Clin. Pharmacol. 23:455.

Eichelbaum M., Bertilsson L., Küpfer A., Steiner E., Meese CO., 1988. Enantioselectivity of 4-hydroxylation in extensive and poor metabolizers of debrisoquine. Brit. J. Clin. Pharmacol. 25:505.

Evans DAP., Manley FA., McKusick VA., 1960. Genetic control of isoniazid metabolism in man. Brit. Med. J. 2:485.

Evans DAP., Mahgoub A., Sloan TP., Idle JR., Smith RL., 1980. A family and population study of the genetic polymorphism of debrisoquine oxidation in a white British population. J. Med. Genet. 17:102.

Evans WE., Relling MV., 1990. Xbal 16-plus 9-kilobase DNA restriction fragments identify a mutant allele for debrisoquin hydroxylase: Report of a family study. Mol. Pharmacol. 37:639.

Gaedigk A., Blum M., Gaedigk R., Eichelbaum M., Meyer UA., 1991. Deletion of the entire cytochrome P450 CYP2D6 gene as a cause of impaired drug metabolism in poor metabolizers of the debrisoquine/sparteine polymorphism. submitted to Amer. J. Hum. Gen..

Gex-Fabry M., Balant-Gorgia AE., Balant LP., Garrone G., 1990. Clomipramine metabolism. Model-based analysis of variability factors from drug monitoring data. Clin. Pharmacokinet. 19:241.

Ghoneim MM., Korttila K., Chiang CK., Jacobs L., Schoenwald RD., Mewaldt SP., Kayaba KO., 1981. Diazepam effect and kinetics in Caucasians and Orientals. Clin. Pharmacol. Ther. 29:749.

Gibaldi M., Perrier D., 1982. "Pharmacokinetics". Marcel Dekker, New York.

Gonzalez FJ., Vilbois F., Hardwick JP., McBride OW., Nebert DW., Gelboin HV., Meyer UA., 1988a. Human debrisoquine 4-hydroxylase (P450IID1): cDNA and deduced amino acid sequence and assignement of the CYP2D locus to chromosome 22. Genomics. 2:174.

Gonzalez FJ., Skoda RC., Kimura S., Zanger UM., Nebert DW., Gelboin HV., Hardwick JP., Meyer UA., 1988b. Characterization of the common genetic defect in humans deficient in debrisoquin metabolism. Nature 331:442.

Gram LF., Brøsen K., 1990. Conditions under which genetic polymorphisms are clinically relevant. In: "European Consensus Conference on Pharmacogenetics" (G. Alvan, LP. Balant, PR. Bechtel, AR. Boobis, LF. Gram and K. Pithan, eds.), page 87. Commission of the European Communities, Luxembourg.

Guengerich F.P., Distlerath I.M., Reilly P.E.B., Wolff T., Shimada T., Umbenhauer D.R. Martin M.N., 1986. Human-liver cytochromes P-450 involved in polymorphisms of drug oxidation. Xenobiotica 16:367.

Gundert-Remy U., 1990. Relevance of genetic polymorphism in metabolism for drug development: the point of view of regulatory agencies. In: "European Consensus Conference on Pharmacogenetics" (G. Alvan, LP. Balant, PR. Bechtel, AR. Boobis, LF. Gram and K. Pithan, eds.), page 129. Commission of the European Communities, Luxembourg.

Heim M., Meyer UA., 1990. Genotyping of poor metabolisers of debrisoquine by allele-specific PCR amplification. Lancet. 336:529.

Horai Y., Nakano M., Ishizaki T., Ishikawa K., Zhou HH., Zhou BJ., Liao CL., Zhang LM., 1989. Metoprolol and mephenytoin oxidation polymorphism in Far Eastern Oriental subjects: Japanese versus mainland Chinese. Clin. Pharmacol. Ther. 46:198.

Ishizaki T., Eichelbaum M., Horai Y., Hashimoto K., Chiba K., Dengler HJ., 1987. Evidence for polymorphic oxidation of sparteine in Japanese subjects. Brit. J. Clin. Pharmacol. 23:482.

Jack DB., Wilkins MR., Quaterman CP., 1983. Lack of evidence for polymorphism in metoprolol metabolism. Brit. J. Clin. Pharmacol. 16:188.

Jackson PR., Tucker GT., Lennard MS., Woods HF., 1986. Polymorphic drug oxidation: Pharmacokinetic basis and comparison of experimental indices. Brit. J. Clin. Pharmacol. 22:541.

Jackson PR., Tucker GT., Lennard MS., Woods HF., 1990. Testing for bimodality in frequency distributions of data suggesting polymorphism of drug metabolism — Statistical methods. In: "European Consensus Conference on Pharmacogenetics" (G. Alvan, LP. Balant, PR. Bechtel, AR. Boobis, LF. Gram and K. Pithan, eds.), page 43. Commission of the European Communities, Luxembourg.

Kagimoto M., Heim M., Kagimoto K., Zeugin T., Meyer UA., 1990. Multiple mutations of the human cytochrome P450IID6 gene (CYP2D6) in poor metabolizers of debrisoquine. Study of the functionnal significance of individual mutations by expression of chimeric genes. J. Biol. Chem. 265:17209.

Kimura S., Umeno M., Skoda RC., Gelboin HV., Meyer UA., Gonzalez FJ., 1989. The human debrisoquine 4-hydroxylase (CYP2D) locus: sequence and identification of the polymorphic CYP2D6 gene, a related gene, and a pseudogene. Am. J. Hum. Genet. 45:889.

Kübli A., Balant L., Dayer P., Balant-Gorgia A., Fabre J., 1982. Influence du polymorphisme génétique de l'oxydation sur les études de biodisponibilité: A propos du bufuralol. J. Pharm. Clin. 1:301.

Kroemer H.K., Mikus G., Kronbach T., Meyer UA., Eichelbaum M., 1989. In vitro characterization of the human cytochrome P-450 involved in polymorphic oxidation of propafenone. Clin. Pharmacol. Ther. **45**:28.

Kumana CR., Lander IJ., Chan M., Ko W., Lin HJ., 1987. Differences in diazepam pharmacokinetics in Chinese and white Caucasians: Relation to body lipid stores. Europ. J. Clin. Pharmacol. **32**:211.

Küpfer A., Preisig R., 1984. Pharmacogenetics of mephenytoin: A new drug hydroxylation polymorphism in man. Europ. J. Clin. Pharmacol. **26**:753.

Lennard MS., Tucker GT., Woods HF., 1986. The polymorphic oxidation of β-adrenoceptor antagonists. Clinical pharmacokinetic considerations. Clin. Pharmacokin. **11**:1.

Lennard MS., Tucker GT., Silas JH., Freestone S., Ramsay LE., Woods HF.,1983. Differential stereoselective metabolism of metoprolol in extensive and poor debrisoquin metabolizers. Clin. Pharmacol. Ther. **34**:732.

Lewis RV., Lennard MS., Jackson PR., Tucker GT., Ramsay LE., Woods HF., 1985. Timolol and atenolol: Relationships between oxidation phenotype, pharmacokinetics and pharmacodynamics. Br. J. Clin. Pharmacol. **19**:329.

Lou YC., Ying L., Bertilsson L., Sjöqvist F.,1987. Low frequency of slow debrisoquine hydroxylation in a native Chinese population. Lancet **II**:852.

Mahgoub A., Idle JR., Lancaster R., Smith RL., 1977. Polymorphic hydroxylation of debrisoquine in man. Lancet **II**: 584.

Mellström B., Bertilsson L., Lou YC., Säwe J., Sjöqvist F.,1983. Amitriptyline metabolism: Relationship to polymorphic debrisoquine hydroxylation. Clin. Pharmacol. Ther. **34**:516.

Meyer UA., Gut J., Kronbach T., Skoda C., Meyer UT., Catin T., Dayer P., 1986. The molecular mechanisms of two common polymorphisms of drug oxidation — Evidence for functional changes in cytochrome P-450 isozymes catalysing bufuralol and mephenytoin oxidation. Xenobiotica **16**:449.

Meyer UA., Zanger UM., Grant D., Blum M., 1990. Genetic polymorphism of drug metabolism. In: "Advances in Drug Research" (B. Testa, ed.), Vol. 19, page 197. Academic Press, London.

Mikus G., Gross AS., Beckmann J., Hertrampf R., Gundert-Remy U., Eichelbaum M., 1989. The influence of the sparteine/debrisoquin phenotype on the disposition of flecainide. Clin. Pharmacol. Ther. **45**:562.

Nakamura K., Goto F., Ray WA., Mc Allister CB., Jacqz E., Wilkinson GR., Branch RA., 1985. Interethnic differences in genetic polymorphism of debrisoquine and mephenytoin hydroxylation between Japanese and Caucasian populations. Clin. Pharmacol. Ther. **38**:402.

Oates NS., Shah RR., Idle JR., Smith RL., 1983. Influence of oxidation polymorphism on phenformin kinetics and dynamics. Clin. Pharmacol. Ther. **34**:827.

Pierce DM., Smith SE., Franklin RA.,1987. The pharmacokinetics of indoramin and 6-hydroxyindoramin in poor and extensive hydroxylators of debrisoquine. Europ. J. Clin. Pharmacol. **33**:59.

Rane A., 1990. Documentation of the metabolism of new chemical entities: future perspective. In: "European Consensus Conference on Pharmacogenetics" (G. Alvan, LP. Balant, PR. Bechtel, AR. Boobis, LF. Gram and K. Pithan, eds.), page 135. Commission of the European Communities, Luxembourg.

Sanz E., Villén T., Alm C., Bertilsson L., 1989. S-Mephenytoin hydroxylation phenotypes in a Swedish population determined after co-administration with debrisoquine. Clin. Pharmacol. Ther. **45**:495.

Schmid B., Bircher J., Preisig R., Küpfer A., 1985. Polymorphic dextromethorphan metabolism: Co-segretation of oxidative O-demethylation with debrisoquin hydroxylation. Clin. Pharmacol. Ther. **38**:618.

Siddoway LA., Thompson KA., McAllister CB., Wang T., Wilkinson GR., Roden DM., Woosley RL., 1987. Polymorphism of propafenone metabolism and disposition in man: Clinical and pharmacokinetic consequences. Circulation **75**:785.

Skoda RC., Gonzalez FJ., Demierre A., Meyer UA., 1988. Two mutant alleles of the human cytochrome P450db1-gene (P450C2D1) associated with genetically deficient metabolism of debrisoquine and other drugs. Proc. Natl Acad. Sci. USA. **85**:5240.

Tucker GT., Jackson PR., Lennard MS., Woods HF., 1990. Pharmacokinetic-pharmacogenetic modelling — A basis for the display and detection of polymorphisms in drug metabolism. In: "European Consensus Conference on Pharmacogenetics" (G. Alvan, LP. Balant, PR. Bechtel, AR. Boobis, LF. Gram and K. Pithan, eds.), page 59. Commission of the European Communities, Luxembourg.

von Bahr C., 1990. Relevance of genetic polymorphism of metabolism for drug development: The point of view of industry. In: "European Consensus Conference on Pharmacogenetics" (G. Alvan, LP. Balant, PR. Bechtel, AR. Boobis, LF. Gram and K. Pithan, eds.) page 119. Commission of the European Communities, Luxembourg.

von Bahr C., Guengerich FP., Movin G., Nordin C., 1989. The use of human liver banks in pharmacogenetic research. In: "Clinical Pharmacology in Psychiatry: From Molecular Studies to Clinical Reality" (SG. Dahl and LF. Gram, eds.), page 163. Springer-Verlag, Heidelberg.

Wang T., Roden DM., Wolfenden HT., Woosley RL., Wood AJJ., Wilkinson GR., 1984. Influence of genetic polymorphism on the metabolism and disposition of encainide in man. J. Pharmacol. Exp. Ther. **228**:605.

Wedlund PJ., Aslanian WS., Mc Allister CB., Wilkinson GR., Branch RA., 1984. Stereoselective metabolism and pharmacogenetic control of 5-phenyl-5-ethylhydantoin. J. Pharmacol. Exp. Ther. **230**:28.

Yue QY., Bertilsson L., Dahl-Puustinen ML., Säwe J., Sjöqvist F., Johansson I., Ingelman-Sundberg M., 1989. Dissociation between debrisoquine hydroxylation phenotype and genotype among Chinese. Lancet **II**:870.

Zanger UM., Vilbois F., Hardwick JP., Meyer UA., 1988. Absence of hepatic cytochrome Pbull causes genetically deficient debrisoquine hydroxylation in man. Biochemistry. **27**:5447.

Zekorn C., Achtert G., Hausleiter HJ., Moon CH., Eichelbaum M., 1985. Pharmacokinetics of N-propylajmaline in relation to polymorphic sparteine oxidation. Klin. Wochenschr. **63**:1180.

Zhou HH., Wood AJJ., 1990. Differences in stereoselective disposition of propranolol does not explain sensitivity differences between white and Chinese subjects: Correlation between the clearance of (–)- and (+)-propranolol. Clin. Pharmacol. Ther. **47**:719.

Zhou HH., Antony LB., Roden DM., Wood AJJ.,1990. Quinidine reduces clearance of (+) -propranolol more than (–)-propranolol through marked reduction in 4-hydroxylation. Clin. Pharmacol. Ther. **47**:686.

THE ROLE OF PHARMACOKINETICS AND PHARMACODYNAMICS IN THE DEVELOPMENT OF THERAPEUTIC PROTEINS

Joyce Mordenti and James D. Green

Department of Safety Evaluation
Genentech, Inc.
South San Francisco, CA 94080

ABSTRACT

Since the degree of preclinical work conducted to address safety issues in animals is usually rate limiting to the initiation of Phase I clinical studies, the preclinical program should be designed to allow rapid entry into the clinic without compromising safety. The establishment of acceptable therapeutic ratios relies on the relationship of doses utilized in toxicology studies and their projected relationship to the clinical dosing regimen. Experience with multiple therapeutic proteins (biomacromolecules) in our laboratory indicates that the pharmacokinetic behavior of many proteins is predictable across species. In selected cases, this information permits extrapolation of preclinical safety and efficacy data to the clinical setting when doses are related on the basis of pharmacokinetic equivalence rather than on a body weight (mg/kg) or body surface area (mg/m^2) basis. With a better understanding of this cross species relationship, the confidence in the safety of a therapeutic agent in initial clinical studies is increased. Based on our experience, an approach is presented that maximizes the relevance of the preclinical information gained, minimizes the scope of the initial toxicology studies, and, therefore, minimizes time to initiation of Phase I. This strategy may be useful for selected classes of compounds. Also, several case studies are presented to illustrate how pharmacokinetics was used to bridge the gap between discovery research, preclinical studies, and clinical trials during the development of two therapeutic proteins (rCD4 and rCD4-IgG) for the treatment of Acquired Immunodeficiency Syndrome (AIDS).

THE ROLE OF PHARMACOKINETICS AND PHARMACODYNAMICS

Whether one is developing a small molecule or a biomacromolecule, the relationship of dose (concentration) to effect is pivotal. The method chosen for selecting preclinical doses depends on particular objectives and, to a great extent, on the data and amount of test material that are available. Defining a 'rational' dose early

New Trends in Pharmacokinetics, Edited by A. Rescigno and A.K. Thakur
Plenum Press, New York, 1991

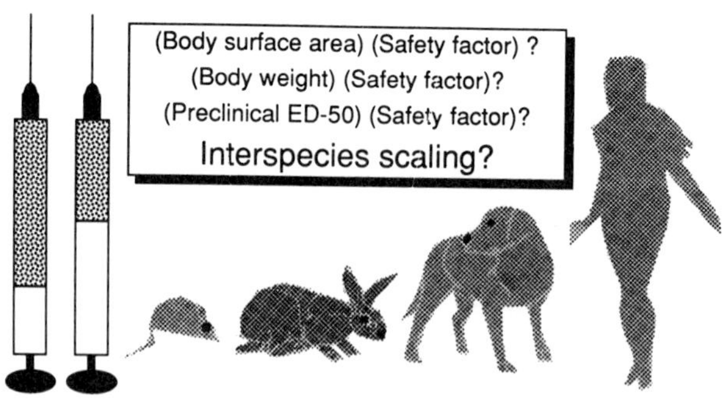

Figure 1. How does one select a dose?

in the development process may expedite subsequent development through a better understanding of how preclinical data will relate to the expected clinical outcome. This may result in a substantial saving in time and resources without a loss of safety.

Figure 1 provides several paradigms for extrapolating doses from animals to humans. The technique of interspecies scaling has been found to be particularly useful for setting and projecting doses for Phase I. Traditional methods of scaling doses on the basis of body weight (mg/kg), body surface area (mg/m^2), or other physiologic parameter are actually specific examples of interspecies scaling and will be discussed in more detail under 'Dose Selection.'

Interspecies scaling techniques may provide predictability, control, increased confidence in dosimetry, and a better understanding of the pharmacological, toxicological, and pharmacodynamic relationships that have been established preclinically [Chappell and Mordenti, 1991]. Establishing the pharmacokinetics early in the development process is important because it provides a framework for an accelerated dose escalation under pharmacokinetic control in Phase I and II clinical studies (Figure 2) [Peck, Collins and Harter, 1990]. The end result can be a decrease

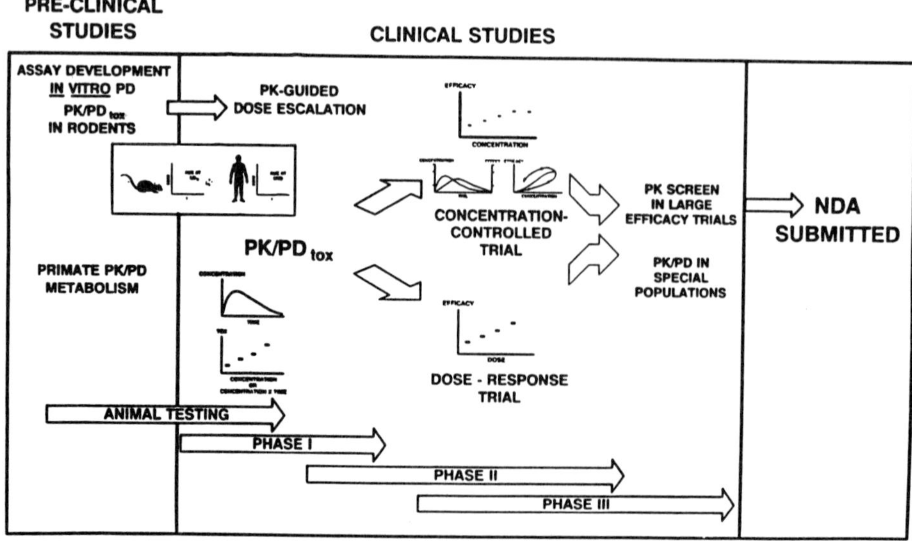

Figure 2. Incorporating pharmacokinetics and pharmacodynamics in drug development. [Peck, Collins and Harter, 1990]. PK = Pharmacokinetics, PD = Pharmacodynamics

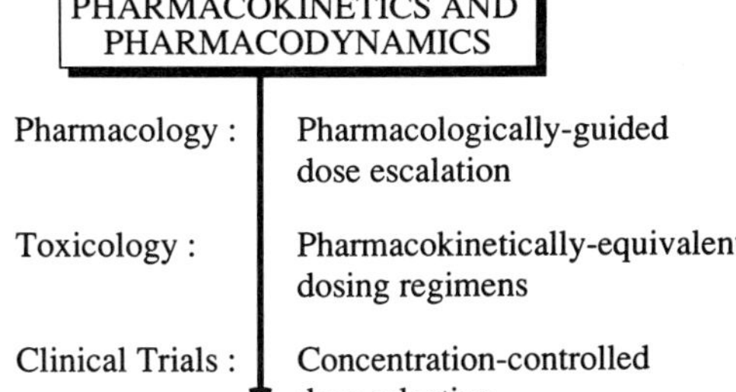

Figure 3. Where does dose selection impact?

in both preclinical and clinical development time and a better understanding of the meaning of dose as early clinical studies begin.

Pharmacokinetics and pharmacodynamics have important roles in selecting doses in preclinical safety and efficacy studies (Figure 3). Integration of this information into drug development programs is the basis of this development strategy. *In vitro* and *in vivo* pharmacology data help to establish the relationship of dose (concentration) to effect and can be used for initial projections of the clinical dose range (concentration) that are desirable. Preclinical pharmacokinetic and pharmacodynamic studies determine drug disposition and effect in the species chosen for the pharmacology and toxicology programs. The doses for the toxicology program are selected to support the projected clinical dosing regimen and to determine the initial therapeutic ratio in pharmacologically active species. When the clinical trials begin, the dose is titrated to the efficacious range based on the pharmacokinetic and pharmacodynamic projections, and then the dose is adjusted based upon the effects that are observed in the initial trials.

UNIQUE POINTS TO CONSIDER FOR THERAPEUTIC PROTEINS

Unlike small, organic molecules, several unique issues need to be considered when performing safety and disposition studies with therapeutic proteins, namely, species specificity, immunogenicity, host cell contaminants, and perhaps most importantly, the small amount of material available for initial preclinical studies. Proteins may not have been highly conserved during evolution; thus, some proteins exhibit activity in a narrow host range, e.g., recombinant human interferon-gamma is active only in higher order primates. When activity is species specific, toxicity and efficacy may be difficult to establish preclinically. Even when proteins exhibit a high degree of sequence homology, parenteral administration to nonhomologous hosts can result in antibody production. Host cell contaminants, such as heterogeneous proteins, endotoxins, viruses [Office of Biologics Research and Review, 1987; Harbour and Woodhouse, 1990], and bovine serum albumin, when present, are routinely removed or inactivated during the manufacturing and purification process and generally do not present major obstacles with current technology. For biomacromolecules, the high cost and limited quantities of test material during the early stages of development must be considered in the overall design of preclinical studies. Typically, only milligram to gram quantities of protein are available (@ $100-1,000K/g); whereas, kilogram quantities of material are often available for the development of small molecules at substantially lower costs (e.g., $1-10K/kg). As a result, the doses for the preclinical development of a biomacromolecule must be selected carefully.

To date, these points have not hindered our ability to evaluate safety in preclinical models and predict safe use in humans.

INTERSPECIES SCALING

Laboratory animals rarely eliminate drugs at the same rate as humans; thus, it is often erroneously concluded that animal data do not provide useful pharmacokinetic information or that preclinical data do not relate to clinical outcome. When species size is taken into account, it is sometimes possible to mathematically extrapolate pharmacokinetic, pharmacodynamic, and toxicologic data across species using interspecies scaling techniques [Chappell and Mordenti, 1991; Mordenti and Chappell, 1989]. The basis for interspecies scaling is the predictability with which anatomical and physiological parameters vary across species — as species decrease in size, they have proportionately larger organs and faster blood circulation time. Since drug clearance is often associated with organ size and blood flow, it is not surprising that small animals frequently eliminate drugs more rapidly than humans, and animal size must be taken into account when doses are extrapolated to humans.

Two methods of interspecies scaling are in widespread use: the allometric approach and the physiologic approach [Mordenti, 1986b]. The method that one chooses will depend on overall goals and, to a great extent, on the data that are available. For selected macromolecules, the allometric approach, which is the easier methodology, may be sufficient. The allometric approach is empiric, not mechanistic; thus it only relates observed phenomena across species without any information on underlying processes. For compounds that undergo species-specific metabolism, that are bound to blood and tissue components with different avidity in different species, that exhibit species-specific binding proteins, or that elicit a significant immune response during one dosing interval (e.g., monoclonal antibodies), a physiological approach may be advisable. A physiologic model can include conversion mechanisms, enzyme activities, binding parameters, and other information that impact on drug disposition and interspecies relationships. The allometric appproach will be discussed briefly in the following section. A comprehensive review of the physiologic approach is presented in another chapter (Krishnan and Andersen, this volume).

The allometric approach utilizes a power function (allometric equation) written as follows:

$$Y = a\, W^b \tag{1}$$

where Y is the parameter of interest, W is body weight, a is the allometric coefficient, and b is the allometric exponent. Equations of this form result whenever an organ or physiologic function change in proportion to total body weight [Kleiber, 1950]. This technique has been used successfully to describe a number of physiological and anatomical properties for birds, reptiles, fish, and mammals [McMahon and Bonner, 1983; Calder, 1984; Schmidt-Nielsen, 1984].

The power function can be linearized by taking its logarithm. The log-transformed allometric equation is written as follows:

$$\log Y = \log a + b \log W \tag{2}$$

where $\log a$ is the y-intercept and b is the slope. The allometric coefficient, a, is the value of the physiological variable (Y) at 1 unit of body weight, i.e, if W is in kg, then a is the value for a 1 kg animal. The sign and magnitude of the exponent (b) indicates how the physiological variable is changing as a function of body weight. In

$$\boxed{\begin{aligned} \text{Clearance} &= aW^X \\ \text{Volume} &= bW^Y \end{aligned}}$$

To produce the same steady-state concentrations:

$$\boxed{D_{human} = D_{animal} \left(\frac{W_{human}}{W_{animal}}\right)^X}$$

To produce the same peak concentrations:

$$\boxed{D_{human} = D_{animal} \left(\frac{W_{human}}{W_{animal}}\right)^Y}$$

$$\boxed{\begin{aligned} &\text{let } X = 0.7 \text{ and } Y = 1 \\ &\text{when no data are available} \end{aligned}}$$

Figure 4. Interspecies extrapolation techniques to estimate equivalent doses. These equations can be used to estimate doses that produce the same average steady-state concentrations or the same peak concentrations, provided the assumptions for allometric scaling apply. Units: CL, volume/time; V, volume; D, dose as total amount (mg) administered (not as mg/kg); and W, weight

general, biological frequencies tend to have an exponent of -0.25; biological time periods tend to have an exponent of 0.25; clearance, physiologic flow rates, and metabolic rate tend to have an exponent of 0.75; and volumes tend to have an exponent of 1.0. Body surface area has an exponent of 0.67.

Allometric scaling techniques have been used to extrapolate pharmacokinetic data for small organic molecules [Boxenbaum, 1984; Boxenbaum and D'Souza, 1990] and, more recently, for biomacromolecules [Mordenti, Chen, et al., 1990; Mordenti, Chen, et al., 1991] from laboratory animals to humans. The allometric exponents for the pharmacokinetic parameters tend to approximate the exponents for corresponding physiologic variables, i.e., half-life (min): $b \sim 0.2$ to 0.4, clearance (mL/min): $b \sim 0.6$ to 0.8, and the volume of distribution (mL): $b \sim 0.8$ to 1.0. (NOTE: If the pharmacokinetic parameters are expressed in different units, then the allometric exponents will change accordingly, e.g. clearance (mL/min/kg): $b \sim -0.4$ to -0.2; volume (mL/kg): $b \sim -0.2$ to 0.) These approximations are usually true for drugs or metabolites that are eliminated solely by physical processes, i.e., biliary, renal, or pulmonary excretion; they are applied cautiously, if at all, to compounds that are metabolized, that have nonlinear pharmacokinetics, or that exhibit saturable or species-specific binding. These caveats are true for small molecules as well as macromolecules.

DOSE SELECTION

If very little is known about the biologic activity and pharmacokinetics of a protein, then $W^{0.7}$ is used as an initial estimate of the allometric scaling factor for clearance, $W^{1.0}$ is used as an initial estimate of the allometric scaling factor for volume of distribution, and the doses are scaled across species accordingly. When efficacy or toxicity are shown to be related to clearance, e.g., related to area under the curve (AUC) or steady-state concentrations, the allometric scaling factor will probably be in the range of $W^{0.6}$ to $W^{0.8}$, which means that smaller animals will require more drug per unit body weight. When efficacy or toxicity appear to be related to peak concentrations (Cmax), the allometric scaling factor may be closer to

$W^{1.0}$, which translates to identical mg/kg across species. Dose extrapolation equations are provided in Figure 4. It should be noted that $W^{0.7}$ predicts a smaller dose for humans than $W^{1.0}$; thus, $W^{0.7}$ provides a more conservative estimate for the human dose than scaling doses on a mg/kg basis. The more conservative approach is recommended when little is known about the safety or efficacy of a compound.

Body surface area is proportional to $aW^{0.67}$. Since this allometric scaling factor resides within the usual range of the allometric scaling factor for clearance, i.e., $aW^{0.6}$ to $aW^{0.8}$, it has served as a reasonable surrogate for cross species dose extrapolation. Despite its success, the use of body surface area as a scaling factor should be discouraged — the calculations are unwieldy in a clinical setting, and it has no biological significance with regards to drug toxicity other than providing an allometric scaling factor that may be similar to drug clearance. The body surface area scaling approach can be replaced with the less cumbersome, straightforward $W^{0.7}$ scaling approach in most cases.

CASE STUDIES

The following two case studies will serve as an example of how interspecies scaling techniques were used to bridge the gap between discovery research, preclinical studies, and clinical trials during the rapid development of two therapeutic proteins for the treatment of Acquired Immunodeficiency Syndrome (AIDS). Complete interspecies scaling data for these two proteins are provided elsewhere [Mordenti, Shaieb, et al., 1991; Mordenti, Chen, et al., 1991].

A. Recombinant Soluble CD4 (rCD4)

CD4 is a surface glycoprotein found primarily on a subset of mature peripheral T cells that recognize antigens presented by class II MHC molecules [Sattentau and Weiss, 1988; Janeway, 1988]; it is also the putative receptor for the AIDS virus, human immunodeficiency virus type 1 (HIV-1) [Maddon, Dalgleish, et al., 1986]. Recombinant soluble CD4 is a 50 kDa glycoprotein produced in Chinese Hamster ovary (CHO) cells; it binds gp120 and blocks HIV-1 infection of T cells and monocytes *in vitro* [Smith, Byrn, et al., 1987; Byrn, Sekigawa, et al., 1989]. It is hypothesized that it will inhibit HIV-1 infection *in vivo* by serving as a decoy and prohibiting interactions between the virus and the T cell; thus it is being developed for the treatment and prevention of AIDS.

Preclinical pharmacokinetic studies with rCD4 were conducted in rats, rabbits and monkeys. Figure 5 shows the allometric plot of rCD4 clearance as a function of body weight. The unweighted log-transformed data were extrapolated to 70 kg by linear regression analysis to predict the clearance of rCD4 in humans. The predicted clearance in humans was used to estimate steady-state rCD4 concentrations in the Phase I continuous intravenous infusion studies, using the simple relationship,

$$C_{ss} = \text{Infusion rate/CL} \tag{3}$$

where Css is steady-state concentration, and CL is the predicted clearance (Table 1). The range of values for clearance was established by assuming the normal variation in this parameter is 1/2- to 2-fold. A similar approach was used to extrapolate the volume of distribution data to 70 kg, and peak serum concentrations of rCD4 following IV bolus administration in humans were estimated using the simple relationship,

$$\text{Peak} = \text{Dose/V} \tag{4}$$

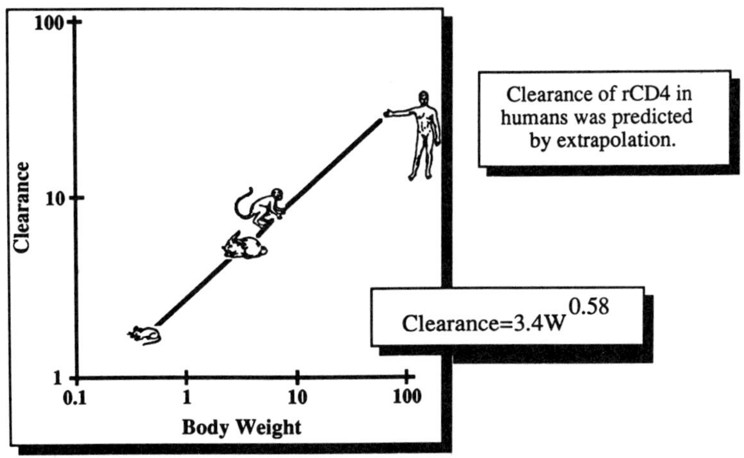

Figure 5. Interspecies scaling of rCD4 clearance (mL/min) *versus* body weight (kg) where V is the predicted volume of distribution (Table 2). Both sets of predictions were included in our Investigational New Drug (IND) application and illustrate the prospective use of interspecies scaling.

Table 1

Predicted and observed steady-state rCD4 concentrations in humans on approximately days 7 and 14 of a 2-week constant IV infusion regimen

Total dose (μg/kg/24 hr)[a]	Predicted concentration (ng/mL)	Observed concentration (ng/mL)[b]
1	<0.8-2.4	1-2.9
10	6-24	10-24
30	18-72	23-31
100	60-240	77-191
300	180-720	279-401
1000	600-2400	422-744

a. n = 3 patients/dose; data not available from all patients.
b. Yarchoan, Thomas, et al. 1989.

Table 2

Predicted and observed peak rCD4 concentrations in humans following a single IV bolus dose

Total dose (μg/kg)[a]	Predicted concentration (ng/mL)	Observed concentration (ng/mL)[b]
1	15-26	18-47
10	150-260	103-452
30	450-790	425-1330
100	1500-2600	1390-1920
300	4500-7900	2430-5020

a. n = 6 patients/dose; data not available from all patients.
b. Approximately ten minutes post-dose [Kahn, Davis, et al., 1989; Kahn, Allan, et al. 1990].

From Phase I clinical trials, it was determined that the average clearance [mean (SD)] of rCD4 in humans was 0.82 (0.35) mL/min/kg, the initial volume of

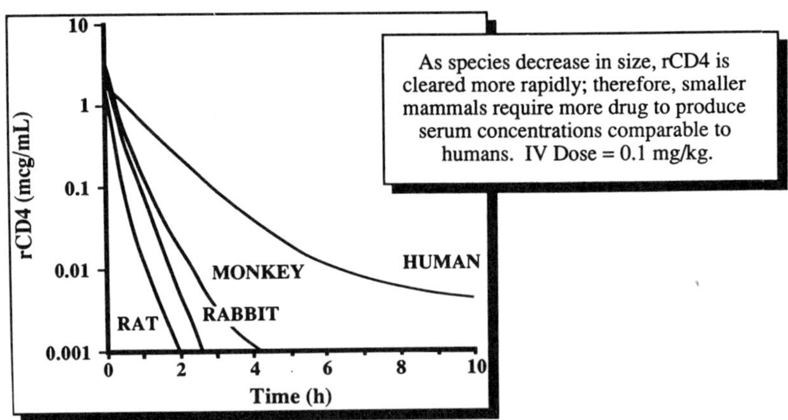

Figure 6. The pharmacokinetics of rCD4 in rat, rabbit, monkey, and human for a 0.1 mg/kg IV dose. Data for laboratory animals are simulated from pharmacokinetic parameters obtained at other doses; data for humans are observed [Kahn, Allan, et al., 1990].

distribution was 52 (28) mL/kg, and the volume of distribution at steady state was 70 (45) mL/kg [Kahn, Allan, et al. 1990].

The allometric equation for rCD4 clearance (mL/min) without the human data was $3.4W^{0.58}$; when the human data were added to the regression analysis, the equation became $3.4W^{0.62}$. These results reinforce two important points: (1) rCD4 clearance scales across species as predicted by theory (i.e., allometric exponent ~ 0.6 to 0.8), and (2) smaller animals eliminate rCD4 more rapidly than larger species. These findings have important implications in the evaluation of safety data and introduce the concept of pharmacokinetic (dose) equivalents; that is, doses or dosing regimens that produce similar serum concentrations in different species. Figure 6 illustrates the pharmacokinetics of rCD4 in various species; as species increase in size, the clearance (mL/min/kg) is slower, and rCD4 remains in the serum for proportionately longer times.

Figure 7 provides dosing regimens for rCD4 in mouse, rat, monkey, and human that are pharmacokinetically equivalent (the mouse was predicted by interspecies extrapolation and later confirmed in a small pilot study; the rabbit was omitted from this comparison because of the similarity to monkey with respect to body weight). The mouse requires 28-times more drug (on a per weight basis) than the human to maintain equivalent steady-state concentrations, because the mouse is eliminating the drug 28-times more rapidly that the human.

A common error in safety assessment programs is to interpret doses on a simple mg/kg basis. In the previous example, 28 mg/kg in the mouse would be erroneously

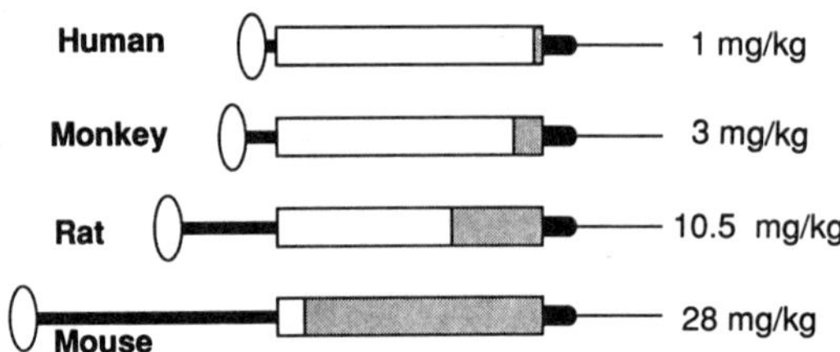

Figure 7. Dose per kg of rCD4 that produces the same average steady-state serum concentrations in each species

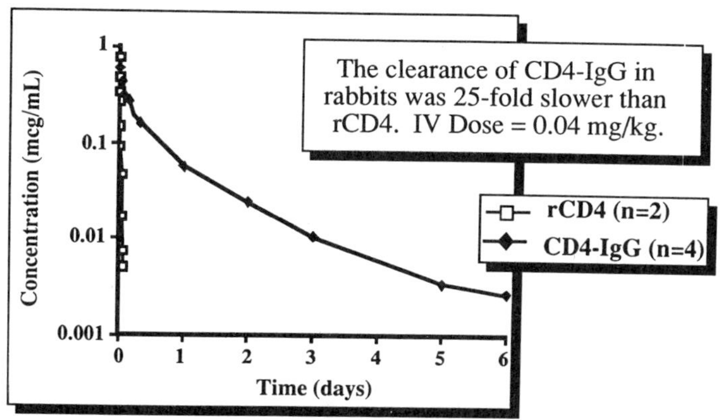

Figure 8. The pharmacokinetics of rCD4 (open square) and CD4-IgG (solid diamond) in rabbits at a dose of 0.04 mg/kg

interpretted to be 28-times larger than the human (1 mg/kg) dose (a considerable safety factor); by knowing the pharmacokinetics, 28 mg/kg in the mouse is correctly evaluated as only equivalent to the human (1 mg/kg) dose — a much different interpretation.

B. Recombinant CD4 Immunoglobulin G (CD4-IgG)

Since interspecies scaling techniques suggested that rCD4 would have a rapid clearance and a short half-life in humans, development began on a second generation molecule that would have a more desirable, prolonged disposition profile. This research culminated in the development of the CD4 immunoadhesin [Capon, Chamow, et al., 1989]. Several CD4 immunoadhesin molecules were produced and made available for *in vitro* and *in vivo* testing. One immunoadhesin, CD4-IgG, was selected for development based on favorable *in vitro* properties and a 25-fold reduction in clearance in rabbits compared to rCD4 (Figure 8).

CD4-IgG is a 98 kDa glycoprotein produced in CHO cells. It is composed of the N-terminal two immunoglobulin-like domains of CD4 joined to the Fc region of human IgG1 [Harris, Wagner, and Spellman, 1990]. Not only does this molecule retain the gp120 binding capabilities of rCD4, but it displays the advantageous characteristics of an IgG antibody, including long serum half-life, placental transfer, and antibody-directed cell-mediated cytotoxicity (ADCC) against HIV-infected cells [Byrn, Mordenti, et al., 1990].

Table 3
Predicted and observed steady-state CD4-IgG concentrations in humans on approximately days 7 and 14 of a 2-week constant IV infusion regimen

Total dose (μg/kg/24 hr)[a]	Predicted concentration (μg/mL)	Observed concentration (μg/mL)[b]
30	0.3-1.2	0.24-0.53
100	1.0-4.0	1.9-2.7
300	3.0-12.0	3.9-6.6
1000	10.0-40.0	15.2-21.6

a. n = 3 patients/dose; data not available from all patients.
b. Yarchoan, Pluda, et al. 1990.

Preclinical pharmacokinetic studies were conducted in rats, rabbits and monkeys. The unweighted log-transformed pharmacokinetic data were extrapolated

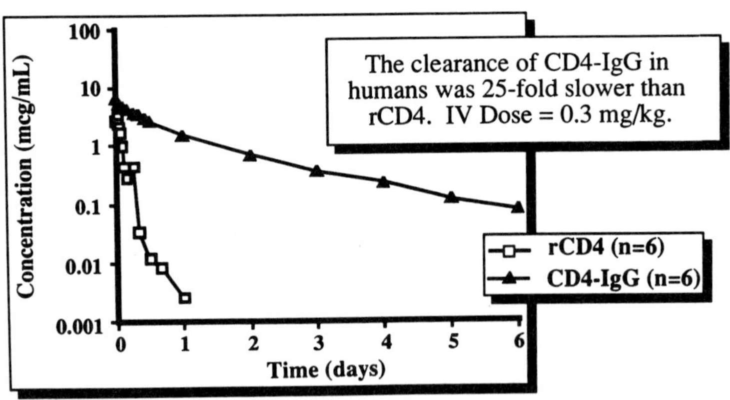

Figure 9. The pharmacokinetics of rCD4 (open square) [Kahn, Allan, et al., 1990] and CD4-IgG (solid triangle) [Collier, Katzenstein, et al., 1990] in humans at a dose of 0.3 mg/kg

to 70 kg by linear regression analysis to predict the clearance and volume of distribution of CD4-IgG in humans. These data were used to estimate steady-state concentrations in the Phase I continuous intravenous infusion studies (Table 3) and peak serum concentrations following IV bolus administration (Table 4).

Table 4
Predicted and observed.peak CD4-IgG concentrations in humans following a single IV bolus dose

Total dose (μg/kg)[a]	Predicted concentration (μg/mL)	Observed concentration (μg/mL)[b]
30	0.26-1.2	0.39-1.4
100	0.87-4.0	0.94-4.0
300	2.6-12.0	4.2-8.6
1000	8.7-40.0	14.5-31.1

a. n = 6 patients/dose; data not available from all patients.
b. Approximately ten minutes post-dose [Collier, Katzenstein, et al. 1990].

From the preclinical data, the clearance of CD4-IgG in humans was expected to be 25-fold slower than rCD4 and the volume of distribution approximately equivalent to rCD4. The actual clearance [mean (SD)] of CD4-IgG from two patients receiving 1 mg/kg as an IV bolus was 1.92 (0.2) mL/hr/kg, the initial volume of distribution was 50 (10) mL/kg, and the volume of distribution at steady state was 76 (16) mL/kg (personal communication, Terri Hodges, M.D., New England Deaconess Hospital, Boston, MA). The pharmacokinetic profiles for rCD4 and CD4-IgG in humans are compared in Figure 9 (Note the similarity to Figure 8). Again, the preclinical data were remarkably accurate in predicting human disposition.

C. Experience with Other Macromolecules

Interspecies scaling techniques, similar to those shown for rCD4 and CD4-IgG, were applied to pharmacokinetic data for Activase® tissue plasminogen activator, recombinant human growth hormone, and relaxin. The clearance and volume of distribution data for these proteins were shown to scale well across species, too [Mordenti, Chen, et al. 1991].

EXPEDITED DEVELOPMENT STRATEGY

Expediting the drug development process has been the subject of much recent discussion [Mordenti, Shaieb, et al., 1991; Fromson, 1989; Lasagna, 1989]. Based on information presented in the previous sections, the pharmacokinetic disposition of several biomacromolecules has been shown to be predictable when extrapolated from animal to man. Furthermore, it has been shown that a meaningful relationship can be established between administered doses on the basis of pharmacokinetic equivalents. Other authors have referred to this relationship as concentration controlled, pharmacologically guided, or pharmacokinetically equivalent dose selection [Mordenti, 1986a; Peck, Collins, and Harter, 1990; Collins, Grieshaber, and Chabner, 1990; Collins, Zaharko, et al., 1986; Collins, Leyland-Jones, and Grieshaber, 1987]; but this approach has not received much support. The ability to better understand the relationship between administered dose, subsequent exposure, and measured effect provides increased confidence in our ability to project safe (and perhaps effective) doses in Phase I and beyond. In selected instances, this increased ability to control and predict serum concentrations, efficacy, and target organ exposure could translate into a more rapid start of Phase I single-dose studies on the basis of less preclinical safety data than is currently submitted [Green and Mordenti, 1990; Green, 1990] .

In the usual course of drug development, single-dose acute studies are followed by repeat-dose studies that usually vary from 1 to 13 weeks in duration. These studies are initiated sequentially, and final, audited reports are included or available at the time of IND submission. The time to filing the IND submission after the initiation of the initial, definitive toxicology studies usually ranges from 15 to 18 months, in some cases being as short as 12 months. Employing an expedited development strategy in 1989 and 1990, our submission time to US regulatory authorities ranged from 3 to 7 months.

A generalized development strategy is presented in Figure 10; the ultimate development plan for individual therapeutics must be tailored on a case-by-case basis. Stage 1 represents the basic core study types: efficacy; absorption, distribution, metabolism, excretion (ADME); and acute and special toxicology studies. On the basis of these data, the IND application is supported to allow single-

MONTH 1	MONTH 2	MONTH 3	MONTH 4	MONTH 5	MONTH 6
Start STAGE 1 testing: Method development and PK-PD with "research" grade material		GMP Material Available	Start Toxicology	Final Reports. Start STAGE 2 testing	File IND

MONTH 7	MONTH 8	MONTH 9	MONTH 10	MONTH 11	MONTH 12
Start Phase I single dose	Multiple dose Phase I studies begin upon completion of histopathologic evaluation and communication of STAGE 2 summary data to FDA & IRB.				End Phase I

Figure 10. sample timeline for an expedited development strategy. Time from rate-limiting toxicology studies to start of clinical Phase I is 12 weeks. KEY: PK-PD, pharmacokinetic and pharmacodynamic studies; GMP, test material made according to Good Manufacturing Practice; IND, Investigational New Drug application; FDA, United States Food and Drug Administration; and IRB, Institutional Review Board (this board approves clinical protocols).

dose human trials, provided no limiting toxicities are observed and a favorable therapeutic ratio for the intended population is determined. Stage 2 testing involves a continuation of preclinical efficacy work in selected models to further refine the PB-PK (physiologically based pharmacokinetic) and PD-PK (pharmacodynamic-pharmacokinetic) correlations. A limited repeat-dose study (4 weeks) is initiated concurrently with the filing of the application. In certain cases, interim or abbreviated communications regarding the status of multi-dose studies can be provided to regulatory authorities and clinical site Investigation Review Boards (IRB) to support extension of clinical protocols to multi-dose regimens. Additional toxicology studies (reproduction, teratology, and longer duration multi-dose studies) would be initiated depending upon the expected duration of clinical treatment (if scientifically warranted) or the nature of the treated population (e.g., women of child bearing age). It should be noted that the success of this approach requires that good communication be maintained with the regulatory authorities that are responsible for approvals. In the United States, our experience with the Food and Drug Administration has been favorable with discussions occurring in the form of pre-IND meetings. Early agreement on the general approach as well as discussion of components of the program that are unique to the specific molecule have allowed expedited filings of IND's and timely approvals. For this approach to be successful, the development program must be closely managed to allow significant savings in time to be realized.

SUMMARY

Drug development programs that include information regarding pharmacokinetic profiles and projected dose equivalents will have a significant impact on the safety evaluation process. For selected molecules, the predictability of clinical pharmacokinetics provides greater confidence in dose selection and interpretation of safety data. This is specially important for biomacromolecules with relatively uneventful profiles or when species specificity is encountered.

The coordination of preclinical pharmacology, toxicology, and disposition studies will enhance the speed, quality and relevance of the preclinical safety data. This information provides the foundation for the subsequent extrapolation to the clinical setting. For well-managed programs, the end result can be a substantial saving in the time and resources without compromising the ability to support safe use in man.

REFERENCES

Boxenbaum H., 1984. Interspecies Pharmacokinetic Scaling and the Evolutionary-Comparative Paradigm. Drug Metab. Rev. **15**:1071.

Boxenbaum H. and D'Souza R.W., 1990. Interspecies Pharmacokinetic Scaling, Biological Design, and Neoteny. In: "Advances in Drug Research," Volume 19, page 139. Academic Press, London.

Byrn R.A., Mordenti J., Lucas C., Smith D., Marsters S., Johnson J., Cossum P., Chamow S., Wurm F., Gregory T., Groopman J., Capon D., 1990. Biological Properties of a CD4 Immunoadhesin. Nature **344**:667.

Byrn R.A., Sekigawa I., Chamow S.M., Johnson J.S., Gregory T.J., Capon D.J., Groopman J.E., 1989. Characterization of In Vitro Inhibition of Human Immunodeficiency Virus by Purified Recombinant CD4. J. Virol. **63**:4370.

Calder III W. A., 1984. "Size, Function, and Life History." Harvard University Press, Cambridge, MA.

422

Capon D. J., Chamow S.M., Mordenti J., Marsters S.A., Gregory T., Mitsuya H., Byrn R.A., Lucas C., Wurm F.M., Groopman J.E., Broder S, Smith D.H., 1989. Designing CD4 Immunoadhesins for AIDS Therapy. Nature **337**: 525.

Chappell W.R. and Mordenti J., 1991. Extrapolation of Toxicological and Pharmacological Data from Animals to Humans. In: "Advances in Drug Research (B. Testa ed.), Volume 20, page 1. Academic Press Inc., San Diego, CA.

Collier A., Katzenstein D., Coombs R., Holodniy M., Mordenti J., Arditti D., Ammann A., Merigan T., Corey L., 1990. Safety and Pharmacokinetics of Intravenous Recombinant CD4 Immunoadhesin (rCD4-IgG) (AIDS Clinical Trials Protocol 121), Abstract S.B.480. Proc. Sixth International Conference on AIDS, San Francisco, CA, Vol 3, page 206. (Serum concentration data subsequently corrected by a factor of 1.44).

Collins J.M., Grieshaber C.K., Chabner B.A., 1990. Pharmacologically-Guided Phase I Clinical Trials Based Upon Preclinical Drug Development. J. Nat.Cancer Inst. **82**:1321.

Collins J.M., Leyland-Jones B., Grieshaber C.K., 1987. Role of Preclinical Pharmacology in Phase I Clinical Trials: Considerations of Schedule-Dependence. In: "Concepts in Cancer Chemotherapy" (F.M. Muggia, ed), page 129. Martinus Nijhoff, Boston.

Collins J.M., Zaharko D.S., Dedrick R.L., Chabner B.A., 1986. Potential Roles for Preclinical Pharmacology in Phase I Clinical Trials. Cancer Treat. Rep. **70**:73.

Fromson J.M., 1989. Perspectives in Pharmacokinetics: A Phased Approach to Drug Development. J. Pharm. Biopharm. **17**:509.

Green J.D., 1990. Preclinical Safety Evaluation Strategy For Biomacromolecules — A Perspective. Abstract in: "From Clone To Clinic. An International Meeting on the Development of Biotechnological Products for Use in Humans." RAI Congress Centre Amsterdam, the Netherlands, page 39.

Green J.D., and Mordenti J., 1990. Rational Basis for Dose Selection in Preclinical and Clinical Safety Assessment, Abstract 876. Toxicologist **10**:219.

Harbour C. and Woodhouse G., 1990. Viral Contamination of Monoclonal Antibody Preparations: Potential Problems and Possible Solutions. Cytotechnology **4**:3.

Harris R. J., Wagner K. L., and Spellman, M. W., 1990. Structural Characterization of Recominant CD4-IgG Hybrid Molecule. *Eur. J. Biochem.* **194**:611.

Janeway C.A.J., 1988. Accessories or Coreceptors? Nature **335**:208.

Kahn J. O., Davis A. J., Groopman J. E., Kaplan L. D., Sherwin S. A., Volberding P. A., 1989. Pharmacokinetic Studies of Recombinant Soluble CD4 in Patients with AIDS and AIDS Related Complex, Abstract Th.B.0.5. Proc. Fifth International Conference on AIDS, Montreal, Cananda, page 212. (Pharmacokinetic data in abstract were preliminary; data in table are final results.)

Kahn J. O., Allan J. D., Hodges T. L., Kaplan L. D., Arri C. J., Fitch H. F., Izu A. E., Mordenti J., Sherwin S. A., Groopman J. E., Volberding P. A., 1990. The Safety and Pharmacokinetics of Recombinant Soluble CD4 (rCD4) in Subjects with the Acquired Immunodeficiency Syndrome (AIDS) and AIDS-related Complex: A Phase I Study. Annals Int. Med. **112**:254.

Kleiber M., 1950. Physiological Meaning of Regression Equations. J. Appl. Physiol. **2**:417.

Lasagna L., 1989. Congress, the FDA, and New Drug Development: Before and After 1962. Persp. Biol. Med. **32**:322.

Maddon P.J., Dalgleish A.G., McDougal J.S., Clapham P.R., Weiss R.A., Axel R., 1986. The T4 Gene Encodes the AIDS Virus Receptor and is Expressed in the Immune System and the Brain. Cell **47**:333.

McMahon T.A. and Bonner J.T., 1983. On Size and Life. Scientific American Books, New York.

Mordenti J., 1986a. Dosage Regimen Design for Pharmaceutical Studies Conducted in Animals. J. Pharm. Sci. 75:852.

Mordenti J., 1986b. Man versus Beast: Pharmacokinetic Scaling In Mammals. J. Pharm. Sci. 75:1028.

Mordenti J. and Chappell W., 1989. The Use of Interspecies Scaling in Toxicokinetics. In: "Toxicokinetics and New Drug Development" (A. Yacobi, J. Skelly, V. Batra, eds.), page 42. Pergamon Press, New York.

Mordenti J., Chen S., Moore J., Ferraiolo B., 1990. Total Body Clearance of Recombinant and Synthetic Proteins are Well Described by an Allometric Relationship, Abstract 955. Toxicologist 10:239.

Mordenti J., Chen S., Moore J., Ferraiolo B., Green, J. D., 1991. Interspecies Scaling of Clearance and Volume of Distribution Data for Five Therapeutic Proteins. Pharmaceutical Research, Vol.11 (Nov.), in press.

Mordenti J., Shaieb D., Chow P., Cossum P., Ferraiolo B., Lewandowski M., Moore J., Green J. D., 1991. Preclinical Safety Evaluation Strategy for Biomacromolecules — A Perspective. In: "Safety Assessment for Pharmaceuticals" (Shayne C. Gad, Ed.). Van Nostrand Reinhold, New York, NY (in press).

Office of Biologics Research and Review, 1987. Points to Consider in the Manufacture and Testing of Monoclonal Antibody Products for Human Use, Office of Biologics Research and Review, Center for Drugs and Biologics, Food and Drug Administration, U.S.A., June 1, 1987.

Peck C., Collins J. Harter J., 1990. Incorporation of Pharmacokinetic and Pharmacodynamic Intelligence into Early Drug Development, Abstract PP-3. Clin. Pharmacol. Ther. 47:126.

Sattentau Q. and Weiss R., 1988. The CD4 Antigen: Physiological Ligand and HIV Receptor. Cell 52:631.

Schmidt-Nielsen K., 1984. Scaling: Why is Animal Size So Important? Cambridge University Press, New York.

Smith D.H., Byrn R.A., Marsters S.A., Gregory T., Groopman J.E., Capon D.J., 1987. Blocking of HIV-1 Infectivity by Soluble, Secreted Form of the CD4 Antigen. Science 238:1704.

Yarchoan R., Thomas R. V., Pluda J. M., Perno C. F., Mitsuya H., Marczyk K. S., Sherwin S. A., Broder S., 1989. Phase I Study of the Administration of Recombinant Soluble CD4 (rCD4) by Continuous Infusion to Patients with AIDS or ARC, Abstract M.C.P. 137. Proc. Fifth International Conference on AIDS, Montreal, Canada, page 564. (Pharmacokinetic data in abstract were preliminary; data in table are final results.)

Yarchoan R., Pluda J. M., Adamo D., Thomas R. V., Mordenti J., Goldspiel B. R., Ammann A., Broder S., 1990. Phase I Study of rCD4-IgG Administered by Continuous Intravenous (IV) Infusion to Patients with AIDS or ARC, Abstract S.B.479. Proc. Sixth International Conference on AIDS, San Francisco, CA, Vol 3, Page 205. (Serum concentration data subsequently corrected by a factor of 1.44).

CONTRIBUTORS

Dr. Melvin E. Andersen
Chemical Ind. Institute of Toxicology
6 Davis Drive
Research Triangle Park, NC 27709

Prof. Luc P. Balant
Psychiatric Univ. Institutions of Geneva
47 rue du XXXI Décembre
1207 Genève
Switzerland

Prof. James S. Beck
4959 Vantage Crescent N.W.
Calgary, Alberta T3A 1X6
Canada

Dr. A. Bertrand Brill
Department of Nuclear Medicine
U.Mass. Medical Center
Worcester, MA 01655

Dr. Michel Eichelbaum
Fischer-Bosch-Institut für
Klinische Pharmakologie
Auerbachstrasse 112
7000 Stuttgart 50
Germany

Dr. Julie Eiseman
Division of Developmental Therapeutics
University of Maryland Cancer Center
655 West Baltimore Street
Baltimore, MD 21201

Dr. Alan Fischman
Division of Nuclear Medicine
Massachusetts General Hospital
Boston, MA 02114

Dr. Daniel R. Krewski
Room 109
Environmental Health Center
Tunney's Pasture
Ottawa, Ontario K1A 0L2
Canada

Prof. Giuliano Mariani
C.N.R. Fisiologia Clinica
via Savi 8
56100 Pisa
Italy

Dr. Bernard M. Mazière
Service Hospitalier "Frédéric Joliot"
4, Place du Général-Leclerc
91406 Orsay
France

Dr. Joyce Mordenti
Genentech, Inc.
460 Point San Bruno Blvd.
South San Francisco, CA 94080

Prof. Aldo Rescigno
School of Pharmacy
University of Parma
Via Massimo d'Azeglio 85
43100 Parma
Italy

Prof. Malcolm Rowland
Department of Pharmacy
University of Manchester
Oxford Road
Manchester M13 9PL
U.K.

Dr. David E. Schafer
Veterans Administration Medical Service
West Haven, CT 06516

Dr. James T. Stevens
Director of Toxicology
Ciba-Geigy Corp.
410 Swing Road
Greensboro, NC 27409

Dr. Alberto Tajana
Recordati Industria Chimica
via M. Civitali 1
20148 Milano
Italy

Prof. Bernard Testa
Ecole de Pharmacie
Université de Lausanne
place du Château 3
1005 Lausanne
Switzerland

Dr. Ajit K. Thakur
Hazleton Laboratories America
9200 Leesburg Turnpike
Vienna, VA 22180

PARTICIPANTS

Gun Almquist
Dept. Human Pharmacokinetics
Explorative Clinical Research
AB Draco, Box 34
S-22100 Lund, Sweden

Dr. Melvin E. Andersen
Chemical Ind. Institute of Toxicology
6 Davis Drive
Research Triangle Park, NC 27709
U.S.A.

Prof. Luc P. Balant
Psychiatric Univ. Institutions of Geneva
47 rue du XXXI Décembre
1207 Genève, Switzerland

Dr. Virgilio Ballabeni
Istituto di Farmacologia e Farmacognosia
Università di Parma
via Massimo d'Azeglio 85
43100 Parma, Italy

Dr. Elisabetta Barocelli
Istituto di Farmacologia e Farmacognosia
Università di Parma
via Massimo d'Azeglio 85
43100 Parma, Italy

Prof. James S. Beck
Division of Medical Biophysics
University of Calgary
2920 - 24th Ave.N.W.
Calgary, Canada

Dr. S. Eralp Bellibas
Ege Tip Fakultesi
Farmakoloji Anabilim Dali
Bornova - Izmir, Turkey

Dr. Riccardo Bellina
Istituto di Patologia Medica I
Università degli Studi
Pisa, Italy

Dr. Walter Bencivelli
CNR Fisiologia Clinica
via Savi 8
I-56100 Pisa, Italy

Dr. Marta Benedetti
Istituto Superiore di Sanità
Laboratorio di Tossicologia Comparata
viale Regina Elena 299
00161 Roma, Italy

Dr. Giuseppina Benoni
Istituto di Farmacologia
Policlinico Borgo Roma
37134 Verona, Italy

Prof. Nazan Bergisadi
Dept. Pharmaceutical Technology
Faculty of Pharmacy
Univ. Istanbul
Beyazit - Istanbul, Turkey

Dr. Donata Bertin
Dipart. Farmacocinetica e Metabolismo
Farmitalia - Carlo Erba
via Imbonati 24
20159 Milano, Italy

Dr. Bianca Maria Bocchialini
Istituto di Farmacologia e Farmacognosia
Università di Parma
via Massimo d'Azeglio 85
43100 Parma, Italy

Eva Bondesson
Department of Human Pharmacokinetics
Explorative Clinical Research
AB Draco, Box 34
S-22100 Lund, Sweden

Dr. A. Bertrand Brill
Department of Nuclear Medicine
U.Mass. Medical Center
Worcester, MA 01655, U.S.A.

Dr. Richard C. Brundage
Department of Pharmaceutics
College of Pharmacy
University of Minnesota
308 Harvard Street S.E.
Minneapolis, MN 55455, U.S.A

Dr. Paola Brusa
Istituto Chimica Farmaceutica Applicata
corso Raffaello 31
10125 Torino, Italy

Harry Bushe
Department of Nuclear Medicine
U. Mass. Medical Center
Worcester, MA 01655, U.S.A.

Dr. Begoña Calvo
Departamento de Farmacia
Avda. del Campo Charro b/n
37008 Salamanca, Spain

Dr. Giovanni Cosmi
Life Science Research
Roma Toxicology Centre SpA
via Tito Speri 12/14
00040 Pomezia (Roma), Italy

Dr. Marco Criscuoli
Laboratori Guidotti S.p.A.
via Livornese 402
56010 S. Piero a Grado (PI), Italy

Dr. Jack Dacre
U. S. Army Biomedical Research and
Development Laboratory
Fort Dietrich
Frederick, MD 21702, U.S.A.

Dr. Peter Terence Daley-Yates
Department of Pharmacy
University of Manchester
Oxford Road
Manchester M13 9PL, U. K.

Dr. Michael David
Byk Gulden
Postfach 6500
D-7750 Konstanz, Germany

Dr. B. E. Davies
SmithKline Beecham Pharmaceuticals
Coldharbour Road
The Pinnacles
Harlow, Essex CM19 5AD, U. K.

Dr. D. Domurado
Université de Technologie de Compiègne
Département Génie Biologique
Centre de Recherches de Royallieu
Compiègne, France

Dr. Franco Dosio
Istituto di Chimica Farmaceutica Applicata
corso Raffaello 31
10125 Torino, Italy

Dr. Constantin Efthymiopoulos
Dipart. Farmacocinetica e Metabolismo
Farmitalia - Carlo Erba
via Imbonati 24
20159 Milano, Italy

Dr. Michel Eichelbaum
Fischer-Bosch-Institut für
Klinische Pharmakologie
Auerbachstrasse 112
7000 Stuttgart 50, Germany

Dr. Julie Eiseman
Division of Developmental Therapeutics
University of Maryland Cancer Center
655 West Baltimore Street
Baltimore, MD 21201, U.S.A.

Dr. William F. Elmquis
Department of Pharmaceutics
College of Pharmacy
University of Minnesota
308 Harvard Street S.E.
Minneapolis, MN 55455, U.S.A.

Prof. Laszlo Endrenyi
Department of Pharmacology
Medical Sciences Building
University of Toronto
Toronto, Ontario M5S 1A8, Canada

Prof. Jorge Errecalde
Catedra de Farmacologia
Facultad de Ciencias Veterinarias
Universidad Nacional de La Plata
CC 296. 1900
La Plata, Argentina

Dr. Luigi Ferrante
Facoltà di Medicina e Chirurgia
Università degli Studi di Ancona
Ancona, Italy

Dr. Franca Ferrari
Dipartimento di Chimica Farmaceutica
Università di Pavia
viale Taramelli 12
27100 Pavia, Italy

Dr. Alan Fischman
Division of Nuclear Medicine
Massachusetts General Hospital
Boston, MA 02114, U.S.A.

Dr. Roberto Foroni
Servizio di Fisica Sanitaria
Ospedale Borgo Trento
37134 Verona, Italy

Dr. Giuliana Galli
Alfa Wassermann S.p.A.
via Ragazzi del '99 n.5
40133 Bologna, Italy

Dr. Maria Angela Girometta
Camillo Corvi S.p.A.
Stradone Farnese 118
29100 Piacenza, Italy

Prof. Gulin Guvendik
Department of Toxicology
Faculty of Pharmacy
University of Ankara
06100 Tandogan - Ankara, Turkey

Dr. Giorgio Iervasi
CNR Fisiologia Clinica
via Savi 8
56100 Pisa, Italy

S. K. Jain
Institut für Pharmakologie
Toxikologie und Pharmazie
Tierärzliche Hochschule
Büntweg 17
3000 Hannover 771, Germany

Dr. Anupam Kaur
School of Life Sciences
Guru Nanak Dev University
Amritsar-143 005, India

Dr. Daniel R. Krewski
Biostatistics and Computer Applications
Environmental Health Center
Tunney's Pasture
Ottawa, Canada

Dr. Ole W. Krogsgaard
Dept. Clin. Physiol. & Nuclear Medicine
University Hospital
9 Blegdamsvej
DK-2100 Copenhagen 0, Denmark

Dr. Helen O. Kwanashie
Department of Pharmacology and
Clinical Pharmacy
Faculty of Pharmaceutical Sciences
Ahmadu Bello University
Zaria, Nigeria

Dr. Frank Larsen
The Royal Danish School of Pharmacy
Dept. of Biological Sciences
Pharmacology and Toxicology
2 Universitetsparken
DK-2100 Copenhagen, Denmark

Dr. Michel Lemaire
Sandoz AG
Pharma Division
Bau 507/801
CH-4002 Basle, Switzerland

Dr. Jean Logan
Department of Chemistry
Brookhaven National Laboratory
Upton, NY 11973, U.S.A.

Marie-Yvette Madrid
Department of Chemical Engineering
Massachusetts Institute of Technology
Cambridge, MA 02139, U.S.A.

Vitus Malerczyk
Hoechst AG
Klinische Forschung
Abteilung für Biometrie und
Dokumentation
Postfach 80 03 20
6230 Frankfurt am Main 80, Germany

Prof. Giuliano Mariani
C.N.R. Fisiologia Clinica
via Savi 8
56100 Pisa, Italy

Dr. Bernard M. Mazière
Service Hospitalier "Frédéric Joliot"
Hôpital D'Orsay
Orsay, France

Dr. Nicola Molea
Servizio di Medicina Nucleare
Clinica Medica II
Università di Pisa
Pisa, Italy

Dr. Paolo Morazzoni
Laboratorio di Farmacocinetica
Inverni della Beffa
via Ripamonti 99
20141 Milano, Italy

Dr. Joyce Mordenti
Genentech, Inc.
460 Point San Bruno Blvd.
South San Francisco, CA 94080, U.S.A.

Dr. Sabine Ott
Sandoz Pharma AG
Human Pharmacology Department
Drug Safety Assessment
Building 386/Office 956
CH-4002 Basle, Switzerland

Dr. Hilal Ozgunes
Ziya Gokalp cad. 78/3
06600 Ankara, Turkey

Prof. Pietro Pavoni
V Clinica Medica
Policlinico Umberto I
00100 Roma, Italy

Prof. José Luis Pedraz
Departamento de Farmacia
Avda. del Campo Charro b/n
37008 Salamanca, Spain

Prof. Aldo Rescigno
School of Pharmacy
University of Parma
Via Massimo d'Azeglio 85
Parma, Italy

Dr. Lillian E. Riad
Department of Pharmaceutics
College of Pharmacy
University of Minnesota
308 Harvard Street S.E.
Minneapolis, MN 55455, U.S.A.

Prof. Malcolm Rowland
Department of Pharmacy
University of Manchester
Oxford Road
Manchester M13 9PL, U.K.

David Saldana
Dep. de Quimica Inorganica y Analisis
E.T.S.I.I.
Univ. de Valladolid
Valladolid, Spain

Prof. Amparo Sanchez
Departamento de Farmacia
Avda. del Campo Charro b/n
37008 Salamanca, Spain

Dr. Ronald J. Sawchuk
Clinical Pharmacokinetics Laboratory
University of Minnesota
308 Harvard Street S.E.
Minneapolis, MN 55455, U.S.A.

Dr. David E. Schafer
Veterans Administration Medical Service
West Haven, CT 06516, U.S.A.

Dr. Enrico Seccamani
Sorin Biomedica S.p.A.
via Crescentino
13040 Saluggia (VC), Italy

Dr. Thomas Skripsky
Ciba-Geigy Corp.
Dept. AG2.51
CH-4002 Basel, Switzerland

Dr. James T. Stevens
Director of Toxicology
Ciba-Geigy Corp.
410 Swing Road
Greensboro, NC 27409, U.S.A.

Dr. Alberto Tajana
Recordati Industria Chimica
via M. Civitali 1
20148 Milano, Italy

Prof. Bernard Testa
Ecole de Pharmacie
Université de Lausanne
place du Château 3
1005 Lausanne, Switzerland

Dr. Ajit K. Thakur
Hazleton Laboratories America
9200 Leesburg Turnpike
Vienna, VA 22180, U.S.A.

Dr. Alessandra Torroni
Ellem Industria Farmaceutica
viale dell'Industria 15/17
20094 Corsico (MI), Italy

Dr. Francis L. S. Tse
Sandoz AG, Pharma Division
Bau 507/909
CH-4002 Basle, Switzerland

Prof. Isik Tuglular
E. U. Tip Fakultesi
Farmakoloji Anabili
Dali, Bornova - Izmir, Turkey

Prof. Frantisek Vitek
Inst. Biophysics & Nuclear Medicine
Charles University, Salmovska 3
12000 Prague 2, Czechoslovakia

Prof. Nevin Vural
Department of Toxicology
Faculty of Pharmacy
University of Ankara
06100 Tandogan, Ankara, Turkey

Dr. Lauren Zech
Laboratory of Theoretical Biology
Building 10, Room 4B-56
National Institutes of Health
Bethesda, MD 20892, U.S.A.

SUBJECT INDEX

435